# RING-FORMING POLYMERIZATIONS

## PART B, 2: HETEROCYCLIC RINGS

This is Volume 13 B, 2  of
ORGANIC CHEMISTRY
A series of monographs
Editors: ALFRED T. BLOMQUIST and HARRY WASSERMAN

A complete list of the books in this series appears at the end of the volume.

# RING-FORMING

# POLYMERIZATIONS

## ROBERT J. COTTER and MARKUS MATZNER

RESEARCH AND DEVELOPMENT DEPARTMENT
CHEMICALS AND PLASTICS
UNION CARBIDE CORPORATION
BOUND BROOK, NEW JERSEY

## PART B, 2
## *Heterocyclic Rings*

1972

ACADEMIC PRESS   New York and London

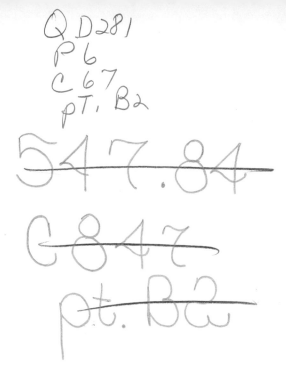
ACADEMIC PRESS, INC.
111 Fifth Avenue, New York, New York 10003

*United Kingdom Edition published by*
ACADEMIC PRESS, INC. (LONDON) LTD.
24/28 Oval Road, London NW1

LIBRARY OF CONGRESS CATALOG CARD NUMBER: 68-26633

PRINTED IN THE UNITED STATES OF AMERICA

To our wives, Barbara and Debora, and children, Patricia, Robert, Jr., Katherine, and Stella. Their patient understanding, interest, and encouragement have contributed immeasurably to the writing of this book.

# Contents

*Preface*                                                                     xiii

*Contents of Part A*                                                            xv

*Contents of Part B, 1*                                                        xvi

**Introduction**                                                              xvii

CHAPTER I.    **Rings Containing Four Carbon Atoms**

    A. Polyimides                                                            1
       Polyimide Homopolymers                                                 1
       Method 1.    Thermal Dehydration of Polyamic Acids                      7
       Method 2.    Chemical Dehydration of Polyamic Acids                    20
       Method 3.    Polyimides via Intermediate Polyiminolactones            22
       Method 4.    Polyimides from Diamines and Tetranitriles              23
       Method 5.    Polyimides from Diisocyanates and Dianhydrides         24
       Method 6.    Miscellaneous Polyimide Syntheses                       26
       Properties                                                            26
       Applications                                                          40
       Poly(Amide-Imides)                                                    41
       Method 7.    Thermal Dehydration of Polyamic Acids                  42
       Method 8.    Chemical Dehydration of Polyamic Acids                 48
       Method 9.    Polymerization of Trimellitic Anhydride with
             Diisocyanates                                           48
       Properties and Applications                                           49
       Poly(Ester-Imides)                                                    52
       Method 10. Thermal Dehydration of Polyamic Acids                      52
       Method 11. Poly(Ester-Imides) from Dianhydrides and Diiso-
             cyanates                                                55
       Method 12. Poly(Ester-Imides) from Trimellitic Anhydride
             and Diamines and Diols                                  56
       Properties of Poly(Ester-Imides)                                      56
       Poly(Imide-Imidazopyrrolones)                                         60
       Method 13. Thermal Dehydration of Polyamic Acids                      61
       Method 14. Polymerization in Polyphosphoric Acid                      62
       Polytriazololactams                                                   63
       Method 15. Thermal Dehydration of Appropriate Polyamic
             Acids                                                   63
       Oxadiazole Copolymers                                                 64

Method 16.  Thermal Dehydration of Polyamic Acids to Give
Copolymers Containing the 1,3,4-Oxadiazole
Ring                                                          64
Method 17.  Thermal Dehydration of Polyamic Acids to Give
Copolymers Containing the 1,2,4-Oxadiazole
Ring                                                          65
Polyiminolactones                                                        66
Method 18.  Chemical Dehydration of Polyamic Acids                        66
Chemical Reactions of Polyiminolactones                                  67
B.  Polymers Containing Pyrrole and Related Rings                          69
Method 1.  Polyteracyanoethylene                                          69
Method 2.  Polyphthalocyanines                                            70
Method 3.  Polymerization of Multifunctional Nitriles and
Amines                                                        71
Method 4.  Polypyrrolines from Diamines                                   72
Method 5.  Polypyrollinones                                               72
C.  Poly(1,3-Oxazinones) and Poly(1,3-Oxazindiones)                       73
Method 1.  Poly(1,3-Oxazinones) from o-Amino Aromatic
Carboxylic Acids                                              73
Method 2.  Poly(1,3-Oxazindiones) from Diisocyanates and
Bis(Hydroxy Acids)                                            75
Method 3.  Poly(1,3-oxazindiones) from Carbonate-Blocked
Bis(Hydroxy Acids) and Diamines                               76
D.  Polydioxins, Polyoxazines, and Polythiazines                          78
Method 1.  Polydioxins from Tetraphenols and Tetrachloro-
quinoxalines                                                  78
Method 2.  Polyoxazines and Polythiazines from Diamino-
diphenols and Tetrachloro or Tetrahydroxy
Compounds                                                     78
Method 3.  Polyoxazines and Polythiazines from Diamino-
diphenols and Quinones                                        79
Method 4.  Condensation of Diaminodithiophenols with Bis-
(α-Haloalkyl Aryl Ketones)                                    80
E.  Polyquinoxalines and Related Types                                     81
Method 1.  Pyrazine Ring-Containing Polymers                              81
Method 2.  Polyquinoxalines from Bis(o-Diamines) and Tetra-
carbonyl Compounds                                            81
Method 3.  Polyquinoxalines by Polycondensation in Poly-
phosphoric Acid                                               83
Properties of Polyquinoxalines                                            84
F.  Polypiperazines and Polydiketopiperazines                             85
Method 1.  Polypiperazines from Bis(β-Hydroxyethyl)Di-
amines                                                        85
Method 2.  Polymerization of Bis(α-Amino Acids) to Poly-
diketopiperazines                                             85
Method 3.  Polydiketopiperazines from Bis(α-Chloroacetam-
ides)                                                         86
G.  Polypyrimidines and Polytetraazapyrenes                               87
Method 1.  Polypyrimidine Syntheses                                       87
Method 2.  Polytetraazapyrene Synthesis                                   87

H. Polypyrimidones, Polyquinazolones, and Polyquinazolinediones 89
    Method 1. Polypyrimidones from Unsaturated Isocyanates 89
    Method 2. Polyquinazolones from Diamines and Bis(benzoxazinones) 89
    Method 3. Polyquinazolones from Aromatic Bis(o-Amino Acids) and Derivatives 91
    Method 4. Polyquinazolinediones by Condensation of Diisocyanates with Aromatic Bis(o-Amino Acids) 93
    Method 5. Poly(isoindoloquinazolinediones) from Condensation of Aromatic Bis(o-Aminoamides) with Dianhydrides 95
    Properties of Polyquinazolones and Polyquinazolinediones 95
I. Polyacetals and Polyketals 96
    Method 1. Condensation of Polyols with Dialdehydes and Diketones 96
    Method 2. Polyacetals by Transacetalization 97
    Miscellaneous Related Syntheses 97
    Properties of Polyacetals and Polyketals 98
J. Polyquinones 98
    Method 1. Polyquinones from Halo-Substituted p-Quinones 98
    Table I.1. Polyimides from Aliphatic and Cycloaliphatic Dianhydrides 102
    Table I.2. Polyimides from Multicyclic Dianhydrides 114
    Table I.3. Polypyromellitimides 120
    Table I.4. Polyimides Based on 3,3',4,4'-Benzophenonetetracarboxylic Acid Dianhydride 145
    Table I.5. Miscellaneous Polyimides 153
    Table I.6. Poly(amide-imides) from Diamino Amides 173
    Table I.7. Poly(amide-imides) from Tricarboxylic Acid Derivatives 181
    Table I.8. Poly(amide-imides) from Amide Anhydrides 187
    Table I.9. Poly(ester-imides) from Ester Anhydrides 189
    Table I.10. Poly(ester-imides) from Diamino Esters 196
    Table I.11. Poly(ester-imides) from Dimethyl Bis(trimellitimidates) 198
    Table I.12. Polyiminolactones 200
    Table I.13. Poly(imide-imidazopyrrolones) 204
    Table I.14. Polymers Containing Pyrrole and Related Rings 205
    Table I.15. Poly(1,3-oxazinones) and Poly(1,3-oxazindiones) 210
    Table I.16. Polydioxins 222
    Table I.17. Polyoxazines 225
    Table I.18. Polythiazines 227
    Table I.19. Polyquinoxalines and Related Polymers 230
    Table I.20. Polypiperazines and Polydiketopiperazines 239
    Table I.21. Polypyrimidines and Polytetraazopyrenes 242
    Table I.22. Polyquinazolones 244
    Table I.23. Polyquinazolinediones 249
    Table I.24. Polyacetals and Polyketals 254
    Table I.25. Polyquinones 260
    References 262

CHAPTER II.  **Intra–Intermolecular Polymerizations Leading to Heterocyclic Rings**

A. Polymerization of Dialdehydes                                    291
   Method 1. Free-Radical Initiation                                291
   Method 2. Ziegler Initiation                                     292
   Method 3. Cationic Initiation                                    292
   Method 4. Anionic Initiation                                     295
B. Polymerization of Diepoxides                                     296
   Method 1. Free-Radical Initiation                                296
   Method 3. Cationic Initiation                                    297
   Method 4. Anionic Initiation                                     298
C. Polymerization of Divinyl Acetals and Ketals                     299
   Method 1. Free-Radical Initiation                                299
   Method 2. Ziegler Initiation                                     306
   Method 3. Cationic Initiation                                    306
D. Polymerization of Unsaturated Esters                             306
   Polymerizations of Unsaturated Glycidyl Esters                   306
   Method 3. Cationic Initiation                                    307
   Method 4. Anionic Initiation                                     307
   Polymerization of Diunsaturated Esters                           307
   Method 1. Free-Radical Initiation                                308
E. Polymerization of Diunsaturated Anhydrides                       317
   Method 1. Free-Radical Initiation                                317
   Method 4. Anionic Initiation                                     320
F. Polymerization of Diunsaturated Germanium Derivatives            321
   Method 2. Ziegler Initiation                                     321
G. Polymerization of Diunsaturated Ammonium Salts, Amine
   Oxides, and Amines                                               322
   Method 1. Free-Radical Initiation                                322
H. Polymerization of Diunsaturated Amides                           328
   Method 1. Free-Radical Initiation                                328
   Method 4. Anionic Initiation                                     331
I. Polymerization of Diisocyanates                                  332
   Method 1. Free-Radical Initiation                                332
   Method 4. Anionic Initiation                                     332
J. Polymerization of Dinitriles                                     335
   Method 1. Free-Radical Initiation                                335
   Method 2. Ziegler Initiation, Miscellany                        336
K. Polymerization of Diunsaturated Phosphorus-Containing
   Compounds                                                        337
   Method 1. Free-Radical Initiation                                337
   Method 4. Anionic Initiation                                     339
L. Polymerization of Diunsaturated Sulfur-Containing Com-
   pounds                                                           339
   Method 1. Free-Radical Initiation                                339
   Method 2. Ziegler Initiation                                     345
   Method 3. Cationic Initiation                                    345
M. Polymerization of Diunsaturated Silicon Derivatives             346
   Method 1. Free-Radical Initiation                                346
   Method 2. Ziegler Initiation                                     349

Method 3. Cationic Initiation     350
Method 4. Anionic Initiation     350
N. Miscellaneous Polymerizations     350
Method 1. Free-Radical Initiation     350
Method 2. Ziegler Initiation     351
Table II.1. Polydialdehydes     352
Table II.2. Polydiepoxides     355
Table II.3. Poly(divinyl acetals) and Poly(divinyl ketals)     357
Table II.4. Polymers from Unsaturated Glycidyl Esters     368
Table II.5. Polymers from Diunsaturated Esters     369
Table II.6. Polymers from Diunsaturated Anhydrides     379
Table II.7. Polymers from Polyunsaturated Ammonium Salts
    and Amine Oxides     382
Table II.8. Copolymers from Diunsaturated Ammonium Salts
    and Sulfur Dioxide     395
Table II.9. Polymers from Diunsaturated Amines     400
Table II.10. Polymers from Diunsaturated Amides     407
Table II.11. Cyclopolymers from Diisocyanates and Diiso-
    thiocyanates     411
Table II.12. Cyclopolymers from Dinitriles     415
Table II.13. Phosphorous-Containing Cyclopolymers     417
Table II.14. Sulfur-Containing Cyclopolymers     421
Table II.15. Silicon-Containing Cyclopolymers     427
Table II.16. Miscellaneous Cyclopolymers     431
References     432

CHAPTER III. **α,β-Unsaturated Aldehyde Polymerizations**

A. Acrolein Polymers     442
Polymerizations     442
Properties     442
Structure     450
Copolymers     455
B. Polymers from α-Substituted Acroleins     456
C. Polymerization of β-Substituted Acroleins     459
D. Miscellaneous Polymers     460
Table III.1. Initiator Systems for Acrolein Polymerization     443
Table III.2. Properties of Various Polyacroleins     446
Table III.3. Polymerization of α-Alkyl Acroleins     462
References     468

CHAPTER IV. **Miscellaneous Ring-Forming Polymerizations**

A. Polydiazadiphosphetidines     473
B. Polytetrazadiborines     473
C. Polytetrazoles     475
D. Polybenzboroxazinones     476
E. Polysulfimides and Related Polymers     476
F. Poly(4-phosphoniapyran salts)     478
G. Poly(amide-indoles)     478

H. Polyxanthones      479
I. Copolymerization of 1,4-Butynediol with Acrolein      480
J. Polymerizations to Fused Pyridine Ring-Containing Polymers      481
K. Polydiazepines      489
L. Polydisalicylides      490
M. Polyferrocenes and Polytitanocenes      491
N. Miscellaneous Polymerizations      492
    References      493

**Supplementary Reference List**      495

**Supplementary Reference List for Part A**      523

*Author Index*      533

*Subject Index*      553

# Preface

These volumes comprise a comprehensive review and compilation of ring-forming polymerization reactions that proceed with heterocyclic ring formation. The phenomenal growth of research and commercial development efforts in this area of polymer chemistry is readily apparent from even the most cursory glance at the table of contents herein. That this growth appears to be continuing unabated is evident from the size of the updated supplementary reference lists at the end of the volumes.

These books are designed to fill the need of polymer and organic chemists for a rapid, easy-to-use key to the available literature on these new materials. Because of the presence of heterocyclic rings of various types in biological macromolecules, the contents of these works will also be of interest to those concerned with the polymer chemistry of living systems.

The large number of heterocyclic ring-containing polymers that have been investigated has resulted in a need for an orderly and systematic presentation of their synthesis and properties. The emphasis in these works is on polymerization reactions that yield polymers containing a ring that is not present in any of the monomers. Additionally, polymerizations leading to linear, high molecular weight products are stressed, although some information on cross-linked materials is included. Polymerizations which propagate via double bond (containing heteroatoms) formation, "a two-membered ring," are discussed initially. Subsequent chapters are arranged according to the number of carbon atoms in the ring that are formed during polymerization. This is followed up through rings containing four carbon atoms. Intra–intermolecular ring-forming polymerizations and $\alpha,\beta$-unsaturated aldehyde polymerizations are treated in separate chapters. Both of these classes generally lead to rings containing five carbon atoms in keeping with the content outline of the other chapters.

The tables of polymers that have been assembled and presented stand alone as a handy index to polymers that have been prepared by ring-forming polymerizations. In most instances they are complete bibliographies for the individual polymers since many have not been prepared by other routes. The monomers from which the polymers in the tables have been prepared are also listed in the Subject Index. Quickly locating a reference wherein the polymerization of a particular monomer has been studied can be a valuable time-saver for those faced with choosing one synthetic path from among several for that particular monomer.

The use of these volumes in conjunction with Part A of this work provides ready access to information about literally hundreds of monomers and polymers. No other source comparable in convenience and scope is presently available.

As mentioned in the Preface to Part A of this endeavor, the continued interest and efforts of many people have eased the task of writing this book. Our sincere appreciation is again extended to those same professional colleagues who have been of invaluable assistance. In addition, we acknowledge the skillful assistance of Rita Matts, who typed most of the manuscript for these two volumes. The aid of Patricia Kralovich and Mary Lou Peters who provided supplementary secretarial help is sincerely appreciated. Anneli DePaolis provided an invaluable assist by redrawing many of the chemical formulas for the manuscript. Finally, we wish to acknowledge and extend our appreciative thanks to the personnel of Academic Press for their thoughtful assistance and cooperation at many points during production.

ROBERT J. COTTER
MARKUS MATZNER

# Contents of Part A

Introduction

CHAPTER I.      Carbon–Carbon Double Bond-Forming Polymerizations

CHAPTER II.     Intra–Intermolecular Polymerizations Leading to
                Homocyclic Carbon Rings

CHAPTER III.    Diels–Alder Polymerizations

CHAPTER IV.     Polymerization of Diisopropenyl Monomers and Other
                Carbocyclic Ring-Forming Polymerizations

CHAPTER V.      Metallorganic Ring-Containing Polymers of Unsaturated
                Carbon Compounds

CHAPTER VI.     Metallorganic Ring Polymers from Nitrogen Chelate Ligands

CHAPTER VII.    Metallorganic Ring Polymers from Nitrogen: Oxygen
                Chelate Ligands

CHAPTER VIII.   Metallorganic Ring Polymers from Nitrogen: Sulfur
                Chelate Ligands

CHAPTER IX.     Metallorganic Ring Polymers from Oxygen Chelate Ligands

CHAPTER X.      Metallorganic Ring Polymers from Sulfur Chelate Ligands

CHAPTER XI.     Polysiloxanes, Polysilazanes, and Polymetalloxanes from
                Ring-Forming Polymerizations

Supplementary Reference List

Author Index—Subject Index

# Contents of Part B, 1

*Introduction*

CHAPTER I.    Multiple Bond-Forming Polymerizations

CHAPTER II.   Rings Containing Two Carbon Atoms

CHAPTER III.  Rings Containing Three Carbon Atoms

*Supplementary Reference List*

*Author Index—Subject Index*

# Introduction

This book is organized according to the specific heterocyclic ring structure that is formed in a ring-forming polymerization. This arrangement was chosen to be of maximum aid to synthetic polymer chemists who are interested in tailormaking polymer repeat units. By classifying his proposed new repeat unit according to the types of rings present in it, he will be able to consult this book to ascertain whether any ring-forming polymerizations are of interest to him. Each chapter is further arranged according to polymer type and the various methods that have been used to prepare that type. The emphasis is on synthetic methods, with the numbers assigned to the methods also being used in the tables to indicate how the tabulated polymers were made. Although the considerations that result in a number being assigned to a method may seem arbitrary at times, they were often necessary to aid in the codification of the data.

Although the term polymer can mean different things to many people, this book is about linear polymers that were made intentionally. Tars, distillation residues, or glops from reactions that "went bad" (unintentional polymerizations) are not included. Ring-forming polymerizations that were planned to be such by their experimenters are what have been considered. The qualitative description of the molecular weight of a polymer can be ambiguous. In this book, the term "high molecular weight" is used to mean that a level has been attained that imparts mechanical properties to the polymer. That is, it is either film- or fiber-forming and mechanical property evaluation is possible. Another way to look at the use of this term is that the polymer is high enough in molecular weight so that further increases do not appreciably change its mechanical properties. "Low molecular weight" is used to designate oligomers, and "moderate molecular weight" describes polymer samples with molecular weights up to the threshold level for property appearance. Since definite molecular weight numbers associated with these various criteria differ from class to class, and, moreover, are generally not available anyway, these qualitative terms were found to be useful and meaningful.

Whether or not a high molecular weight, linear polymer was obtainable from a particular ring-forming polymerization was used as the main criterion for judging the merits of the method. The data in the tables were also selected to indicate at a glance if the polymer was obtained at a high molecular weight. However, being well aware of how fickle polymerizations can be with respect to whether high, moderately high, or low molecular weight products are obtained, all ring-forming polymerizations that were found have been included. Too many examples exist in the authors' experiences of "unsuccessful" polymerizations that were made to work when greater attention was paid to experimental techniques, monomer purification procedures, or other details. Finally, a word of warning is in order to those who may be uninitiated in the difficulties attendant to proving the structure of a polymer. The structures reported for the polymers described in this book are those suggested by the workers whose research is cited. In some cases, the evidence presented for the assigned structures hinged mainly on the reactants employed and an elemental analysis of the product. Of course, where evidence was found that disputed an assigned structure, it is included and discussed. The newer, non-destructive, spectrophotometric methods for structure determination have been used on many ring-containing polymers, and their wider use is to be encouraged. In any event, we hope this alert will serve those who may not be aware of this problem.

### EXPLANATION OF TABLES

The tables of polymers that have been assembled and presented in this book stand alone as a handy index to those polymers that have been prepared by ring-forming polymerizations. In most instances, they are complete bibliographies for the individual polymers, since many have not been prepared by other routes. The general guidelines that were used in assembling these tables from the literature data are given below.

### Number

The first column in every table contains an arabic number that has been assigned to each polymer or entry in the table. The first entry in a table is No. 1, with the following ones being numbered consecutively.

### Structure

The tables contain the structure of the repeating unit of the polymer in the second column. Within a table, the polymers are arranged according to increasing complexity and/or empirical formula of the repeat unit. In general, the arrangement that has been employed allows one to readily compare the effect of changes in structure on polymer properties. Some guidelines that were used are illustrated by the following examples. Polymer classes such as polyazines which have a constant structural feature are written with this feature at the left, viz., $\{N-N=CH-R-CH\}_n$. The examples are then tabulated in order of increasing complexity of R according to the general guidelines given below. A polymer class that possesses a variable "constant structural feature" such as poly(Schiff Bases) is listed similarly, viz., $\{N-R'-N=CH-R-CH\}_n$. The polymers are then listed with R' held constant according to increasing R complexity. Then R' is changed according to the same guidelines used for R, and the polymers listed again according to increasing R complexity.

The guidelines employed to adjudge increasing complexity of R and R' radicals are as follows.

1. Alkylene.
2. Alkylene-containing heteroatoms ($(CH_2)_2O(CH_2)_2$) follow the all carbon-containing alkylene ($(CH_2)_5$). Oxygen precedes nitrogen precedes sulfur heteroatoms.
3. Cycloalkylene.
4. Aralkylene, e.g. xylylene, $-CH_2-C_6H_4-CH_2-$.
5. Aromatic Radicals.
   i. One Ring. Para precedes meta precedes ortho substitution.
   ii. One Heteroaromatic Ring. Para precedes meta precedes ortho substitution.
   iii. Two Rings. Para substituted rings first. If connected by other atoms, carbon atoms precede heteroatoms. One connecting atom precedes two, etc.
   
      After both para substituted rings; one para—one meta, one para—one ortho; two meta; one meta—one ortho; etc.
   iv. Naphthalene Rings. 1,4-Subthtalene precedes 1,5 precedes 2,7, etc.
   v. Three Rings. Similar hierarchy as outlined above.
   vi. Fused Three-Ring Systems.
   vii. Four Rings, etc.

Copolymers are listed under the homopolymers to which they are most closely related. In some instances, the name of the monomer from which the polymer was prepared is given under the repeat unit structure. If a structure has been presented in the text preceding the table, only a reference to that page and structure number may be listed. Abbreviations that are used in the structural formulas are always explained under the structure. The symbol $\{(\ )_x(\ )_y\}_n$ denotes a copolymer, not necessarily a block copolymer.

*Method*

The tables all contain a Method column which contains at least one number for each entry. This number refers to the synthetic method described in the preceding text of that chapter that was used to prepare the polymer. Since some polymers have been prepared by more than one method, one entry can have more than one method number. Data that are subsequently reported on the same line as the method number were obtained on a polymer sample that was prepared by the indicated method.

### Solubility

All of the tables contain a column relating solubility information. This column generally lists the solvents that have been reported for the polymer. If nothing was said about the solubility of the polymer in the original article, this column is left blank. If a polymer was described as being "insoluble," this has been reported in the table. If data reported in the Molecular Weight column were obtained by solution methods, the first solvent listed under Solubility is the solvent that was used.

### Molecular Weight

All tables contain a column that is entitled "Molecular Weight." The data listed herein can be of several types and are what the original article said about the molecular weight of the polymer that was obtained. If a range of molecular weight data was reported, only the highest values are listed. Abbreviations for the various types of viscosities are as follows: $\eta$, unspecified; $[\eta]$, intrinsic; $\eta_{red}$, reduced; $\eta_{sp}$, specific; $\eta_{inh}$, inherent; $z\eta$, viscosity number. Actual molecular weights are also reported in this column with the method used following in parentheses. Other abbreviations that appear in this column are $\bar{M}$-, number average molecular weight; $\bar{M}_w$, weight average molecular weight; DP, degree of polymerization; and VPO, vapor-phase osmometry. The first solvent that is listed under Solubility is generally the one that was used to obtain molecular weight data by a solution method.

### Tg ($°C$)

This is the glass or second-order transition temperature. These data were only infrequently available and are sometimes reported under the Remarks and Property Data column.

### Melting Point, Tm ($°C$)

This is the melting point of the polymer, but since this term is still used loosely in the polymer literature, the reported data are not always true melting points. When the presence or absence of crystallinity has been determined, the result has also been recorded, sometimes under Remarks and Property Data. Thus, when a material is noted as being crystalline, the reported melting point is quite apt to be a true crystalline melting point. When the method of determining the "melting point" has been reported it has been recorded in the table.

Some abbreviations used in this column are d, decomposition; PMT, polymer melt temperature; TGA, thermogravimetric analysis; and DTA, differential thermal analysis.

### Remarks and Property Data

Most tables contain this column to record other pertinent data that give a more complete picture of the particular polymer. Abbreviations that can be found in these columns include psi, pounds per square inch; IR, infrared; EPR, electron paramagnetic resonance; NMR, nuclear magnetic resonance; and some of the ones already defined previously.

### References

The reference numbers listed are those where preparation of the particular polymer entry has been described or where the polymer is mentioned. They refer only to the list at the end of the chapter in which they appear.

### Abbreviations

| | |
|---|---|
| AIBN | azobis(isobutyronitrile) |
| DMA | dimethylacetamide |
| DMF | dimethylformamide |
| DMSO | dimethylsulfoxide |
| TCNE | tetracyanoethylene |
| THF | tetrahydrofuran |

## CHAPTER I

# Rings Containing Four Carbon Atoms

## A. Polyimides

### 1. POLYIMIDE HOMOPOLYMERS

Aromatic polyimides [1] are a class of polymers that display excellent

$$\left[ \begin{array}{c} N \underset{\underset{O}{\overset{\overset{O}{\parallel}}{C}}^{\overset{O}{\parallel}}{\diagdown} Ar_1 \underset{\underset{O}{\overset{\overset{O}{\parallel}}{C}}}{\overset{\overset{O}{\parallel}}{C}} N - Ar_2 \end{array} \right]_n \qquad (I\text{-}1)$$

**[1]**

mechanical and electrical properties and very high temperature stability. They have matured to commercial status. A commercially available representative is Polymer H [4] derived from pyromellitic anhydride [2] and 4,4'-diaminodiphenyl ether [3]. It is produced and sold by the Dupont Company.

$$n \; \text{[2]} \quad + \quad n \; H_2N\text{—}\langle\bigcirc\rangle\text{—}O\text{—}\langle\bigcirc\rangle\text{—}NH_2 \quad \longrightarrow$$

**[2]**          **[3]**

$$\left[ \begin{array}{c} N \underset{\underset{O}{\overset{\overset{O}{\parallel}}{C}}}{\overset{\overset{O}{\parallel}}{C}} \bigcirc \underset{\underset{O}{\overset{\overset{O}{\parallel}}{C}}}{\overset{\overset{O}{\parallel}}{C}} N\text{—}\langle\bigcirc\rangle\text{—}O\text{—}\langle\bigcirc\rangle \end{array} \right]_n \qquad (I\text{-}2)$$

**[4]**

## METHODS

The reaction of a dianhydride [5] with a diamine [6] in a solvent such as *N,N*-dimethylformamide proceeds smoothly at room temperature and yields a polyamic acid [7] [Eq. (I-3)].

$$n \text{[5]} + n\ H_2NAr_2NH_2 \xrightarrow[\text{25°C}]{\text{DMF}} \left[ \text{HNCOAr}_1\text{CONHAr}_2 \atop \text{COOH COOH} \right]_n \qquad \text{(I-3)}$$

[5]              [6]                    [7]

The dehydration of [7] leads to the polyimide [1].

$$\left[ \text{HNCOAr}_1\text{CONHAr}_2 \atop \text{COOH COOH} \right]_n \xrightarrow{-\,2n\ H_2O} \left[ \text{N} \atop \right]_n \qquad \text{(I-4)}$$

[7]                                      [1]

The transformation shown in Eq. (I-4) can be performed by heat or by the use of a chemical reagent. Thermal dehydration and related reactions are the subject of Method A,1. Method A,2 discusses the use of chemical reagents. The preparation of polyamic acids is common to both methods and will be considered first.

The most commonly used conditions for the preparation of polyamic acids are those that are depicted in Eq. (I-3) (*4*). In addition to DMF, *N,N*-dimethylacetamide, DMSO, *N*-methylpyrrolidone (*4, 36, 49, 53, 58, 139, 208, 218, 465, 468, 562*), and mixtures of these solvents are most frequently employed. The first step of the reaction consists in the formation of a charge-transfer complex between the dianhydride and the solvent. This was shown with pyromellitic anhydride [2] (*36*) and 3,3′,4,4′-diphenyl ether tetracarboxylic acid dianhydride [8] (*360*). The structure of the complex derived from [8]

[8]

and DMSO is presumed to be [9]. Species [9] is essentially an activated form

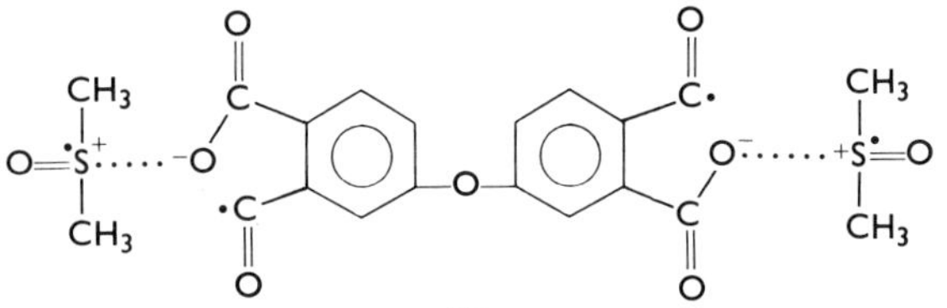

[9]

of the dianhydride. Polymerization of [9] proceeds as described in Eq. (I-5).

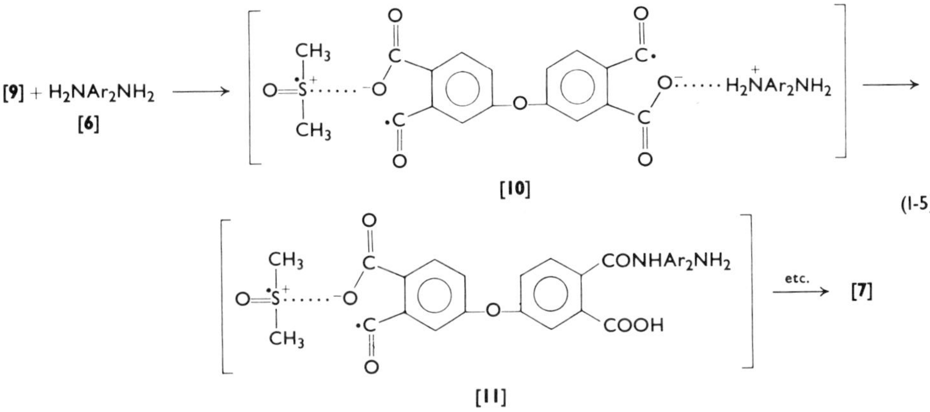

[9] + H₂NAr₂NH₂ →

[6]

[10]

(I-5)

[11]

→ [7]

The synthesis of polyamic acids was studied by several investigators (*121, 122, 172, 198, 200, 207, 208, 213, 351, 620, 721*). The effects of experimental variables upon the molecular weight of the polymer were prime objectives of these studies. There are two conditions that must be met if a high molecular weight material is desired. These are (1) the absence of moisture and (2) relatively low reaction temperatures. Polyamic acids are very sensitive to water and hydrolyze easily. The ease of hydrolysis stems from the fact that the materials are polymeric *o*-carboxy-substituted amides. Literature data (*720*) indicate that amides [12] hydrolyze $10^5$ times faster than their unsubstituted analogs [13]. The effect of water is more complicated, however.

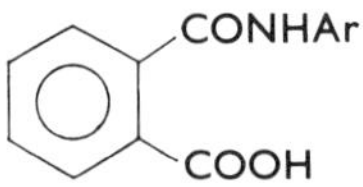

[12]

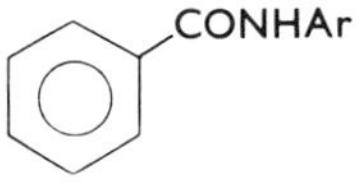

[13]

It was shown (*86, 717, 718*) that water actually catalyzes the formation of the polyamic acid. Simultaneously, water also hydrolyzes the polymer. The rate of the hydrolysis reaction increases with temperature.

Another deleterious effect of water is the hydrolysis of the dianhydride monomer and/or the anhydride end groups (*198, 200, 361*). Both reactions (I-6) and (I-7) will lead to chain termination and consequently low molecular

$$O \overset{\overset{O}{\underset{\|}{C}}}{\underset{\underset{\|}{C}}{}} Ar_1 \overset{\overset{O}{\underset{\|}{C}}}{\underset{\underset{\|}{C}}{}} O + H_2O \longrightarrow O \overset{\overset{O}{\underset{\|}{C}}}{\underset{\underset{\|}{C}}{}} Ar_1 \overset{COOH}{\underset{COOH}{}} \qquad (I\text{-}6)$$

$$[5] \qquad\qquad [14] \qquad\qquad [15]$$

$$\text{Growing polymer} \cdots\cdots Ar_1 \overset{\overset{O}{\underset{\|}{C}}}{\underset{\underset{\|}{C}}{}} O + H_2O \longrightarrow \cdots\cdots Ar_1 \overset{COOH}{\underset{COOH}{}} \qquad (I\text{-}7)$$
$$\text{chain}$$

$$[16] \qquad\qquad [14] \qquad\qquad [17]$$

weight polymer. Studies pertaining to the formation of polyamic acids from pyromellitic anhydride (*198, 200*) have shown that the highest molecular weights resulted when the solid anhydride was added to a solution of the diamine in the solvent. The rationale given was that no premature hydrolysis by moisture prior to reaction was taking place under these conditions.

Reaction temperatures are usually kept below 35°C. Higher temperatures have several deleterious effects. First, if fortuitous water is present, the rate of hydrolysis of both the polyamic acid and the anhydride end groups increases (*718*). Also, at higher temperatures imidation starts and additional water is formed (*276, 277*). Finally, if the imidation attains an appreciable degree, low

$$\bigcirc \overset{CONH\cdots\cdots}{\underset{COOH}{}} \longrightarrow \bigcirc \overset{\overset{O}{\underset{\|}{C}}}{\underset{\underset{C}{}}{}} N\cdots\cdots + H_2O \qquad (I\text{-}8)$$

$$[18] \qquad\qquad [19] \qquad\qquad [14]$$

molecular weight polyimide precipitates and further chain growth is effectively stopped (*620*). The formation of polyamic acids [Eq. (I-3)] is temperature dependent. Its rate is higher as the temperature is increased. It was claimed (*720, 721*) that for each dianhydride/diamine pair there is an optimum temperature at which best results are obtained.

The solvents and reaction conditions that were described above are most commonly employed for the synthesis of polyamic acids. The literature, and in particular the patent literature, describes a multitude of variations. In many instances the solutions of the polyamic acids are applied directly to wire or other substrates and yield a polyimide coating after baking. The accompanying table summarizes these modified procedures.

Experimental Conditions for the Preparation of Polyamic Acids

| Reaction conditions | Remarks | References |
|---|---|---|
| 1. DMA + molecular sieves | Water is eliminated. Very high molecular weight polymer results | 353 |
| 2. DMA + anhydrous alcohol | — | 54, 467 |
| 3. DMA (or DMF) + phenolic ether | The dianhydride forms an activated complex with the ether. Very high molecular weight polymer is obtained (M.wt. ~6–7,000,000) | 68 |
| 4. THF, acetone, methyl ethyl ketone, cyclohexanone and other nonsolvents. They may contain up to 10% of DMF, DMA, DMSO, or N-methyl-pyrrolidone | Polyamic acid of high molecular weight precipitates and is easily isolated as a solvent-free powder | 83, 200 |
| 5. DMF, DMA, DMSO, N-methyl-caprolactam, or N-methyl pyrrolidone in admixture with hydrocarbon solvents such as toluene or other alkylated benzenes | Solutions of the polyamic acids are useful as wire enamels. Addition of the hydrocarbon solvent improves several features of the enamel. Addition of silicones further increases the smoothness of the coating | 45, 150, 206–208, 213, 267 |
| 6. Cresols, xylenols, halogenated phenols, optionally admixed with a hydrocarbon diluent | Better process economics. Applicable to selected diamines and dianhydrides only. Solutions of the polyamic acids are useful in coatings | 148, 328, 614–618 |
| 7. DMF, DMSO, or N-methyl-pyrrolidone in admixture with phenol or cresol | Better economics | 282 |
| 8. Mixtures of cresol and quinoline | Useful in coatings | 240 |
| 9. Mixtures of phenol, cresylic acid, and pyridine | Useful in coatings | 246 |
| 10. Pyridine, at reflux | Partially imidized, powdery polyamic acid results from coagulation in acetone. Heating to 300°–325°C completes the imidation | 243 |

Experimental Conditions for the Preparation of Polyamic Acids—*continued*

| Reaction conditions | Remarks | References |
| --- | --- | --- |
| 11. Pyridine, optionally admixed with DMA | — | *486, 488* |
| 12. Mixtures of dimethyl sulfone with Cellosolve acetate | — | *300* |
| 13. Nitrobenzene, tetramethylene sulfone, sulfolane, α-chloro-naphthalene, or ditolylmethane | Heating the solution (or suspension) of the polyamic acid is often a good way to the polyimide | *371, 374, 375* |

Aromatic polyimides are generally infusible materials that cannot be processed by the techniques employed for thermoplastic polymers. It was claimed (*524*) that the polyimide from 3,3',4,4'-diphenyl ether tetracarboxylic acid dianhydride [8] and 4,4'-diaminodiphenyl ether [3] is processable and soluble in Fomal (a mixture of 10 parts of phenol with 7 parts of 2,4,6-trichloro-

[8]                                     [3]

phenol) when it is prepared at sufficiently low molecular weight. Molecular weight control was accomplished at the polyamic acid stage by using a mono-anhydride or a monoamine (*218, 524*). The use of the olefinically unsaturated monoanhydride [20] was described (*506, 507*). Polyamic acids that were end-capped with [20] cross-linked during thermal imidation. The technique is useful in composites.

[20]

Several studies of the aging behavior of solutions of polyamic acids were performed. Viscosity changes of DMA solutions containing the polymers from pyromellitic anhydride [2] and primary aromatic diamines were investigated (*276, 277*). A rapid early drop in viscosity at room temperature was observed. It was apparently due to two factors: (a) some adventitious water and (b) exchange reactions of free amine or free anhydride with the *o*-carboxy-amide linkage leading to chain cleavage. Slow decline in viscosity took place on longer standing, presumably a consequence of hydrolysis by the water formed from partial cyclization to the imide. A number of additives that reportedly prevent these viscosity changes were described in patents (*173, 174*).

With a view toward understanding the processes that occur during the thermal cyclization of polyamic acids to polyimides, the behavior of the former at high temperatures in air was studied by infrared techniques (*600*). Results indicated that cleavage to amine and anhydride took place simultaneously with the imidation.

The use of specially designed solvent systems allowed the preparation of polyamic acids in the form of organosols (*75*). Polyamic acids were also characterized by dilute solution techniques (*674, 677*).

Two examples of bulk reactions are of interest. It was reported (*547*) that the reaction of pyromellitic anhydride [2] with 1,3-diamino- or 1,2-diamino-benzene to give the corresponding polyamic acids could be performed without solvent at 200°C, under 89,000 psi pressure. Note that [2] and 1,2-diamino-benzene do not form a polymer under usual solution conditions. Other reports (*466, 671*) claimed the copolymerization of a dianhydride with a diamine in bulk to give the polyimide directly. An example is shown (*466*).

METHOD 1. THERMAL DEHYDRATION OF POLYAMIC ACIDS

The thermal dehydration shown in Eq. (I-4) and several related reactions comprise this method. Most commonly, reaction (I-4) is performed by heating a film of the polyamic acid at elevated temperatures *in vacuo* or an inert atmosphere (*4, 15, 62*). The time/temperature cycles vary with the particular polyamic acid. In some instances, the heating cycle lasts from 30 seconds to $\frac{1}{2}$ hour at 300°–350°C (*15, 208*); in other cases heating was continued for

$$(I\text{-}4)$$

several hours at lower temperatures, i.e., 150°C, optionally followed by a few minutes at 300°–350°C (*121, 122*). Generally, longer reaction times are employed at lower temperatures.

Thermal imidation of polyamic acids has been the subject of numerous investigations (*224, 225, 388, 405, 690*). The following mechanism was proposed for the cyclization of the polymer derived from pyromellitic anhydride [2] and 1,3-diaminobenzene [24] (*690*):

$$(I\text{-}10)$$

The processes that take place during the thermal imidation are more complicated, however, than those represented in Eq. (I-10). The particular polyamic
acid that is being cyclized and the experimental conditions employed play a
predominant role in determining what kind of reactions will occur. For example, three types of free radicals were observed in an ESR study (*112*) of the
thermal dehydration of polymer [30].

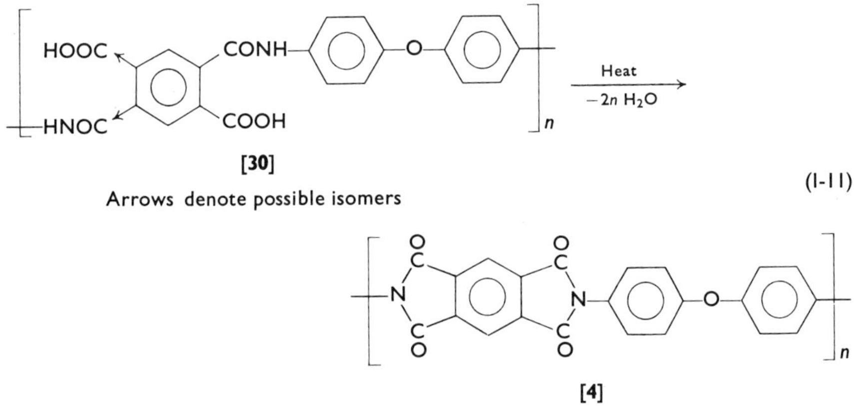

[30]

Arrows denote possible isomers

(I-11)

[4]

The formation of the imide structure may proceed intra- or intermolecularly
(*322, 361*). The former mode is preferred since it leads to a linear chain.

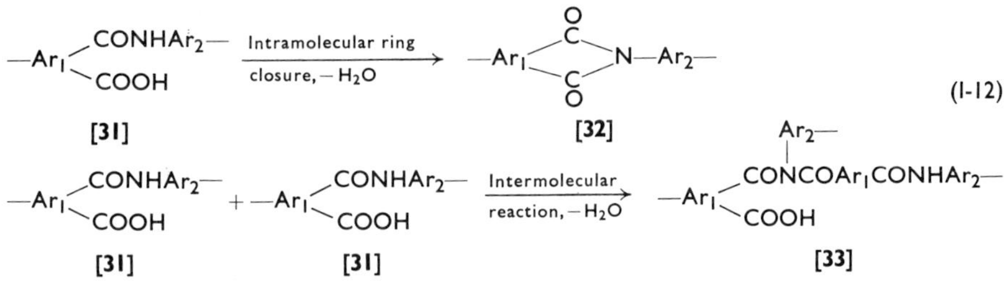

(I-12)

Intermolecular reactions give branched and cross-linked materials. It was
shown in one instance (*361*) that the activation energy of the intramolecular
process was higher than that of the intermolecular process. Therefore, high
dehydration temperatures were recommended in order to obtain a linear
polymer (*361*).

Various experimental modifications of the imidation depicted in Eq. (I-4)
have been described in the literature. They are as follows:

(a) Dehydration of the polyamic acids by refluxing their DMSO, DMF, or
DMA solutions (*64, 73, 483, 551, 558*). A fabrication technique whereby

most of the solvent is driven off from these solutions by shear at high temperature, followed by complete devolatilization and conversion to polyimide at a more elevated temperature, was reported (306).

(b) Compression molding of the free polyamic acids at high temperatures and pressures (27, 63, 198, 200, 248).

(c) A modification of (b) above. Two precondensates are first prepared. One is made using a dianhydride/diamine mole ratio of 2:1; the other is prepared from the same reagents but at a mole ratio of 1:2. The precondensates are mixed to give a molding powder. The latter can be filled optionally and is then compressed to the polyimide (78, 487).

The addition of adjuvants to the polyamic acids prior to dehydration proved useful for the preparation of a variety of composites. Foams were made by bubbling nitrogen, or adding solid carbon dioxide or other appropriate blowing agents to the solutions of the polymeric acids (14, 312, 314). The use of metal acetylacetonates led to metal-filled compositions useful in decorative and electrical applications (15, 217). The addition of powdered perfluorocarbon polymers yielded materials possessing a low friction coefficient and good thermal and chemical resistance (294, 295, 443). Products resistant to high-temperature corona discharge were similarly obtained when a variety of organometallics were added to the polyamic acids (428). Several silicone polymers led to compositions useful as solvent-resistant coatings and adhesives (475).

When the imidation is carried out under well-defined solution conditions, fine polyimide powders consisting of particles having uniform dimensions are obtained. Surface areas as high as 500 $m^2$/gm have been achieved (283, 284, 286, 477). These powders can be coalesced into shaped objects under high temperature (200°–500°C) and pressure (10,000–30,000 psi). The coalesced materials display good mechanical properties. Generally, these preparations consist in heating the polyamic acids in solution in the presence of a base. A fine, partially imidated precipitate results (269, 270). Further heating of the latter under nitrogen atmosphere at 325°C for 8–16 hours yields the final products. Modifications of this procedure have been described (287, 495).

Dianhydrides and diamines are not the only known starting materials for the preparation of polyamic acids and polyimides. Aminoanhydrides can also be utilized (42, 79, 80, 152). An example is shown in Eq. (I-13) (152).

Arrows denote isomers

Note that as early as 1908, Bogert and Renshaw (*110*) prepared the anhydride [34]. They stated: "It does not melt but gives off more water at higher temperatures, perhaps with formation of a polymolecular imide." Bogert's and Renshaw's paper is probably the first literature reference to an aromatic polyimide!

The polymerization depicted in Eq. (I-13) is catalyzed by acids ($BF_3$, $CH_3COOH$, HBr, etc.) and bases (triethylamine and pyridine) (*153*). Salts of aminoanhydrides (*152*) and the *N*-acyl derivatives (*94, 504*) were also useful.

The literature describes a number of variations of Method A,1. Some follow:

(*a*) *Polymerization of Tetraacids with Diamines* (*109, 126, 212, 213, 322*)

The polymerization of tetraacids with diamines is performed in bulk, preferably in an inert atmosphere, optionally under reduced pressure. The mixture of the tetraacid and diamine is heated in stages to temperatures as high as 300°–325°C. The method is of particular utility for the preparation of the fusible polyimides based on aliphatic diamines. In one instance (*307*), polyphosphoric acid was used as both the solvent and condensing agent.

(*b*) *Polycondensation of Ester-Amide Salts* (*95, 97, 124, 214, 216*)

An illustration is provided in Eq. (I-14). The ester-amide salt [39] can be

Diester-diacid;
arrows denote isomers

(I-14)

Ester-amide salt $\xrightarrow{\text{300°C, N}_2}$ Polyimide

[39] [40]

recrystallized prior to imidation (*216*). It was claimed (*391*) that the salt route led to a more linear polyimide than the polyamic acid route. The polyimide can be formed in a bulk reaction, for example, compression molding (*216*); or by heating a suspension of the amide-ester salt in diphenyl ether (*251, 484, 485*).

*c. Cyclization of Poly(ester-amides)*

Heating of poly(ester-amides) [41] at elevated temperatures yields the corresponding polyimides [1].

$$
\left[\begin{array}{c} \text{HNCOAr}_1\text{CONHAr}_2 \\ \diagup \qquad \diagdown \\ \text{COOR} \quad \text{COOR} \end{array}\right]_n \xrightarrow{200°-350°C} \left[\text{N}\underset{\underset{\text{C}}{\overset{\text{C}}{}}}{\overset{\overset{\text{O}}{\text{C}}}{\diagup\diagdown}}\text{Ar}_1\underset{\underset{\text{O}}{\text{C}}}{\overset{\overset{\text{O}}{\text{C}}}{\diagdown\diagup}}\text{N}-\text{Ar}_2\right]_n \quad (\text{I-15})
$$

[41]                                          [1]

The ester derivatives [41] of the polyamic acids can be obtained by several methods:

(1) Esterification of the corresponding acid (*18, 20, 548*).
(2) The reaction of a diacid-diester with a diamine (*190*).

$$
n\begin{array}{c}\text{HOOC}\diagdown\qquad\diagup\text{COOH}\\ \qquad\text{Ar}_1\\ \text{ROOC}\diagup\qquad\diagdown\text{COOR}\end{array} + n\,\text{H}_2\text{NAr}_2\text{NH}_2 \xrightarrow{\sim150°C} \left[\begin{array}{c} \text{HNCOAr}_1\text{CONHAr}_2 \\ \diagup\qquad\diagdown \\ \text{COOR}\quad\text{COOR}\end{array}\right]_n \quad (\text{I-16})
$$

[42]                    [6]                              [41]

(3) The condensation of a tetraester with a diamine (*427, 476, 647, 660*).

$$
n\begin{array}{c}\text{ROOC}\diagdown\qquad\diagup\text{COOR}\\ \qquad\text{Ar}_1\\ \text{ROOC}\diagup\qquad\diagdown\text{COOR}\end{array} + n\,\text{H}_2\text{NAr}_2\text{NH}_2 \xrightarrow{200°-280°C} \left[\begin{array}{c} \text{HNCOAr}_1\text{CONHAr}_2 \\ \diagup\qquad\diagdown \\ \text{COOR}\quad\text{COOR}\end{array}\right]_n
$$

[43]                    [6]                              [41]

(I-17)

(4) The coupling of a diacid halide–diester with a diamine. Reaction (I-18) can be performed interfacially (*519*) or via low-temperature polycondensation (*590, 611, 612*).

$$
n\begin{array}{c}\text{ROOC}\diagdown\qquad\diagup\text{COOR}\\ \qquad\text{Ar}_1\\ \text{ClOC}\diagup\qquad\diagdown\text{COCl}\end{array} + n\,\text{H}_2\text{NAr}_2\text{NH}_2 \longrightarrow \left[\begin{array}{c} \text{HNCOAr}_1\text{CONHAr}_2 \\ \diagup\qquad\diagdown \\ \text{COOR}\quad\text{COOR}\end{array}\right]_n \quad (\text{I-18})
$$

[44]                    [6]                              [41]

(5) Treatment of polyiminolactones with alcohols or thiols (*642*). Poly-iminolactones [45] are isomeric with the polyimides. Their preparation and

$$
\left[\text{O}\underset{\underset{\text{N}}{}}{\overset{\overset{\text{O}}{\text{C}}}{\diagup\diagdown}}\text{Ar}_1\underset{\underset{\text{N}-\text{Ar}_2}{}}{\overset{\overset{\text{O}}{\text{C}}}{\diagdown\diagup}}\text{O}\right]_n \xrightarrow[\text{RSH}]{\text{ROH or}} \left[\begin{array}{c} \text{NHCOAr}_1\text{CONHAr}_2 \\ \diagup\qquad\diagdown \\ \text{COOR(SR)} \\ \text{COOR(SR)}\end{array}\right]_n \quad (\text{I-19})
$$

[45]                                          [46]

properties are described in a separate section of this chapter. Treated with alcohols or thiols they readily give the poly(ester-amides) or polythioester-amides [46]. The thio analogs are also easily converted to the polyimides on heating.

(6) The use of trimethylsilyl esters [48] was described (111); N-silylamines [47] were the starting materials. Using the relatively inexpensive tetrahydro-furan as solvent was claimed as an advantage of this modification (111).

$$n\ (CH_3)_3SiHNAr_2NHSi(CH_3)_3\ +\ n\ \underset{[5]}{\text{O}}\overset{\text{O}}{\underset{\text{O}}{\text{C}}}Ar_1\overset{\text{O}}{\underset{\text{O}}{\text{C}}}O \xrightarrow[\text{25°C}]{\text{THF}}$$

[47]          [5]

(I-20)

$$\left[ \begin{array}{c} \text{HNCOAr}_1\text{CONHAr}_2 \\ \diagup \quad \diagdown \\ \text{COOSi(CH}_3)_3 \\ \text{COOSi(CH}_3)_3 \end{array} \right]_n$$

[48]

The ester-amide approach was also successful in the preparation of polyimides from amino anhydrides (681, 682).

$$n\ \underset{\substack{\text{NH}_2\cdot\text{HCl}\\ [49]}}{\overset{\text{COOC}_6H_5}{\underset{\text{COOC}_6H_5}{\bigcirc}}} \xrightarrow{115°\text{–}200°C} \underset{[50]}{\text{Poly(amide-ester)}} \xrightarrow[\text{Vacuum}]{190°\text{–}280°C} \underset{[51]}{\text{Polyimide}} \qquad (I-21)$$

### d. Dehydration of Ammonium Salts of Polyamic Acids (28, 62, 221, 259, 561)

The ammonium salts [52] of polyamic acids are prepared by treating the latter with an amine. They are often soluble in water. The aqueous solutions are useful in coatings (62) and in the preparation of composites (259).

$$\left[ \begin{array}{c} \text{HNCOAr}_1\text{CONHAr}_2 \\ \diagup \quad \diagdown \\ \text{COOH} \quad \text{COOH} \end{array} \right]_n \xrightarrow[\text{H}_2\text{O/solvent}]{2n\ R_3N} \left[ \begin{array}{c} \text{HNCOAr}_1\text{CONHAr}_2 \\ \diagup \quad \diagdown \\ \overset{-}{\text{COO}}\overset{+}{\text{NHR}_3} \\ \overset{-}{\text{COO}}\overset{+}{\text{NHR}_3} \end{array} \right]_n \xrightarrow{\text{Heat}}$$

[7]                              [52]                (I-22)

$$\left[ \begin{array}{c} \text{N}\overset{\text{O}}{\underset{\text{O}}{\text{C}}}Ar_1\overset{\text{O}}{\underset{\text{O}}{\text{C}}}\text{N}-Ar_2 \end{array} \right]_n$$

[1]

*e. Thermal Imidation of Poly(amide-amides)* (*29, 59*)

The amides of polyamic acids [53] are prepared by heating a polyimino-lactone [45] with an amine or ammonia. Heating [53] at 300°C for a few minutes

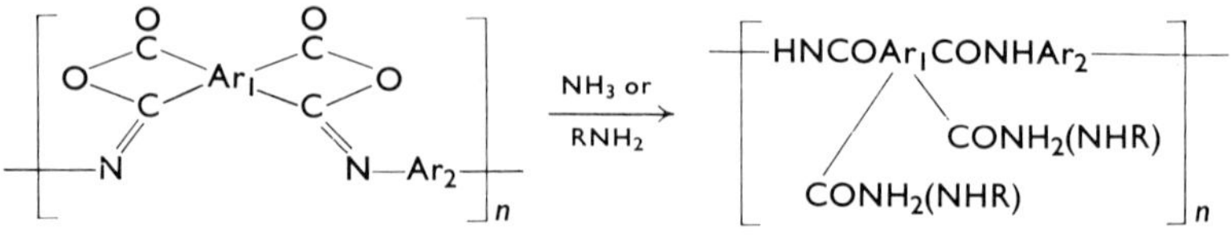

(I-23)

[45]           [53]

yields the polyimide [1]. A variation of this procedure makes use of a poly-tetrazole intermediate [54]. The sequence is shown in Eq. (I-24).

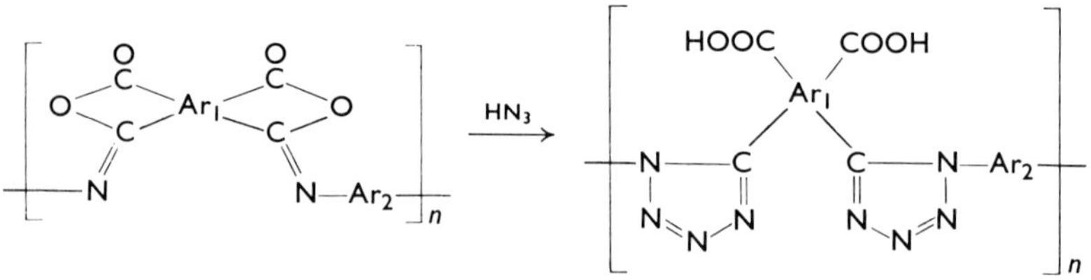

[45]           [54]

(I-24)

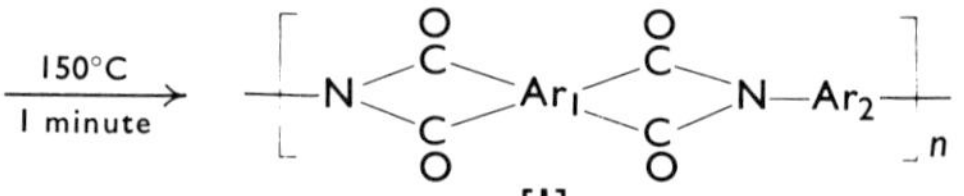

[1]

*f. Other Thermal Imidation Reactions*

Method A,1 is useful for the preparation of high-molecular-weight aromatic, aliphatic, and cycloaliphatic polyimides (*208*). It is also successful for the preparation of polymers that contain heterocyclic nuclei, e.g., derived from diamines such as [55] and [56] (*275*). An alternate preparation of the 1,3,4-oxadiazole-containing materials is discussed in Section A,4.

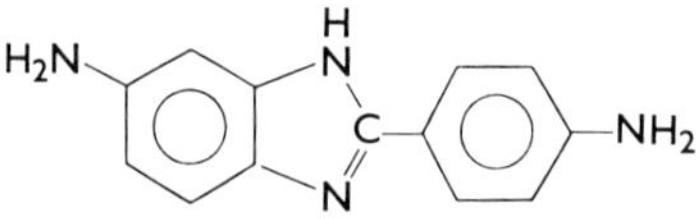

[55] 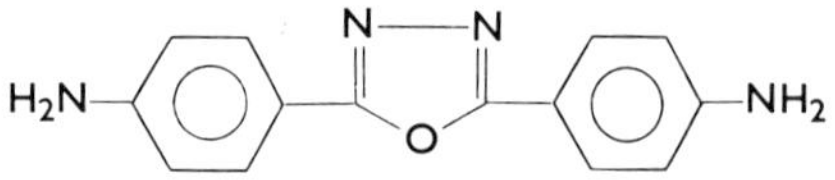           [56]

The reaction of the dianhydride [57] with a variety of diamines gave co-polymeric poly(amide-imides) due to decarboxylation during the thermal imidation (*40*).

[57]

The preparation of fibers by dry-spinning of polyamic acids followed by imidation has been reported (*337, 479*).

The use of dianhydride [58] led to products that contained seven-membered imide rings (*296, 298*).

[58]  + *n* H$_2$NAr$_2$NH$_2$  [6]  ⟶  [59]                    (I-25)

As mentioned before, the great majority of polyimides are insoluble and infusible materials. It is of interest to note that proper variation of the structure did lead to moldable and soluble polymers. For instance, the dianhydride [62], prepared as shown in Eq. (I-26), was condensed with a variety of diamines. Some of the products were soluble and tractable (*488, 494*).

[60]       [61]   + 2   Heat ⟶   [62]                          (I-26)

The condensation of hydrazines with dianhydrides has been described by several investigators (*199, 366, 368*). The behavior of the parent compound, hydrazine, with pyromellitic anhydride was studied by Korshak *et al.* (*368*). It is summarized in Eq. (I-27). The polyamic acid [64] was prepared under typical conditions in DMF, pyridine, or hexamethylphosphoramide. Its thermal dehydration proceeded differently when carried out with or without a solvent. In solution, polymer [66] was obtained. Heating the neat polyacid

$$(1 \cdot 27)$$

*in vacuo* gave the polyimide [65]. The structures of both were supported by infrared evidence. Isomerization of [65] into [66] was easily accomplished at 200°C in the presence of glycerine. Interestingly, a copolymer containing both rings could be prepared by stopping the reaction of [64] to [66] before completion, isolating the material, and continuing the imidation in the solid state (*368*).

Apparently, only the expected polyimides [69] are obtained from arylene dihydrazines (*366*). An example is shown in Eq. (I-28).

$$(I-28)$$

The condensation of *N,N'*-diaminopyromellitimide [70] with pyromellitic anhydride has also been studied (*199*). Infrared investigation of the product did not allow differentiation between the two possible structures [71] and [72].

Stille and Morgan (*633*) investigated the polymerization of 4,8-diphenyl-1,5-diazabicyclo[3,3,0]octane-2,3,6,7-tetracarboxylic dianhydride [73] with aliphatic and aromatic diamines. The structures of the polyimides [75] were confirmed by their independent preparation from benzalazine [76] and the bismaleimides [77] via a 1,3-addition polymerization [Eq. (I-30)].

An interesting thermal rearrangement followed by decarboxylation was observed with the polyimide from 4,4′-diamino-3,3′-dihydroxylbiphenyl [78] and pyromellitic anhydride (354). The reaction sequence is shown in scheme (I-31). Poly[2,2′-(p-phenylene)-6,6′-dibenzoxazole] [81] was the end product at 450°C.

Elastomeric polyimides were prepared from siloxane diamines and poly-oxyethylene diamines (186, 330). An example is shown (186).

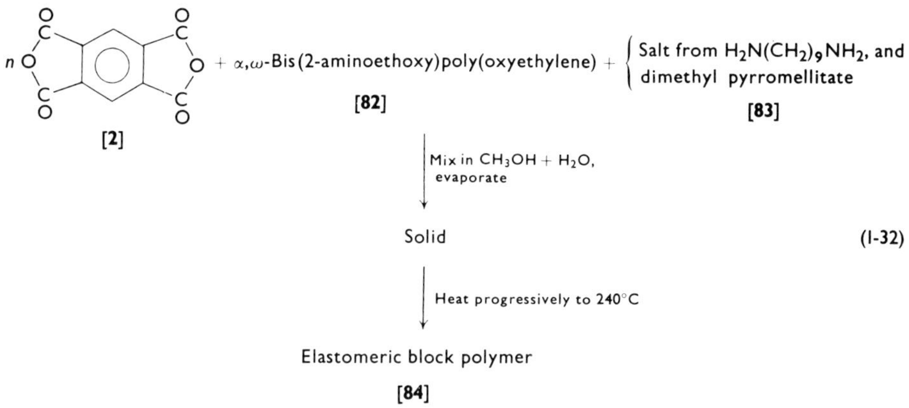

$n$ [2] + α,ω-Bis(2-aminoethoxy)poly(oxyethylene) + { Salt from $H_2N(CH_2)_9NH_2$, and dimethyl pyrromellitate

[82]   [83]

Mix in $CH_3OH + H_2O$, evaporate

Solid   (I-32)

Heat progressively to 240°C

Elastomeric block polymer

[84]

The preparation of rigid block polymers has been described (659). Two oligomeric segments, e.g., [87] and [90], possessing mutually reactive end groups were first prepared. Their condensation led to the poly(ester-imide) [91]. Good thermal stability and improved solubility characteristics were claimed for this product (659).

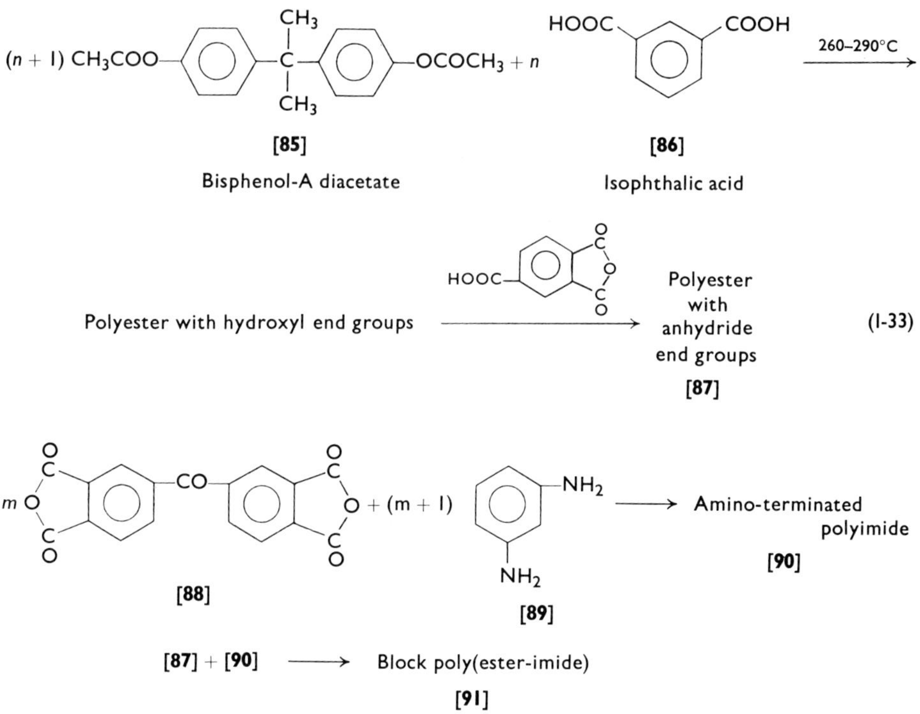

$(n + 1)$ $CH_3COO$—⬡—C(CH₃)₂—⬡—$OCOCH_3$ + $n$ HOOC—⬡—COOH  →  260–290°C

[85]   [86]
Bisphenol-A diacetate   Isophthalic acid

Polyester with hydroxyl end groups  →  Polyester with anhydride end groups   (I-33)

[87]

$m$ [88] + $(m + 1)$ [89] →  Amino-terminated polyimide

[90]

[87] + [90]  ⟶  Block poly(ester-imide)

[91]

METHOD 2. CHEMICAL DEHYDRATION OF POLYAMIC ACIDS

In addition to heat, several reagents are capable of promoting the dehydration of a polyamic acid to the corresponding polyimide. [Eq. (I-4)]. The most frequently used system is a mixture of acetic anhydride and pyridine (26, 27, 49, 465). An inert solvent such as benzene or cyclohexane may be added (27, 93, 209, 210). A variety of other bases are useful in lieu of pyridine. These include quinoline, isoquinoline, their substituted derivatives, triethylamine, N-methylmorpholine, and several others (47, 61, 145, 208, 325, 474). In addition to acetic anhydride, the usefulness of numerous other anhydrides has been claimed (145, 311, 313, 325).

Typically, the reaction is performed in the following way. A cast film of the polyamic acid is dried and then soaked in the dehydrating mixture, generally at room temperature. Depending on the particular case, reaction times may vary from a few minutes to several days. The film is then extracted with benzene or dioxane and dried at elevated temperatures. In many instances the extraction is omitted and the film dried directly after the treatment. Several modifications of this procedure have been described (145, 311, 325). For instance, acetic anhydride and pyridine may be added to the polyamic acid solution prior to casting.

Examination of the literature indicates that treatment at elevated temperatures (up to 300°–400°C) is usually given to the polyimide article after dehydration (209, 210, 218). Heating times vary, depending on the temperature. Improvement in electrical and other properties was claimed to occur under these conditions. Possibly, additional cyclization takes place and yields a polymer that displays better performance characteristics.

The combination of anhydride/base is not the only known reagent that transforms polyamic acids to polyimides. Amino anhydrides [92] have also been claimed (81). N-Acylazoles, where the azole group is an imidazole,

$$\left( \begin{array}{c} R \\ R_1 \end{array} \!\!\!> N - R_2 C \!\!\! \underset{\displaystyle \parallel}{\overset{\displaystyle O}{}} \!\!\! \right)_2 \!\!\! - O$$

[92]

benzimidazole, pyrazole, 1,2,3-triazole, benzotriazole, or tetrazole, have also been described (82). Finally, a host of "miscellaneous" reagents are given in reference (222). Some examples are $SOCl_2$/pyridine, $C_6H_5P(O)Cl_2$/pyridine, $CH_3COCl$/pyridine, $P_2O_5$/pyridine, $(ClCH_2CH_2CO)_2O$/pyridine, and polyphosphoric acid/pyridine. It should be noted that many of these reagents also promote the cyclization of polyamic acids to poly(iminolactones) (29, 59, 386, 387, 470) as shown in Eq. (I-34). The latter isomerize upon heating into the corresponding polyimides. The isomerization reaction [Eq. (I-35)] is Method

$$(I-34)$$

$$(I-35)$$

A,3 for the preparation of polyimides. Dehydrations with the above reagents are generally followed by a heating step. Thus, the sequence of products that form in these cases is probably

$$\text{Polyamic acid} \xrightarrow{\text{Dehydration}} \text{Polyiminolactone} \xrightarrow{\text{Heat}} \text{Polyimide} \quad (I-36)$$

$$[7] \qquad\qquad [45] \qquad\qquad [1]$$

The reaction of biphenyl-2,2′, 6,6′-tetracarboxylic acid dianhydride with a variety of aromatic diamines [Eq. (I-25)] was studied (296, 298). Dehydration of the intermediate polyamic acid was performed both via Methods A,1 and A,2. Interestingly, in this case, cyclization occurred by refluxing the acid in acetic anhydride without added base.

Method A,2 utilizes relatively mild reaction conditions. Therefore, secondary reactions such as cleavage and cross-linking are minimized (675, 676, 678). Cyclization of the polyamic acid based on pyromellitic anhydride and 2,4-diaminoisopropylbenzene [93] yields the polyimide [94], which is soluble in several solvents. The solubility allowed for the determination of the

$$(I-37)$$

degree of polymerization. It was shown that relatively minor changes in the number average molecular weight took place on chemical dehydration (675, 676, 678).

Method A,2 was used for the preparation of a wide range of polyimides (2, 26). Under well-controlled experimental conditions, polyimide powders consisting of particles possessing uniform dimensions and high surface areas were made. They could be coalesced into shaped objects under high temperature and pressure (219). Method A,2 has also been used for the preparation of foams (312, 314) and other cellular polyimides (13, 14). Filled compositions were obtained by adding fillers to the polyamic acids prior to imidation (138, 210).

An interesting reaction was observed with the polyimide from pyromellitic anhydride and 3,3'-bis(carbamido)-4,4'-diaminobiphenyl [95]. When the polyimide [96] was refluxed in acetic anhydride for 10 hours a polypyrimido-pyrrolone [97] was obtained (551).

$$(I\text{-}38)$$

METHOD 3. POLYIMIDES VIA INTERMEDIATE POLYIMINOLACTONES

Under certain conditions, polyamic acids can be dehydrated to polymers that are isomeric with the polyimides (15, 19, 29, 169). These polymers are referred to as poly(iminolactones) or poly(isoimides) [45] and are discussed in a separate section of this chapter. Upon heating, polyiminolactones re-

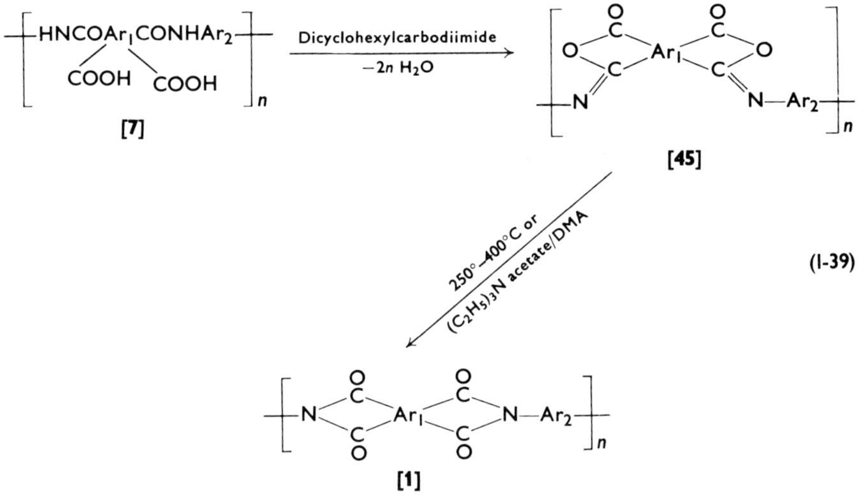

(I-39)

arrange into the corresponding polyimides. The reaction is rapid and some-times complete within seconds (25). Adequate adjustment of time and tem-perature allows the preparation of poly(imide-iminolactone) copolymers (387).

## METHOD 4. POLYIMIDES FROM DIAMINES AND TETRANITRILES

The following reaction was reported (90).

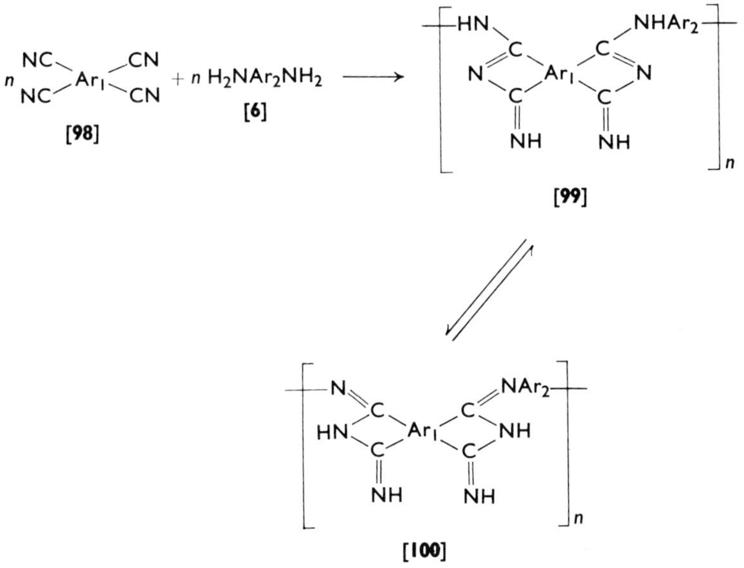

(I-40)

## METHOD 5. POLYIMIDES FROM DIISOCYANATES AND DIANHYDRIDES

The reaction of a diisocyanate with a dianhydride gives a polyimide and carbon dioxide. The synthesis is illustrated in Eq. (I-41). Both aromatic and aliphatic diisocyanates were used (*96, 151, 223, 447, 493*). The reactions are

$$n\,[\mathbf{5}] + n\,OCNAr_2NCO\;[\mathbf{101}] \xrightarrow{\text{Heat}} 2n\,CO_2\;[\mathbf{102}] + [\mathbf{1}]_n \qquad (\text{I-41})$$

preferably carried out in a solvent such as DMF, DMA, DMSO, or *N*-methylpyrrolidone (*151, 223, 447, 452, 493*). Polymerization temperatures vary as a function of the system. For instance, in DMF the preferred temperature is about 130°C. A secondary condensation shown in Eq. (I-42) takes place to a significant extent when stronger heating is applied (*447*). In other solvents, satisfactory results were obtained at temperatures up to 160°C.

$$HCON(CH_3)_2\;[\mathbf{103}] + OCN \cdots\;[\mathbf{104}] \xrightarrow{150°C} CO_2\;[\mathbf{102}] + \cdots N{=}CHN(CH_3)_2\;[\mathbf{105}] \qquad (\text{I-42})$$

Bulk preparation of a polyimide (190°C, *in vacuo*) was also reported (*447*). The following mechanism was postulated for this polymerization (*223, 447*).

$$[\mathbf{106}] + OCN\cdots\;[\mathbf{104}] \longrightarrow [\mathbf{107}] \xrightarrow[-CO_2]{\text{Heat}} [\mathbf{19}] \qquad (\text{I-43})$$

In the first step, the isocyanate function interacts with the anhydride group to give a polymer that possesses structural units [107]. The latter are unstable to heat, lose carbon dioxide, and yield the polyimide [19]. Reaction of pyromellitic anhydride [2] and 4,4'-diisocyanatodiphenylmethane [108] was performed in DMF at 130°C (447). Both the polyimide [110] and its precursor [109] were isolated. The structure of [109] was supported by its infra-

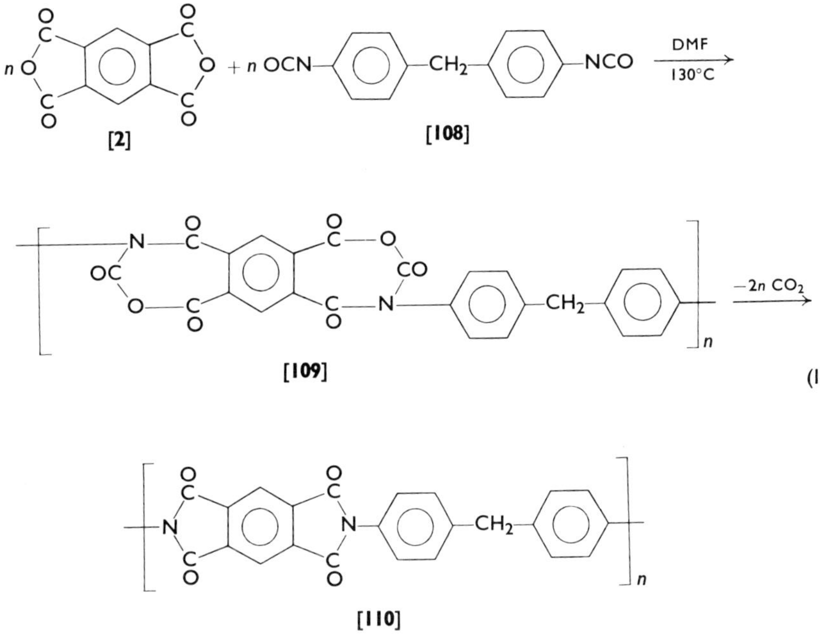

[2]                    [108]

[109]                                                        (I-44)

[110]

red spectrum; moreover, heating at 130°C in DMF resulted in loss of carbon dioxide and formation of [110]. Interestingly, when the reaction was performed under $CO_2$ pressure, the yield of [109] increased, as expected from LeChatelier's principle (447).

The condensation of pyromellitic anhydride with a variety of diisocyanates to isocyanato-terminated polymers was reported (452). The end groups were masked via reaction with a phenol. The products were claimed to be useful as adhesives and auxiliary agents for rubber (452). In another application, a polyfunctional isocyanate (polyaniline-polyisocyanate, PAPI) was condensed with 3,3',4,4'-benzophenonetetracarboxylic acid dianhydride to give a rigid foam. The latter possessed excellent thermal stability, fire, solvent, and chemical resistance (223).

An interesting synthesis of a polyimide that combines Methods A,1 and A,5 has been described (589). The synthetic scheme is shown in Eq. (I-45).

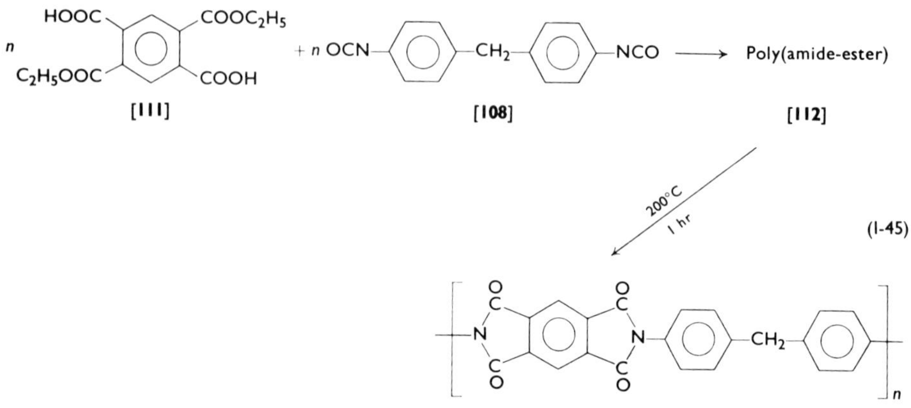

[III]                          [108]                          [112]

(I-45)

[110]

## METHOD 6. MISCELLANEOUS POLYIMIDE SYNTHESES

The following ring-opening/ring-forming polymerization has been described (559):

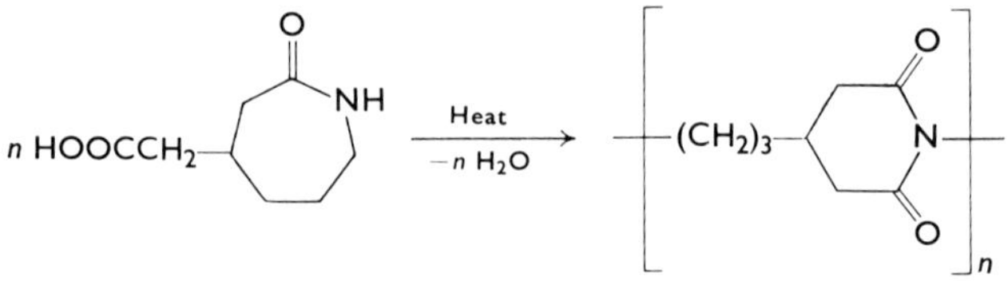

[113]                          [114]

(I-46)

High molecular weight polymers possessing reduced viscosities of up to 4.5 (m-cresol) were obtained. Their structure was supported by infrared and nuclear magnetic resonance spectral evidence (601). An interesting phenomenon was observed. When the formation of polymer [114] was conducted at temperatures *above* its melting point (~280°C), an amorphous material was obtained; on the other hand, polymerizations at temperatures below the melting point of [114] gave a crystalline product. Annealing of the amorphous resin above its glass transition temperature (~90°C) resulted in rapid crystallization (559).

## PROPERTIES OF POLYIMIDES BASED ON ALIPHATIC DIAMINES

The properties of polyimides based on aliphatic diamines have not been of significant practical interest. Due to the aliphatic residues, the glass transition temperatures and oxidative stabilities of these polymers are low. Typical data are listed in the accompanying table (*620*).

Properties of Polypyromellitimides Based on Aliphatic Diamines

| Diamine | Stability[a] (hr) | $T_g$ (°C) |
|---|---|---|
| 3-Methyl-1,7-diaminoheptane | 8–10 | 135 |
| 4,4-Dimethyl-1,7-diaminoheptane | 20–30 | 135 |
| 1,9-Diaminononane | 20–25 | 110 |

[a] Time during which sample retains toughness when heated in air, at 175°C.

The polypyromellitimides were moldable and in many instances, soluble. For example, it was reported that the polymer from 1,9-diaminononane can be compression molded at 340°C (*212, 213*) and that it is soluble in *m*-cresol (*322*). Another polypyromellitimide based on diamine [115] "could be pressed

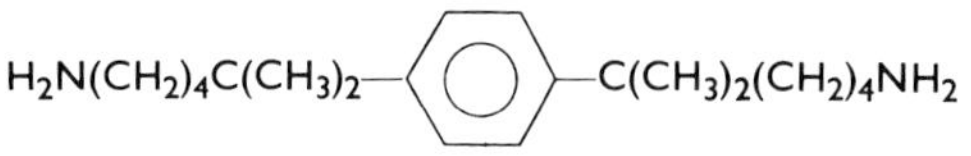

$$H_2N(CH_2)_4C(CH_3)_2 - \!\!\!\left\langle\bigcirc\right\rangle\!\!\! - C(CH_3)_2(CH_2)_4NH_2$$

[115]

at 310° to a tough film" (*215*). Copolyimides from mixtures of pyromellitic anhydride [2], 2,2-bis(3,4-dicarboxyphenyl)propane dianhydride [116] and 1,4-bis(aminomethyl)benzene [117] were compression and injection moldable (*251, 484, 485*) at very high temperatures and pressures.

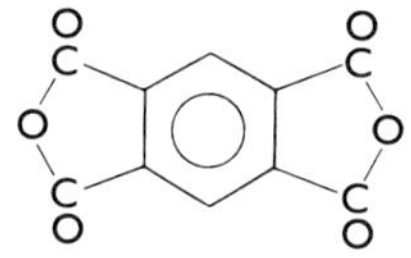

[2]

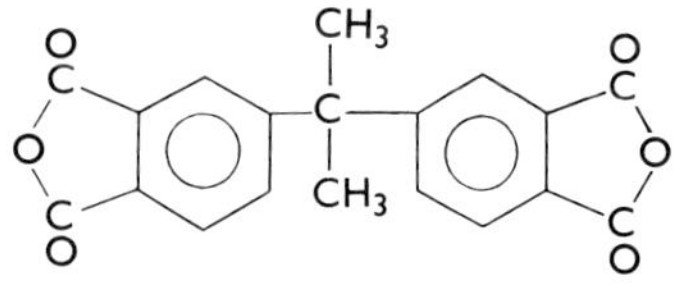

[116]

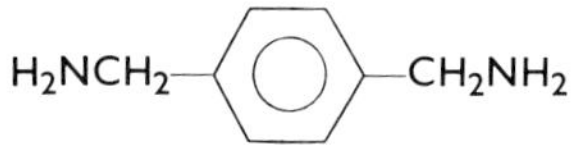

[117]

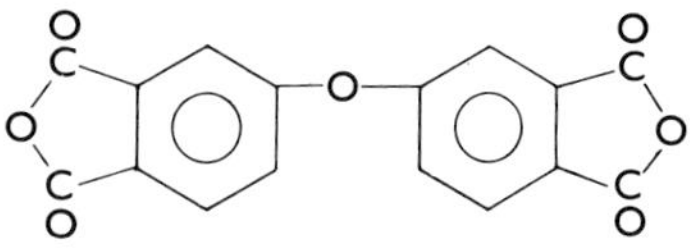

[8]

Moldable products were also prepared from other dianhydrides. Polymers derived from [8] and aliphatic diamines could be thermally processed when the amino functions were separated by at least four carbon atoms (*620*). A series

of polyimides utilizing both aliphatic and aromatic diamines was prepared
from the dianhydride [118] (39). The aromatic diamines yielded polymers that

[118]

[114]

did not soften or melt; softening points of the aliphatic derivatives were in
the range of 110°–230°C. The polyimide [114] was soluble in several solvents
including formic acid, m-cresol, trifluoroethanol, and sulfuric acid (559).
It melted at 280°C and could be fabricated above that temperature.

The relatively poor thermal-oxidative stabilities of aliphatic pyromellitimides were illustrated earlier. Several thermal stability studies comparing a
variety of aliphatic and aromatic polyimides were performed (509, 510, 567).
In all instances, the aliphatic materials degraded faster and at lower temperatures than the aromatic materials.

## PROPERTIES OF POLYIMIDES BASED ON AROMATIC DIAMINES

Polyimides prepared from aromatic diamines possess several outstanding
properties. They are the most thermally stable, commercially available
polymers. As an illustration, Polymer H [4], which is based on pyromellitic
anhydride [2] and 4,4′-diaminodiphenyl ether [3], possesses the following
unique combination of properties (11, 12).

[4]

Polymer H

(1) Good mechanical properties over the temperature range of liquid
helium (b.p. −269°C) to +500–600°C. Its "zero-strength temperature"* is
about 800°C, well above the value of 550°C for aluminum.

(2) High electrical resistivity, high dielectric strength, and low loss. The
dielectric properties are relatively constant over a wide range of temperature
and frequency.

---

* Maximum temperature at which a film of the polymer sustains a stress of 20 psi.

(3) Mechanical and electrical properties maintained at high levels after prolonged exposure to high temperature. Based on experimental data that were obtained on films aged at 300°–400°C, it was predicted that Polymer H could last for several years at 250°C.

(4) Polymer H is flameproof and infusible; there are practically no known solvents for it. In addition, its radiation resistance is excellent (*359*).

The polymer represented by formula [4] is sold by the DuPont Company. The name "Polymer H" stems from the fact that it has the ability to exceed the insulation requirements of class H (180°C). The material is also sold under such names as "Kapton Film," "Polymer SP," etc. (*107, 582*).

The properties and applications of aromatic polyimides are discussed in many reviews (*11, 12, 33, 34, 166, 183, 188, 189, 197, 389, 414, 520, 638, 643, 648, 649, 669*).

### a. Solubility and Melt-Processibility

Aromatic polyimides are generally insoluble, infusible materials that cannot be processed by the usual techniques for thermoplastics. Films or other objects are made from the soluble polyamic acid precursor stage. Solubilities of some typical pyromellitimides are listed in the accompanying table.

Solubilities of Typical Polypyromellitimides

| Diamine | Solvent | References |
|---|---|---|
| H$_2$N—⟨○⟩—NH$_2$ | Amorphous: conc. H$_2$SO$_4$ <br> Crystalline: insoluble | *620, 622* |
| NH$_2$ ⟨○⟩ NH$_2$ | Amorphous: conc. H$_2$SO$_4$ <br> Crystalline: insoluble | *620, 622* |
| H$_2$N—⟨○⟩—⟨○⟩—NH$_2$ | Fuming HNO$_3$ | *620, 622* |
| H$_2$N—⟨○⟩—CH$_2$—⟨○⟩—NH$_2$ | Conc. H$_2$SO$_4$ | *620, 622* |
| H$_2$N—⟨○⟩—C(CH$_3$)$_2$—⟨○⟩—NH$_2$ | Conc. H$_2$SO$_4$ | *620, 622* |

Solubilities of Typical Polypyromellitimides—*continued*

| Diamine | Solvent | References |
|---|---|---|
| 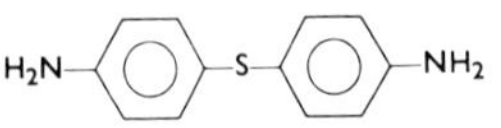 | Fuming HNO$_3$ | *620, 622* |
| 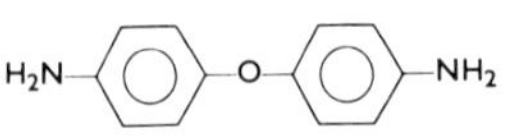 | Fuming HNO$_3$, molten SbCl$_3$ mixtures AsCl$_3$+ SbCl$_3$ | *181, 620, 622* |
| 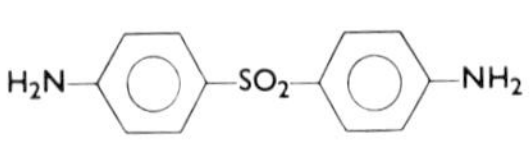 | Conc. H$_2$SO$_4$ | *620, 622* |
| 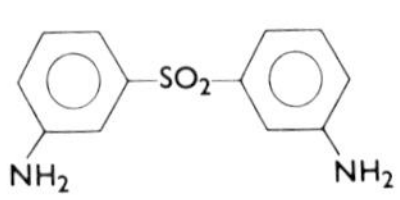 | Conc. H$_2$SO$_4$ | *620, 622* |

Varying the structures of the dianhydride and diamine moieties did yield some soluble and melt-processible polymers. For example, Russian workers (*370, 374*) have shown that the introduction of bulky substituents into the chain enhances solubility. Polyimides from pyromellitic anhydride and the

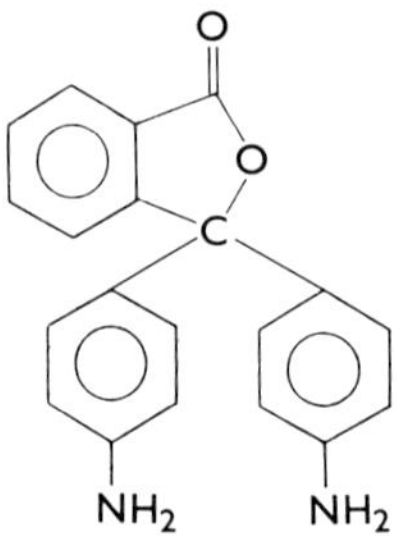

[119]

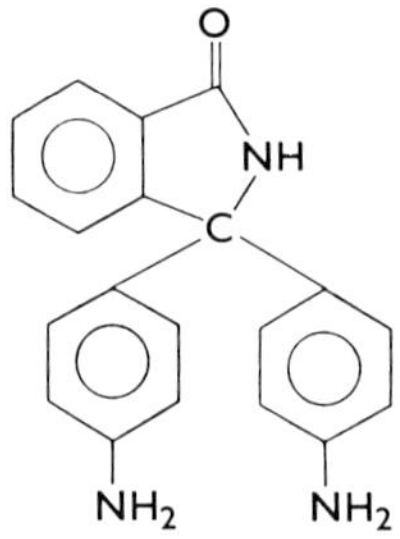

[120]

diamines [119] and [120] were soluble in nitrobenzene. Additional soluble and/or melt-processible materials are tabulated in the accompanying table (pp. 31–32).

### b. Crystallinity

Some polyimides are crystalline as made. Several others could be crystallized via annealing at high temperature (*609*). The crystalline behavior of a few representative aromatic polyimides is described in the accompanying table (p. 33).

Soluble and/or Melt-Processible Polyimides Based on Aromatic Diamines

| Dianhydride | Diamine | Solubility and/or melt-processibility | References |
|---|---|---|---|
| | | Soluble in *m*-cresol | *671* |
| | | Soluble in DMF | *558* |
| | | Soluble in DMF Moldable at 320°C | *137, 245  488* |

1:1 molar ratio

Soluble and/or Melt-Processible Polyimides Based on Aromatic Diamines—*continued*

| Dianhydride | Diamine | Solubility and/or melt-processibility | References |
|---|---|---|---|
| (pyromellitic dianhydride structure) | $H_2N$—(phenyl)—$CH(CH_3)_2$, $NH_2$ | Soluble in DMF and DMA | 56, 647 |
| (pyromellitic dianhydride structure) | $H_2N$—(phenyl, $COOC_2H_5$)—(phenyl, $COOC_2H_5$)—$NH_2$ | Soluble in DMSO and DMA | 551 |
| (dianhydride with $C(CF_3)_2$ linkage) | $H_2N$—(phenyl)—$C(CF_3)_2$—(phenyl)—$NH_2$ | Soluble in $CHCl_3$, $C_6H_6$, dioxane, and acetone | 564 |
| Various | Triphenylmethane-bis(alkoxy)-diamines | Soluble in DMA | 26 |

Crystalline Behavior of Aromatic Polyimides

| Dianhydride | Diamine | Crystallinity | References |
|---|---|---|---|
| | | Crystallizable | *377, 620, 622* |
| " | | Crystallizes readily | *377, 620, 622* |
| " | | Highly crystalline | *377, 620, 622* |
| " | | Slightly crystalline | *377, 620, 622* |
| " | | Crystallizable with difficulty | *377, 620, 622* |
| " | | Crystallizable | *377, 620, 622* |
| " | | Crystallizable | *377, 620, 622* |
| | | Amorphous as made | *603* |
| " | | Crystallizable at 250°C | *609* |

It should be noted that crystal melting was not observed on heating the crystalline pyromellitimides (*620, 622*).

### c. Hydrolytic Stability

The hydrolytic stability of aromatic polyimides varies depending on the structure of the polymer. The pyromellitimides based on 4,4′-diamino-diphenyl ether and 4,4′-diaminodiphenyl thioether retained toughness after 1 year and 3 months, respectively, in boiling water. On the other hand, the pyromellitimide from *m*-phenylenediamine embrittles within a week under these conditions (*620, 622*).

### d. Thermal Stability

Aromatic polyimides display outstanding thermal stability. The latter was the object of many investigations which utilized a variety of techniques (*157–164, 290, 293, 309, 310, 316, 323, 338, 369, 432, 513–517, 581, 582, 603*). The best way to illustrate the excellent high-temperature properties of these materials is to list the times during which properties are retained upon high-temperature aging as well as the "zero strength temperatures". Such data are tabulated in the accompanying table (*377*).

The outstanding high-temperature stability of the pyromellitimides is further illustrated by the thermogravimetric data shown in Fig. 1. It is readily seen that the subject polymers are stable in an inert atmosphere up to 500°C without noticeable weight loss (*620*).

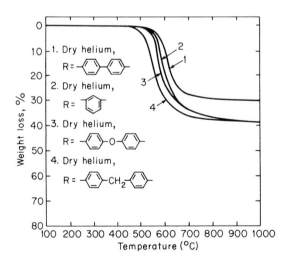

FIG. 1. Thermogravimetric analyses of polypyromellitimides (3°C/min temperature rise). [From Sroog, *J. Polymer Sci.*, Part C, No. 16, 1191–1209 (1967).]

The thermal degradation of polymer H [4] was investigated both in air and vacuum (157–164, 432), It was stable in air up to 420°C. At 485°C, complete volatilization took place within about 5 hours. Degradation rates were determined and a value of 33 kcal/mole was calculated for the energy of activation (157, 158, 162, 432). It was assumed that under these conditions cleavage occurs at the imide bond.

Thermal Stability of Pyromellitimides

| Diamine | Stability[a] at | | Zero strength temperature (°C) |
|---|---|---|---|
| | 275°C (air) | 300°C (air) | |
| $H_2N$—⬡—$NH_2$ | 1 year | 1 month | 900 |
| ⬡ with $NH_2$ (1,4) | 1 year | 1 month | 900 |
| $H_2N$—⬡—⬡—$NH_2$ | — | 1 month | 900 |
| $H_2N$—⬡—$CH_2$—⬡—$NH_2$ | — | 7–10 days | 800 |
| $H_2N$—⬡—$C(CH_3)_2$—⬡—$NH_2$ | — | 15–20 days | 580 |
| $H_2N$—⬡—$S$—⬡—$NH_2$ | — | 6 weeks | 800 |
| $H_2N$—⬡—$O$—⬡—$NH_2$ | 1 year | 1 month | 800 |

[a] Measured by retention of the creasability of the film.

Stability up to 500°C was observed *in vacuo*. Volatilization began above this temperature, leaving ultimately about 45% of the polymer as a brittle carbonized residue. Again, rates and the energy of activation were determined. The latter was equal to 74 kcal/mole. Simultaneously, analysis of the volatile products was performed. Based on these data, the mode of cleavage indicated by the dashed lines in [4] in Eq. (I-47) was postulated (159, 161, 162, 432).

The remaining carbonized residue displayed semiconducting properties (*160, 163, 432*). Their magnitude increased with the increase in the pyrolysis temperature. This is in agreement with the proposed pyrolysis mechanism depicted in Eq. (I-47).

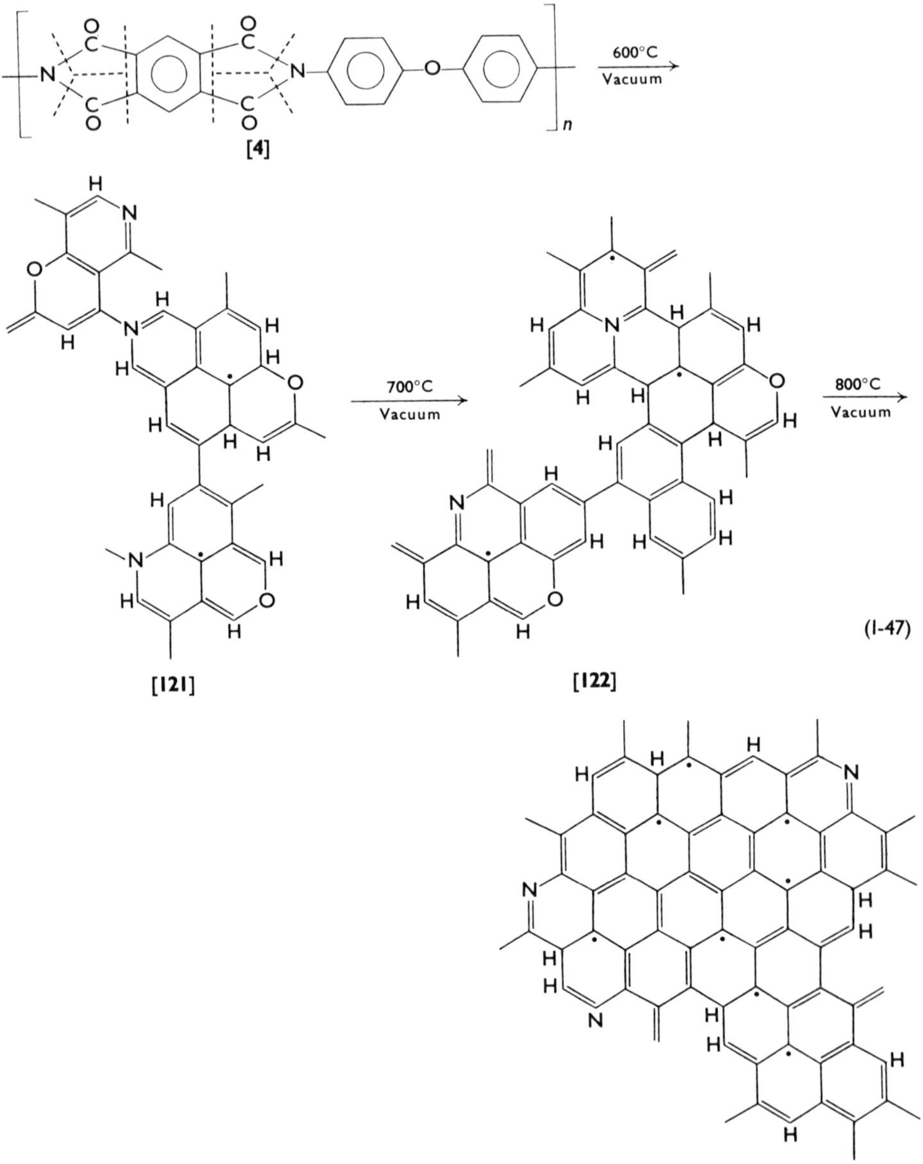

[4]

[121]                [122]

(I-47)

[123]

In another interesting study (324), the thermal stabilities of the two poly-imides [124] and [125] were compared. Polymer [124] has no hydrogen atoms in its backbone, while [125] has two hydrogen atoms per repeat unit. The study

[124]

[125]

indicated that [124] was stable for at least 25 hours in air at 400°C, whereas [125] charred extensively at 320°C both in air and in nitrogen. The thermal stabilities of polyimide copolymers that contained oxadiazole and benzimid-azole residues were investigated (275). Weight loss data showed that both heterocyclic nuclei had an adverse effect on stability.

### e. Glass Transition Temperatures

Several studies performed on Polymer H (107, 157, 158, 430, 431) un-covered no glass transitions ($T_g$) or melting point within the temperature range of $-100°$ to $+500°C$. In addition, the physical property data showed that the commercial film is probably cross-linked. In one instance, however, a $T_g$ of 385°C was determined using a different method (234).

The glass transition temperatures of some aromatic polyimides are listed in the accompanying table. It is noteworthy that annealing results in increased $T_g$'s. Examination by X-ray showed that no crystallinity developed during that step. It was speculated (182) that cross-linking and/or further cyclization of the amic acid units were responsible for the observed behavior.

### f. Mechanical and Electrical Properties

The mechanical and electrical properties of polymer H [4] are given in the accompanying table (309, 310, 323).

Glass Transition Temperatures of Aromatic Polyimides

| Dianhydride | Diamine | $T_g$ (°C) | References |
|---|---|---|---|
| | | None[a]; 385[b] | 182, 234 |
| | | 250[c] | 603 |
| | | 240[a]; 340[b]; 217–243[d] | 182, 234 |
| " | | 290[a]; 294–305[d] | 182 |
| " | | 220[a]; 345[b]; 207–223[d] | 182, 234 |
| " | | 265[a]; 273–284[d] | 182 |
| | | 340[f] | 564 |

[a] From modulus/temperature measurements.
[b] From electrical dissipation factor/temperature measurements.
[c] Via coefficient of thermal expansion.
[d] By differential scanning calorimetry (DSC).
[e] After annealing for 4 minutes, at 400°C, in vacuo.
[f] Method of determination not given in the reference.

Properties of Polymer H

| | |
|---|---|
| Density (gm/cm$^3$) | 1.42 |
| Tensile strength (psi) | |
| 23°C | 24,000 |
| 200°C | 17,000 |
| Tensile modulus (psi) | |
| 23°C | 415,000 |
| 200°C | 260,000 |
| Elongation (%) | |
| 23°C | 65 |
| 200°C | 90 |
| Dielectric constant, 23°C (10$^3$ cps) | 3.5 |
| Dissipation factor, 23°C (10$^3$ cps) | 0.002 |
| Volume resistivity, 23°C (ohm-cm) | 10$^{18}$ |
| Surface resistivity, 23°C (ohms) | >10$^{16}$ |
| Dielectric strength, 23°C, $\frac{1}{4}$-inch electrode (volts/mil) | 7,000 |
| Cut-through temperature (°C) | >435 |
| Moisture absorption (%) | |
| 50% RH | 1.3 |
| 100% RH | 3.0 |

Examination of the data shows that polymer H is a tough material that retains its toughness at high temperatures. Its mechanical and electrical characteristics remain excellent after long periods of exposure to heat. For instance, 80% of the dielectric strength was retained after aging for 8 weeks at 300°C in air. Under similar conditions, the dielectric constant and the dissipation factor showed slight increases of 8% and 12%, respectively (309).

Fibers of the polymer displayed equally good mechanical and thermal behavior (338).

The excellent characteristics of polymer H are further evidenced by the comparative data shown in the accompanying table (p. 40) (11, 12, 30).

Polymer H possesses good resistance to creep and cold flow (30). Films of the material can be oriented via drawing (67) to yield products with higher tensile strengths and moduli.

Mechanical properties of other aromatic polyimides have also been determined (220, 377). Generally, they are rigid polymers with elongations varying from very low (<5%) to high values (~100%). Several are comparable in toughness to polymer H (377).

g. *Miscellaneous Properties*

Aromatic polyimides can be hydrolyzed by aqueous base. This reaction was claimed to be useful for the recovery of monomers from scrap polymer

Comparison of Polymer H with Other Polymers

| Property | Unit | Temp. (°C) | Polymer H | Tedlar[a] | Teflon[b] | Mylar[c] |
|---|---|---|---|---|---|---|
| Density | gm/cm³ | 23 | 1.42 | 1.37 | 2.15 | 1.39 |
| Tensile strength | psi | 23 | 24,000 | 15,000 | 3,000 | 23,000 |
| | | 200 | 17,000 | 1,000 | 400 | 7,000 |
| | | 500 | 5,000 | — | — | — |
| Tensile modulus | psi | 23 | 415,000 | 300,000 | 43,000 | 500,000 |
| | | 200 | 260,000 | 1,000 | 2,000 | 50,000 |
| Ultimate elongation | % | 23 | 65 | 150 | 300 | 100 |
| | | 200 | 90 | 200 | — | 125 |
| Melting point | °C | — | — | — | 290 | 260 |
| Zero strength temperature | °C | — | 800 | 300 | 250 | 250 |
| Solvent resistance | % | — | Excellent | Excellent | Excellent | Good |

[a] Poly(vinyl fluoride).
[b] Poly(tetrafluoroethylene).
[c] Poly(ethylene terephthalate).

(*77*). Polyimides are similarly degraded by reaction with hydrazine. This phenomenon was applied in the development of an etching technique for pyromellitimide films (*226*). It is also useful as an analytical tool for the determination of the polymer structure (*201, 349*).

Several polyimides were found to possess useful catalytic properties. Specifically, catalytic activity was demonstrated for the vapor-phase decomposition of formic acid (*288*), and the high-temperature (400°C) decomposition of nitrous oxide (*289*).

## APPLICATIONS OF POLYIMIDES

One of the most outstanding characteristics of polyimides is their excellent thermal stability. Consequently, they are useful in applications requiring high heat resistance. Laminating operations are one example (*275, 414*). Several polyimide solutions are offered for lamination purposes by the DuPont Company. Other companies are also active in this field. A formulation is sold by the Monsanto Plastics Division under the name of Skygard 700. Some of the properties of the glass laminates are lower than those obtained with other resin systems (epoxies, phenolics, etc.). Typical values are 60,000 psi for flexural strength and 2% for elongation. However, heat stability is superior.

Another area in which polyimides are employed is as adhesives (*27, 166, 227, 232, 235, 275, 321, 476*). Again, room temperature properties can be below those obtained with other structural adhesives, but the utility of polyimides becomes apparent at high temperatures (*414*). It was claimed (*415, 416, 481, 482*) that the treatment of polyimide surfaces with bases enhances their adhesive characteristics.

The use of the polymers in structural coatings was reported (*190, 209, 414*). Resistance to high temperatures in aerospace applications was the prime motivation in these studies.

The excellent electrical properties of polyimides were discussed in the preceding sections. Their usefulness as insulators is described in numerous reports (*53, 54, 135, 147, 148, 175, 299, 467, 468, 501, 503*). Composite insulators based on polymer H and other resins have also been evaluated (*407–409, 594, 610*). Several varnishes including "Pyre ML" which are designed for magnet wire coating are offered commercially.

Special solid systems were successfully molded by the DuPont Company and are sold under the designation "Vespel." Another commercial grade, SP-2, is a 15% graphite-filled composition. Note that the literature reports a variety of other filled polyimide materials. Fillers such as silica fibers (*623*), molybdenum disulfide (*165*), and copper powder (*165*) were studied. Good friction and wear characteristics were claimed (*165*).

Laminates of polyimides with other resins were prepared. Good adhesion between polymer H and a tetrafluoroethylene/hexafluoropropylene copolymer was achieved by preirradiation of the polyimide surface (*70, 492*). Laminates of polymer H with novolaks displayed good shear strength and thermal stability (*85*). Materials consisting of two different adjacent polyimide layers and a process for welding the two layers together were also described (*202, 203*). Conductive films that possessed high resistance to permeation by water vapor were prepared by metal or metal oxide coating of the polymer (*352, 584*). Polyimides were also shown to be superior as seals in hand-operated valves for ultrahigh vacuum (*304*).

Polyimides are discussed in many reviews (*30, 31, 194, 196, 230, 280, 308, 350, 369, 377, 383, 436, 440, 441, 448, 450, 536, 544, 565, 568, 571, 588, 619–622, 645, 668, 670, 679, 680, 692*).

## 2. POLY(AMIDE-IMIDES)

Poly(amide-imides) are copolymeric materials that contain both the amide [126] and the imide [127] linkages. As expected, their behavior reflects these structural features. For example they possess very good mechanical properties,

—ArNHCOAr′

[126]

$$\begin{array}{c} \text{—CO} \\ \phantom{xxx}\diagdown \\ \phantom{xxxx}\text{N—} \\ \phantom{xxx}\diagup \\ \text{—CO} \end{array}$$

[127]

but their high-temperature stability is inferior to that of polyimide homo-polymers (*351, 369*). It decreases with an increase in amide content (*518*). Poly(amide-imides) have been the subject of several reviews (*35, 113, 233, 413*).

## METHODS

The methods that are useful for the preparation of poly(amide-imides) are very similar to those used for the synthesis of polyimides. The condensation of an appropriate anhydride and diamine yields a polyamic acid that contains amide groups in its chain. Its dehydration leads to the copolymer amide-imide.

The synthesis of the polyamic acids and their cyclization is accomplished in essentially the same ways as described in the section on polyimides, by heating at high temperature (Method A,7) or by using chemical reagents (Method A,8). Both methods are briefly discussed.

### Method 7. Thermal Dehydration of Polyamic Acids

A variety of starting materials were used for synthesis of the polymers. In the first approach, a diamino amide is condensed with a dianhydride, as shown in Eq. (I-48) (*121, 122, 238, 274*).

Diamino-terminated polyamides could also be used in lieu of [128] (*121, 122*). In a different modification, a diamine, a dibasic acid chloride, and a dianhydride were reacted simultaneously (*393*) [Eq. (I-49)]. Another variation utilized a diester-dichloride instead of the dianhydride [Eq. (I-50).

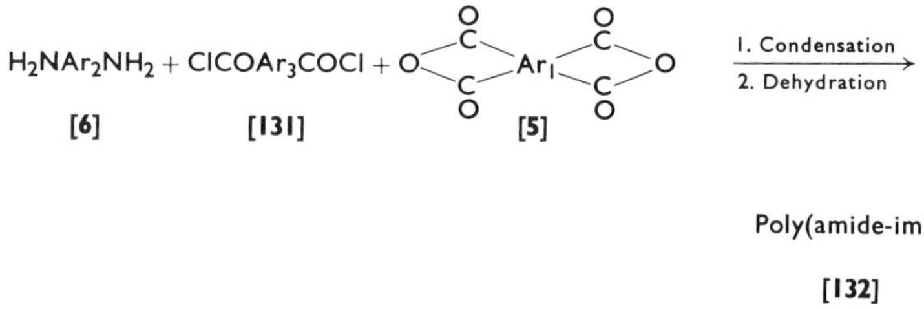

$$H_2NAr_2NH_2 + ClCOAr_3COCl + \underset{[5]}{O} \quad \xrightarrow[\text{2. Dehydration}]{\text{1. Condensation}}$$

[6]   [131]

(I-49)

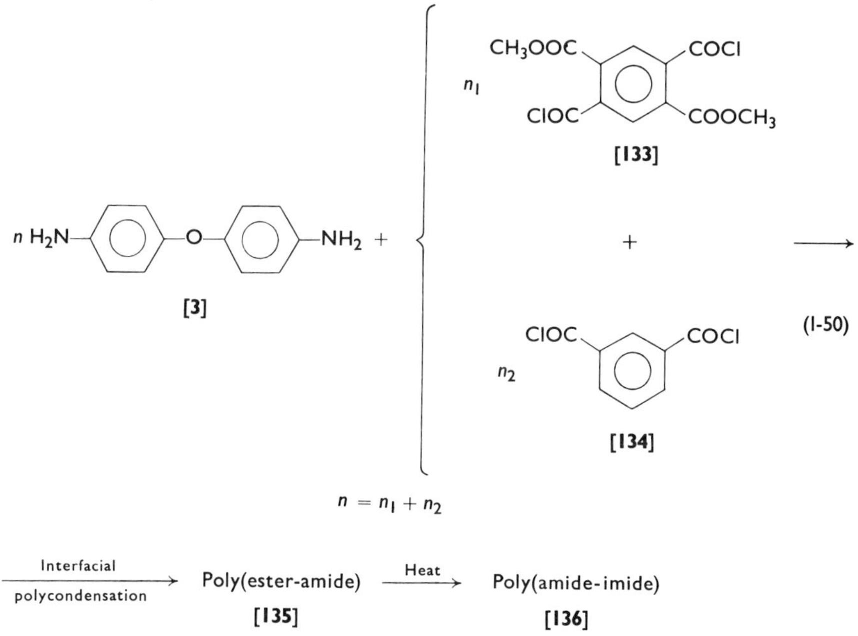

Poly(amide-imide)

[132]

$$n\,H_2N-\!\!\!\left\langle\bigcirc\right\rangle\!\!-O-\!\!\left\langle\bigcirc\right\rangle\!\!-NH_2 +$$

[3]

(I-50)

$$n = n_1 + n_2$$

$$\xrightarrow[\text{polycondensation}]{\text{Interfacial}} \text{Poly(ester-amide)} \xrightarrow{\text{Heat}} \text{Poly(amide-imide)}$$

[135]   [136]

An interesting synthesis of the poly(imide-benzimidazole) [140] from the triamino amide [137] was reported (*275*). It is shown in Eq. (I-51). Note that it is very important to use 1 mole of pyromellitic anhydride [2] per mole of the triamino derivative [137]. If an excess of the dianhydride is employed, a gelled product results due to interaction with the third amino group. The reactivity of the latter is low, however, and it does not enter into condensation when a 1:1 stoichiometric ratio of [2] to [137] is maintained.

[137]  [2]  [138]  [139]  [140]

(I-51)

The condensation of several dihydrazides with dianhydrides was reported (*334, 367, 655–657*). Infrared data support structure [141] for the polymers, and not structure [142].

[141]  [142]

The second approach makes use of tricarboxylic acid derivatives. An example is shown in Eq. (I-52) (50, 461). The solutions of the intermediate poly(amide-amic acids) obtained in this and in the preceding approach are useful in coating wire and other substrates. Treatment of these solutions prior to use by ethylene or propylene oxides has been described (624). The oxides act as hydrochloric acid scavengers; this decreases the corrosiveness of the enamel and improves the quality of the coating.

Other solvent systems that are useful for the reactions of the type shown in Eq. (I-52) include DMA (331), DMA/xylene/triethylamine (136, 242), N-methylpyrrolidone/xylene/triethylamine (140), DMA/N-methylpyrrolidone (114), and DMA/pyridine (331). It was claimed that coatings possessing excellent surface characteristics were obtained when acetamide or acetanilide were added to poly(amide-amic acid) solutions in DMA or mixtures of DMA/N-methylpyrrolidone (114).

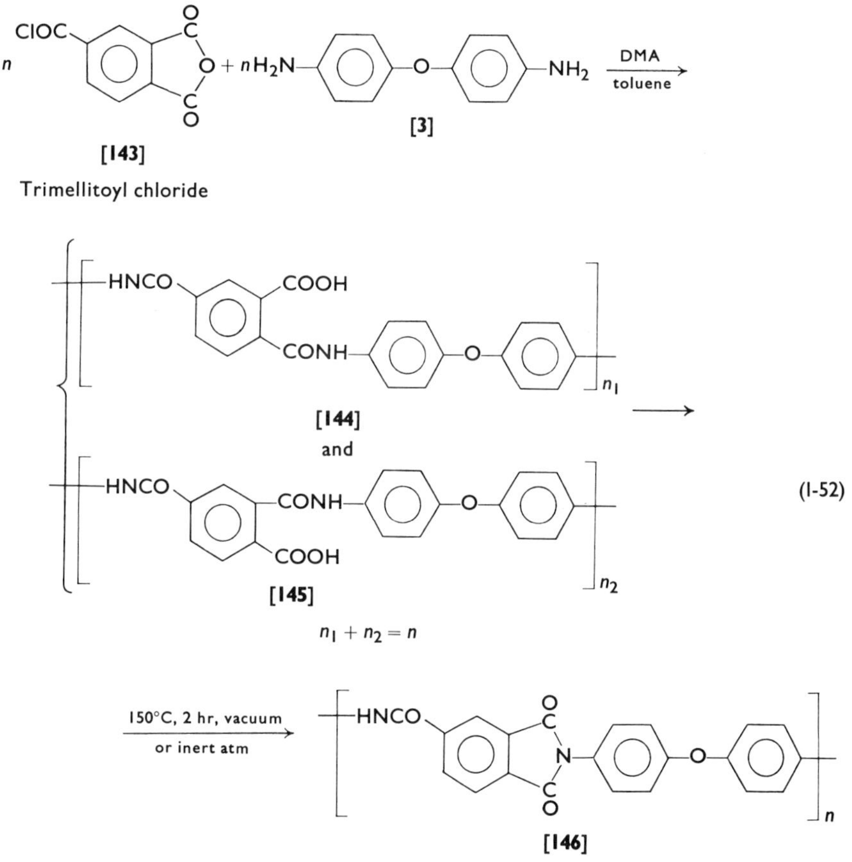

[143]
Trimellitoyl chloride

[3]

[144]
and

[145]

$n_1 + n_2 = n$

(I-52)

[146]

The condensation described in Eq. (I-52) can also be performed using interfacial or bulk polymerizations. Examples are shown in Eqs. (I-53) and (I-54). The reactions of diamino-terminated polyamides with tricarboxylic acid derivatives have been reported (*136*, *242*). Copolymeric amide-imides were also prepared by the reaction of a dianhydride and a tricarboxylic acid monoanhydride with a diamine (*146*).

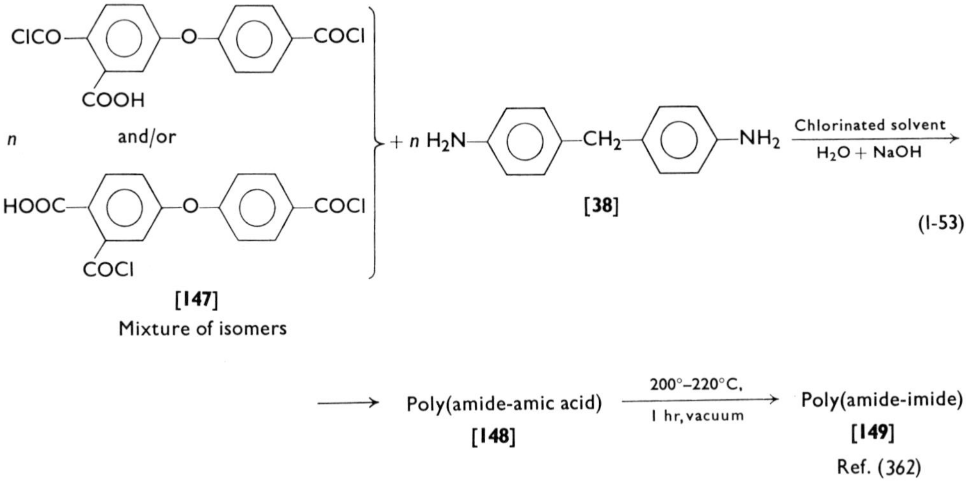
ClCO
O
COCl
COOH
n
and/or
HOOC
O
COCl
COCl
[147]
Mixture of isomers
+ n H₂N
CH₂
NH₂
[38]
Chlorinated solvent
H₂O + NaOH
(I-53)
Poly(amide-amic acid)
[148]
200°–220°C,
1 hr, vacuum
Poly(amide-imide)
[149]
Ref. (362)

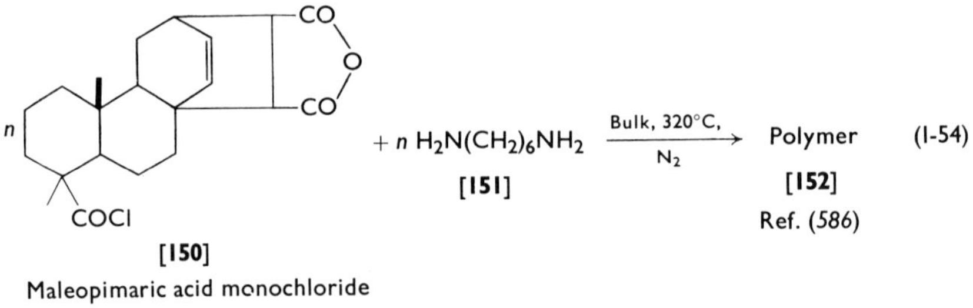
CO
O
CO
n
COCl
[150]
Maleopimaric acid monochloride
+ n H₂N(CH₂)₆NH₂
[151]
Bulk, 320°C,
N₂
Polymer
[152]
Ref. (586)
(I-54)

A third approach consists in conducting the polymerization with an amide-containing dianhydride (*423*, *496*, *497*).

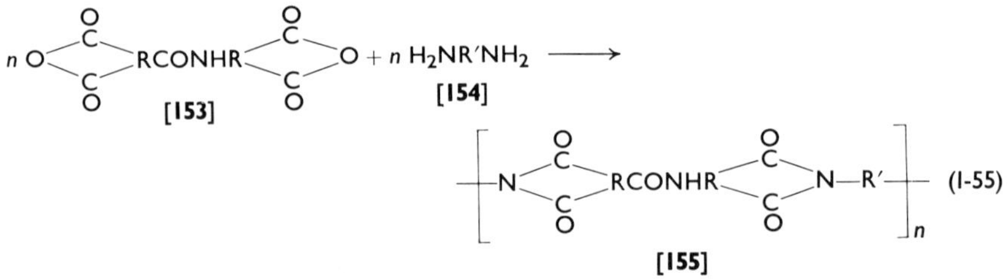
n O
RCONHR
O + n H₂NR′NH₂
[153]
[154]
N
RCONHR
N—R′
n
(I-55)
[155]

The preparation of the dianhydrides is straightforward. The condensation of a tricarboxylic acid monochloride with a diamine is the route that is most frequently utilized (*156, 329, 332, 497, 586*). An example is given in Eq. (I-56) (*156*). The dianhydride does not have to be isolated; it can be reacted further *in situ* to give high polymer (*155*).

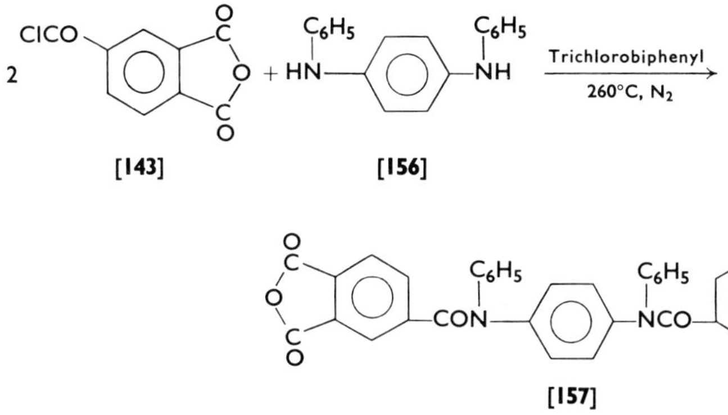

[143]                          [156]

(I-56)

[157]

Another synthesis of the intermediate [153] makes use of the interaction of a tricarboxylic acid monoanhydride with a diamide (*419, 420*). This is illustrated in Eq. (I-57). This reaction is applicable to both aromatic and aliphatic diacylamino compounds.

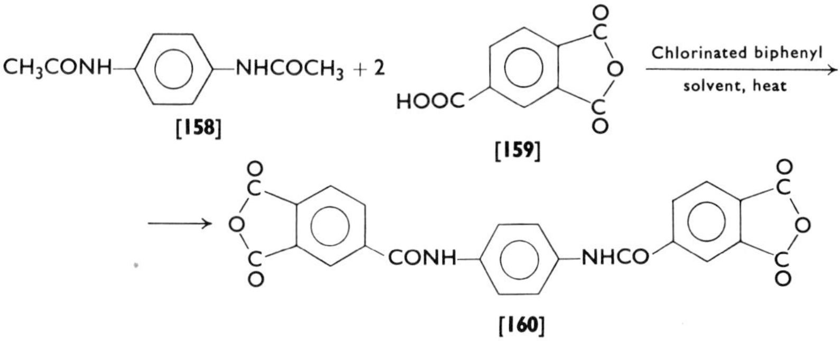

[158]                          [159]                          (I-57)

[160]

Other approaches to poly(amide-imides) are reported in the literature. The use of diamino-terminated polyamic acids is described (*93, 121, 122, 238*). It is claimed, for example, that the reaction of the latter with an aromatic diester at elevated temperature gave high polymer (*93*). The reaction of a dianhydride with an amino acid (*572, 596*) or an amino alcohol (*69, 74, 247*) gives derivatives of the types [161] and [162]. Compound [161] can be poly-condensed with diamines and lead to copolymeric amide-imides (*572, 596*).

The diol [162] was reacted with diisocyanates and dihydroxyl-terminated polyesters or polyethers in the presence of chain-extending agents such as 1,4-butanediol, to give elastomeric poly(imide-urethanes) (69, 74, 247). These reactions are not discussed in detail because they are not ring-forming polymerizations.

[161]                    [162]

## METHOD 8. CHEMICAL DEHYDRATION OF POLYAMIC ACIDS

The results are very similar to those reported earlier for polyimides (334).

## METHOD 9. POLYMERIZATION OF TRIMELLITIC ANHYDRIDE WITH DIISOCYANATES

The reaction of a tricarboxylic acid monoanhydride with a diisocyanate gives a poly(amide-imide) (91, 252, 256, 578). An example is shown in Eq. (I-58). The polymerization is preferably conducted in solvents; DMF and

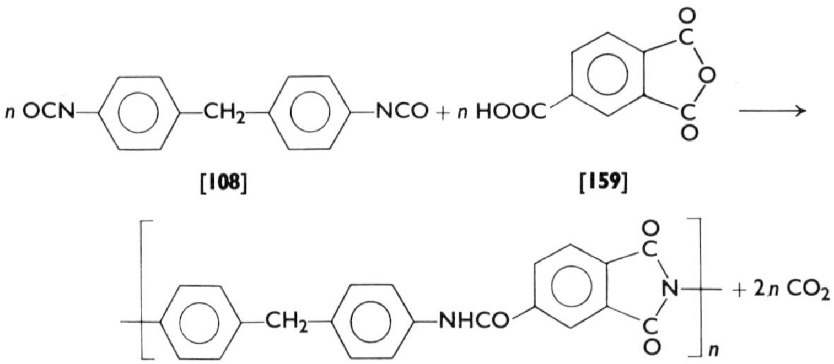

[108]                    [159]

(I-58)

[163]

N-methylpyrrolidone are most commonly used. Reaction temperatures vary within the range of 80°-160°C. The polymer solution can be utilized directly for coatings and wire insulation (256). The method was successfully applied to the preparation of poly(amide-imide)–polyimide block copolymers (91).

The reaction of trimellitic anhydride with a variety of diisocyanato-terminated materials was reported (452). The condensations were performed in solvents such as N-methylpyrrolidone, o-dichlorobenzene, glycol monomethylether acetate, and cresol. The end groups were masked via reaction with phenols. The products were claimed to be useful as adhesives and auxiliary agents for rubber (452).

## PROPERTIES AND APPLICATIONS OF POLY(AMIDE-IMIDES)

A very wide variety of poly(amide-imides) have been prepared. The properties of the polymers varied depending on their structure. As an illustration, the poly(amide-imide) that was synthesized from 4,3'diaminobenzanilide [164] and pyromellitic anhydride [2] had the following properties (*182, 234*).

[164]                                          [2]

$T_g$ (°C) = 205; 360
Tensile strength (psi) = 11,000
Tensile modulus (psi) = 270,000
Elongation at break (%) = 15

Properties of Poly(amide-imides) Based on Pyromellitic Anhydride[a]

| Diamine | Inherent viscosity[b] | Film properties[c] | Weight loss at 325°C[c] (%) | | | |
|---|---|---|---|---|---|---|
| | | | 100 hr | 200 hr | 300 hr | 400 hr |
| 4,4'-Diaminobenzanilide | 2.20 | Brit. | 5.7 | 8.4 | 11.9 | 12.1 |
| 4,3'-Diaminobenzanilide | 1.55 | Flex. | 4.3 | 7.8 | 10.8 | 11.9 |
| 3,4'-Diaminobenzanilide | 2.19 | Flex. | 2.0 | 4.2 | 6.9 | 9.8 |
| 3,3'-Diaminobenzanilide | 1.18 | Flex. | 3.2 | 6.5 | 9.8 | 11.2 |
| 3,5'-Diaminobenzanilide | 1.29 | No film, just crumbs | — | — | — | |
| Isophthal (4-aminoanilide) | 1.46 | Flex. | 6.9 | 9.4 | 14.4 | 20.4 |
| $N,N'$-*m*-Phenylenebis(4-amino-benzanilide) | 1.48 | Flex. | 6.0 | 9.2 | 12.5 | 15.6 |
| Isophthal (3-aminoanilide) | 1.48 | Flex. | 6.8 | 8.1 | 10.5 | 13.2 |
| $N,N'$-*m*-Phenylenebis(3-amino-benzanilide) | 1.07 | Flex. | 6.2 | 8.3 | 14.0 | 20.3 |
| $MPD_{7.23}IP^e_{6.23}$ | 1.16 | Flex. | 6.7 | 16.5 | 29.8 | — |
| $MPD_{7.6}IP_{4.98}TP^e_{1.66}$ | 1.21 | Flex. | 8.5 | 22.4 | 44.4 | — |

[a] From Bowen and Frost, *J. Polymer Sci.* Part A, 1, 3135–3150 (1963).

[b] Measured on the polyamic acid precursor in DMA, 0.5% solution, at 25°C.

[c] The samples consisted of films approximately 1 mil thick adhering to aluminum. The film was judged to be flexible if it could be creased without cracking; brittle if it cracked on creasing. Brit., brittle; Flex., flexible.

[d] In air, in a forced-draft oven. The values at 100 hour intervals were obtained from plots of weight loss versus time.

[e] MPD, *m*-phenylenediamine units; IP, isophthalic units; TP, terephthalic units.

It is of interest that, depending on the method of determination, two values were obtained for the glass transition temperature. One, at 205°C, resulted from measurements of the electrical dissipation factor (*234*). The other was obtained by measuring the modulus as a function of temperature (*182*). No satisfactory rationale was offered to explain this discrepancy. After the polymer was annealed at 250°C *in vacuo* (24 hours), modulus measurements indicated

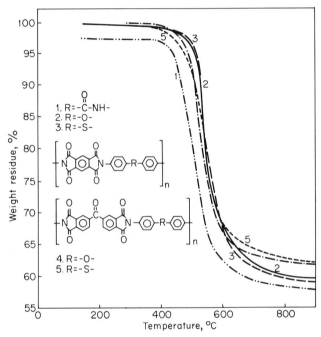

FIG. 2. Thermogravimetric analysis for aromatic poly(amide-imides) and polyimides in nitrogen. T = 150°C per hour. (Reprinted from Freeman, *et. al.*, Polymer Engineering and Science, Vol. 9, No. 1, January, 1969.)

a glass transition temperature of 390°C. Cross-linking or a more complete imidation were thought responsible for this phenomenon (*182*).

Poly(amide-imides) display lower thermo-oxidative stability than polyimides (*413*). The stabilities of these materials were investigated by several workers (*122, 581, 582*). The mechanical behavior and weight loss data for several poly(amide-imides) derived from pyromellitic anhydride and a variety of amide-containing diamines are shown in the accompanying table (p. 49) (*122*).

The data show that there is an appreciable weight loss at 325°C in air. As a comparison, the poly(pyromellitimide) from 4,4′-diaminodiphenyl ether displayed a loss of 6.6 wt. % only after 400 hours under similar conditions

*(122)*. The better high-temperature properties of polyimides are also clearly demonstrated by the results of thermogravimetric analysis (Fig. 2); glass transition temperature data as measured by a power factor (Fig. 3) is an additional indication of their behavior *(413)*.

In spite of the fact that the high-temperature behavior of the poly(amide-imides) is not as good as that of the polyimides, the former display impressive performance at somewhat lower temperatures. Excellent quality laminates

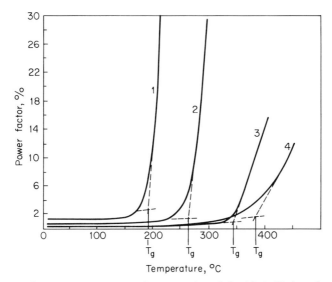

FIG. 3. Power factor vs. temperature for aromatic poly(amide-imides) and polyimides; showing $T_g$ (1) $MPD_{16}$–$IP_9$–$TP_3$–$PMDA_4$; (2) (MAB–PPD)–PMDA; (3) DAPS–PMDA; (4) DAPE–PMDA; where: MPD, *m*-phenylenediamine; IP, isophthalic acid; TP, terephthalic acid; PMDA, pyromellitic anhydride; MAB, *m*-aminobenzoic acid; PPD, *p*-phenylene-diamine; DAPS, 4,4′-diaminodiphenyl sulfide; DAPE, 4,4′-diaminodiphenyl ether. (Reprinted from Freeman, *et al.*, SPE TRANSACTIONS, Vol. 5, No. 2, p. 80, April, 1965.)

were prepared *(35, 119, 120, 233, 293, 652, 653)*. An example of a laminating resin is the Amoco Chemical Company's AI-10 material, represented by formula [**163**]. Aging studies of these laminates indicated good retention of properties at temperatures as high as 275°–300°C *(120, 413)*.

Poly(amide-imides) are valuable in electrical applications for the following reasons *(413)*:

(a) High dielectric strength which is retained at high temperatures in humid environment; it remains high after long aging.

(b) Good thermal stability.

(c) Excellent chemical and abrasion resistance.

(d) Ability to form tough, smooth surfaces. Resistance of film to heat and pressure.

The combination of these characteristics makes the polymers useful in magnet wire coatings and insulating varnishes (*140, 146, 233, 412, 413*). In addition to their present use in these areas, poly(amide-imides) show promise in a variety of other applications including binder materials for composite slot liners, sleeving for high-temperature electrical insulation applications, slot-sticks, and phase insulation (*413*). In one instance, an aliphatic polyamide containing imide groups was claimed to be useful in ink binders for flexographic inks (*84*).

### 3. POLY(ESTER-IMIDES)

Poly(ester-imides) are copolymeric materials that contain both imide [**127**] and ester structures [**165**]. These polymers possess good adhesive and

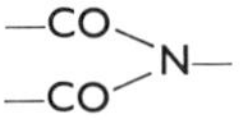

[**127**] 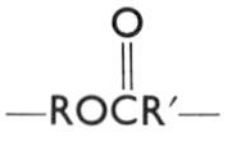                    [**165**]

electrical properties. They are the subject of two excellent reviews (*413, 569*).

METHOD 10. THERMAL DEHYDRATION OF POLYAMIC ACIDS

This method involves the condensation of an ester-containing dianhydride with a diamine. A polyamic acid is formed which on thermal dehydration yields the poly(ester-imide). The reaction sequence is shown in Eq. (I-59). The formation of the polyamic acids is conducted at room temperature in typical

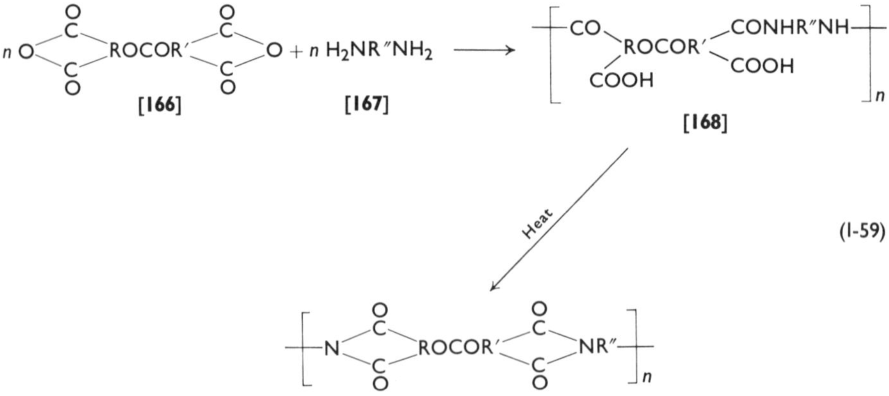

[**166**]          [**167**]                    [**168**]

(I-59)

[**169**]

solvents, including DMF, DMA, DMSO, and *m*-cresol (*422*). In contrast to the observations in the homopolyimide area (*121, 122, 198, 200*), the order of addition of the reagents has no effect on the molecular weight of the poly-amic acid (*422*). Imidation is performed by heating the polymeric acids to temperatures of up to 250°C for variable periods of time that are a function of the particular system.

The ester dianhydrides [**166**] are prepared by the reaction of a tricarboxylic acid monoanhydride with an aliphatic or aromatic diester. Two techniques were developed:

(*a*) Exchange in solution (*419, 420, 422*). The preparation is shown in Eq. (I-60).

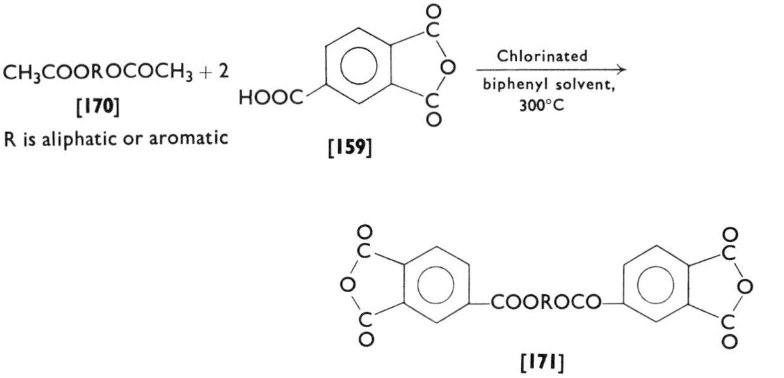

CH$_3$COOROCOCH$_3$ + 2

[**170**]

R is aliphatic or aromatic

HOOC

[**159**]

Chlorinated biphenyl solvent, 300°C

(I-60)

—COOROCO—

[**171**]

(b) Exchange in bulk (*144*). An example is given in Eq. (I-61).

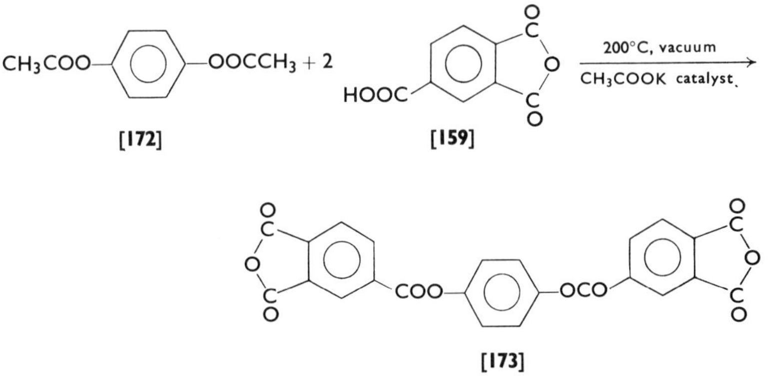

CH$_3$COO—⟨ ⟩—OOCCH$_3$ + 2

[**172**]

HOOC

[**159**]

200°C, vacuum

CH$_3$COOK catalyst.

(I-61)

—COO—⟨ ⟩—OCO—

[**173**]

The reaction of ester-dianhydrides with hydrazides was investigated by Loncrini *et al.* (*426*). A typical polymerization is presented in Eq. (I-62).

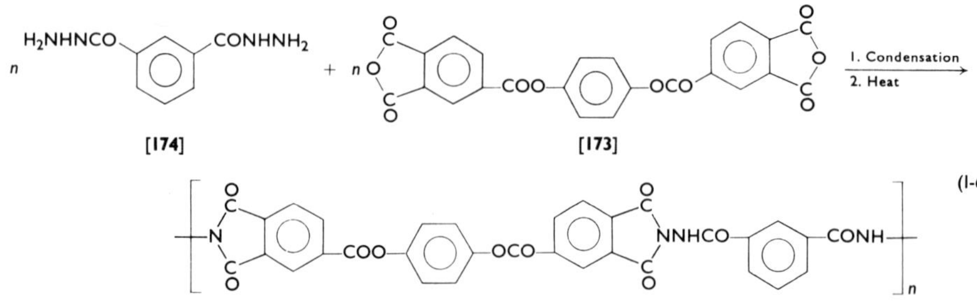


[174] [173] [175] (I-62)

Structure [**175**] was supported by infrared data. No evidence was found for the presence of the six-membered ring [**142**] in the polymer.

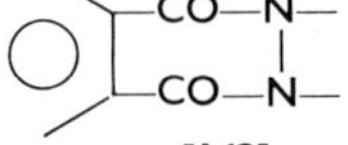

[142]

The rearrangement shown in Eq. (I-63) is well known from the literature.

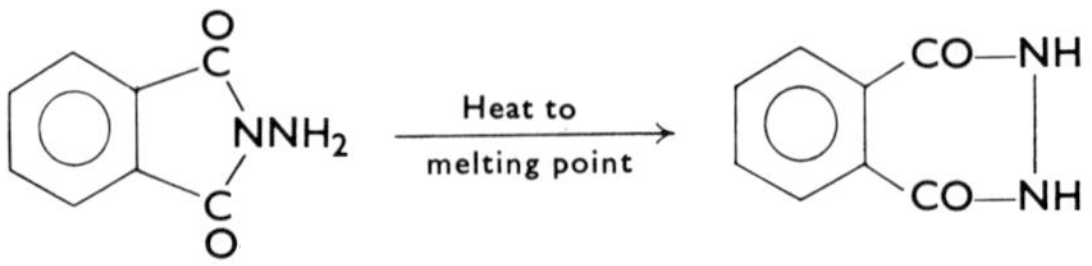


[176] [177] (I-63)

Therefore, the absence of structure [**142**] was rather unexpected and indicated that *N*-aroylaminophthalimides must be stable under the conditions at which

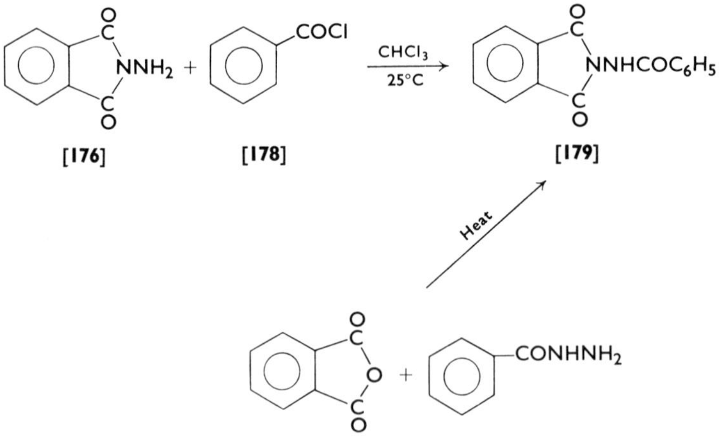


[176] [178] [179] [180] [181] (I-64)

the unsubstituted amino derivative [176] rearranges. This was demonstrated by the synthetic sequence in Eq. (I-64) (426). Both routes yielded the same product [179], an N-aroylaminophthalimide, which was recovered unchanged after heating for 5 minutes at 350°C.

Poly(ester-imides) are good electrical insulating materials. There are numerous patents (51, 52, 133, 134, 141, 143, 168, 239, 241, 244, 249, 253, 254, 265, 291, 445, 446, 491, 583, 593, 613) in which a variety of poly(ester-imide) formulations useful in coating electrical conductors are described. Cyclization to the imide is heat-induced and often takes place on the substrate. Depending on the particular formulation, other cross-linking reactions may accompany the imidation. An example is shown in scheme (I-65) (118).

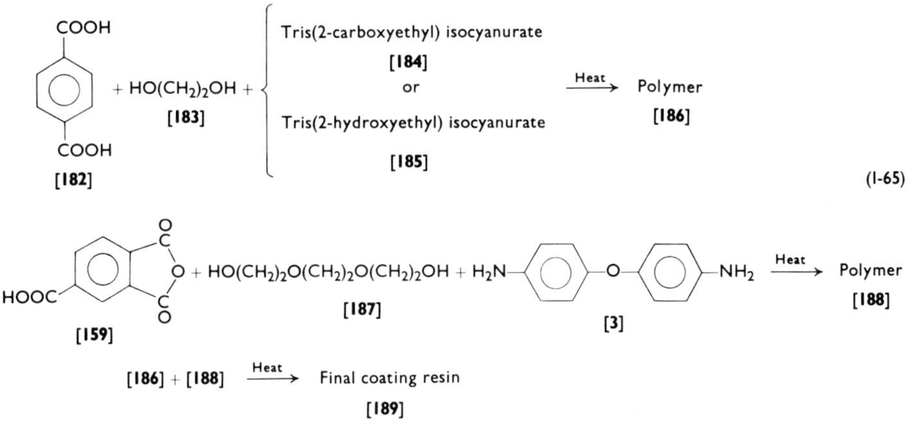

(I-65)

[186] + [188] $\xrightarrow{\text{Heat}}$ Final coating resin
[189]

## METHOD 11. POLY(ESTER-IMIDES) FROM DIANHYDRIDES AND DIISOCYANATES

The polymerization shown in Eq. (I-66) was reported (453). The ester-imide structure [192] was supported by infrared evidence. The material was useful in wire-coating formulations.

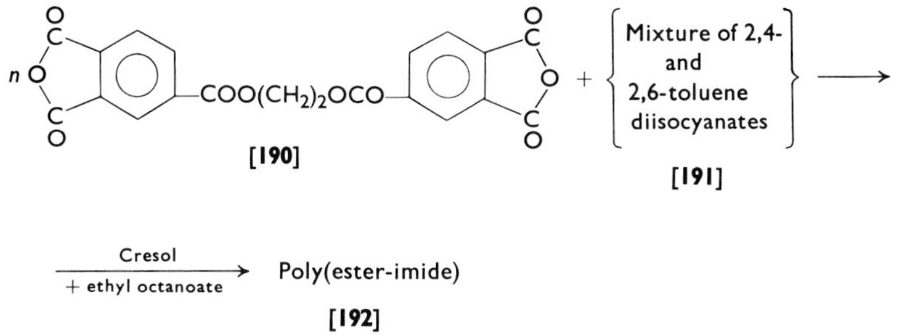

(I-66)

$\xrightarrow[\text{+ ethyl octanoate}]{\text{Cresol}}$ Poly(ester-imide)
[192]

METHOD 12. POLY(ESTER-IMIDES) FROM TRIMELLITIC ANHYDRIDE AND DIAMINES AND DIOLS

The following preparation of poly(ester-imides) was reported (356).

Ring formation occurs in the first step, i.e., during the formation of [193], but not during the actual polymerization. The method is included because of the $T_g$ versus structure data that are available for a series of polymers [195]. The values are given in the accompanying table; a decrease in the glass transition temperatures is observed with progressively longer chains between the ester groups.

| R″ | R | $T_g$ (°C) |
|---|---|---|
| $(CH_2)_6$ | $(CH_2)_2$ | 98 |
| $(CH_2)_6$ | $(CH_2)_4$ | 80 |
| $(CH_2)_6$ | $(CH_2)_5$ | 73 |
| $(CH_2)_6$ | $(CH_2)_6$ | 67 |
| $(CH_2)_6$ | $(CH_2)_7$ | 45 |

PROPERTIES OF POLY(ESTER-IMIDES)

The properties of a series of poly(ester-imides) that were prepared from the ester-dianhydrides [196] and aromatic diamines are given in the accompanying table (p. 57) (422).

Mechanical properties were measured on films based on the dianhydrides in which the Ar group of [196] was [197] and [198]. The data are presented in

[196]

## Properties of Poly(ester-imides)[a]

| Anhydride, Ar = | Diamine | Inherent viscosity[b] | Properties of cured films[c] | Polymer melt temperature (°C) |
|---|---|---|---|---|
| | p-Phenylenediamine | 2.80 | Flexible | >500 |
| | m-Phenylenediamine | 1.35 | Flexible | >500 |
| | 3,3'-Dimethoxy-4,4'-diaminodiphenyl | 1.04 | Flexible | >500 |
| | Benzidine | 3.00 | Flexible | >500 |
| | 4,4'-Diaminodiphenyl ether | 1.00 | Flexible | >500 |
| | 4,4'-Diaminodiphenyl methane | 0.81 | Flexible | >500 |
| | 4,4'-Diaminodiphenyl sulfone | 0.45 | Brittle | >500 |
| | 50% 4,4'-Diaminodiphenyl ether–50% benzidine | 0.78 | Flexible | >500 |
| | Durenediamine | 0.46 | Brittle | >500 |
| | 4,4'-Diaminodiphenyl ether | 0.35 | Flexible | 390 |
| | Benzidine | 0.41 | Flexible | >500 |
| | p-Phenylenediamine | 1.03 | Flexible | >500 |
| | 4,4'-Diaminodiphenyl ether | 0.34 | Flexible | 235 |
| | p-Phenylenediamine | 0.31 | Brittle | >500 |
| | m-Phenylenediamine | 0.31 | Brittle | >500 |
| | 4,4'-Diaminodiphenyl ether | 0.53 | Brittle | >500 |
| | 4,4'-Diaminodiphenyl methane | 0.45 | Brittle | >500 |

[a] From Loncrini, J. Polymer Sci., Part A–I, 4, 1531–1541 (1966).

[b] Inherent viscosities were determined in N-methyl-2-pyrolidone at 0.5% concentration on the polyamic acid precursor.

[c] The film was judged to be flexible if it would take a full 180° bend without cracking.

[197]                    [198]

Tensile Properties of Poly(ester-imide) Films[a]

| Anhydride from | Amine | Cure[b] conditions | Tensile strength × 10³ psi | | Tensile modulus × 10³ psi | | Elongation (%) | |
|---|---|---|---|---|---|---|---|---|
| | | | Room temp. | 200°C | Room temp. | 200°C | Room temp. | 200°C |
| [197] | p-Phenylenediamine | A | 22.3 | 11.3 | — | — | 5 | 16 |
| [197] | m-Phenylenediamine | A | 12.7 | 4.4 | — | — | 5 | 40 |
| [197] | 4,4′-Diaminodiphenyl ether | A | 13.1 | 3.2 | — | — | 5 | 50 |
| — | ML[c] | A | 13.7 | 6.8 | — | — | 11 | 20 |
| [197] | m-Phenylenediamine | B | 14.8 | 7.5 | 542 | 276 | 7 | 20 |
| [197] | m-Phenylenediamine | C | 15.8 | 6.5 | 454 | 255 | 9 | 15 |
| [197] | p-Phenylenediamine | C | 16.8 | 11.3 | 1310 | 549 | 3 | 3 |
| [197] | 4,4′-Diaminodiphenyl ether | C | 15.0 | 6.3 | 440 | 193 | 14 | 23 |
| [197] | 50% 4,4′-Diamino- diphenyl ether– 50% benzidine | C | 22.7 | 9.9 | 790 | 354 | 6 | 11 |
| — | ML[c] | C | 14.8 | 9.7 | 468 | 172 | 28 | 45 |
| [197] | 4,4′-Diaminodiphenyl- methane | D | 14.5 | 6.1 | 544 | 191 | 6 | 21 |
| [198] | 4,4′-Diaminodiphenyl ether | D | 9.9 | 6.3 | 360 | 152 | 4 | 7 |
| [198] | Benzidine | D | 13.4 | 6.2 | 471 | 208 | 6 | 10 |

[a] From Loncrini, J. Polymer Sci., Part A-1, 4, 1531–1541 (1966).
[b] A, 1 hr at 100°C + 1 hr at 200°C; B, $\frac{1}{2}$ hr at 100°C + $\frac{1}{2}$ hr at 200°C + 1 hr at 260°C; C, 1 hr at 100°C + $\frac{1}{2}$ hr at 200°C + 1 hr at 240°C; D, 1 hr at 100°C + $\frac{1}{2}$ hr at 200°C + 4 hr at 225°C.
[c] ML, from the DuPont Company, is a polyamic acid reaction product of pyromellitic anhydride and 4,4′-diaminodiphenyl ether.

the accompanying table (422). Examination of the data indicates that poly-(ester-imides) display tensile strengths that are within the expected range. Moduli are surprisingly high, however. In fact they are higher than the values obtained under comparable curing conditions for Polymer H. Incidentally, the effect of imidation conditions upon polymer properties is of interest.

Poly(ester-imides) show lower thermal-oxidative stabilities than poly-imides. This is due to the presence of ester groups in the macromolecular

Weight Loss of Polymers Heated at Various Temperatures in Air[a]

| Structure | Type | Wt. loss during 100 hr of heating (wt. %) | | | | | Total wt. loss (wt. %) |
|---|---|---|---|---|---|---|---|
| | | 260°C | 280°C | 300°C | 325°C | 350°C | |
| | Ester-imide | 1.20 | 1.38 | 3.42 | 7.95 | 29.0 | 42.95 |
| | Amide-imide[b] | 6.17 | 1.58 | 6.49 | 10.92 | 71.8 | 96.93 |
| | Ester | 3.90 | 1.39 | 10.44 | 22.92 | 60.4 | 99.01 |
| | Imide[c] | 1.16 | 0.07 | 1.38 | 2.40 | 12.0 | 17.01 |

[a] From Loncrini, J. Polymer Sci., Part A–1, 4, 1531–1541 (1966).
[b] Amoco Chemical Co. AI-10.
[c] DuPont, Polymer H.

chains. Typical stabilities determined via weight loss for some representative polymers are shown in the accompanying table (p. 59) (*422*). It is clearly seen that the poly(ester-imide) is inferior to Polymer H. On the other hand, its stability is better than that of the Amoco poly(amide-imide), AI-10. The weakness of the ester linkage is evidenced by the practically total volatilization of poly(hydroquinone isophthalate) under the experimental conditions. It is to be noted that both the poly(ester-imide) and Polymer H retained flexibility after the treatment at 300°C, while the two other materials were completely embrittled. It was also observed that the polymers described in this section displayed very strong adhesion to glass (*422*).

## 4. HETEROCYCLIC IMIDE-CONTAINING COPOLYMERS

Copolymers that contain an imide ring as well as another heterocyclic ring are discussed in this section. Several such copolymers have already been described in the sections dealing with polyimides and poly(amide-imides). However, with very few exceptions the preparation of most of the polymers discussed earlier involved the use of a "ready-made" heterocyclic co-monomer; the ring-forming process in the great majority of the examples involved the imide structure only. Condensation of 2,5-bis(4-aminophenyl)-1,3,4-oxa-diazole with pyromellitic anhydride [Eq. (I-68)] is an example. Copolymers in

[56]          [2]

(I-68)

[199]

which both heterocyclic rings are formed simultaneously are considered in this section.

## POLY(IMIDE-IMIDAZOPYRROLONES)

Poly(imide-imidazopyrrolone) copolymers contain units of [200] and [201] in their chains. Both polyimides and polyimidazopyrrolones possess outstanding thermal stability. The incorporation of the imide structure into the very rigid imidazopyrrolone was expected to increase the flexibility of the material without affecting its stability at high temperature.

[200]

[201]

"Imidazopyrrolone"

METHOD 13. THERMAL DEHYDRATION OF POLYAMIC ACIDS

Reaction of an aromatic triamine with an aromatic dianhydride in a polar solvent such as $N,N$-dimethylacetamide yields the intermediate soluble poly(amic-acid) [203]. The acid is cast to a film which is dehydrated by heating at elevated temperatures. Poly(imide-imidazopyrrolones) [204] are obtained (*185*). The reaction is exemplified in Eq. (I-69) using 1,2,4-triaminobenzene [202] and pyromellitic anhydride [2] (*185*).

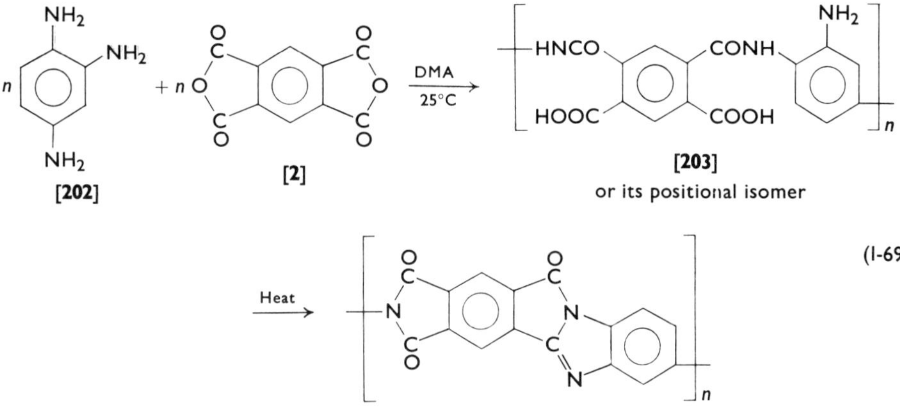

[202]     [2]     [203]
or its positional isomer

(I-69)

[204]

Formula [204] is an idealized representation of the copolymer. Infrared studies indicated the presence of units [200] and [201] in the product (*185*).

In a slight variation, poly(imide-imidazopyrrolones) were prepared by coreacting a mixture of an aromatic tetramine and diamine with an aromatic dianhydride (*98, 255*) [Eq. (I-70)]. Proper adjustment in the tetraamine/diamine ratio led to copolymers possessing the desired proportion of structures [200] and [201]. The general conditions that were used were similar to those described above. The success of the present method is predicated upon the solubility of the intermediate poly(amic acid). If the poly(amic acid) is insoluble, only low molecular weight materials are obtained (*185*).

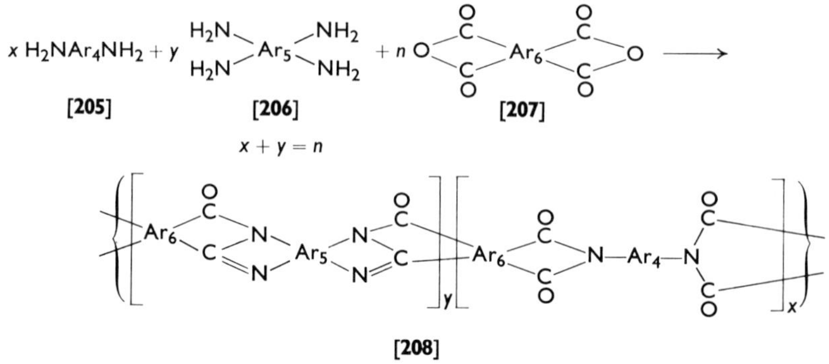

$$x + y = n$$

(I-70)

**[208]**

## METHOD 14. POLYMERIZATION IN POLYPHOSPHORIC ACID

A 3–5% solution of 1,2,4-triaminobenzene **[202]**, and 1,4,5,8-naphthalene-tetracarboxylic acid dianhydride **[209]** in polyphosphoric acid was heated at 180°C for 12–14 hours. The reaction mixture was hydrolyzed in ice water. A high molecular weight copolymer, presumably **[210]** ([$\eta$] 0.88 in 97% $H_2SO_4$) was isolated (*185*). The method was applied successfully to other diamines.

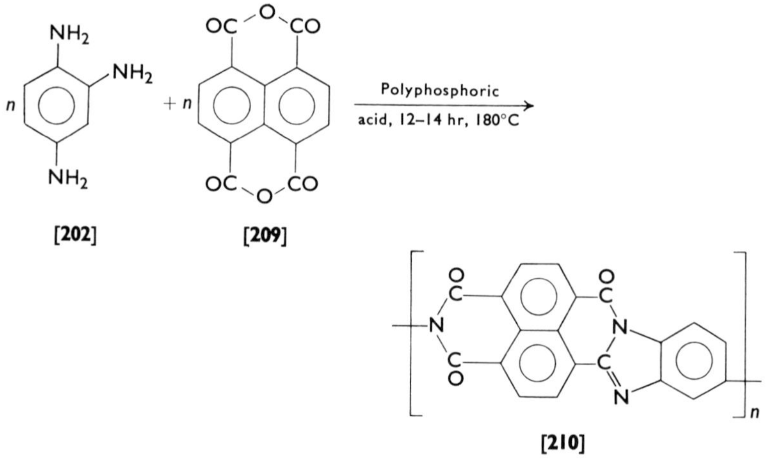

(I-71)

**[202]**          **[209]**

**[210]**

It is of particular interest in those cases where Method A,13 fails due to the insolubility of the amic acid intermediate. The reaction was also successfully utilized in the synthesis of poly(imide-imidazopyrrolones) starting with mixtures of aromatic tetra- and diamines, instead of the triamine (*100*).

The poly(imide-imidazopyrrolones) are infusible materials of limited solubility. The only solvent that is mentioned in the literature is concentrated sulfuric acid. Poly(imide-imidazopyrrolones) display excellent thermal stability. Thermogravimetric analysis of polymers **[204]** and **[210]** at a heating rate of 10°C/minute showed that both were stable beyond 540°C. At 1000°C, the weight losses were 32% for [204] and 26% for [210]; at 1176°C the weight losses were 38 and 31%, respectively (*185*).

The poly(imide-imidazopyrrolones) are tabulated in Table I-13. (See also Chapter III of Part I; Section C-2).

## POLYTRIAZOLOLACTAMS

METHOD 15. THERMAL DEHYDRATION OF APPROPRIATE POLYAMIC ACIDS

The synthesis of polytriazololactams is shown in Eq. (I-72) (*89*).

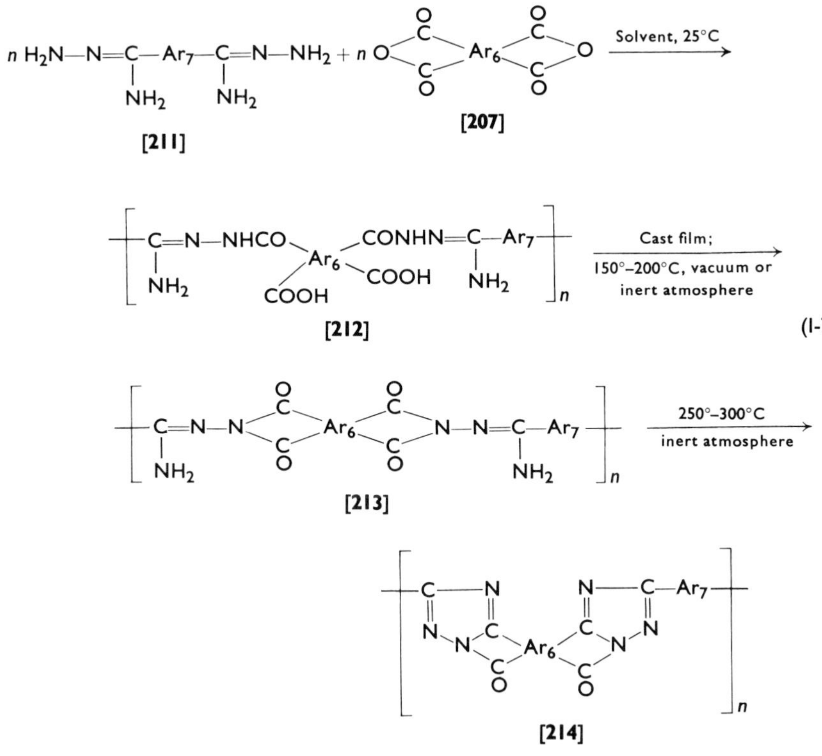

[211]    [207]

[212]    (I-72)

[213]

[214]

Reaction of the bis(amidrazone) [211] with the dianhydride [207] is performed at room temperature in a polar solvent such as DMF, DMSO, or *N*-methyl-2-pyrrolidone. The intermediate poly(amic acid) [212] is cast to a film and then thermally dehydrated. The dehydration proceeds in two steps: at temperatures of 150°–200°C the polyimide [213] is obtained; further heating at 250°–300°C yields the poly(triazololactam) [214].

Excellent mechanical properties, good thermal resistance (up to 400°–500°C), as well as good resistance to γ-rays were claimed for these polymers (*89*).

OXADIAZOLE COPOLYMERS

METHOD 16. THERMAL DEHYDRATION OF POLYAMIC ACIDS TO GIVE COPOLYMERS CONTAINING THE 1,3,4-OXADIAZOLE RING

Several copolymers that contain an imide ring and the 1,3,4-oxadiazole ring were discussed in the section on polyimides (Section A,1). The majority of these copolymers are prepared (see Tables I.3 and I.4) from oxadiazole diamines via reaction with a dianhydride. However, an alternative method is also available (*250, 275*). It consists in reacting a diaminohydrazide with a dianhydride followed by thermal dehydration of the intermediate poly(hydrazide-amic acid). An example is shown in Eq. (I-73) (*275*). The two heterocyclic

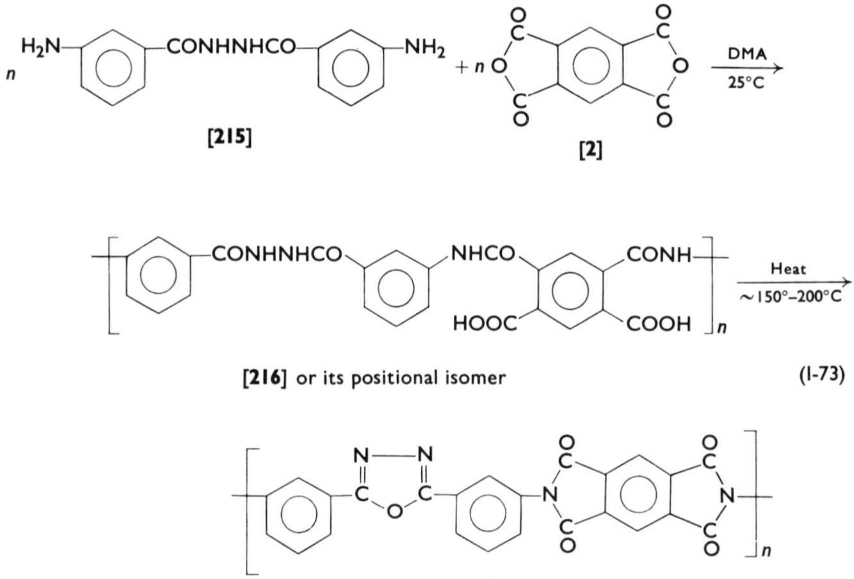

[215]     [2]

[216] or its positional isomer     (I-73)

[217]

structures (oxadiazole and imide) form at different rates; under proper experimental conditions it is possible to isolate the intermediate poly(imidehydrazide) [218] (*275*).The condensation of 2,5-bis(3-aminophenyl)-1,3,4-

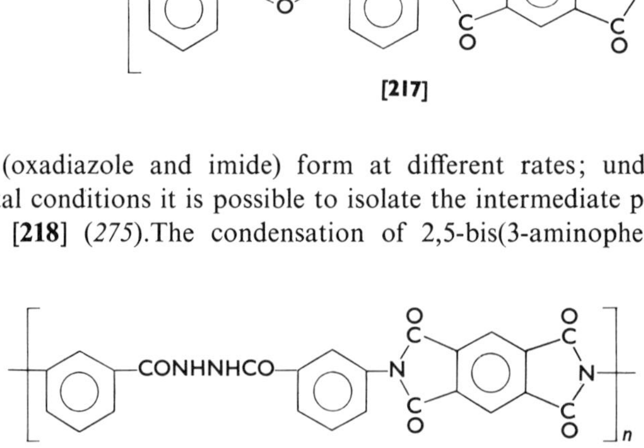

[218]

oxadiazole [219] with pyromellitic anhydride [2] leads to the same polymer [217] (275).

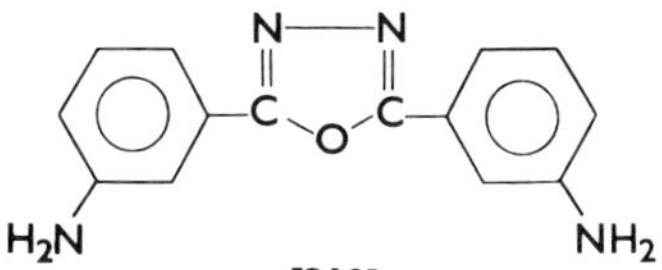

[219]

A somewhat different approach to the synthesis of poly(1,3,4-oxadiazole-imides) is shown in Eq. (I-74) (275). An imide-containing diacid chloride [220] is used as the starting material.

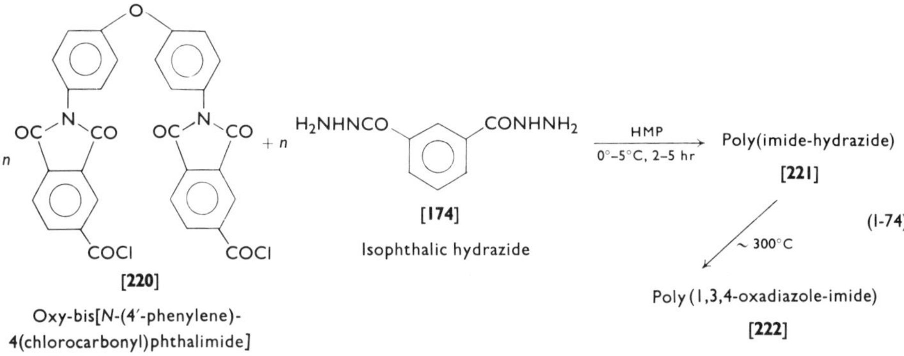

[220]

Oxy-bis[N-(4'-phenylene)-
4(chlorocarbonyl)phthalimide]

[174]

Isophthalic hydrazide

Poly(imide-hydrazide)

[221]

(I-74)

~ 300°C

Poly(1,3,4-oxadiazole-imide)

[222]

Polyimides containing the 1,3,4-oxadiazole ring and their properties are tabulated in Tables I.3 and I.4. Generally, the introduction of the 1,3,4-oxadiazole moiety into the chain results in a decrease of the polymer's thermal stability.

METHOD 17.  THERMAL DEHYDRATION OF POLYAMIC ACIDS TO GIVE COPOLYMERS CONTAINING THE 1,2,4-OXADIAZOLE RING

The synthetic route whereby these copolymers can be prepared is shown in Eq. (I-75) (345, 597). Condensation of isophthalbis(amidoxime) [223] with trimellitic acid chloride [143] in DMA gave the intermediate dianhydride [224]. The latter was condensed in DMA with 4,4'-diaminodiphenyl ether; dehydration of the poly(amic-acid) at 200°C under reduced pressure led to the poly(1,2,4-oxadiazole-imide) [225]. Polymer [225] was claimed to be stable up to 400°C.

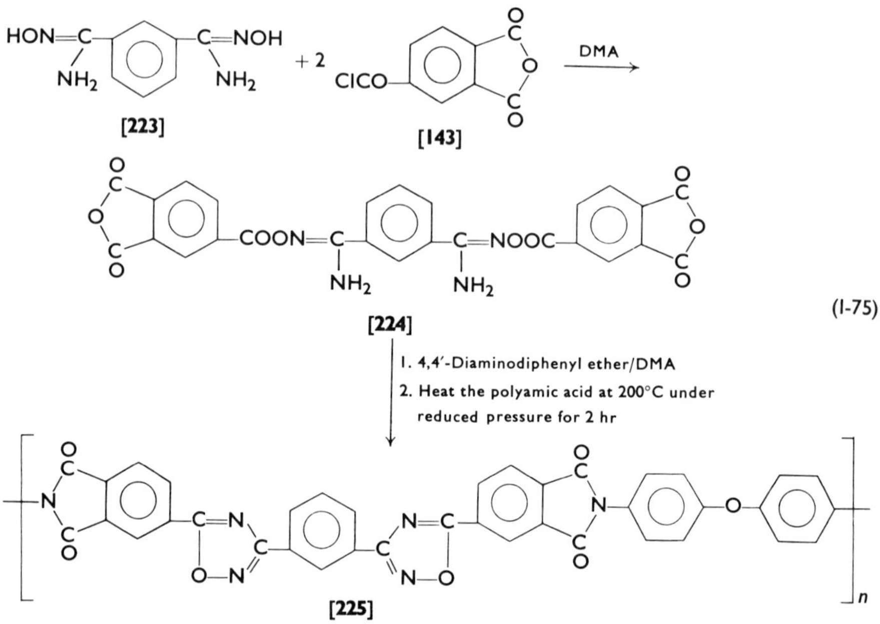

$$[223] + 2 \ [143] \xrightarrow{DMA}$$

$$[224]$$

(I-75)

1. 4,4'-Diaminodiphenyl ether/DMA
2. Heat the polyamic acid at 200°C under reduced pressure for 2 hr

$$[225]$$

## 5. POLYIMINOLACTONES

Polyiminolactones constitute a class of polymers that are isomeric with polyimides. The description of their preparation and chemical reactions follows.

### METHOD 18. CHEMICAL DEHYDRATION OF POLYAMIC ACIDS

Polyiminolactones or polyisoimides [45] are prepared by the sequence shown in Eq. (I-76).

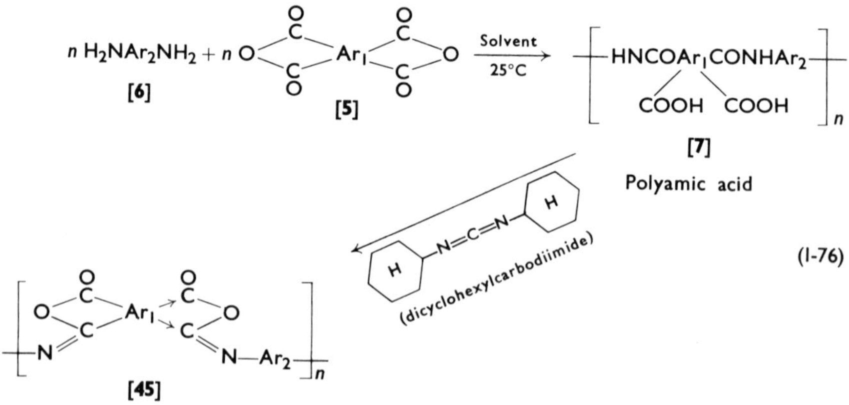

$$n \ H_2NAr_2NH_2 + n \ [5] \xrightarrow[25°C]{Solvent} [7]$$

[6]          [5]

Polyamic acid

(dicyclohexylcarbodiimide)

(I-76)

$$[45]$$

The diamine [6] reacts with the dianhydride [5] at room temperature. Dimethyl-formamide and dimethylacetamide are preferred solvents (*15, 19, 29, 169*). The polyamic acid solution is dehydrated with dicyclohexylcarbodiimide at room temperature to yield the polyiminolactone [45] (*15, 19, 29, 169*), The polymer may be a mixture of isomers (hence, the arrows in formula [45]).

Several types of reagents are suitable for this dehydration. Carbodiimides appear to be one of the best (*15, 19, 29, 169, 471*). In some instances the combination of a carbodiimide with pyridine was found beneficial (*19*). Other excellent reagents include monochloroacetic anhydride/2-methylpyrazine (*29, 386*), phenylphosphonyl dichloride/pyridine (*29, 386*), and trifluoracetic anhydride (*29, 57, 59, 386, 470*). In addition, the use of pyridine or triethyl-amine with a variety of organic and inorganic acid halides (e.g., propionyl fluoride and thionyl chloride) also led to polyiminolactones (*386, 471*). The dehydration is preferably carried out at room temperature (*19, 386*). The polyamic acid may be used as a solution in DMF or DMA. Alternately, a film of the acid can be first cast and then soaked in a benzene or cyclohexane solution of the dehydrating agent (*19, 57, 386*). Other techniques have also been developed (*145, 325*).

## REACTIONS OF POLYIMINOLACTONES

Upon heating, polyiminolactones rearrange into the corresponding poly-imides. The reaction is rapid and sometimes completed within seconds (*25*). Adjustment of time and temperature can lead to poly(imide-iminolactone) copolymers (*387*). The isomerization can also be performed by chemical means. Triethylammonium acetate in DMA solution is the reagent of choice (*386, 387*). Again, proper control of the reaction conditions leads to poly(imide-iminolactone) copolymers of well-defined composition (*387*).

Polyiminolactones react readily with alcohols, thiols, ammonia, and amines and give the corresponding derivatives of the polyamic acids. These reactions are summarized in Eq. (I-78).

[45]   (I-77)

[I]

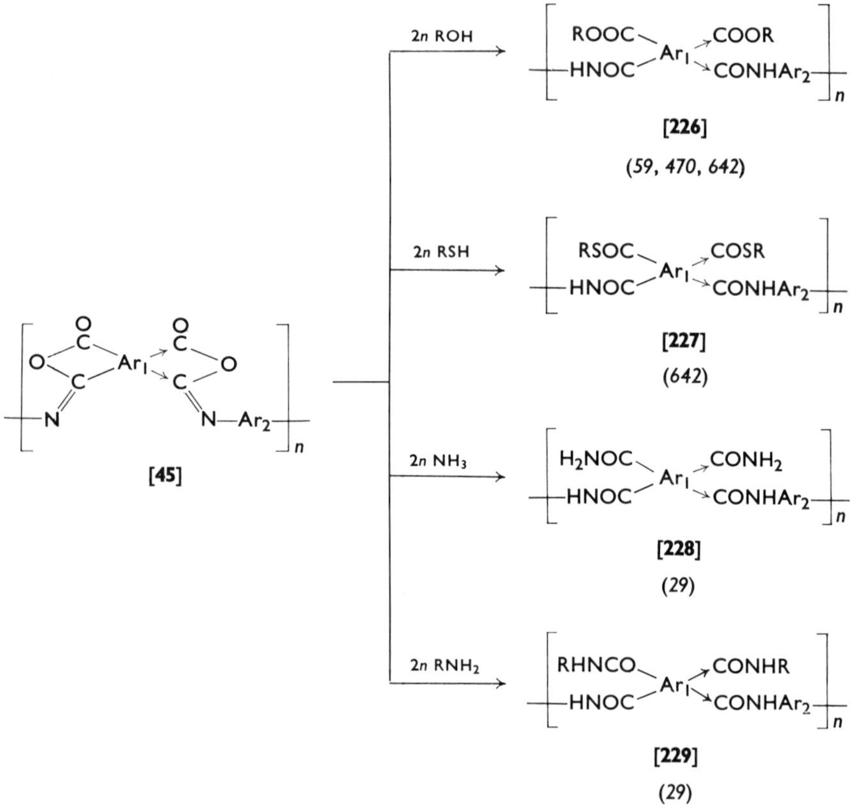

$$\textbf{[45]} \quad \xrightarrow{2n\ \text{ROH}} \quad \begin{bmatrix} \text{ROOC} & \text{Ar}_1 & \text{COOR} \\ \text{HNOC} & & \text{CONHAr}_2 \end{bmatrix}_n \quad \textbf{[226]} \quad (59,\ 470,\ 642)$$

$$\xrightarrow{2n\ \text{RSH}} \quad \begin{bmatrix} \text{RSOC} & \text{Ar}_1 & \text{COSR} \\ \text{HNOC} & & \text{CONHAr}_2 \end{bmatrix}_n \quad \textbf{[227]} \quad (642)$$

$$\xrightarrow{2n\ \text{NH}_3} \quad \begin{bmatrix} \text{H}_2\text{NOC} & \text{Ar}_1 & \text{CONH}_2 \\ \text{HNOC} & & \text{CONHAr}_2 \end{bmatrix}_n \quad \textbf{[228]} \quad (29)$$

$$\xrightarrow{2n\ \text{RNH}_2} \quad \begin{bmatrix} \text{RHNCO} & \text{Ar}_1 & \text{CONHR} \\ \text{HNOC} & & \text{CONHAr}_2 \end{bmatrix}_n \quad \textbf{[229]} \quad (29)$$

(I-78)

Polymers [226]–[229] yield polyimides upon heating at 250–400°C (29, 642). A similar reaction takes place when polyiminolactones are treated with hydrazoic acid (644).

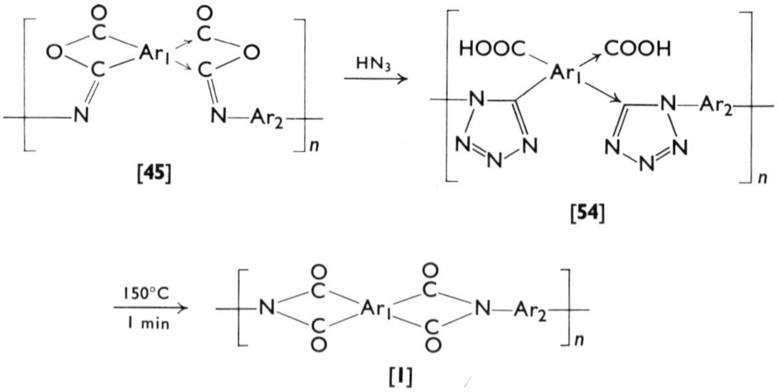

(I-79)

The polyiminolactones that have been prepared and their properties are listed in Table I.12. Care is recommended in interpreting those properties that were obtained as a result of testing at elevated temperatures. They may in fact describe the poly(imide-iminolactone) copolymers rather than the poly-iminolactones.

## B. Polymers Containing Pyrrole and Related Rings

METHOD 1. POLYTETRACYANOETHYLENE

Metal-tetracyanoethylene polymers were discussed in the first volume of this book (Part A, Chapter IV, pp. 150–154). As was noted then, homopolymerization of tetracyanoethylene occurs on heating at 200°C. Nucleophiles, such as alcohols, phenols, and amines (104), or conjugated polymers (105) are reported to catalyze polymerization. Using 1–2 mole % of a nucleophile like alcohol gave a 43% yield of polymer in 10 hours at 200°C. Polyphenyl and polyanthryl are examples of the conjugated type of polymers that were employed. Black, infusible, powdery products are obtained with both types of catalysts. The scheme shown in Eq. (I-80) depicts a probable reaction path for this

(I-80)

polymerization (105, 106). Studies of the thermal stability and electrical conductivity of these products have been reported (104).

METHOD 2. POLYPHTHALOCYANINES

Pyromellitonitrile (1,2,4,5-tetracyanobenzene) polymerizes to polyphthalocyanines on being heated at 250°–450°C (654). Related polymers containing

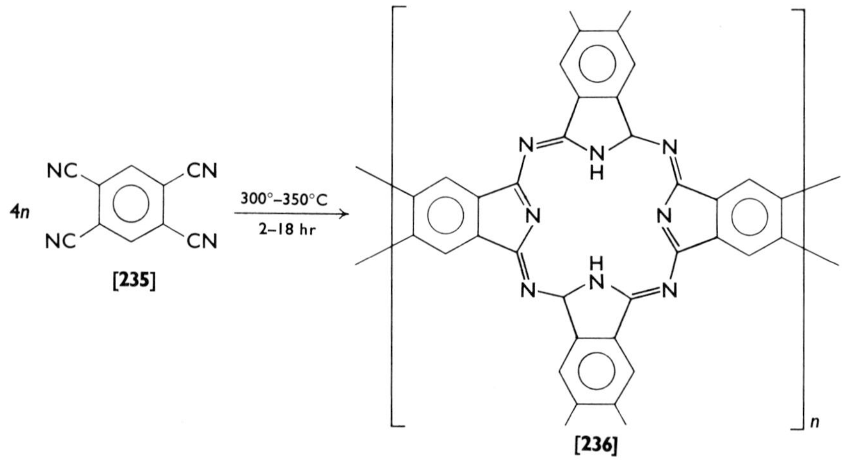

[236]

(I-81)

metals were described in Chapter VI of Part A of this book (p. 156). In addition to the polymerization conditions shown in Eq. (I-81), heating [235] with

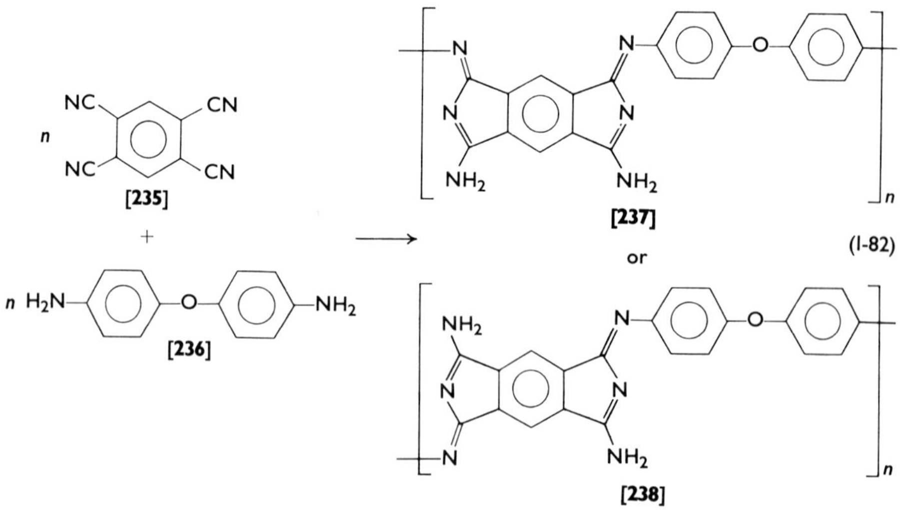

(I-82)

methanol or ammonia at 100°C yields an intermediate product (*177, 683*). Subsequent heating of this material, which is probably related to pyromellitimide, at 250°–300°C results in polyphthalocyanine formation. Removal of the magnesium from magnesium polyphthalocyanine with concentrated sulfuric acid is an additional route to polymers with structure [236] (*176*). All of the products prepared by these various conditions were dark, purple to black materials that have been investigated extensively for their electrical properties (*176, 433, 654, 683*).

METHOD 3. POLYMERIZATION OF MULTIFUNCTIONAL NITRILES AND AMINES

Polymers possessing pyrrole rings are formed when diamines react with tetranitriles. By careful control of reaction conditions, this polymerization can be made to yield linear structures such as those shown in Eq. (I-82) (*534*). A reaction time of 1 hour at reflux in methoxyethanol containing sodium yields soluble polymers of moderately high molecular weights. Further reaction of [237] or [238] with additional diamine is slow. The complete ladder structure polymer shown in Eq. (I-83) requires reaction times of days for its formation (*500, 505, 533, 535*). The uncatalyzed, high-temperature condensation of

(I-83)

diamines with tetranitriles has also been carried out in solvents such as α-chloronaphthalene (b.p. 263°C), or in bulk at 200°–300°C (*434*). Products obtained by these high-temperature routes are less tractable than those prepared by reaction at lower temperature. Compounds such as [241] and [242] also polymerize with amines to polymers possessing structures similar to those shown above (*505, 533*).

The structure of the polymers shown above are most likely idealized, inasmuch as many were intractable. They are all colored and have been investigated for their electrical conductivity properties (*434, 535*). Some members of this polymer family are reported to be thermally stable in air to 350°C (*434*).

## METHOD 4. POLYPYRROLINES FROM DIAMINES

A Japanese patent describes the polymerization of diamines with 1,2,4,5-tetrakis(chloromethyl)benzene (*596*). Triethylamine is used to scavenge the hydrogen chloride. Dichloroethane, dimethylformamide, dimethyl sulfoxide, and dimethylacetamide are useful solvents at 80°–120°C. Although evidence for the polymer's molecular weight is not available, elemental analysis indicated that cyclization of the pyrroline rings was incomplete.

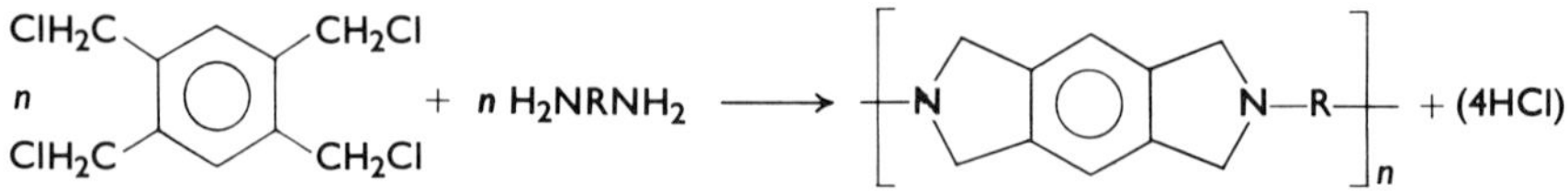

[243]        [244]        (I-84)

## METHOD 5. POLYPYROLLINONES

Condensation of bis(*o*-aminocarboxylic acids) with *p*-xylylene dichloride is reported to proceed as shown in Eq. (I-85). Polyphosphoric acid at 80°–160°C yields moderately high molecular weight polymer (*697*). Heating

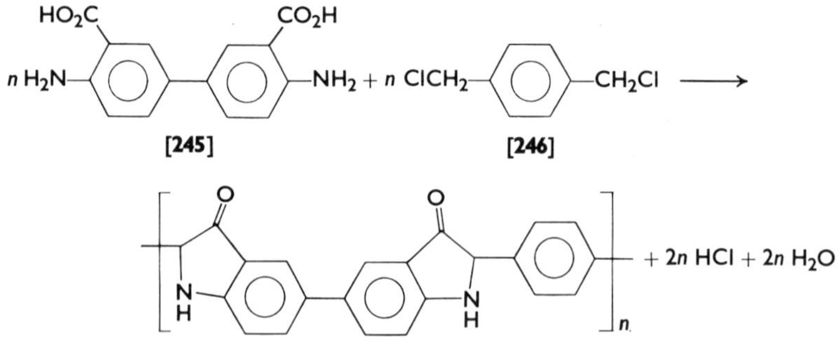

[245]        [246]

(I-85)

[247]

the reactants in dimethylformamide or pyridine at 160°C yields the same polymer (*698*). Condensation of dialdehydes with [245] has yielded very high molecular weight polymers that may possess structures analogous to [247] (*693*).

Examples of the polymer classes discussed in this section are listed in Table I.14.

## C. Poly(1,3-Oxazinones) and Poly(1,3-Oxazindiones)

METHOD 1. POLY(1,3-OXAZINONES) FROM o-AMINO AROMATIC CARBOXYLIC ACIDS

Although many poly(1,3-oxazinones) are prepared by cyclizing suitably substituted polyamides, several direct routes have also been described. Heating a bis(o-amino aromatic acid) in polyphosphoric acid is reported to yield a high molecular weight polymer containing 1,3-oxazinone groups (*396*). The example shown in Eq. (I-86) was effected in 2–3 hours at 140°–160°C. The mechanism of this condensation has been studied and phosphoryl-

ated intermediates postulated. An optimum polymerization temperature of 150°C was found (*398, 400*). The use of polyphosphoric acid for preparing precursor polyamides for poly(1,3-oxazinone) synthesis has been described. Adding lithium chloride to the PPA gives higher molecular weight polyamides (*704*). Dinitriles (*699*), diesters (*701*), and diamides (*702*) can be used as diacid component precursors for polyamide synthesis with polyphosphoric acid. One-step polymerization to a polyoxazinone is also effected by fuming sulfuric acid containing lithium chloride (*704*). The reported example [Eq. (I-87)] utilized phenyl esters for reactants.

As mentioned above, the synthesis of a high molecular weight polyamide containing cyclizable carboxylic acid groups is frequently used for polyoxazinone synthesis. Low-temperature, solution polycondensation in aprotic, polar solvents is preferred over interfacial or bulk procedures (*401, 576, 577,*

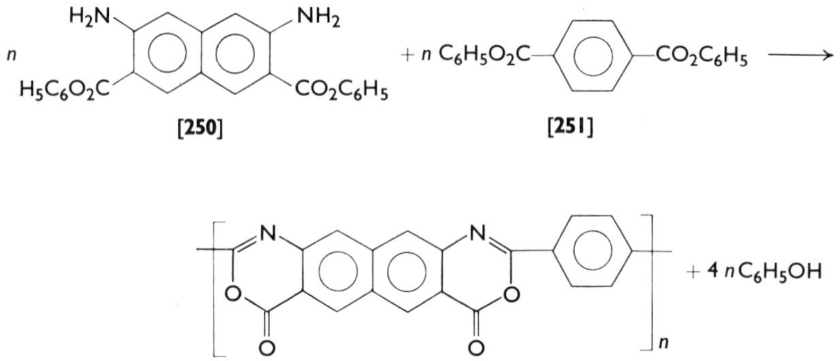

[250]          [251]

(I-87)

[252]

$\eta_{inh}$ 2.02

$+ 4n C_6H_5OH$

706, 707). N-Methylpyrrolidone containing lithium chloride (303, 508, 694, 703) or triethylamine in dimethylacetamide (508, 576, 577) are illustrative reaction systems for polymerization of diacid halides with bis(o-amino acids). A bis(ketene) was used in one instance for polyamide synthesis (303). Very

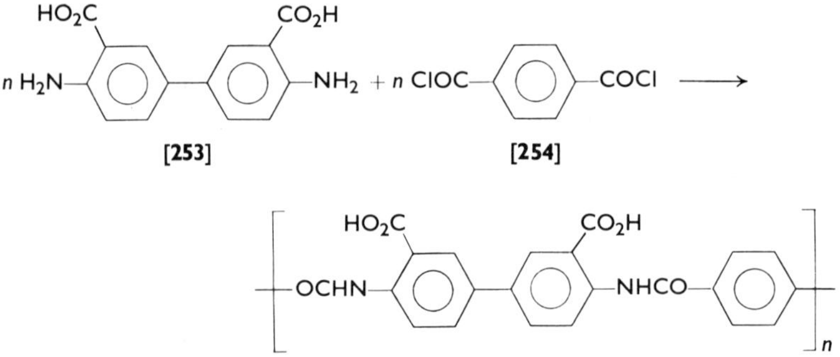

[253]          [254]          (I-88)

[255]

high molecular weight polyamic acids [255] are obtainable. Cyclization of this type polymer is carried out thermally in vacuo. Cyclodehydration begins at

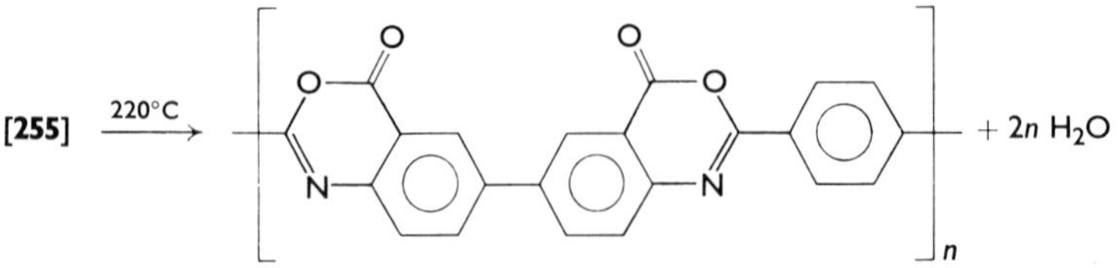

[255] $\xrightarrow{220°C}$          $+ 2n H_2O$          (I-89)

[256]

about 160°–200°C (*401, 508, 695, 710*), with final temperatures up to 360°C being employed. This reaction is generally carried out on films cast from the polyamic acid. The oxidative cyclization shown in Eq. (I-90) occurred on heating the polyamide [**257**] in air at 380°–400°C for 1 hour (*694*). Chemical methods for dehydrating polyamic acids to polyoxazinones that have been

[**257**]

(I-90)

[**258**]

reported include polyphosphoric acid at 200°–250°C for several hours (*401*), and acetic anhydride containing pyridine (*348, 401, 695, 705, 708, 710*) or isoquinoline (*705*). Cyclization in the presence of a phosphite ester has also been described (*709*).

Many of the poly(1,3-oxazinones) prepared by the variations of this method are insoluble. However, the data in Table I.15 show that some of the soluble members possess very high molecular weights. Only meager, additional characterization data have been reported.

METHOD 2. POLY(1,3-OXAZINDIONES) FROM DIISOCYANATES AND BIS(HYDROXY ACIDS)

The direct synthesis of poly(1,3-oxazindiones) from diisocyanates and diphenyl esters of bis(*o*-hydroxycarboxylic acids) has been reported (*108, 260*). The reaction shown in Eq. (I-91) was carried out in dimethyl sulfoxide at 100°C for several hours. Added amines, such as triethylenediamine, are catalysts. Very high molecular weights are obtainable. When aliphatic esters of the bis(hydroxy acids) are polymerized with diisocyanates, the reaction is generally carried out in two steps (*16, 17*). The intermediate polyurethane [**262**] is synthesized in inert, polar solvents with a basic catalyst. Subsequent heating at 200°–300°C in vacuum converts it to the polyoxazindione [**263**].

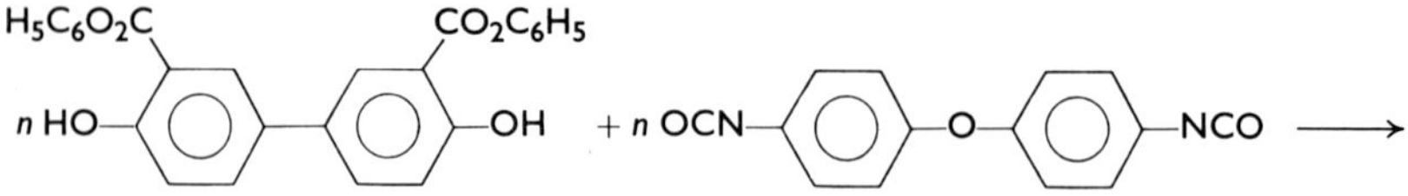

[259]          [260]

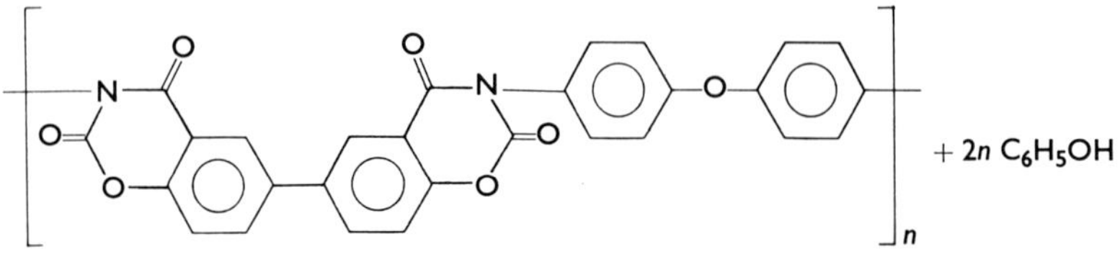

(I-91)

+ 2n C6H5OH

[261]

Similar polymerizations with dihydroxymonocarboxylic esters yield poly-oxazindione-urethanes (22). Some aliphatic members of this family based on tartaric acid have also been reported. Films prepared from polymers based

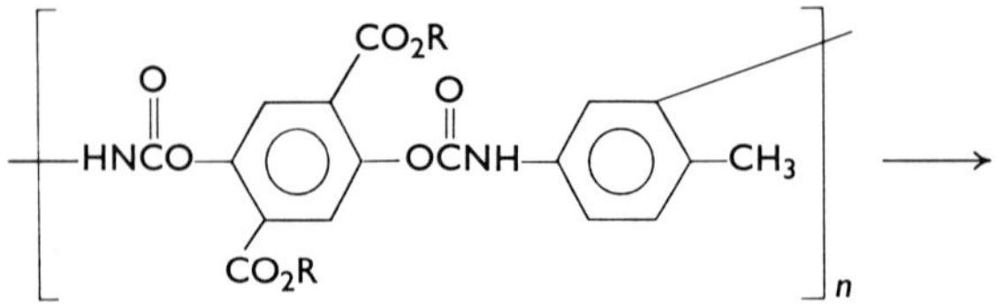

[262]

(I-92)

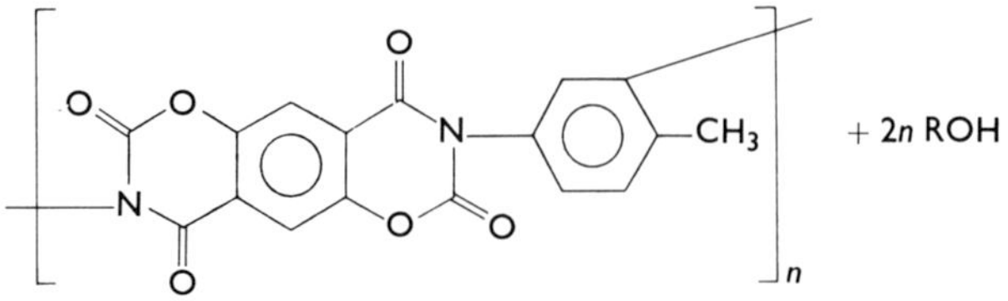

+ 2n ROH

[263]

on diphenyl ether diisocyanates are useful as electrical insulating materials (557).

METHOD 3. POLY(1,3-OXAZINDIONES) FROM CARBONATE-BLOCKED BIS-(HYDROXY ACIDS) AND DIAMINES

Polyamides prepared from carbonate-blocked bis(o-hydroxy acid chlorides) cyclize to polyoxazindiones when heated. An example is shown in Eq. (I-93)

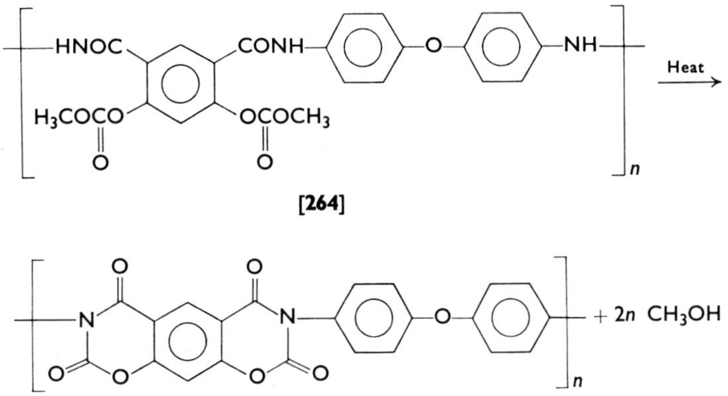

**[264]**

(I-93)

+ 2n CH₃OH

**[265]**

(*16*). The precursor polyamides are prepared from diacid halides and diamines by solution polycondensation at room temperature. Copoly(amide-oxazin-diones) are prepared analogously from monohydroxy diacids (*21*). High molecular weights have been obtained for many members of this family of polymers.

A variant of the above synthetic procedure uses phenyl chloroformate/pyridine to cyclize a preformed polyamide possessing free *o*-hydroxyl groups (*115*). The polyamides were prepared either in the melt, or in phenols or cresols as solvents.

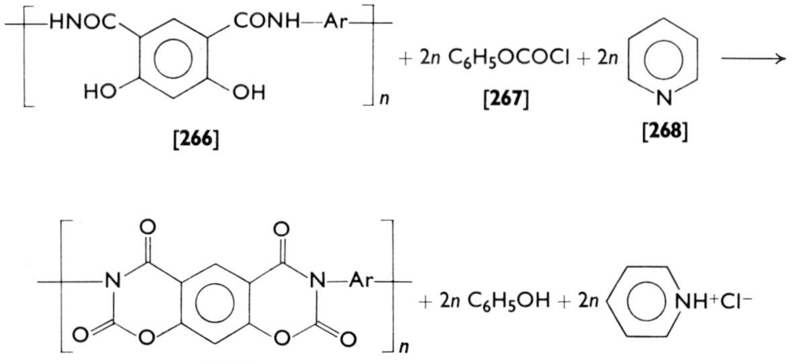

+ 2n C₆H₅OCOCl + 2n

**[266]**          **[267]**          **[268]**

(I-94)

+ 2n C₆H₅OH + 2n

**[269]**

The property data available on poly(1,3-oxazinones) and poly(1,3-oxazin-diones) are quite meager. Essentially all of the data are presented in Table I.15. No one compound has been singled out for scale up and intensive evalua-tion. Reviews have mentioned this family of polymers (*350, 692*).

## D. Polydioxins, Polyoxazines, and Polythiazines

METHOD 1. POLYDIOXINS FROM TETRAPHENOLS AND
TETRACHLOROQUINOXALINES

The condensation of aromatic tetrahydroxy compounds with tetrachloro-quinoxaline derivatives leads to polydioxins (685). The example shown in Eq. (I-95) was carried out in refluxing pyridine for 16–24 hours under nitrogen. Melt polymerization and the use of naphthalene or nitrobenzene as solvents

[270]          [271]

(I-95)

[272]          + 4n HCl

are additional variants. The polymers obtained (Table I.16) are not always completely soluble. The soluble products possessed viscosities indicative of low to moderately high molecular weight polymers. The lack of reported mechanical property data is also indicative of their low molecular weights. Thermal stability data for these materials were indistinguishable from those obtained on many other classes of ring-containing polymers.

METHOD 2. POLYOXAZINES AND POLYTHIAZINES FROM DIAMINODIPHENOLS
AND TETRACHLORO OR TETRAHYDROXY COMPOUNDS

The polymerization of diaminodiphenols and diaminodithiophenols has been reported and yields very high molecular weight polyoxazines and poly-thiazines. The only coreactants found to be useful to date are either tetra-chloro- or tetrahydroxyquinoxaline type compounds. An overall schematic representation of this polymerization is shown in Eq. (I-96). Path B, employing the tetrahydroxy coreactant in polyphosphoric acid, yields the highest molecular weight polyoxazines (X = O). Reaction conditions are 200°–250°C. for 24 hours (442, 521, 522). The synthesis of polyoxazines from the tetrachlorobis (quinoxaline) derivative (Path A) is generally carried out in refluxing pyridine. Naphthalene can also be used as a medium for this path.

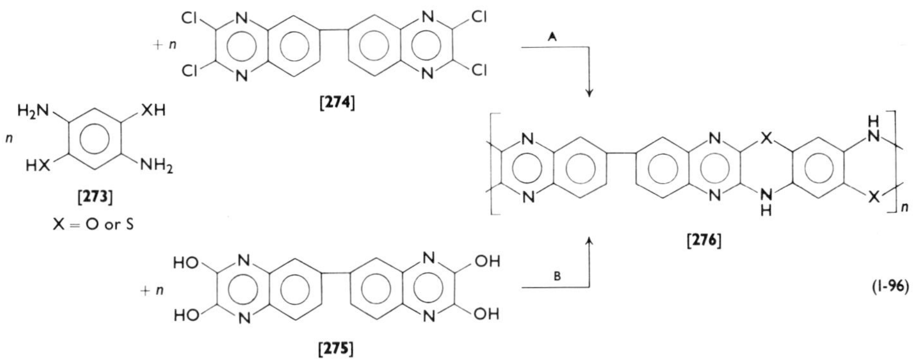

(I-96)

However, the highest molecular weight polythiazines are obtained by polymerizing the tetrachloroquinoxaline derivative [274] with the diamino-dithiophenol [273] (Path A) (442, 521, 522). Dimethylacetamide is a preferred solvent for carrying out the polymerization at 150°C for 16–24 hours. Poly-thiazines are obtainable by Path B using polyphosphoric acid at 200°–240°C, but their molecular weights are generally lower. With both the polyoxazines and polythiazines synthesized by the above routes, a heat treatment at high temperatures *in vacuo* is sometimes necessary to complete cyclization of the polymers (442). All of the polymers that were prepared possess dark colors.

METHOD 3. POLYOXAZINES AND POLYTHIAZINES FROM DIAMINODIPHENOLS AND QUINONES

Diaminodiphenols (628, 632) and diaminodithiophenols (442, 523) poly-merize with quinones to yield polyoxazines or polythiazines. The quinones that have been studied are 2,5-dihydroxy-*p*-benzoquinone, 2,5-diacetoxy-*p*-benzoquinone, 2,5-dichloro-*p*-benzoquinone, and 2,5-dichloro-2,5-dihydroxy-*p*-benzoquinone. One-step polymerization to a ladder polyoxazine occurs in hexamethylphosphoramide at 120°C as shown in Eq. (I-97) (628). The inter-mediate poly(Schiff base) is isolable if the monomers are reacted in tetra-

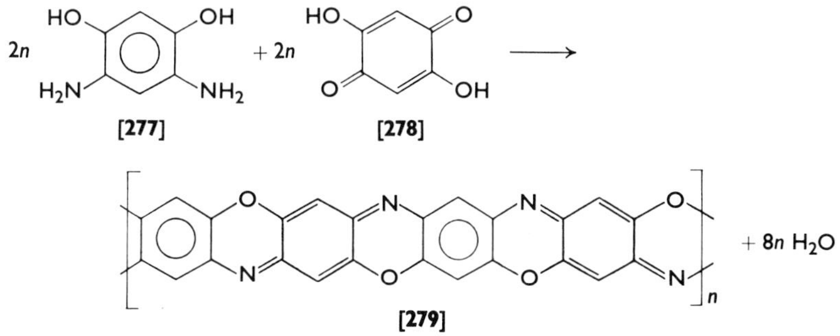

(I-97)

hydrofuran. The polymerization of the diaminodithiophenol type monomer was conducted in dimethylacetamide at 150°C for 24 hours.

The use of 2,5-dichloro-2,5-dihydroxy-*p*-benzoquinone in this polymerization yields an intermediate that can cyclize with elimination of either water or hydrogen chloride (628). Both routes do occur so that the polymer obtained possesses structure [280]. The data in Tables I.17 and I.18 show that moder-

**[280]**

X = —OH, —Cl

ately high molecular weights have been obtained for the few polymers that have been prepared by this method. The structures drawn for these polymers are no doubt idealized, because of the possibilities that exist for isomerism.

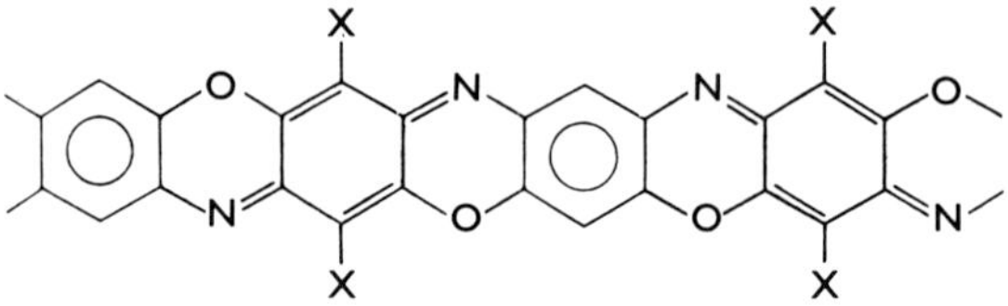

METHOD 4. CONDENSATION OF DIAMINODITHIOPHENOLS WITH BIS(α-HALOALKYL ARYL KETONES)

Polythiazines form from condensation of diaminodithiophenols with bis(α-haloalkyl aryl ketones) (116, 117). The reaction was carried out in refluxing dimethylformamide for up to 20 hours. The molecular weights of the products were determined by bromine analysis and ranged up to 25,000. No mechanical property data were reported.

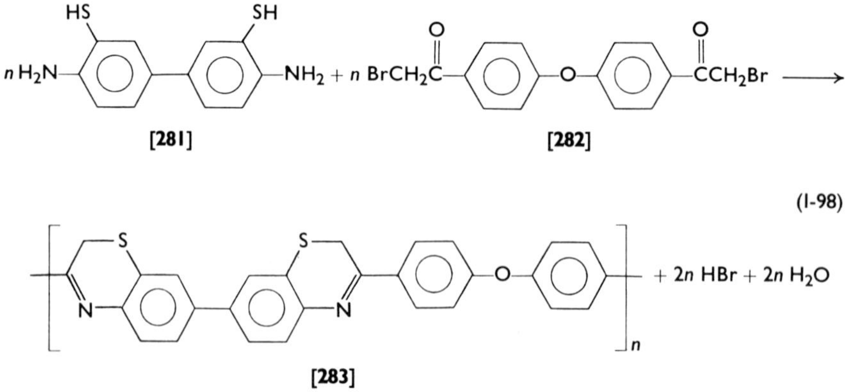

[281]          [282]

(I-98)

+ 2n HBr + 2n H₂O

[283]

## E. Polyquinoxalines and Related Types

METHOD 1. PYRAZINE RING-CONTAINING POLYMERS

Polymerization of the anion in silver cyanide is reported to yield the "ladder" type polymer shown in Eq. (I-99) (691). Precise conditions that were used to effect this polymerization are not available.

$$6n \text{ AgCN} \longrightarrow \qquad (I\text{-}99)$$

[284]

Another report describes the polymerization of the bis(α-bromoketone) [285] with ammonia to yield the polyphenylenepyrazine [286]. Dimethylformamide at temperatures up to reflux was the reaction solvent (335). Properties for polymer [286] were not reported.

$$n \text{ BrCH}_2\text{CO} \!-\!\!\left\langle \bigcirc \right\rangle\!\!-\! \text{COCH}_2\text{Br} + 2n \text{ NH}_3 \longrightarrow$$

[285]

$$+ 2n \text{ H}_2\text{O} + 2n \text{ HBr} \qquad (I\text{-}100)$$

[286]

METHOD 2. POLYQUINOXALINES FROM BIS(o-DIAMINES) AND TETRACARBONYL COMPOUNDS

Aromatic tetraamines have been condensed with a variety of bis(1,2-dicarbonyl compounds) to yield polyquinoxalines. As the data in Table I.19 show, many examples possessing very high molecular weights have been prepared. Solution polymerization (followed by a sintering step) and melt polymerization are employed. Generation of the amine reactant *in situ* via a Leuckart reaction has also been described. The molecular weight of the product [288] was not reported.

On balance, solution polymerization to a prepolymer that is advanced in molecular weight by a sintering operation is the preferred procedure. Solvents for the initial reaction include dimethylformamide, dimethylacetamide, N-methylpyrrolidone, and m-cresol. Although reflux temperatures are used,

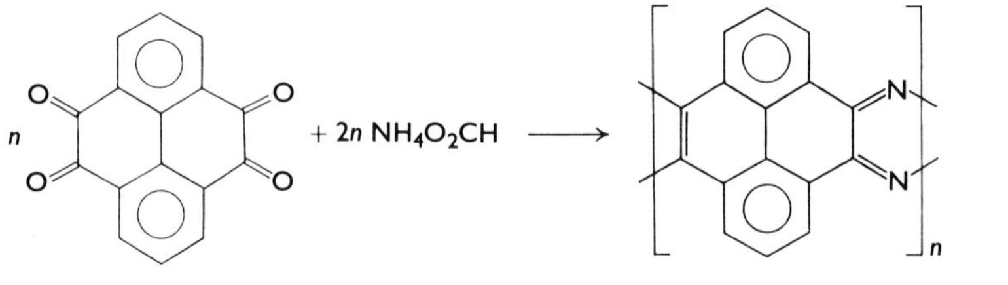

[287]                                    [288]

m-cresol at room temperature yielded very high molecular weight polymers
from a group of bis(benzils) (687–689). An example is shown in Eq. (I-102).

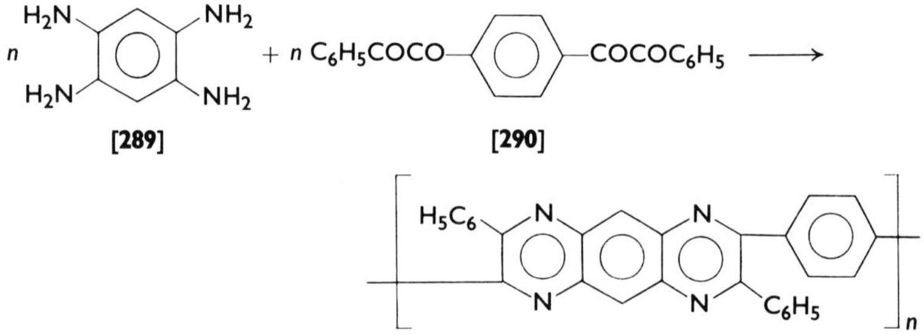

[289]                    [290]

(I-102)

[291]

$[\eta]$ 0.97

The solid state, or sintering operation is carried out at temperatures of 300°–
400°C in vacuo for periods of about an hour (204, 315, 318, 626). Polyphosphoric acid has been used as the solvent and condensing agent for polyindoloquinoxaline synthesis (598, 599). Moderately high molecular weight
polymers were obtained [Eq. (I-103)].

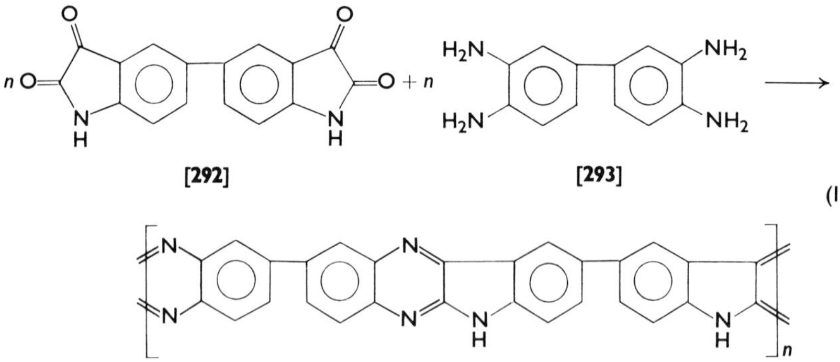

[292]                              [293]

(I-103)

[294]

$[\eta]$ 0.30

Direct, melt polymerization of the amine and ketone monomers can also yield high molecular weight polymer. Temperatures of up to 400°C for several hours are used (*37, 317, 627*). Melt polymerization conditions were required to effect reaction between 1,2,4,5-tetraaminobenzene and 1,2,6,7-tetraketopyrene, since no reaction occurred in solution (*630*). In another study, bulk, melt procedures yielded higher molecular weight polymers than solution procedures (*315*).

Disubstituted quinones have been polymerized with amines to yield polymers that most probably possess quinoxaline units. They are listed in Table I.19. The monomers studied are 2,3-dichlorobenzoquinone (*336*), 2,5-dihydroxy-benzoquinone (*429, 570, 629, 631, 632*), and 3,6-dichloro(difluoro)-2,5-dihydroxybenzoquinone (*631*).

## METHOD 3. POLYQUINOXALINES BY POLYCONDENSATION IN POLYPHOSPHORIC ACID

It has been reported that aromatic tetraamines self-condense in poly-phosphoric acid to moderately high molecular weight polymers containing quinoxaline units (*44*). Temperatures from 250° to 340°C appear necessary to effect this reaction. The reaction shown in Eq. (I-104) did not proceed at 290°C. Polymer [295] probably still contains some precursor dihydrophen-azine units, although most are probably oxidized to [295].

$$ \text{[293]} \xrightarrow[\text{9 hr}]{340°C} \text{[295]} + 2n\,NH_3 \qquad (I\text{-}104) $$

Polyphosphoric acid also effects the condensation of aromatic tetraamines with tetrasubstituted tetraazaanthracenes (*195, 342, 343*). Temperatures of 250°C are employed for the polymerization of tetrachloro-, tetrahydroxy-, or tetraphenoxy tetraazaanthracenes with tetraamines. The ring initially formed in the polymerization is in the dihydro form. Monomers that self-polymerize in an analogous manner are known. An example is shown in Eq. (I-105).

$$ \text{[296]} \xrightarrow[\text{20 hr}]{\substack{PPA \\ 250°C}} \text{[297]} \qquad (I\text{-}105) $$

PROPERTIES OF POLYQUINOXALINES

Polyquinoxalines are highly colored materials, with many of the examples listed in Table I.19 being black. Although some are insoluble or only partially soluble, others are completely soluble in more powerful solvents such as sulfuric acid, *m*-cresol, or hexamethyl phosphoramide. Excellent resistance to refluxing, aqueous potassium hydroxide has been reported for some members of this polymer family (*317*). The molecular weights of many poly-quinoxalines have been quite high, as evidenced by viscosity data and their ability to yield tough fabrications. Flexible cast films are reported in several instances (*317, 627, 687–689*). Spinning of polyquinoxaline fibers from sulfuric acid solution has been patented (*579*).

The condensation of a tetraamine with a tetracarbonyl compound does not necessarily lead to a polymer with a uniform repeating unit. In the case of polymer [**298**] and others, evidence was obtained to show that the 2,2′-structure is formed exclusively (*191, 626, 637*). Just how general this finding may be is

[**298**]

not known, because most studies did not pin down polymer structure. Glass transition temperatures for a number of polyquinoxalines are reported in Table I.19. They range from about 200° up to 400°C. Crystallinity has been detected in some of these polymers. Phenyl substituents on the quinoxaline rings raise the $T_g$, whereas ether linkages in the backbone cause 20°–70°C decreases.

The thermal stability of most of the polyquinoxalines has been investigated. As a class, they are one of the most thermally stable. Numerous members do not exhibit significant decomposition until about 500°C in air, as determined by thermogravimetry. In inert atmospheres, stability to 40°–50°C higher temperatures is found. Phenyl substituents on the quinoxaline ring improve thermal stability (*317, 687*). Polymers derived from aliphatic tetraketones are less stable than those possessing an all aromatic structure (*627*). Surprisingly, some examples possessing "ladder" structures were not significantly more stable than "single-stranded" structures.

Interest in polyquinoxalines for adhesives and laminates has been described (*316, 320*). Electrical property data have also been of interest (*336, 537*).

Reviews that have mentioned polyquinoxaline-type polymers are quite numerous (*33, 196, 230, 319, 350, 369, 436–438, 440, 441, 451, 537, 607, 625, 645, 679, 692*).

## F. Polypiperazines and Polydiketopiperazines

### METHOD 1. POLYPIPERAZINES FROM BIS($\beta$-HYDROXYETHYL)DIAMINES

The dehydration of bis($\beta$-hydroxyethyl)diamines is reported to yield polypiperazines (149). In the examples reported, low molecular weight polymer apparently was desired, since substantially less than the theoretical amount of water was taken off. Pyrophosphoric acid was employed as a catalyst for the polymerization shown in Eq. (I-106), which was conducted at 260°–380°C.

$$n \text{ HOCH}_2\text{CH}_2\text{NH(CH}_2)_{12}\text{NHCH}_2\text{CH}_2\text{OH} \longrightarrow$$

[299]

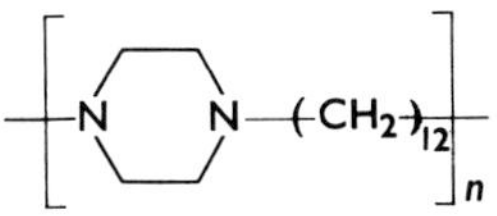

[300]

### METHOD 2. POLYMERIZATION OF BIS($\alpha$-AMINO ACIDS) TO POLYDIKETOPIPERAZINES

Bis($\alpha$-amino acids) polymerize to polydiketopiperazines. At least in a couple of examples, very high molecular weight polymers have been obtained. Polymerization conditions primarily involve heating in bulk or in a solvent such as $m$-cresol. The free diacid or diethyl ester can be used (10, 228). An example is shown in Eq. (I-107). Bulk polymerization is usually conducted at temperatures near 200°C for periods of several hours.

$$n \text{ HOOC}-\underset{\underset{\text{NH}_2}{|}}{\text{CH}}-(\text{CH}_2)_6-\underset{\underset{\text{NH}_2}{|}}{\text{CH}}-\text{COOH} \longrightarrow$$

[301]

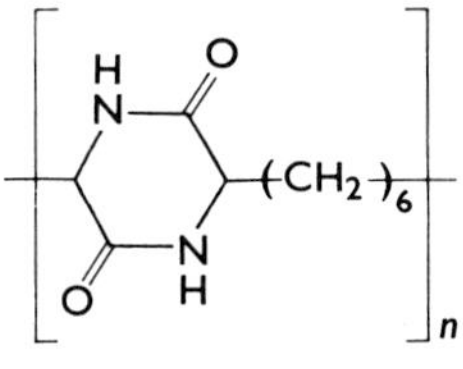

[302]

Hexamethylenebis(iminoacetic acid) [303] polymerizes to very high molecular weight polymer possessing $N,N'$-disubstitution (10, 228).

Phosphoric acid catalyzes the polymerization, which was carried out in *m*-cresol at temperatures up to 218°C.

$$n \, HOOCCH_2NH-(CH_2)_6-NHCH_2COOH \longrightarrow$$

[303]

(I-108)

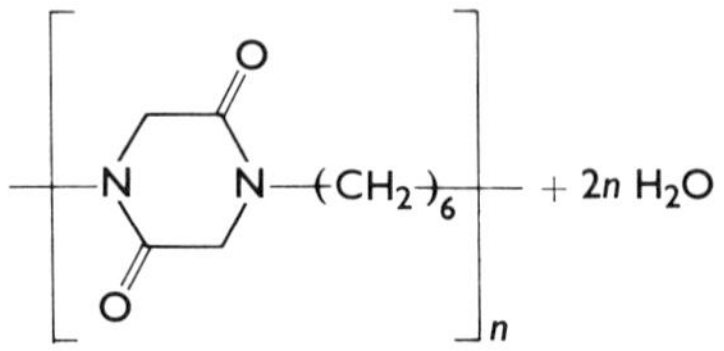

[304]

## METHOD 3. POLYDIKETOPIPERAZINES FROM BIS(α-CHLOROACETAMIDES)

A recent, unique route to polydiketopiperazines employs aromatic bis-(α-chloroacetamides) (*339*). Heating this class of monomer with an epoxide, such as phenyl glycidyl ether, yields moderately high molecular weight poly-diketopiperazines. Quaternary ammonium salts catalyze this polymerization; phenyl glycidyl ether acts as the hydrogen chloride acceptor. Reaction can be carried out in phenyl glycidyl ether as solvent or in dimethylacetamide containing lithium chloride. In the latter medium, reaction required 50 hours at 100°C.

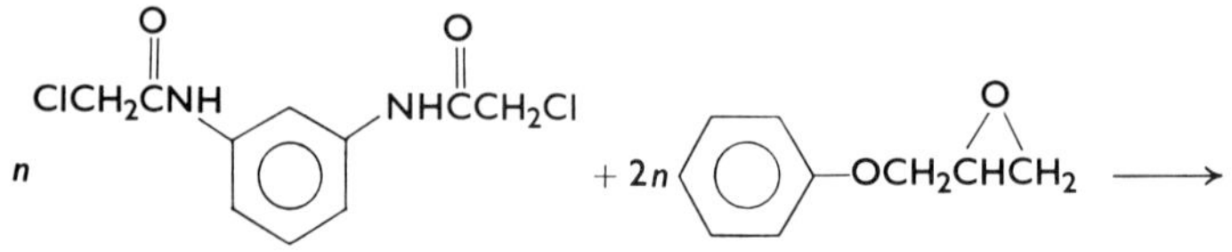

[305]

(I-109)

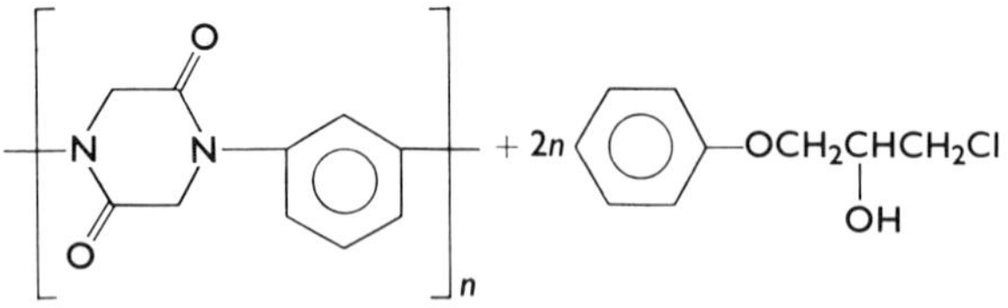

[306]                    [307]

Polymers discussed in this section are tabulated in Table I.20.

## G. Polypyrimidines and Polytetraazapyrenes

METHOD 1. POLYPYRIMIDINE SYNTHESES

Two different ring-forming polymerization reactions have been reported for polypyrimidine synthesis. Alkali metal-catalyzed polymerization of dinitriles yields polyaminopyrimidines (333). Sodium methoxide, sodium hydride, butyllithium and potassium *tert*-butoxide are some of the catalysts that were used. The polymers, which had variously disubstituted 4-amino-pyrimidine rings, possessed the physical characteristics of low molecular weight polymers. Using acetonitrile as a co-monomer improved their tractability.

Condensation of diamidines with bis(β-keto esters) yields poly(4-hydroxy-pyrimidines) possessing moderate molecular weights (639). The diamidine was freed from its hydrochloride with sodium methoxide just before reaction. The condensation was conducted in piperidine at 65°–70°C for 4 hours. Polymers possessing inherent viscosities of 0.15–0.21 were obtained. Infrared spectral studies confirmed the polymer structures. Thermal stability data for these materials are reported in Table I.21.

METHOD 2. POLYTETRAAZAPYRENE SYNTHESIS

Marvel and co-workers have pioneered the condensation of 1,4,5,8-tetra-aminonaphthalene with diacid derivatives. Diphenyl esters were superior to diacid chlorides because they ultimately resulted in higher molecular weight polymer. The postulated sequence of reactions that led ultimately to the polytetraazapyrenes is shown in Eq. (I-111) (187). Initial reaction was carried out at 170°–250°C for a few hours, followed by a sinter-type advancing step at

(I-111)

300°C *in vacuo*. The soluble portions of the products possessed moderately high molecular weights. The thermal stability and electrical properties of these polymers were investigated. They were found to be slightly less stable thermally than polybenzimidazoles; electrical properties were typical of insulators.

Polymerization of 1,4,5,8-tetraaminonaphthalene with dianhydrides leads to ladder-type polymers of structure similar to the polypyrrones discussed in Chapter III of Part B, Vol. 1 (*38*).

## H. Polypyrimidones, Polyquinazolones and Polyquinazolinediones

### Method 1. Polypyrimidones from Unsaturated Isocyanates

Vinyl isocyanate can be polymerized either through the isocyanate or vinyl groups to yield different polymers. Subsequent cyclization of the pendant groups in [**317**] or [**318**] yields the ladder polypyrimidone [**319**] (*32, 529–532*). The *N*-vinyl-1-nylon polymer [**317**] is obtained by cyanide-initiated poly-

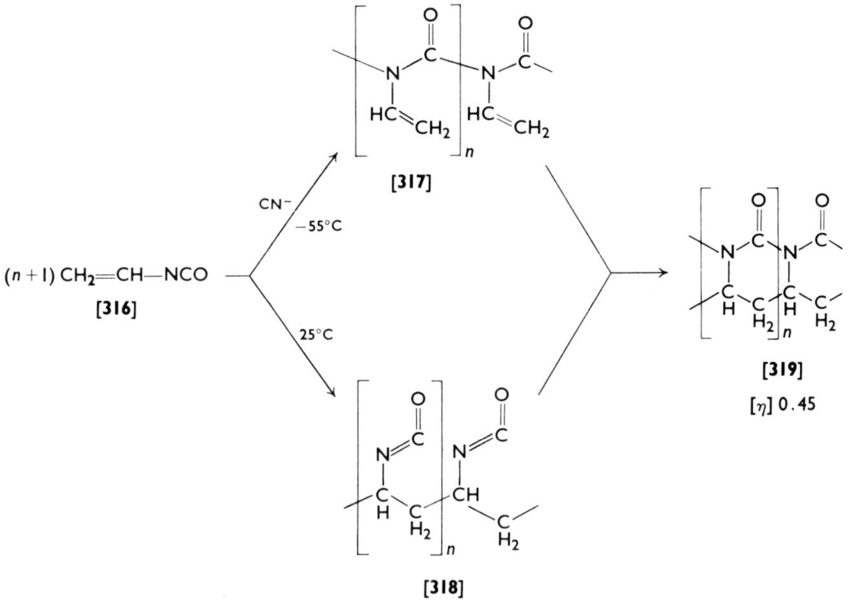

CN⁻
−55°C
25°C
(n + 1) CH2=CH—NCO
[316]
[317]
[318]
[319]
[η] 0.45


(I-112)

merization in dimethylformamide at −55°C (*520, 532*). Vinyl polymerization to [**318**] occurs at room temperature. Formation of the ladder polymer [**319**] is effected by radical or ionic agents, and samples possessing moderately high molecular weights ($[\eta] = 0.45$) have been obtained. The polymer was soluble in pyridine and melts with decomposition at about 390°C (*532*). Investigations of other unsaturated isocyanates have been performed, but the results are not as well defined. Propenyl isocyanate polymerizes to the 1-nylon, whereas isopropenyl isocyanate does not (*529*). $\beta$-Styryl isocyanate has also been investigated.

### Method 2. Polyquinazolones from Diamines and Bis(benzoxazinones)

Some very high molecular weight polyquinazolones have been synthesized by polycondensation of aromatic diamines with bis(benzoxazinones).

Specifically, 2,2'-dimethyl-[6,6'-bibenzoxazine]-4,4'-dione **[320]** condenses with diamines as shown in Eq. (I-113) (297, 406). Solution polymerization gave better results than melt polymerization (193). *m*-Cresol was preferred as

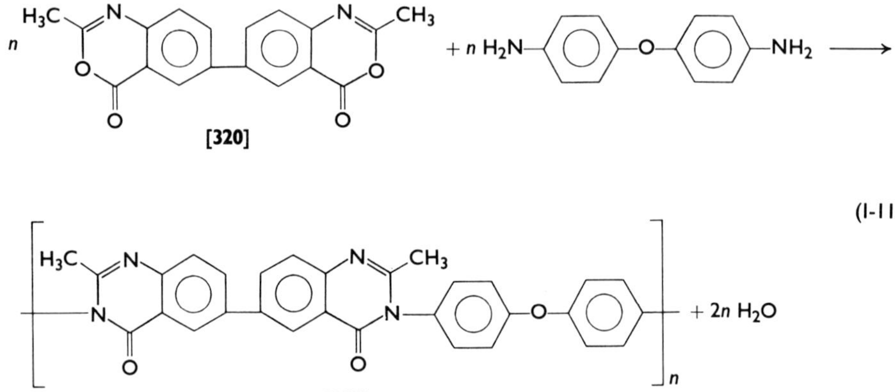

**[320]**

(I-113)

**[321]**

$\eta_{inh}$ 2.34

solvent over dimethylformamide or dimethyl sulfoxide. Subsequent heating of the polymer formed in solution at 200–325°C was used to advance the molecular weights. The analogous 2,2'-diphenyl derivative has been prepared and polymerized (604, 605). It was less reactive than **[320]**, but did yield high molecular weight polymers when polymerization was conducted in *p*-chlorophenol (605). An extensive study of various polymerization conditions showed this to be the optimum solvent (605). An attempt to polymerize 2,2'-trifluoromethyl-[6,6'-bibenzoxazine]-4,4'-dione with a diamine was unsuccessful due to the reaction of the amine with the trifluoromethyl groups (193). Isomeric polymers (Nos. 5 and 6, Table I.22) were obtained by using the isomeric bis(benzoxazinones) ([**322**] and [**323**]) for polymerization with diamines (592). The structures of most of these polymers was confirmed by infrared spectral studies.

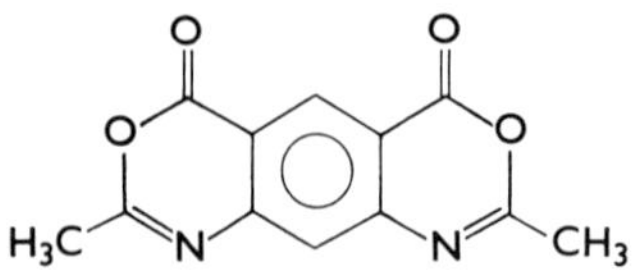

**[322]**

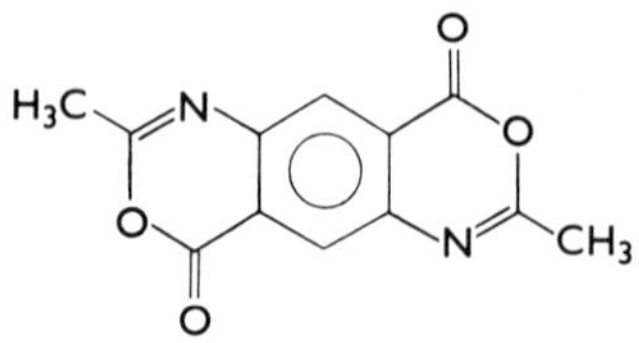

**[323]**

METHOD 3. POLYQUINAZOLONES FROM AROMATIC BIS($o$-AMINO ACIDS) AND DERIVATIVES

Aromatic bis($o$-amino acids) have been used to synthesize polyquinazolones by a variety of reactions. Poly(2-alkylquinazolones) possessing very high molecular weights are obtained by direct condensation of bis($o$-amino acids) with aromatic or aliphatic bis($N$-acylamido) compounds (344, 700). Polyphosphoric acid or fuming sulfuric acid is used as the medium and condensing agent [Eq. (I-114)]. A similar condensation leading to 2-arylquinazolones

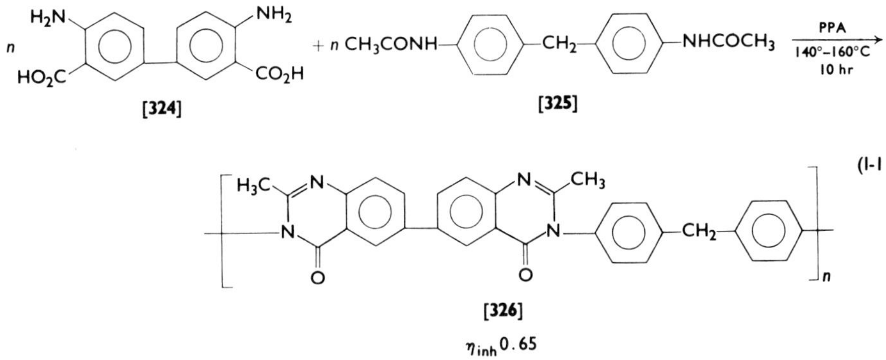

[324]     [325]     (I-114)

[326]

$\eta_{inh}$ 0.65

occurs between bis($N$-aroylamino acids) and diamines (99). Melt, bulk procedures were employed [Eq. (I-115)].

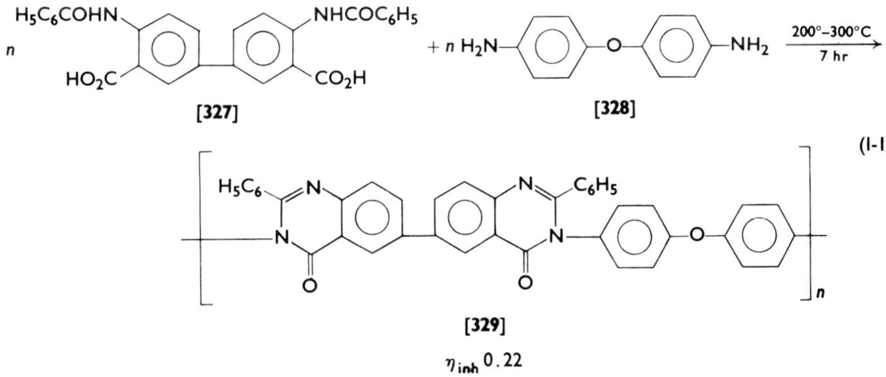

[327]     [328]     (I-115)

[329]

$\eta_{inh}$ 0.22

Post-treatments of some polymers prepared from aromatic bis($o$-amino acids) also yields polyquinazolones. Heating polymers prepared from bis-(iminoesters) and aromatic bis($o$-amino acids) has yielded very high molecular weight polymer (*595, 640*). The connecting link in the polymers formed from this route is via the 2-position in the quinazolone ring, as shown in Eq. (I-116).

$$n \quad [330] \quad + n \ H_5C_2OC\text{—}(CH_2)_8\text{—}COC_2H_5 \quad [331] \quad \xrightarrow[100°-165°C/4 \ hr]{DMA}$$

[330] structure: $H_2N$, $NH_2$, $HO_2C$, $CO_2H$ on biphenyl

[331] structure: with NH groups

[Polymer] $\xrightarrow[2 \ hr]{250°C/1 \ mm.}$                                        (I-116)

[332] structure $\quad + 2n \ C_2H_5OH + 2n \ H_2O$

$(CH_2)_8$, HN, NH, O

[332]

$[\eta]$ 0.82

A similar conversion can be effected on polyamides prepared from bis(*o*-amino amides) [333] (*268, 552*). This cyclization can also proceed to form

$$[333] \longrightarrow [334] \tag{I-117}$$

[333] structure: NHC=O, CNH$_2$=O

[334] structure: N, NH, O

[333]          [334]

intermolecular imide groups with attendant ammonia liberation. This path is reportedly minimized when potassium carbonate/methanol is used for cyclization (*268*). Finally, treatment of polyamides from bis(*o*-amino acids) and diacids with aniline is reported to form poly(3-phenylquinazolones) (*397*). Molecular weight data on the product are not available.

In addition to the above reports, several references describe the direct, self-condensation of bis(*o*-amino acids) in polyphosphoric acid (*399, 400, 403*). Although the structures of the products were not elucidated in detail, they could contain quinazolone moieties. Polymerization times of 3 hours at 150°–160°C under nitrogen have yielded polymer with inherent viscosities ($H_2SO_4$) up to 0.73 (*399*).

METHOD 4. POLYQUINAZOLINEDIONES BY CONDENSATION OF DIISOCYANATES WITH AROMATIC BIS(o-AMINO ACIDS)

Polymerization of diisocyanates with aromatic bis(o-amino acids) has been extensively investigated for polyquinazolinedione synthesis. It is generally agreed that the initial step is the formation of a poly(ureido acid) [336] (24, 327, 402, 713). The initial cyclization product from [336] is an iminobenzoxazinone which then rearranges to the quinazolinedione [338] at higher tem-

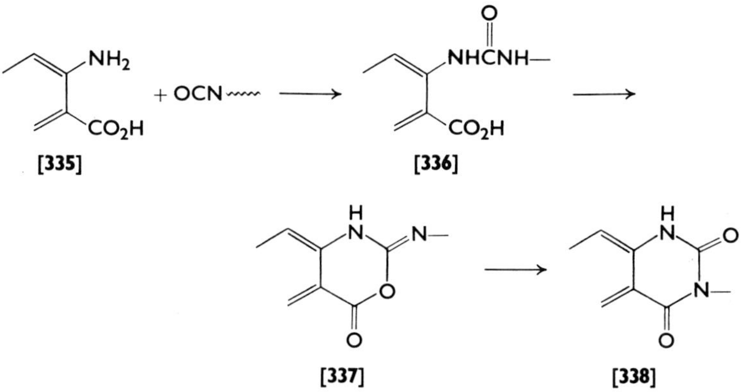

[335]   [336]

(I-118)

[337]   [338]

peratures. Esters of the aromatic bis(o-amino acids) react similarly (24, 458), as do isothiocyanates (502).

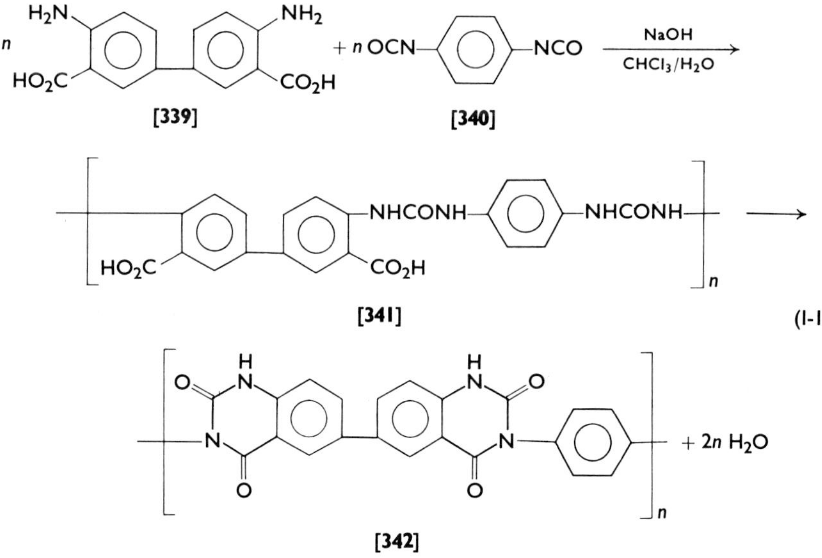

[339]   [340]

[341]

(I-119)

[342]

$\eta_{inh}$ 1·91

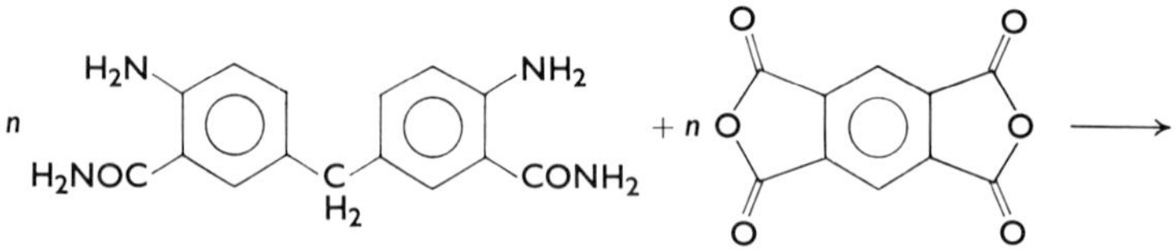

[343] [344]

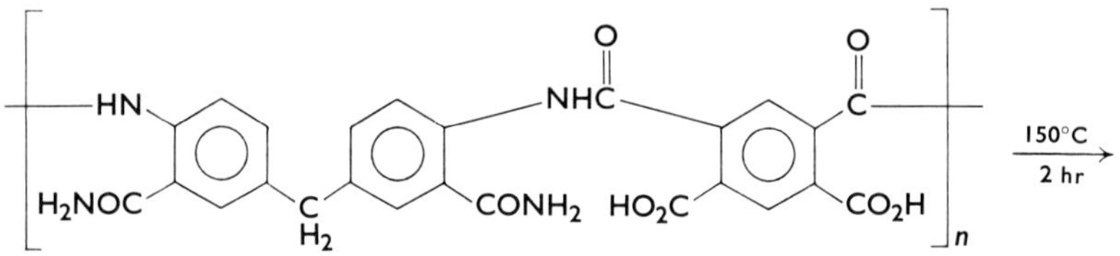


[345]

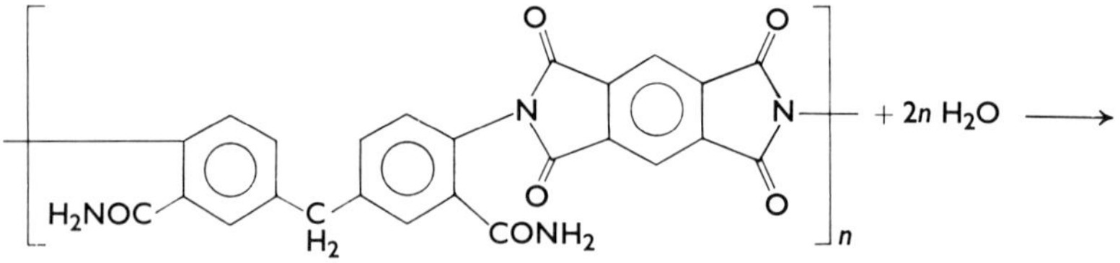

[346] (I-120)

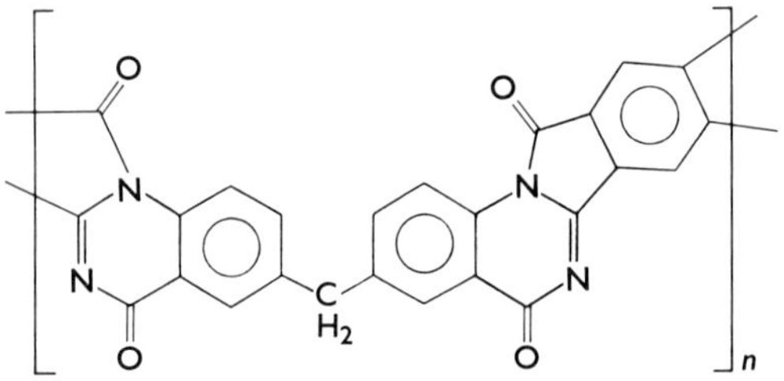

[347]

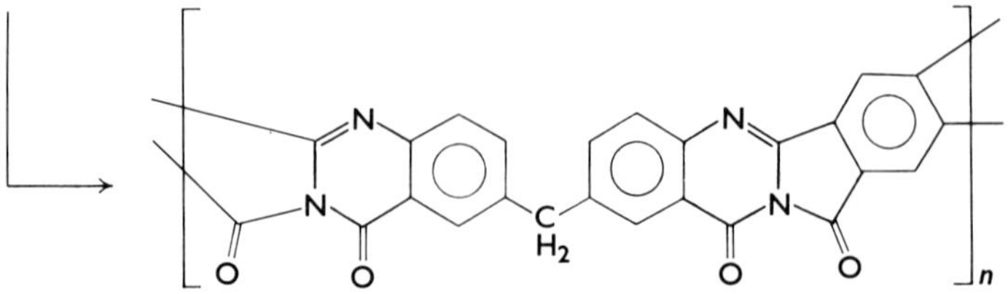

[348]

Initial polymerization to the poly(ureido acid) has been carried out in (1) polar, aprotic solvents such as dimethylacetamide or N-methylpyrrolidone with and without catalysts (23, 24, 650, 714); (2) in polyphosphoric acid or mixtures of polyphosphoric acid with fuming sulfuric acid (142, 402, 458, 696); and (3) interfacially in a chloroform/water system (711, 713). All methods are satisfactory with respect to giving final polymers with very high molecular weights. Cyclization of the precursor polymers can be effected by heating in polyphosphoric acid or in vacuo (300°–400°C). Cyclization to the oxazinone [337] occurs at 80°–100°C in PPA; the oxazinone rearranges to the quinazoline-dione [338] at about 150°C. Thermally induced cyclization occurs at temperatures of 160°–180°C and 230°–250°C for the respective steps (713). Numerous examples of polymers possessing usefully high molecular weights are listed in Table I.23. An example is shown in Eq. (I-119) (711).

METHOD 5. POLY(ISOINDOLOQUINAZOLINEDIONES) FROM CONDENSATION OF AROMATIC BIS(o-AMINO AMIDES) WITH DIANHYDRIDES

Reaction of a dianhydride with a bis(o-amino amide) can be directed to yield polyisoindoloquinazolinediones (550). Initial reaction is carried out in polar, aprotic solvents to yield the poly(o-amido amide) [345] (404). Cyclization to the polyimide [346] occurs by heating to 150°C, or chemically, with dicyclohexylcarbodiimide (549) and acetic anhydride–pyridine (551). Further cyclization to the polyisoindoloquinazolinediones [348] occurs at 250°C. Extensive studies of model compounds confirmed this reaction path. Moderately high molecular weight polymers are obtainable as shown by the data in Table I.23.

## PROPERTIES OF POLYQUINAZOLONES AND POLYQUINAZOLINEDIONES

Polyquinazolones and polyquinazolinediones possess many of the typical characteristics of aromatic heterocyclic polymers. They are soluble, but primarily only in acidic-type solvents. Most are colored materials that melt with decomposition. Crystallinity has been detected in some of the poly-quinazolinediones, but not the polyquinazolones. The polyquinazolinediones are reported to possess excellent resistance to acids and bases (402, 404, 650).

The thermal stability of these polymers has been investigated. They decompose before melting and a great deal of thermoanalytical data is available (see Tables I.22 and I.23). In the quinazolone series, the polymers substituted in the 2-position with phenyl groups were more thermally stable than those similarly substituted with methyl groups (*604*). The phenyl-substituted poly-mers decomposed at the same temperature (~425°C) in either air or argon.

Mechanical property data for the highest molecular weight polymers are reported in Tables I.22 and I.23. The polyquinazolones are described as being thermally stable coating materials (*72, 406*), as well as useful laminating resins (*268*). The polymer classes described in this section have been discussed in reviews (*196, 350, 451, 607, 692*).

Polyphthalazines, which are isomeric with polyquinazolinediones, have been synthesized from reaction of dianhydrides and diamines or dihydrazides (*418*). In addition, the synthesis of very high molecular weight polyhydro-uracils and polyiminoimidazolidiones by post-cyclization reactions has also been reported (*205*).

## I. Polyacetals and Polyketals

### METHOD 1. CONDENSATION OF POLYOLS WITH DIALDEHYDES AND DIKETONES

Polyacetals and polyketals can be synthesized by ring-forming polymeriza-tions from polyols and dicarbonyl compounds. Since pentaerythritol has been investigated most extensively, many of the polymers contain the *spiro* acetal or ketal structural feature (see Table I.24).

Polymerization of the free dialdehydes (*1, 179, 180, 292, 385*) or diketones (*43, 672*) has been reported. Acid catalysts are used and the reaction is generally carried out while removing water. This is accomplished merely by heating in bulk polymerizations (*1*) or by azetrope formation with a solvent such as benzene (*43, 179, 180, 672*). However, the condensation has also been effected in water at reflux, with precipitation of the polymer as formed serving as the driving force for polymerization (*179, 180*). In general, higher molecular

$$n \begin{array}{c} HOH_2C \\ HOH_2C \end{array} \!\!\!\! C \!\!\!\! \begin{array}{c} CH_2OH \\ CH_2OH \end{array} + n \ (C_2H_5O)_2CH(CH_2)_3CH(OC_2H_5)_2 \xrightarrow[\substack{o\text{-Phenylanisole,} \\ 125°–197°C, \ 3 \ hr}]{HCl}$$

[349]                        [350]

(I-121)

$$\left[ \begin{array}{c} OCH_2 \\ OCH_2 \end{array} \!\!\!\! C \!\!\!\! \begin{array}{c} CH_2O \\ CH_2O \end{array} \!\!\!\! CH \!\!-\!\! (CH_2)_3 \!\!-\!\! CH \right]_n + 4n \ C_2H_5OH$$

[351]

$\eta_{sp}$ 0.71

weight polymers are obtained by using the acetals or ketals of the dicarbonyl compounds. Reaction conditions are similar, with acid catalysts such as *p*-toluenesulfonic (*43, 179*), hydrochloric (*1, 171, 279*), or phosphoric (*1, 560*) acids being employed. Bulk reaction (*46*), or reaction in solution (*46, 171*), has yielded high molecular weight polymer. An example is shown in Eq. (I-121). 2-Alkoxydihydropyran and 2,6-dialkoxytetrahydropyran have been used as precursors to dialdehydes in this polymerization (*279, 560*). Alkylene oxides (*1*) and pentaerythritol diformal (*385*) are useful polyol precursors. Pentaerythritol diformal was preferred over pentaerythritol for its water solubility in a study of its polycondensation with aldehydes directly on textile fabrics (*385*).

METHOD 2. POLYACETALS BY TRANSACETALIZATION

The self-polycondensation of [352] by acid-catalyzed transacetalization has been reported (*449*). Use of hydrogen chloride as catalyst in a benzene/water medium at $11°$–$12°C$ gave polymer with a molecular weight of about 1500. The same polymer was prepared at higher molecular weights ($\sim$6000) by a metal oxide-catalyzed ester exchange reaction.

$$(I\text{-}122)$$

$$(I\text{-}123)$$

MISCELLANEOUS RELATED SYNTHESES

Several reports have appeared that describe polymerization reactions that could have yielded products structurally similar to those described above. Since details are lacking, they are briefly mentioned here.

Stannic chloride-catalyzed condensation of the bis(oxetane) [356] with terephthalaldehyde or cyclohexanone yields polymer (528) (I-124). A variety

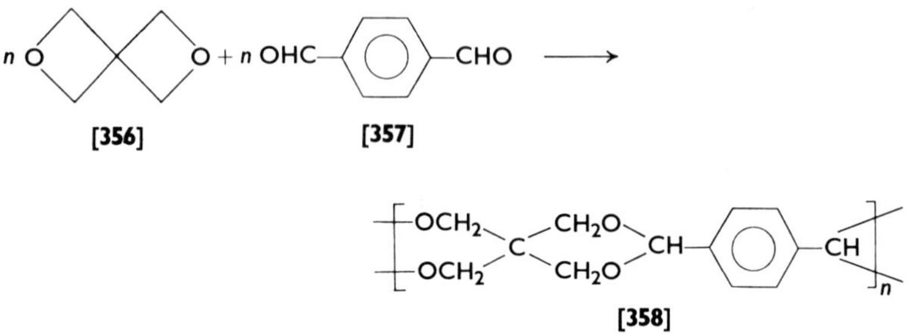

[356]          [357]

(I-124)

[358]

of silicon derivatives have been condensed with pentaerythritol in attempts to synthesize [359] (125, 672). Ill-defined materials were obtained. Phosphorous-containing related materials have been studied (539).

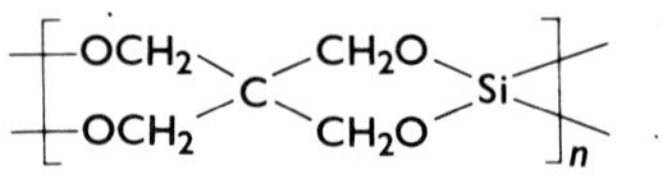

[359]

Low molecular weight polythioketals have been prepared by condensing tetrakis(mercaptomethyl)methane with diketones (127, 229). The reaction was carried out with hydrogen chloride catalyst in dioxane solution.

## PROPERTIES OF POLYACETALS AND POLYKETALS

The properties that have been reported for the polyacetals and polyketals are listed in Table I.24. Enough data to establish clear-cut trends are not available. Applications that have been reported for some of these products include magnet wire coatings (178, 179, 292), plasticizers for cellulosics (1), titanium coatings (167), crease-proofing (385) and shrinkproofing (390) agents for textiles. Reviews describing these products have appeared (376, 587).

## J. Polyquinones

METHOD 1. POLYQUINONES FROM HALO-SUBSTITUTED p-QUINONES

A number of interesting ring-forming polymerizations have been reported for the synthesis of polymers containing quinone rings. Even though high molecular weight polymers were not obtained, their syntheses are described herein. Since all the reactions utilize halo-substituted quinone derivatives, they are all described under this single method.

Both the 2,5-dichloro- and 2,5-dibromo-3,6-dihydroxyquinones have been self-condensed to yield the poly(quinone ether) [361]. Polymerization in

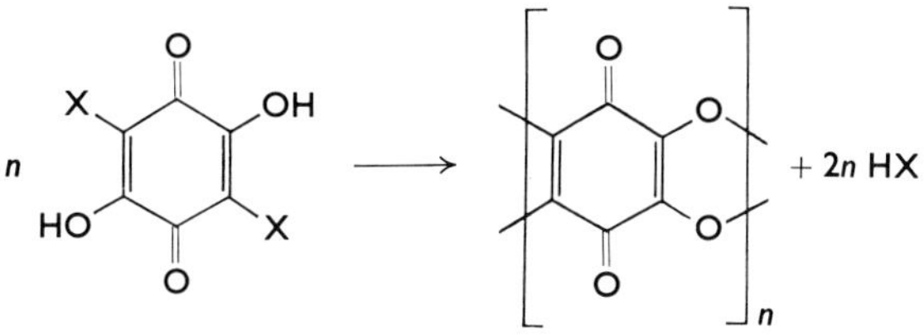

[360]                [361]

X = Cl, Br

(I-125)

solution in diphenyl gave the highest molecular weight polymer (302). Hydrogen chloride and hydrogen bromide are liberated at 218° and 190°C from the respective monomers, when polymerization is conducted in diphenyl. Bulk polymerization was conducted at 250°–310°C, with hydrogen chloride and hydrogen bromide evolution occurring at 285° aod 250°C respectively (102). The spectral and electrical properties of the polymers were of special interest.

Condensation of chloranil with tetrahydroxy-$p$-quinone is another route to [361] (455). Other tetrahydroxy ring compounds react similarly, as shown in Eq. (I-126). Basic solvents or inorganic bases such as sodium carbonate were used at elevated temperatures.

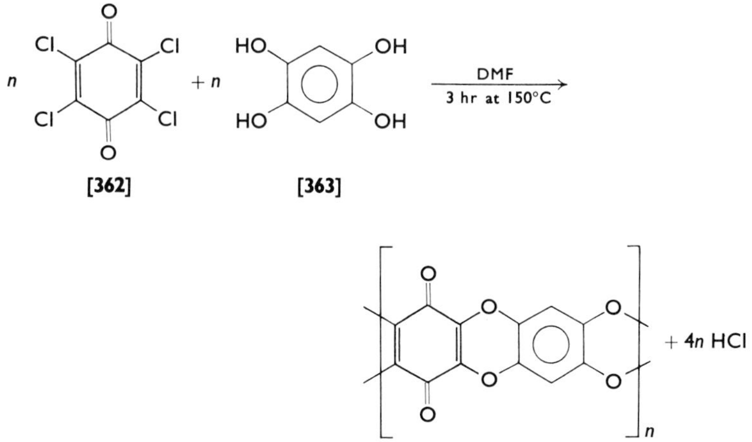

[362]                [363]

(I-126)

[364]

Condensation of ammonia with chloranil under pressure yields the poly-(quinone imine) [365]. Use of potassium amide gave the highest molecular weight polymer (456). It was a dark brown, semiconducting material. Tetra-amines react similarly with chloranil (457).

$$+ 2n \text{ NH}_3 \longrightarrow \quad\quad + 4n \text{ HCl} \qquad (\text{I-127})$$

[362]                    [365]

The poly(quinone thioether) [366] is formed from chloranil and sulfides. Potassium sulfide in xylene yielded polymer with a molecular weight of 5500 (454). Sulfur-containing polyquinones are also reported from the action

$$+ 2n \text{ K}_2\text{S} \longrightarrow \quad\quad + 4n \text{ KCl} \qquad (\text{I-128})$$

[362]                    [366]

of sulfur on polymers [367] and [369]. The precursor polymers are prepared from haloquinones and hydroxyquinones (101), or p-benzoquinone and ammonia (236, 417).

$$\xrightarrow[\substack{250°C \\ 10\ hr}]{S/I_2} \qquad\qquad (\text{I-129})$$

[367]                    [368]

The properties of the polymers described in this section are given in Table I.25. They are dark-colored materials, some of which possess semiconducting properties. Others are reported to be thermally stable pigments.

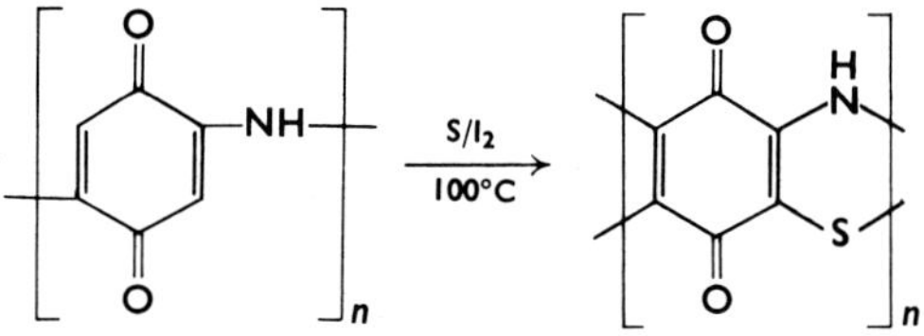

(I-130)

[369]              [370]

TABLE I.1

Polyimides from Aliphatic and Cycloaliphatic Dianhydrides

| No. | Structure | Method | Remarks and property data | References |
|---|---|---|---|---|
| | <br>1,1,2,2-Ethanetetracarboxylic acid dianhydride and | | | |
| 1 | R = $+CH_2+_2$<br>Ethylenediamine | A,1 | Gradual weight loss starts at ~250°C in air (TGA). Contains amide linkages due to decarboxylation during imidation step | 40 |
| 2 | R = $-CH_2-CH-$<br>  $CH_3$<br>1,2-Propanediamine | A,1 | Gradual weight loss starts at ~250°C in air (TGA). Contains amide linkages due to decarboxylation during imidation step | 40 |
| 3 | R = $+CH_2+_3$<br>1,3-Propanediamine | A,1 | Gradual weight loss starts at ~250°C in air (TGA). Contains amide linkages due to decarboxylation during imidation step | 40 |
| 4 | R = $+CH_2+_6$<br>1,6-Hexanediamine | A,1 | Gradual weight loss starts at ~250°C in air (TGA). Contains amide linkages due to decarboxylation during imidation step | 40 |
| 5 | R = $+CH_2+_7$<br>1,7-Heptanediamine | A,1 | Gradual weight loss starts at ~250°C in air (TGA). Contains amide linkages due to decarboxylation during imidation step | 40 |

| # | R | | Description | Ref |
|---|---|---|---|---|
| 6 | $R = -(CH_2)_2-CH=CH(CH_2)_3-$ <br> 1,7-Diaminoheptene-3 | A,1 | Gradual weight loss starts at ~250°C in air (TGA). Contains amide linkages due to decarboxylation during imidation step | 40 |
| 7 | $R = -(CH_2)_3-C(CH_3)_2-(CH_2)_3-$ <br> 1,7-Diamino-4,4-dimethylheptane | A,1 | Gradual weight loss starts at ~250°C in air (TGA). Contains amide linkages due to decarboxylation during imidation step | 40 |
| 8 | $R = -(CH_2)_9-$ <br> 1,9-Diaminononane | A,1 | Gradual weight loss starts at ~250°C in air (TGA). Contains amide linkages due to decarboxylation during imidation step | 40 |
| 9 | $R = -(CH_2)_{10}-$ <br> 1,10-Diaminodecane | A,1 | Gradual weight loss starts at ~250°C in air (TGA). Contains amide linkages due to decarboxylation during imidation step | 40 |
| 10 | $R =$ <br> 1,4-Diaminobenzene | A,1 | Gradual weight loss starts at ~250°C in air (TGA). Contains amide linkages due to decarboxylation during imidation step | 40 |
| 11 | $R =$ <br> 1,3-Diaminobenzene | A,1 | Gradual weight loss starts at ~250°C in air (TGA). Contains amide linkages due to decarboxylation during imidation step. | 40 |
| 12 | $R =$ <br> 1-Methyl-2,4-diaminobenzene | A,1 | Gradual weight loss starts at ~250°C in air (TGA). Contains amide linkages due to decarboxylation during imidation step | 40 |

TABLE I.1—*continued*

Polyimides from Aliphatic and Cycloaliphatic Dianhydrides

| No. | Structure | Method | Remarks and property data | References |
|---|---|---|---|---|
| 13 | 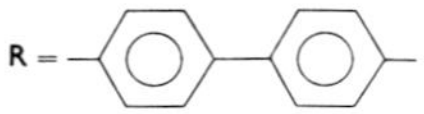 R = [biphenyl] <br> 4,4'-Diaminobiphenyl | A,1 | Gradual weight loss starts at ~250°C in air (TGA). Contains amide linkages due to decarboxylation during imidation step | 40 |
| 14 | 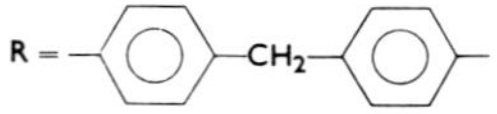 R = —CH₂— [diphenylmethane] <br> 4,4'-Diaminodiphenylmethane | A,1 | Gradual weight loss starts at ~250°C in air (TGA). Contains amide linkages due to decarboxylation during imidation step | 40 |

* * *

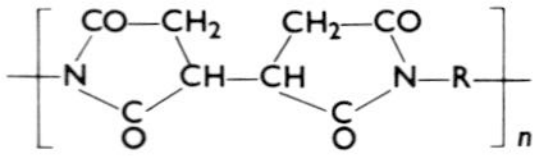

1,2,3,4-Butanetetracarboxylic acid dianhydride and

| No. | Structure | Method | Remarks and property data | References |
|---|---|---|---|---|
| 15 | R = $-(CH_2)_2-$ <br> 1,2-Diaminoethane | A,1 | Does not melt or soften | 39 |
| 16 | 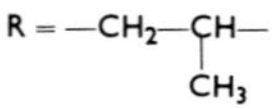 R = $-CH_2-CH-$ <br>         $CH_3$ <br> 1,2-Propanediamine | A,1 | — | 39 |
| 17 | R = $-(CH_2)_3-$ <br> 1,3-Propanediamine | A,1 | — | 39 |

| 18 | $R = +CH_2)_6$<br>1,6-Hexanediamine | A,1 | — | *39* |
| 19 | $R = +CH_2)_7$<br>1,7-Heptanediamine | A,1 | — | *39* |
| 20 | $R = +CH_2)_2 CH=CH(CH_2)_3$<br>1,7-Diaminoheptene-3 | A,1 | — | *39* |
| 21 | $R = +CH_2)_3 C(CH_3)_2 +CH_2)_3$<br>1,7-Diamino-4,4-dimethylheptane | A,1 | — | *39* |
| 22 | $R = +CH_2)_9$<br>1,9-Diaminononane | A,1 | — | *39* |
| 23 | $R =$ 1,4-Diaminobenzene | A,1 | Does not melt or soften | *39, 381* |
| 24 | $R =$ 1,3-Diaminobenzene | A,1 | Does not melt or soften | *39, 381* |
| 25 | $R =$ CH$_3$<br>1-Methyl-2,4-diaminobenzene | A,1 | Does not melt or soften | *39, 381* |

TABLE I.1—*continued*

Polyimides from Aliphatic and Cycloaliphatic Dianhydrides

| No. | Structure | Method | Remarks and property data | References |
|---|---|---|---|---|
| 26 | R = <br> 4,4'-Diaminobiphenyl | A,1 | Stable in air up to 480°C. Does not melt or soften | *39, 381* |
| 27 | R = <br> 4,4'-Diaminodiphenylmethane | A,1 | Does not melt or soften | *39, 381* |
| | *meso*-1,2,3,4-Butanetetracarboxylic acid dianhydride and | | | |
| 28 | R = <br> 1,4-Diaminobenzene | A,1 | Optimum imidation temperature (IR) = 220°C. Does not soften. Properties: Density = 1.405 $g/cm^3$; tensile strength (20°C) = 1000 $kg/cm^2$; at 350°C = 400 $kg/cm^2$. Elongation: (20°C) = 8%; (350°C) = 15%. Amorphous. Stable to acids; unstable to bases. Insoluble. Decomposes in conc. $H_2SO_4$ and in fuming $HNO_3$ | *225, 382 394* |
| 29 | R = <br> 1,3-Diaminobenzene | A,1 | Optimum imidation temperature (IR) = 220°C. Amorphous. Stable to acids; unstable to bases. Insoluble. Decomposes in conc. $H_2SO_4$ and in fuming $HNO_3$ | *225, 382, 394* |

| № | R = | Method | Properties | Ref. |
|---|---|---|---|---|

30

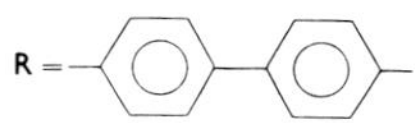

4,4-Diaminobiphenyl

A,1

Optimum imidation temperature (IR) = 250°C; Insoluble. $T_m$ = >500°C. Thermomechanical curve shows no $T_g$. Density = 1.375 g/cm³; tensile strength: (20°C) = 1000 kg/cm²; (350°C) = 350 kg/cm²; elongation: (20°C) = 13%; (350°C) = 60%. Amorphous. Stable to acids; unstable to bases. Decomposes in conc. $H_2SO_4$ and in fuming $HNO_3$

*225, 261, 382, 394*

31

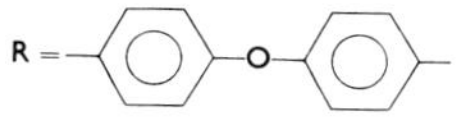

4,4'-Diaminodiphenyl ether

A,1

Optimum imidation temperature (IR) = 280°C; $T_g$ = 290°C; Density = 1.358 g/cm³; tensile strength: (20°C) = 1200 kg/cm²; (300°C) = 300 kg/cm²; elongation: (20°C) = 15%; (300°C) = 20%. Amorphous. Stable to acids; unstable to bases. Insoluble. Decomposes in conc. $H_2SO_4$ and fuming $HNO_3$

*225, 382, 394, 395*

32

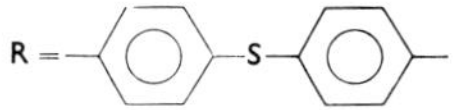

4,4'-Diaminodiphenyl sulfide

A,1

Amorphous. Insoluble. Stable to acids; unstable to bases. Decomposes in conc. $H_2SO_4$ and fuming $HNO_3$

*394, 395*

33

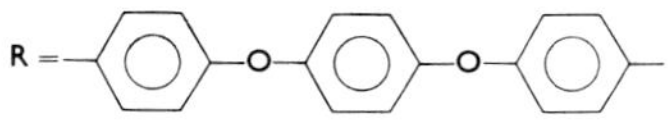

Hydroquinone bis(4-aminophenyl) ether

A,1

$T_g$ = 265°C; Density = 1.335 g/cm³; tensile strength: (20°C) = 1300 kg/cm², (250°C) = 400 kg/cm²; elongation: (20°C) = 50%; (250°C) = 180%. Amorphous. Stable to acids; unstable to bases. Insoluble. Decomposes in conc. $H_2SO_4$ and in fuming $HNO_3$

*394, 395*

* * *

TABLE I.1—*continued*
Polyimides from Aliphatic and Cycloaliphatic Dianhydrides

| No. | Structure | Method | Remarks and property data | References |
|---|---|---|---|---|
| 34 | <br>From $\beta$-(carboxymethyl)-$\epsilon$-caprolactam | A,6 | $T_g \cong 90°C$. $T_m = 281°$–$283°C$. Mol. wt., $\eta_{red}$ up to 4.5 Solubility: $m$-Cresol, HCOOH, $CF_3CF_2OH$, $H_2SO_4$. When polymerization is carried out below polymer melting point crystalline material results. Amorphous material is obtained at polymerization temperatures above polymer $T_m$. Annealing of the amorphous material above the $T_g$ yields crystalline polymer. Fibers and filaments. NMR and IR support polymer structure | 559, 601 |
| | <br>1,2-*trans*-3,4-Cyclobutanetetracarboxylic acid dianhydride and | | | |
| 35 | $R = +CH_2+_6$<br>1,6-Diaminohexane | A,1 | Tough films | 641 |
| 36 | $R = +CH_2+_8$<br>1,8-Diaminooctane | A,1 | Tough films | 641 |
| 37 | $R = +CH_2+_9$<br>1,9-Diaminononane | A,1 | Tough films | 641 |

| | | | | |
|---|---|---|---|---|
| **38** | R = $-(CH_2)_{10}-$<br>1,10-Diaminodecane | A,1 | Tough films | | *641* |
| **39** | R = $-(CH_2)_{12}-$<br>1,12-Diaminododecane | A,1 | Tough films | | *641* |
| **40** | R = [benzene]$-CH_2-$[benzene]<br>4,4'-Diaminodiphenylmethane | A,1 | | — | *641* |

* * *

1,2,3,4-Cyclopentanetetracarboxylic acid dianhydride and

| | | | | |
|---|---|---|---|---|
| **41** | R = $-HNCO-(CH_2)_4-CONH-$<br>Adipic acid dihydrazide | A,1 | | — | *444* |
| **42** | R = $-HNCO-(CH_2)_7-CONH-$<br>Azelaic acid dihydrazide | A,1 | | — | *444* |
| **43** | R = [benzene]<br>1,4-Diaminobenzene | A,1 | Optimum imidation temperature, argon atmosphere (IR) = 260°C | | *224* |

TABLE I.1—*continued*
Polyimides from Aliphatic and Cycloaliphatic Dianhydrides

| No. | Structure | Method | Remarks and property data | References |
|---|---|---|---|---|
| 44 | R = 1,3-Diaminobenzene | A,1 | Optimum imidation temperature, argon atmosphere (IR) = 260°C | 224 |
| 45 | R = —HNCO—CONH— Terephthalic acid dihydrazide | A,1 | — | 444 |
| 46 | R = —HNOC—CONH— Isophthalic acid dihydrazide | A,1 | — | 444 |
| 47 | R = 4,4'-Diaminobiphenyl | A,1 | Optimum imidation temperature, argon atmosphere (IR) = 350°C | 224 |
| 48 | R = 4,4'-Diaminodiphenyl ether | A,1 | Optimum imidation temperature, argon atmosphere (IR) = 300°C | 224 |

* * *

1,2,3,4-cis,cis,cis,cis-Cyclopentane-
tetracarboxylic acid dianhydride and

| | | | | |
|---|---|---|---|---|
| 49 | R = 4,4'-Diaminodiphenyl ether | A,1 | — | 395 |
| 50 | R = 4,4'-Diaminodiphenyl sulfide | A,1 | — | 395 |
| 51 | R = Hydroquinone bis(4-aminophenyl) ether | A,1 | — | 395 |

* * *

1,2,4,5-Cyclohexanone tetrapropionic acid
dianhydride (tetraacetyl derivative) and

| | | | | |
|---|---|---|---|---|
| 52 | R = 4,4'-Diaminobiphenyl | A,1 | — | 660 |

TABLE 1.1—*continued*
Polyimides from Aliphatic and Cycloaliphatic Dianhydrides

| No. | Structure | Method | Remarks and property data | References |
|-----|-----------|--------|---------------------------|------------|
| 53 | R = 4,4'-Diaminodiphenyl ether | A,1 | — | 660 |
|  | * * * 3,4-Dicarboxy-1,2,3,4-tetrahydro-1-naphthalene succinic dianhydride and | | | |
| 54 | R = 1,3-Diaminobenzene | A,1 | Soluble in DMF. $\eta_{inh}$ 0.26. $T_m = 230°C$ | 65, 66, 488 |

**55**

$R = \quad$ —CH₂— structure

**4,4'-Diaminodiphenylmethane**

A,1

Soluble in DMF. $\eta_{inh}$ 0 . 16. Can be formed under pressure at 300°C. Properties: Izod notched shock resistance = 0.78 ft. lb/inch; Rockwell hardness = 112; bending modulus: (23°C) = 33,600 psi, (200°C) = 23,800 psi; tensile strength: (23°C) = 812 psi; elongation: 6·9%; dielectric constant: $10^3$ cp/10 cp = 3.78/3.65; dissipation factor: $10^3$ cp/$10^6$ cp = 0.0031/0.015. Useful in composites, glass laminates and in fibers

488, 494

**56**

$R = \quad$ —O— structure

**4,4'-Diaminodiphenyl ether**

A,1

Izod notched shock resistance = 0.33 ft lbs/inch; Rockwell hardness = 112; bending modulus: (23°C) = 37,800 psi, (200°C) = 19,600 psi; tensile strength (23°C) = 392 psi; elongation (23°C) = 2.6%; dielectric constant: $10^3$ cp/$10^6$ cp = 3.80/3.63; dissipation factor: $10^3$ cp/$10^6$ cp = 0.0034/0.017

488

* * * *

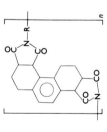

1,2,3,4,5,6,7,8-Octahydrophenanthrene-1,2,5,6-tetracarboxylic acid dianhydride and

**57**

$R =$ Various amines

A,1

Tough, flexible materials. Good heat and oxidation stability. Useful in coatings, as molding resins, adhesives, and laminating resins

65

TABLE I.2
Polyimides from Multicyclic Dianhydrides

| No. | Structure | Method | Solubility | Molecular weight | $T_m$ (°C) | Remarks and property data | References |
|---|---|---|---|---|---|---|---|
| 1 | 3,6-endo-Ethylene-1,2,4,5-cyclohexanetetracarboxylic acid dianhydride and R = p-Phenylenediamine | A,1 | — | — | — | Heat-stable amorphous powder. Useful for the manufacture of building cements and protective coatings | 466 |
| 2 | Bicyclo[2.2.2]oct-7-ene-2,3,5,6-tetracarboxylic acid dianhydride and R = 4,4'-Diaminodiphenylmethane | A,1 | m-cresol | $\eta_{red}$ 0.49 | — | Good electrical properties. Useful in the manufacture of lacquered coatings, moldings, and films. Konig hardness = 191, Erichsen index = 10 | 671 |

3    R = 

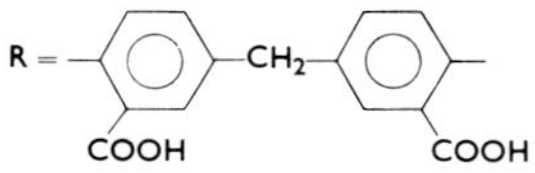

*  *  *

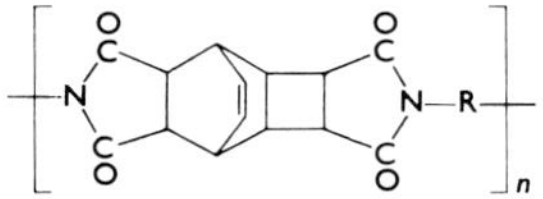

Tricyclo[4.2.2.0$^{2,5}$]dec-7-ene-3,4,9,10-tetracarboxylic acid dianhydride and

| No. | R | | DMF | $\eta_{inh}$ 0.4 | | Remarks | Ref. |
|---|---|---|---|---|---|---|---|
| 3 | | A,1 | — | — | — | Reaction with a diisocyanate or a diepoxide followed by heat treatment yields a tough, cross-linked elastic material useful in coatings. Copolymers are also described | 558 |
| 4 | R = $-(CH_2)_6-$  1,6-Diaminohexane | A,1 | — | — | — | — | 641 |
| 5 | R = $-(CH_2)_8-$  1,8-Diaminooctane | A,1 | — | — | — | — | 641 |
| 6 | R = $-(CH_2)_9-$  1,9-Diaminononane | A,1 | — | — | — | — | 641 |
| 7 | R = $-(CH_2)_{10}-$  1,10-diaminodecane | A,1 | — | — | — | —; | 641 |
| 8 | R = $-(CH_2)_{12}-$  1,12-Diaminododecane | A,1 | — | — | — | — | 641 |

TABLE 1.2—*continued*
Polyimides from Multicyclic Dianhydrides

| No. | Structure | Method | Solubility | Molecular weight | $T_m$ (°C) | Remarks and property data | References |
|-----|-----------|--------|------------|------------------|------------|---------------------------|------------|
| 9 | R = <br> 4,4'-Diaminodiphenylmethane | A,1 | — | — | — | — | 641 |
| 10 | R = <br> 4,4'-Diaminodiphenyl ether | A,1 | — | — | — | Tensile strength = 805 kg/cm²; elongation = 9.5%; dielectric disruptive discharge = 160,000 V/min; Dissipation value = 0.36%; 25°C, at 1000 cycles; dielectric constant = 1.5–2.0 | 480 |
| 11 | R = Various amines | A,1 | — | — | — | — | 538, 553 |

\* \* \*

Tricyclo[4.2.2.0²·⁵]dec-7-ene-7-alkyl-3,4,9,10-tetracarboxylic acid dianhydride and

12  R₁ = C₂H₅

R =

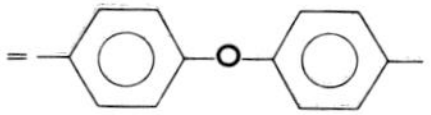

4,4'-Diaminodiphenyl ether

A,1    Dimethyl-octamide    300,000 (viscosity)    >600    Brittle    *123*

13  R₁ = —HC(CH₃)₂

R =

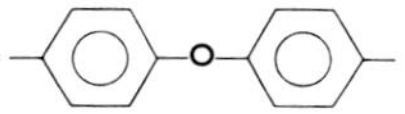

4,4'-Diaminodiphenyl ether

A,1    —    430–450    Tensile strength = 1083 psi; tensile modulus = 20,830 psi; elongation at break = 25%    *123*

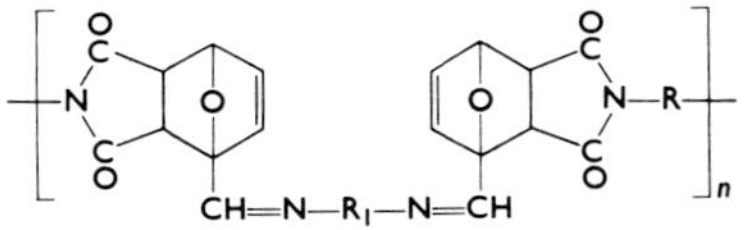

* * * *

Dianhydride prepared by Diels–Alder addition of a Schiff's base from furfural and a diamine on maleic anhydride

14  R₁ =

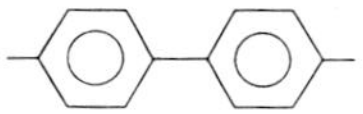

R =

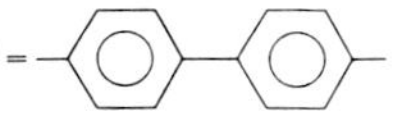

Benzidine

A,1    —    —    Stable in air up to 350°C    *103*

**TABLE 1.2—** *continued*
Polyimides from Multicyclic Dianhydrides

| No. | Structure | Method | Solubility | Molecular weight | $T_m$ (°C) | Remarks and property data | References |
|---|---|---|---|---|---|---|---|
| 15 | $R_1$ = (3,3'-Dimethylbenzidine); R = (Benzidine) | A,1 | — | — | — | Thermal stability (air) data | *103* |
| 16 | $R_1$ = (4,4'-Diaminodiphenyl ether); R = (Benzidine) | A,1 | — | — | — | Thermal stability (air) data | *103* |

* * *

4,8-Diphenyl-1,5-diazabicyclo[3.3.0]octane-
2,3,6,7-tetracarboxylic acid dianhydride and

| 17 | $R = -(CH_2)_6-$ 1,6-Diaminohexane | A,1 | DMF | $\eta_{inh}$ 0.60 | >350 | Dark brown. Decomposes on heating above 350°C. Also prepared by a 1,3-addition of hexamethylenebis(maleimide) across benzalazine, $C_6H_5CH=N-N=CHC_6H_5$. See text | 633 |
| --- | --- | --- | --- | --- | --- | --- | --- |
| 18 | $R =$ m-Phenylenediamine | A,1 | DMF | $\eta_{inh}$ 0.10 | >350 | Dark brown. Decomposes on heating above 350°C. Also prepared by a 1,3- addition of m-phenylenebis(maleimide) across benzalazine $C_6H_5CH=N-N=CHC_6H_5$. See text | 633 |

TABLE I.3
Polypyromellitimides

Pyromellitic anhydride or derivative and

| No. | Structure | Method | Solubility | Molecular weight | $T_m$ (°C) | Remarks and property data | References |
|-----|-----------|--------|------------|------------------|------------|---------------------------|------------|
| 1 | R = — <br> hydrazine | A,1 | — | — | — | Structure of polymer depends on imidation conditions. See text | 368 |
| 2 | R = —CH₂Si(CH₃)₂OSi(CH₃)₂CH₂— <br> 1,3-Bis(aminomethyl)tetramethyldisiloxane | — | — | — | — | Thermal degradation studies | 509 |
| 3 | R = (CH₂)₆ <br> 1,6-Diaminohexane | A,1; A,2; A,5 | — | — | 370 | Tough flexible film. May be cross-linked | 151, 218 322, 493, 513, 514 |
| 4 | R = (CH₂)₂CH(CH₃)(CH₂)₄ <br> 1,7-Diamino-3-methylheptane | A,1 | — | — | — | $T_g$ = 135°C. Oxidatively stable (air, 175°C) for 8–10 hr | 620 |

| No. | R / Name | Method | Solvent | $\eta_{inh}$ | | Properties | Refs. |
|---|---|---|---|---|---|---|---|
| 5 | R = $-(CH_2)_3-C(CH_3)_2-(CH_2)_3-$ <br> 1,7-Diamino-3,3-dimethylheptane | A,1; A,2 | m-Cresol | 0.7 | — | $T_g$ = 135°C. Oxidatively stable (air, 175°C) for 20–30 hr. Compression molding at 340°C yields a tough film; useful in glass laminates. Degree of crystallinity (X-ray) = 10 to 20% | 124, 214, 216, 218, 620 |
| | Copolyimides | A,1 | — | — | — | — | 214 |
| 6 | R = $-(CH_2)_3-OSi(CH_3)(C_6H_5)O-(CH_2)_3-$ <br> Bis(2-aminoethoxy)methylphenylsilane | A,1 | — | — | — | Yields flexible coatings with good thermal and electrical properties | 473 |
| | Cross-linked silicon-containing copolymers | A,1 | — | — | — | Heat-stable and solvent-resistant coatings | 475 |
| 7 | R = $-(CH_2)_2-OSi(C_6H_5)_2O-(CH_2)_2-$ <br> Bis(2-aminoethoxy)diphenylsilane | A,1 | — | — | — | Yields flexible coatings with good thermal and electrical properties | 473 |
| 8 | R = $-(CH_2)_9-$ <br> 1,9-Diaminononane | A,1 | m-Cresol | — | — | $T_g$ = 110°C. Oxidatively stable (air, 175°C) for 20–25 hr. Can be compression-molded at 340°C. Degree of crystallinity (X-ray) = 50%. Useful in laminates | 124, 212, 213, 216, 322, 620 |

TABLE 1.3—*continued*

Polypyromellitimides

| No. | Structure | Method | Solubility | Molecular weight | $T_m$ (°C) | Remarks and property data | References |
|---|---|---|---|---|---|---|---|
| 9 | R = $-(CH_2)_9-$ and bis(β-aminoethoxy) polyoxyethylene residue | A,1 | — | $\eta_{inh}$ 1.04 | — | Vicat softening point = 167°C. Elastomeric; can be melt-spun at 275°C. Properties of filaments: extensibility >300%; elastic recovery (100% extension) = 90%; elastic recovery (200% extension) = 85%; modulus = 0.35 g/denier | 186 |
| 10 | R = $-(CH_2)_3-Si(CH_3)_2-(CH_2)_3-$ 1,3-Bis(γ-aminopropyl)tetramethyldisiloxane | A,1 | Nitro-benzene, chloro-benzene, DMF | $\eta_{red}$ 0.86 | ~180 | Specific gravity = 1.216 g/cm³. Shape stability to heat (Martens) = 62°–63°C; decomposition temperature: 375°C. | 391 |
| 11 | R = $-(CH_2)_{10}-$ 1,10-Diaminodecane | A,1 | m-Cresol | $\eta_{inh}$ 1.0 | — | Compression-moldable at 390°C | 216 |
| 12 | R = $-CH(CH_3)-(CH_2)_8-CH(CH_3)-$ 1,10-Diamino-1,10-dimethyldecane | A,1 | m-Cresol | $\eta_{inh}$ 0.8 | — | Compression-moldable at 200°C to a tough film | 214, 216 |
| 13 | R = $-(CH_2)_3-O-(CH_2)_2-O-(CH_2)_3-$ 1,10-Diamino-4,7-dioxadecane | A,1 | m-Cresol | $\eta_{inh}$ 0.8 | — | Compression-moldable at 325°C | 216 |

| No. | Structure | | | | | Ref. |
|---|---|---|---|---|---|---|
| 14 | R = —(CH$_2$)$_2$—O[Si(CH$_3$)(C$_6$H$_5$)OSi(C$_6$H$_5$)$_2$OSi(CH$_3$)(C$_6$H$_5$)O]—(CH$_2$)$_2$—<br>α,ω-Bis(2-aminoethoxy)-mixed methyl phenyl trisiloxane | A,1 | — | — | — | 300 |
| 15 | R = —(CH$_2$)$_6$—NHP(O)NH—(CH$_2$)$_6$—<br>$\,$ $\,$ $\,$ $\,$C$_6$H$_5$<br>Reaction product of phenylphosphonyl dichloride (1 mole) with 2 moles of 1,6-diaminohexane | A,1 | — | — | — | 715 |
| 16 | R = —(CH$_2$)$_4$—O[Si(CH$_3$)(CF$_3$CH$_2$CH$_2$)O]$_{10}$—(CH$_2$)$_4$—<br>α,ω-Bis(4-aminobutoxy)poly(methyl-γ,γ,γ-trifluoropropyl)siloxane | A,1 | — | — | — | 300 |
| | Copolymers from mixtures of dianhydrides and silicon-containing diamines | | — | — | — | 300 |
| 17 | R = —(CH$_2$)$_3$—[OSi(CH$_3$)$_2$]$_{1-4}$OSi(CH$_3$)$_2$O—(CH$_2$)$_3$—<br>α,ω-Bis(3-aminopropoxy)polydimethylsiloxane | A,1 | — | — | Viscous liquid. Useful in coatings and adhesives | 580 |
| 18 | R = —(CF$_2$)$_2$—O[Si(CH$_3$)$_2$O]$_{100}$—CF$_2$CF$_2$—<br>α,ω-Bis(2-perfluoroethoxy)polydimethylsiloxane | A,1 | — | — | — | 300 |
| 19 | Various aliphatic diamines | A,1 | — | — | — | 212, 213 |
| 20 | R = —(CH$_2$)$_4$—C(CH$_3$)$_2$—⟨benzene⟩—C(CH$_3$)$_2$—(CH$_2$)$_4$—<br>1,4-Bis(5'-amino-1',1'-dimethyl)benzene | A,1 | m-Cresol | $\eta_{inh}$ 1.5–1.8 | Compression-moldable at 310°C. Properties of film: color—pale yellow; stick temp. = 260°C; yield strength = 9390 psi; ultimate strength = 6739 psi; elongation = 70%; flexural modulus (70 mil film) = 606,000 psi (23°C) | 215 |

## TABLE I.3—continued
### Polypyromellitimides

| No. | Structure | Method | Solubility | Molecular weight | $T_m$ (°C) | Remarks and property data | References |
|---|---|---|---|---|---|---|---|
| 21 | R = —CH₂—⬡—O—⬡—CH₂—<br>4,4'-Di(aminomethyl)diphenyl ether | — | — | — | — | Thermal degradation studies | 509 |
| 22 | R = ⬡<br>1,4-Diaminobenzene | A,1; A,2 | conc. $H_2SO_4$ $\eta_{inh}$ 0.5 | | — | Film properties: Density = 1.41 g/cm³; tensile modulus = 520,000 psi; elongation = 5.5%; tensile strength = 14,000 psi. Catalyzes the decomposition of formic acid and the dehydration of cyclohexanol | 218, 220, 237, 288, 621 |
| 23 | R = ⬡<br>1,3-Diaminobenzene | A,1; A,2 | conc. $H_2SO_4$ $\eta_{inh}$ 0.33 | | — | Structure supported by IR. Film properties: density = 1.43 g/cm³; tensile modulus = 400,000 psi; tensile strength = 15,000 psi; stable for 100 hr in boiling water and 18 hr to steam at 180°C; stable for more than 30 days at 300°C in air; zero strength temperature = 800°C. Good insulator (electrical volume resistivity >10⁵ ohm cm). Coalescible powder prepared | 121, 122, 218, 220, 221, 237, 286, 351, 427, 590, 611, 612, 621, 647 |
| | Copolymers | A,1 | — | — | — | — | 220 |

| No. | | | | | | Remarks | Ref. |
|---|---|---|---|---|---|---|---|
| 24 | 4-Isopropyl-1,3-diaminobenzene, R = CH(CH₃)₂ | A,1 | DMF, DMA | $\eta_{inh}$ 1.17 | — | Structure supported by IR. Cast film properties: tensile strength = 320,000 psi; elongation = 13.7%; resistance to traction = 162,400 psi; dielectric constant = 3.3; electrical resistivity >10$^{15}$ ohm cm. Solution properties studied | 56, 647, 676 |
| | Copolymers of 24 | — | — | — | — | — | 56 |
| 25 | 1,3-Diaminoperfluorobenzene | A,1 | — | — | — | Useful as insulating enamel for electrical conductors and as bonding resin in laminates that contain refractory fillers | 184 |
| 26 | 1,4-Dihydrazinobenzene, R = —HN— NH— | A,1 | DMSO | $\eta_{sp}$ 0.17 | — | — | 366 |
| 27 | 1,3-Dihydrazinobenzene, R = —HN— NH— | A,1 | DMSO | $\eta_{sp}$ 0.11 | — | — | 366 |

## TABLE 1.3—*continued*
### Polypyrromellitimides

| No. | Structure | Method | Solubility | Molecular weight | $T_m$ (°C) | Remarks and property data | References |
|---|---|---|---|---|---|---|---|
| 28 | <br>4,4'-Diaminodiphenyl | A,1; A,2 | Fuming $HNO_3$ | $\eta_{inh}$ 0.3 | — | Flexible films. Properties: Density = 1.43 g/cm³; tensile modulus = 850,000 psi; elongation = 14%; tensile strength = 14,900 psi; wt. loss at 325°C (air) after 400 hr = 6.5%. TGA data (air, argon) also indicate excellent thermal stability. Useful in composites with asbestos and barite. Catalyzes dehydration of cyclohexanol. Kinetics of thermal imidation were studied. | *68, 121, 122, 209, 218, 237, 288, 305, 351, 364, 621* |
| 29 | <br>3,3'-Dihydroxy-4,4'-diaminobiphenyl | A,1 | — | — | — | Structure supported by IR. On heating to ~450°C, rearranges and decarboxylates to poly-[2,2'-(p-phenylene)-6,6'-dibenzoxazole] | *354* |
| 30 | <br>3,3'-Dimethoxy-4,4'-diaminobiphenyl | A,1 | — | — | — | Structure supported by IR | *354* |

| # | R = | Method | Solvent | Viscosity | | Comments | Ref. |
|---|-----|--------|---------|-----------|---|----------|------|
| 31 | 3,3-Di(carbethoxy)-4,4'-diaminobiphenyl | A,1 | DMSO, DMA | $\eta_{inh}$ 0.13 | — | Reaction with $NH_3$ in DMA yields the amide-containing polymer. The latter, refluxed in $(CH_3CO)_2O$, gives a poly(pyrimidopyrrolone). See text | 551 |
| 32 | 3,3'-Bis(carbamido)-4,4'-diaminobiphenyl | A,2 | — | — | — | Refluxing in $(CH_3CO)_2O$ gives a poly(pyrimido-pyrrolone). See text | 551 |
| 33 | 3,3'-Diamino-4,4'-dihydroxybiphenyl | A,1 | $H_2SO_4$ | $\eta_{red}$ 0.4 | — | Structure supported by IR. Decomposes at > 380°C (TGA, air, 4°C/min) | 371, 372 |
| 34 | 3,3'-Diamino-4,4'-dianilinobiphenyl | A,1 | Sparingly in boiling DMF, DMA and in strongly acidic solvents | — | — | Structure supported by IR. Begins to decompose in air at ≥ 400°C | 370 |
| 35 | 4,4'-Dihydrazinobiphenyl | A,1 | DMSO | $\eta_{sp}$ 0.16 | — | — | 366 |
| | Copolymers | A,1 | — | — | — | Thermal stability studies | 366 |

TABLE I.3—*continued*

Polypyromellitimides

| No. | Structure | Method | Solubility | Molecular weight | $T_m$ (°C) | Remarks and property data | References |
|---|---|---|---|---|---|---|---|
| 36 | <br>4,4'-Diaminodiphenylmethane | A,1;<br>A,2;<br>A,5 | conc. $H_2SO_4$ | $\eta_{inh}$ 1.7 | >500 | Properties of film: Density = 1.362 g/cm³; modulus: (23°C) 350,000 psi, (200°C) 180,000 psi; elongation: (23°C) 14%, (200°C) 22%; tensile strength: (23°C) 11,900 psi, (200°C) 7100 psi; impact strength = 2.01 kg cm/mil; tear strength 4.9 g/2 inch tear/mil; hydrolytic stability: >100 hr in boiling water, >24 hr in steam. Chemical stability: excellent; thermal stability: >90 hr at 250°C in air; >24 hr at 310°C in air; zero strength temperature: 785°C. Volume resistivity: >2.5 × 10¹⁵ ohm-cm (23°C) >4.1 × 10¹¹ ohm-cm (250°C); dielectric constant ($K$) [table below] Fiber formation and their properties were studied. Useful | 15, 18, 20, 58, 60, 121, 122, 128, 209, 217, 218, 221, 237, 273, 282, 283, 284, 287, 305, 311, 313, 364, 447, 472, 526, 527, 589, 621 |

dielectric constant ($K$)

| 3.96 | 3.86 | 3.19 | 3.17 |
|---|---|---|---|

Dissipation factor ($D_f$)

| 0.0044 | 0.0086 | 0.0109 | 0.0019 |
|---|---|---|---|
| f 10² | 10⁴ | 10² | 10⁵ |

Temperature (°C)

| 23 | 23 | 250 | 250 |
|---|---|---|---|

129

in films, coatings, and composites. Coalescible powder possessing a crystallinity index of 39% (X-ray) was prepared. Compositions containing finely dispersed metals useful in electrical applications were made. Kinetics of thermal imidation were studied

—

---

37

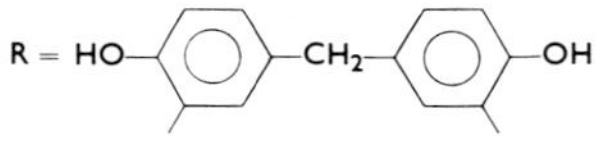

R = HO-⬡-CH₂-⬡-OH

3,3'-Diamino-4,4'-dihydroxydiphenylmethane

A,1

DMF, DMA, DMSO, tetramethylene sulfone, mixtures of the latter with acetone and THF

$\eta_{red}$ up to 1.0

Structure supported by IR. Decomposition begins at >380°C (TGA, air, 4°C/min)

371, 372

---

38

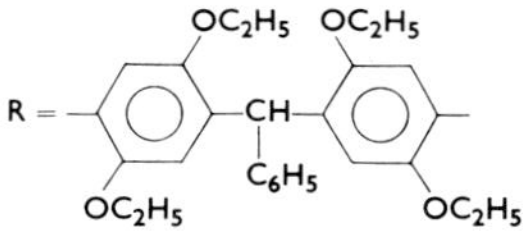

4,4'-Diamino-2,2',5,5'-tetraethoxytriphenylmethane

A,2

DMA

$\eta_{inh}$ 0.80–1.00

Cherry red. Tensile modulus = 374,000 psi (23°C), 157,000 psi (200°C); elongation = 5% (23°C), 24% (200°C); tenacity = 11,900 psi, (23°C), 3500 psi (200°C); Density = 1.388 g/cm³; zero strength temperature ≅ 438°C; volume resistivity: 3.3 × 10¹³ ohm-cm at 23°C, 6.0 × 10¹² ohm-cm at 200°C. Dissipation factor: 0.003–0.005 at 23°C, 0.002–0.007 at 200°C; dielectric constant: 3.4 at 23°C, 2.5 at 200°C

26

---

Copolymers

A,2

—

—

—

26

TABLE I.3—continued
Polypyromellitimides

| No. | Structure | Method | Solubility | Molecular weight | $T_m$ (°C) | Remarks and property data | References |
|---|---|---|---|---|---|---|---|
| 39 | R = [structure] —C(CH$_3$)$_2$ <br> 2,2-Bis(4-aminophenyl)propane | A,1; A,2 | conc. H$_2$SO$_4$ | $\eta_{inh}$ 0.8 | — | Density = 1.30 g/cm$^3$; tensile modulus = 370,000 psi; elongation = 7.4%; tensile strength = 11,900 psi; wt. loss (400 hr, 325°C, air) = 36%. Coalescible powder prepared | 121, 122, 209, 218, 221, 237, 273, 283, 284, 611, 612, 621 |
| | Copolymers of 39 | A,2 | — | — | — | — | 218 |
| 40 | R = HO—[structure]—C(CH$_3$)$_2$—[structure]—OH <br> 2,2-Bis(3-amino-4-hydroxyphenyl)propane | A,1 | DMF, DMA DMSO, tetra-methylene sulfone, mixtures of the latter with acetone and THF | $\eta_{red}$ up to 1.5 | — | Structure supported by IR | 371, 372 |
| 41 | R = [structure with CF$_3$—C—CF$_3$] <br> 2,2-Bis(4-aminophenyl)hexafluoropropane | A,2 | — | — | — | Good thermal stability. Gives tough, slightly colored films. Coalescing powder was prepared | 47, 48 463, 564 |

| No. | R = | Structure | Method | Solvent | Viscosity | | Properties | Ref. |
|---|---|---|---|---|---|---|---|---|
| 42 | 2,2-Bis(4-amino-3-methyl)hexafluoropropane | | A,2 | — | — | — | — | *564* |
| 43 | 2,2-Bis(4-aminophenyl)chloropentafluoropropane | | A,2 | — | — | — | — | *564* |
| 44 | 2,2-Bis(4-aminophenyl)-1,3-dichlorotetrafluoropropane | | A,2 | — | — | — | — | *564* |
| 45 | 9,9-Bis(4-aminophenyl)fluorene | | A,1 | $H_2SO_4$ | $[\eta]$ 0.65 | — | TGA (5°C/min); air: wt. loss starts at 460°C. TGA (5°C/min), helium: wt. loss starts at >500°C | *374, 375* |

TABLE 1.3—*continued*
Polypyromellitimides

| No. | Structure | Method | Solubility | Molecular weight | $T_m$ (°C) | Remarks and property data | References |
|---|---|---|---|---|---|---|---|
| 46 | <br>Anilinephthalein<br>R = | A,1 | DMF, nitrobenzene | $[\eta]$ 1.80<br>$\bar{M}_w$ (light scattering) = 125,000 | >500 | Tear strength = 1300–1400 kg/cm²; elongation at break = 30–40%; TGA (5°C/min), air: wt. loss starts at 410°C, TGA (5°C/min), helium: wt. loss starts at 460°C. Chemical resistance was studied in nitrobenzene solution. Annealing and fractionation studies were also performed | *374, 375, 661, 662, 664–667, 673* |
| 47 | <br>Anilinephthaleinimide<br>R = | A,1 | DMF | $[\eta]$ 1.18 | >500 | Tear strength = 1300–1400 kg/cm²; elongation at break = 30–40%; wt. loss (air): 25 hr at 300°C—none, 5 hr at 400°C—none, 5 hr at 420°–430°C —30% | *374, 375, 661, 662, 666* |
| 48 | <br>N-Benzoylanilinephthaleinimide<br>R = | A,1 | — | — | — | Wt. loss (air): 5 hr at 400°C—none; 5 hr at 420°–430°C—41% | *374* |

| 49 | R = | | | |
|---|---|---|---|---|

**4,4'-Diaminodiphenyl ether**

| A,1; A,2; A,3 | conc. H₂SO₄, fuming HNO₃, SbCl₃, mixtures of AsCl₃ with SbCl₃ | $[\eta]$ up to 2.57 | — | Known (among others) under the names of "Polymer H" and "Vespel" (DuPont). Kinetics of imidation studied. Optimum mechanical properties obtained after heat treatment at 300°–400°C which leads to slight cross-linking. Tensile strength = 25,000 psi; tensile modulus ≅ 400,000 psi; elongation at break = 70–120%. Mechanical properties remain good within the temperature range of −200° to +400°C. Slightly crystalline (X-ray). No $T_g$ indicated by modulus/temperature curve; $T_g$ determined via electrical dissipation factor at 1 kc as a function of temperature was 385°C. Possesses superior electrical properties within the temperature range of −60° to +350°C. Excellent thermal stability: after 5000 hr at 250°C mechanical and electrical properties are still excellent. Displays very good chemical, radiation, and abrasion resistance, excellent dielectric stability at ⩾300°C, good resistance to cracking and good adherence to metals. Useful as a high-temperature film, insulator, coating, and fiber; other uses are in cellular articles, composites, laminates, and foams. Coalescible powders were prepared. Several investigations were performed on this polymer. They include: (a) Thermal degradation studies; | 13, 14, 28, 41, 59 62, 63, 67, 97, 121, 122, 128, 131, 138, 145, 181, 182, 209, 211, 218, 221, 234, 237, 240, 248, 267, 269, 270, 283, 284, 287, 305, 311, 312–314, 325, 337, 351, 364, 383, 384, 405, 407– 410, 415, 416, 428, 470, 478, 479, 481, 482, 489, 490, 519, 561, 566, 579, 611, 612, 621, 623<br><br>157–159, 161, 162 |

TABLE I.3—*continued*
Polypyromellitimides

| No. | Structure | Method | Solubility | Molecular weight | $T_m$ (°C) | Remarks and property data | References |
|---|---|---|---|---|---|---|---|
| 49 | R = 4,4′-Diaminodiphenyl ether | A,1; A,2 | — | — | — | (b) Pyrolysis studies (yields a semi-conducting material); | 160, 163, 164 |
| | | | | | | (c) Studies of the dynamic-mechanical behavior within the temperature range of 60° to 800°K; | 107 |
| | | | | | | (d) Thermographic and dilatometric studies; | 602 |
| | | | | | | (e) Study of the transimidation reaction which leads to a 3-dimensional system, during the thermal imidation; | 603 |
| | | | | | | (f) Study of the alkaline hydrolysis of the polymer | 512 |
| | | | | | | For supplementary information about this polymer, see text | |
| | Copolymers of 49 | A,1; A,2 | — | — | — | — | 27, 61, 128, 285, 474, 486 |
| 50 | R = 3,4′-Diaminodiphenyl ether | A,1 | — | — | — | Flexible film. Wt. loss after 400 hr at 325°C in air = 7.2% | 121, 122, 273 |

135

| No. | Structure | Code | Reagent | $\eta_{inh}$ | Properties | References |
|---|---|---|---|---|---|---|
| 51 | R = 4,4'-Diaminodiphenyl sulfide | A,1; A,2 | Fuming HNO₃ | $\eta_{inh} \geqslant 0.3$ | Film properties: Density = 1.41 g/cm³; tensile modulus = 320,000 psi; elongation = 7–10%; tensile strength = 10,000 psi; wt. loss after 400 hr at 325°C (air) = 7.9%. Fibers prepared and evaluated | 121, 122, 209, 218, 237, 273, 337, 621, 716 |
| | Copolymers of 51 | A,1 | — | — | Good oxidative stability. The sulfide linkage oxidizes to sulfoxide and/or sulfone. Oxygen is consumed and chain breakage prevented | 209, 651 |
| 52 | R = 4,4'-Diaminodiphenyl sulfone | A,1; A,2 | H₂SO₄ | $\eta_{inh} \geqslant 0.3$ | Density = 1.43 g/cm³; tensile modulus = 350,000 psi; elongation = 14%; tensile strength = 8500 psi; useful in composites | 209, 218, 237, 305, 611, 612, 621 |
| 53 | R = Bis(3-aminophenyl)alkyl phosphine oxide | A,1; A,2 | — | — | — | 95 |
| | Copolymers of 53 | A,1; A,2 | — | — | — | 95 |
| 54 | R = Bis-(4-aminophenyl)diethylsilane | A,1 | H₂SO₄ | $\eta_{inh} >0.1$ | Maintains strength up to 500°C | 55, 132 |
| 55 | R = 4,4'-Diaminostilbene | A,2 | — | — | — | 490 |

TABLE 1.3—*continued*
Polypyromellitimides

| No. | Structure | Method | Solubility | Molecular weight | $T_m$ (°C) | Remarks and property data | References |
|---|---|---|---|---|---|---|---|
| 56 | 4,4'-Diaminodiphenyl disulfide | A,1 | — | — | — | — | 5 |
| 57 | 4,4'-Diaminoazobenzene | A,2 | — | — | — | Red. Semicrystalline. Tensile modulus = 1,478,000 psi (25°C), 591,000 psi (200°C); tensile strength = 124,000 psi (25°C), 11,300 psi (200°C); elongation = 3.2% (25°C), 4.9% (200°C); dielectric constant (100 Hz) = 3.67 (25°), 3.18 (200°C); volume resistivity = $1.3 \times 10^{17}$ ohm-cm (25°C), $1.3 \times 10^{14}$ ohm-cm (200°C); dissipation factor (100 Hz) = 0.0023 (25°C), 0.0016 (200°C). Crystalline regions have *trans* configuration. Amorphous regions have *cis* configuration. Shrinks reversibly on heating or UV treatment. Phenomenon possibly due to *trans* ⇌ *cis* interconversion. | 87, 93, 498 |
| | Copolymers of 57 | A,2 | — | — | — | Control of the amount of azo linkages in the copolymers yields materials with coefficients of thermal dilation = 0 | 87, 498 |

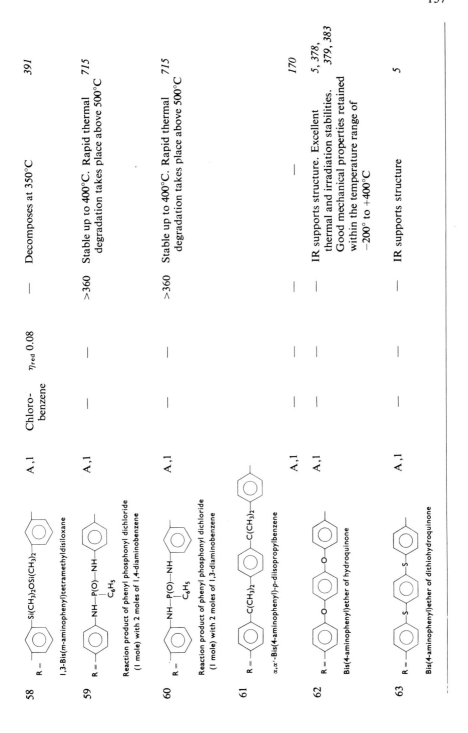

| No. | | Method | Solvent | $\eta_{red}$ | | | Ref. |
|---|---|---|---|---|---|---|---|
| 58 | R = [structure] Si(CH₃)₂OSi(CH₃)₂<br>1,3-Bis(m-aminophenyl)tetramethyldisiloxane | A,1 | Chloro-benzene | $\eta_{red}$ 0.08 | — | Decomposes at 350°C | 391 |
| 59 | R = [structure] NH—P(O)—NH with C₆H₅<br>Reaction product of phenyl phosphonyl dichloride (1 mole) with 2 moles of 1,4-diaminobenzene | A,1 | — | — | >360 | Stable up to 400°C. Rapid thermal degradation takes place above 500°C | 715 |
| 60 | R = [structure] NH—P(O)—NH with C₆H₅<br>Reaction product of phenyl phosphonyl dichloride (1 mole) with 2 moles of 1,3-diaminobenzene | A,1 | — | — | >360 | Stable up to 400°C. Rapid thermal degradation takes place above 500°C | 715 |
| 61 | R = [structure] C(CH₃)₂ — C(CH₃)₂<br>α,α'-Bis(4-aminophenyl)-p-diisopropylbenzene | A,1 | — | — | — | — | 170 |
| 62 | R = [structure] O<br>Bis(4-aminophenyl)ether of hydroquinone | A,1 | — | — | — | IR supports structure. Excellent thermal and irradiation stabilities. Good mechanical properties retained within the temperature range of −200° to +400°C | 5, 378, 379, 383 |
| 63 | R = [structure] S — S<br>Bis(4-aminophenyl)ether of dithiohydroquinone | A,1 | — | — | — | IR supports structure | 5 |

TABLE I.3—*continued*
Polypyromellitimides

| No. | Structure | Method | Solubility | Molecular weight | $T_m$ (°C) | Remarks and property data | References |
|---|---|---|---|---|---|---|---|
| 64 | R = 2,5-Bis(4-aminophenyl)-1,3,4-oxadiazole | A,1; A,2 | — | — | — | Yields tough films. Drawn films are moderately crystalline (X-ray). Does not melt in the conventional sense. Fiber and UV stability data. Decomposition temperature (TGA, N$_2$, 15°C/min) = 525°C. Endothermic transition temperature (DTA, N$_2$, 20°C/min) = 520°C. Relatively unaffected after 35 days at 300°C (air) | 49, 465, 541–543, 545, 573 |
| 65 | R = 2,5-Bis(3-aminophenyl)-1,3,4-oxadiazole | A,1; A,2 | — | — | — | Endothermic transition temperature (DTA, N$_2$, 10°C/min) = 500°C. Does not melt in the conventional sense. Decomposition temperature (TGA, N$_2$,15°C/min) = 500°C. Fiber and UV stability data | 49, 275, 465, 543, 545 |

| No. | R = | | | | | Ref. |
|---|---|---|---|---|---|---|
| 66 | 3,5-Bis(m-aminophenyl)-4-phenyl-1,2,4-triazole | A,1; A,2 | — | — | Endothermic transition temperature (DTA, $N_2$, 10°C/min)=450°C. Film exposed in air on a hot surface softened at 350°C. Decomposition temperature (TGA, $N_2$, 15°C/min) = 510°C | 49, 465, 541–543 |
| 67 | Bis(p-aminophenyl) ether of 2,2-bis(4-hydroxyphenyl)propane | A,1 | — | — | Yields elastic films. Excellent adhesive, thermal, and dielectric properties. Useful in wire coatings | 135 |
| 68 | Reaction product of phenyl phosphonyl dichloride (1 mole) and 2 moles of 4,4'-diaminobiphenyl | A,1 | — | >360 | Stable up to 400°C. Rapid thermal degradation takes place above 500°C | 715 |
| 69 | Reaction product of phenyl phosphonyl dichloride (1 mole) and 2 moles of 3,3'-dimethyl-4,4'-diaminobiphenyl | A,1 | — | >360 | Stable up to 400°C. Rapid thermal degradation takes place above 500°C | 715 |
| 70 | Reaction product of phenyl phosphonyl dichloride (1 mole) and 2 moles of 3,3'-dimethoxy-4,4'-diaminobiphenyl | A,1 | — | >360 | Stable up to 400°C. Rapid thermal degradation takes place above 500°C | 715 |

TABLE I.3—*continued*
Polypyromellitimides

| No. | Structure | Method | Solubility | Molecular weight | $T_m$ (°C) | Remarks and property data | References |
|---|---|---|---|---|---|---|---|
| 71 | R = —⟨⟩—CH₂—⟨⟩—NH—P(O)NH—⟨⟩—CH₂—⟨⟩— with C₆H₅ Reaction of product of phenyl phosphonyl dichloride (1 mole) and 2 moles of 4,4'-diaminodiphenyl methane | A,1 | — | — | >360 | Stable up to 400°C. Rapid thermal degradation takes place above 500°C | 715 |
| 72 | R = —⟨⟩—SO₂—⟨⟩—NHP(O)NH—⟨⟩—SO₂—⟨⟩— with C₆H₅ Reaction product of phenyl phosphonyl dichloride (1 mole) and 2 moles of 4,4'-diaminodiphenyl sulfone | A,1 | — | — | >360 | Stable up to 400°C. Rapid thermal degradation takes place above 500°C | 715 |
| 73 | R = thiazole structure *m*-Bis[4-(*p*-aminophenyl)thiazole-2-yl]-benzene | A,2 | — | — | — | Endothermic transition temperature of films hot drawn at 350° and 425°C (DTA, N₂, 10°C/min) = 565°C. Decomposition temperature (TGA, N₂, 15°C/min) = 550°C | 49, 465, 541–543 |

| 74 | R= | A,1 | — | — | Properties of film: tensile strength = 25,600 psi; elongation = 1.5%. Strength retained for at least 10 hr at 300°C (air). Oxidation temperature = 568°C; TGA (air, 5°C/min) = 15% wt. loss at 490°–535°C; TGA (vac., 5°C/min) = 15% wt. loss at 620°–625°C | *355* |

2,2'-Di-*p*-aminophenyl-2,6-imidazobenzimidazole

| 75 | R= | A,1 | — | — | Stable for at least 10 hr at 300°C (air). Oxidation temperature = 577°C; TGA (air, 5°C/min) = 15% wt. loss at 490°–535°C; TGA (vac., 5°C/min) = 15% wt. loss at 620°–625°C | *355* |

2,2'-Di-*m*-aminophenyl-2,6-imidazobenzimidazole

| 76 | R= | A,1 | — | — | Does not melt in the conventional sense. Fiber properties: yellow-orange; Density = 1.43 g/cc; high crystallinity and medium orientation (X-ray). Tenacity = 4.7 g/den., elongation = 19.3%; initial modulus = 57 g/den. Endothermic transition temperature (DTA, $N_2$, 20°C/min) = 595°C. Decomposition temperature (TGA, $N_2$, 15°C/min) = 615°C. Performance of fibers at high temperatures, their heat aging, and UV stability are described | *546* |

2,2'-Di-*m*-aminophenyl-6,6'-dibenzothiazole

TABLE 1.3—*continued*

Polypyromellitimides

| No. | Structure | Method | Solubility | Molecular weight | $T_m$ (°C) | Remarks and property data | References |
|---|---|---|---|---|---|---|---|
| 77 | 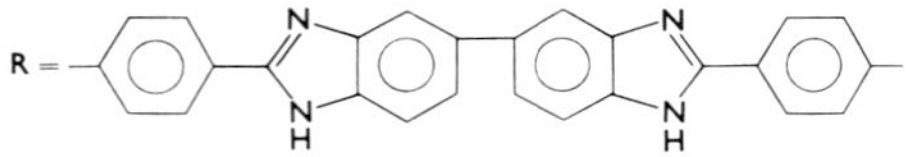 2,2′-Di-*p*-aminophenyl-5,5′-dibenzimidazole | A,1 | — | — | — | Properties of film: tensile strength = 24,750 psi; elongation = 2.0%. Retains strength for at least 10 hr at 300°C (air). Oxidation temperature = 568°C. TGA and DTA data | *340, 355, 543* |
| 78 | 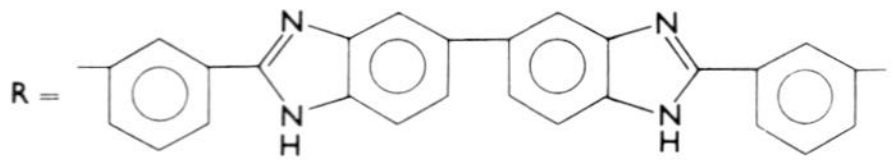 2,2′-Di-*m*-aminophenyl-5,5′-dibenzimidazole | A,1 | — | — | — | Properties of film: tensile strength = 17,600 psi; elongation = 3.5%. Retains strength for at least 10 hr at 300°C (air). Oxidation temperature = 577°C. TGA and DTA data | *355, 541–543* |
| 79 | 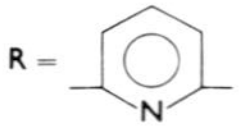 2,6-Diaminopyridine | A,1 | — | — | — | — | *459* |

| No. | Structure / Name | Method | | | Remarks | Ref. |
|---|---|---|---|---|---|---|
| 80 | <br>N,N'-Diaminopyromellitimide | A,1; A,2 | — | — | May contain polymer resulting from rearrangement:<br> | 199 |
| 81 | <br>2-(m-Aminophenyl))-5-aminobenzimidazole | A,1 | — | — | Clear, yellow, brittle film<br><br>Oxidation studies performed | 275 |
| | Copolymers of 81 | A,1 | — | — | Useful as laminating and adhesive resins | 275 |
| 82 | <br>N,N'-Diaminonaphthalene-1,4,5,8-tetracarboxydiimide | A,1 | — | — | Intractable, insoluble, gray-brown powder. Oxidation studies performed | 199 |
| 83 | <br>p-Bis(5-aminobenzoxazol-2-yl)benzene | A,1 | — | — | Fiber properties: color—copper; medium crystallinity and orientation (X-ray); density = 1.49 g/cc; tenacity = 4.8 g/den.; elongation = 5.1%; initial modulus = 168 g/den. Performance of fibers at high temperatures, their heat aging characteristics, and UV stability are described. DTA data also given | 546 |

TABLE 1.3—*continued*
Polypyromellitimides

| No. | Structure | Method | Solubility | Molecular weight | $T_m$ (°C) | Remarks and property data | References |
|-----|-----------|--------|------------|------------------|------------|---------------------------|------------|
| 84 | 4,4'-Bis(5-aminobenzoxazol-2-yl) diphenyl ether | A,1 | — | — | — | Fiber properties: color—bronze; medium crystallinity and orientation (X-ray); tenacity = 5.2 g/den.; elongation = 4.0%; initial modulus = 148 g/den. Performance of fibers at high temperatures, their heat aging characteristics and UV stability are described. TGA and DTA data also given | *546* |

TABLE I.4

Polyimides Based on 3,3′,4,4′-Benzophenonetetracarboxylic Acid Dianhydride

| No. | Structure | Method | Remarks and property data | References |
|---|---|---|---|---|
| | 3,3′,4,4′-Benzophenone tetracarboxylic acid dianhydride or derivative and | | | |
| 1 | R = $+(CH_2)_2+$ Ethylenediamine | A,1 | Clear, flexible, yellow film | 148, 328 |
| 2 | R = $+(CH_2)_6+$ 1,6-Diaminohexane | A,1 | Polymer can be molded at 350°C "under pressure" to a clear tan film | 148, 328 |
| 3 | R = $+(CH_2)_3-Si(CH_3)_2OSi(CH_3)_2+(CH_2)_3+$ 1,3-Bis(γ-aminopropyl)tetramethyldisiloxane | A,1 | — | 330 |
| 4 | R = $+(CH_2)_4-Si(CH_3)_2OSi(CH_3)_2+(CH_2)_4+$ 1,3-Bis(δ-aminobutyl)tetramethyldisiloxane | A,1 | Flexible and abrasion resistant. Excellent insulating coating. Retains flexibility after aging at 250°C for 16 hr; wt. loss (16 hr, 250°C) = 6.2% | 330 |
| 5 | R = $+(CH_2)_3+[Si(CH_3)_2O]_{30}-Si(CH_3)_2+(CH_2)_3+$ α,ω-Bis(γ-aminopropyl)polydimethylsiloxane | A,1 | — | 330 |
| | Copolymers of 1,3,4, and 5 | A,1 | Linear and cross-linked compositions | 330 |
| 6 | R = 1,4-Diaminobenzene | A,1 | Thermal imidation studied. Excellent thermal stability. Weight loss (air, 300°C, 672 hr) = 5.7% | 224, 411 |
| | Copolymers of 6 | A,1 | — | 286 |

TABLE I.4—continued

Polyimides Based on 3,3',4,4'-Benzophenonetetracarboxylic Acid Dianhydride

| No. | Structure | Method | Remarks and property data | References |
|---|---|---|---|---|
| 7 | R= (benzene ring)  1,3-Diaminobenzene | A,1; A,2 | Thermal imidation studied. Westinghouse deisgnation: I-8. Properties: Amorphous (X-ray); $T_s$ = 240°C (modulus–temperature curve), 217°–243°C(DSC). 340°C (via measurements of electrical dissipation factor); tensile strength = 18,000 psi; tensile modulus = 300,000 psi; elongation at break = 11%. Useful in wire coatings. Excellent thermal stability. Coalescible powders were prepared | 148, 182, 224, 234, 266, 269, 270, 281, 282, 284, 286, 287, 293, 316, 328, 411 |
| | | | Polymers end-blocked with p-aminoacetanilide and phthalic anhydride (Westinghouse designation I-40) or with p-aminoacetanilide only (Westinghouse I-66) are excellent high-temperature adhesives for glass and steel. I-40 is useful up to 700°F, retains more than 50% of its original lap shear tensile strength at 550°F (>2000 psi) after 1000 hr at 550°F | 166 |
| | Copolymers of 7 | A,1 | High-temperature adhesives | 148, 166, 328 |
| 8 | R= (benzene ring)  1,3-Diaminobenzene | A,1 | Block-copolymer polyimide/bisphenol-A-isophthalic polyester. Excellent thermal stability and improved solubility in phenolic solvents claimed | 659 |
| 9 | R= (benzene ring) CH(CH$_3$)$_2$  Cumene-2,4-diamine | A,1 | Cross-linking with di-tert-butyl peroxide reported | 326 |

| No. | R = (structure / name) | Method | Properties | Ref. |
|---|---|---|---|---|
| 10 | 1,3-Diamino-5-(2'-imidazolyl)benzene | A,1 | Clear, yellow, brittle films | 275 |
| 11 | 4,4'-Diaminobiphenyl | A,1 | Thermal imidation studied. Stable up to 400°C | 6, 224 |
| 12 | 3,3'-Diamino-4,4'-dianilinobiphenyl | A,1 | Structure supported by IR. Decomposition starts at ⩾400°C (air). Sparingly soluble in boiling DMF, DMA, and strong acidic solvents | 370 |
| 13 | 4,4'-Diaminodiphenylmethane | A,1 | Excellent wire insulator. Coalescible powders were prepared. Cross-linking with di-tert-butyl peroxide was reported | 54, 148, 246, 281, 287, 326, 328, 467, 591 |
| 14 | 2,2-Bis(4-aminophenyl)hexafluoropropane | A,1; A,2 | — | 564 |
| 15 | 4,4'-Diaminobenzophenone | A,1 | — | 148, 328 |

TABLE I.4—continued

Polyimides Based on 3,3',4,4'-Benzophenonetetracarboxylic Acid Dianhydride

| No. | Structure | Method | Remarks and property data | References |
|---|---|---|---|---|
| 16 | R = <br>4,4'-Diaminodiphenyl ether | A,1 | Thermal imidation studied. Westinghouse designation I-7. Properties of film: amorphous (X-ray); $T_g = 220°C$ (modulus/temperature curve), 207°–223°C (DSC), 345°C (via measurement of electrical dissipation factor). Annealing (250°C vac., 20 hr, or 400°C, vac., 4 min) increases $T_g$ to 265°–270°C (modulus/temperature curve) and to 273°–284°C (DSC). Reason: cross-linking and/or further imidation; no crystallinity was induced on annealing (X-ray). Tensile strength = 14,000–15,000 psi; tensile modulus = 300,000 psi; elongation at break = 15%; excellent thermal stability. Wt. loss = 1.2% (air, 300°C, 230 hr); 3.1% (air, 300°C, 717 hr). Has good hydrolytic stability, good adhesive, and good electrical properties. Excellent wire insulator. Coalescible powders were prepared. Coalescing at 400°C, under 3000–4000 psi pressure gave material with following properties: strength index = 6.1; tensile strength = 23,400 psi; elongation = 10.6%; tensile impact strength = 194 ft lb/inch³. Excess diamine may be used to effect cross-linking via reaction: amine + carbonyl group of dianhydride. Cross-linked compositions are good insulators and good adhesives for fiber glass, aluminum, and stainless steel laminates | 6, 182, 221, 224, 234, 647<br><br><br><br><br><br><br><br><br><br><br><br>139, 411, 476, 525, 562<br>148, 269, 270, 282, 283, 284, 287, 328<br><br><br><br><br>53, 147, 175, 468 |
| | Copolymers of 16 | A,1 | — | 137, 245 |
| 17 | R = <br>4,4'-Diaminodiphenyl sulfone | A,1 | — | 148, 328 |

| No. | Structure | A,1 | Notes | Ref. |
|---|---|---|---|---|
| 18 | R = <br> 4,4'-Diaminodiphenyl-N-methylamine | A,1 | Coalescible powders were prepared | 283, 284 |
| 19 | R = <br> 2,5-Bis(4-aminophenyl)-1,3,4-oxadiazole | A,1 | Moderately crystalline (X-ray). Weight loss (air, 325°C, 332 hr) = 4.1%. Decomposition temperature (TGA, N$_2$, 15°C/min) = 500°C. Endothermic transition temp. (DTA, N$_2$ 20°C/min) = 485°C. Fiber property and UV stability data | 275, 545 |
| 20 | R = <br> 2-(p-Aminophenyl)-5-(m-aminophenyl)-1,3,4-oxadiazole | A,1 | Wt. loss (air, 325°C, 332 hr) = 3.6%. Also prepared by condensation of N-(m-aminobenzoyl)-N'-(p-aminobenzoyl)-hydrazine with 3,3',4,4'-benzophenonetetracarboxylic acid dianhydride followed by thermal dehydration. See text in section on heterocyclic copolymers | 275 |
| 21 | R = <br> 2,5-Bis(m-aminophenyl)-1,3,4-oxadiazole | A,1 | Useful in glass laminates. Wt. loss (air, 325°C, 332 hr) = 4.4%. Also prepared by condensation of N,N'-bis(m-aminobenzoyl)hydrazine with 3,3',4,4'-benzophenone-tetracarboxylic acid dianhydride followed by thermal dehydration. See text in section on heterocyclic copolymers | 272, 275 |
| 22 | R = <br> 2,2'-Bis[5-(m-aminophenyl)-1,3,4-oxadiazolyl] | A,1 | Wt. loss (air, 325°C, 232 hr) = 18.8% | 275 |

TABLE 1.4—*continued*

Polyimides Based on 3,3',4,4'-Benzophenonetetracarboxylic Acid Dianhydride

| No. | Structure | Method | Remarks and property data | References |
|---|---|---|---|---|
| 23 | R = *m*-Bis[5-(p-aminophenyl)-1,3,4-oxadiazole-2-yl]benzene | A,1 | Prepared directly from the oxadiazole precursor: (isophthalbis[*N*'-(*p*-aminobenzoyl)hydrazide]) by condensation with 3,3',4,4'-benzophenonetetracarboxylic acid dianhydride followed by thermal dehydration. See text in section on heterocyclic copolymers | 275 |
| 24 | R = *m*-Bis[5-(m-aminophenyl))-1,3,4-oxadiazole-2-yl]benzene | A,1 | Wt. loss (air, 325°C, 332 hr) = 9.1% | 275 |
| 25 | R = 2,2'-Di-p-aminophenyl-2,6-imidazobenzimidazole | A,1 | Tensile strength = 27,200 psi; elongation = 3.3%. Retains strength for at least 10 hr at 300°C (air). Oxidation temperature: 590°C; TGA (air, 5°C/min) = 15% wt. loss at 490°–535°C; TGA (vac., 5°C/min) = 15% wt. loss at 620°–625°C. | 355 |
| 26 | R = 2,2'-Di-m-aminophenyl-2,6-imidazobenzimidazole | A,1 | Tensile strength = 17,600 psi; elongation = 3.3%. Retains strength for at least 10 hr at 300°C (air). Oxidation temperature: = 590°C; TGA (air, 5°C/min) = 15% wt. loss at 490°–535°C; TGA (vac., 5°C/min) = 15% wt. loss at 620°–625°C | 355 |

| No. | R = (structure) | | Properties | Ref. |
|---|---|---|---|---|
| 27 | 2,2'-Di-p-aminophenyl-5,5'-dibenzimidazole | A,1 | Tensile strength = 24,000 psi; elongation = 3.8%. Retains strength for at least 10 hr at 300°C (air). Oxidation temperature = 586°C; TGA (air, 5°C/min) = 15% wt. loss at 520°–550°C; TGA (vac., 5°C/min) = 15% wt. loss at 625°–630°C | 355 |
| 28 | 2,2'-Di-m-aminophenyl-5,5'-dibenzimidazole | A,1 | Tensile strength = 17,500 psi; elongation = 3.5%. Retains strength for at least 10 hr at 300°C (air). Oxidation temperature = 590°C; TGA (air, 5°C/min) = 15% wt. loss at 520°–550°C; TGA (vac., 5°C/min) = 15% wt. loss at 625°–630°C | 355 |
| 29 | 2-(p-Aminophenyl)-6-aminobenzoxazole | A,1 | Tensile strength = 15,000 psi; tensile modulus = 200,000 psi; elongation = 5–20%. Retains toughness after "several" hours in air at 370°C | 563 |
| 30 | 2-(m-Aminophenyl)-6-aminobenzoxazole | A,1 | Tensile strength = 12,000 psi; tensile modulus = 200,000 psi; elongation = 5–10%. Retains toughness and strength for 2 hr at 370°C in air | 563 |
| 31 | 2-(p-Aminophenyl)-6-aminobenzothiazole | A,1 | — | 563 |
| 32 | 5,4'-Diamino-2-phenylbenzimidazole | A,1 | Westinghouse designation: BI-1; Wt. loss (325°C, air, 332 hr) = 8.3%. Is an outstanding adhesive for titanium. At 600°F, bond strength = 1300 psi; after 1000 hr at 600°F, bond strength at 600°F = 800 psi | 166, 275 |

TABLE I.4 —*continued*

Polyimides Based on 3,3'4,4'-Benzophenonetetracarboxylic Acid Dianhydride

| No. | Structure | Method | Remarks and property data | References |
|-----|-----------|--------|---------------------------|------------|
| 33 | R = 5,3'-Diamino-2-phenylbenzimidazole | A,1 | Clear, yellow, flexible film. Tensile strength = 15,000 psi; tensile modulus = 200,000 psi; elongation = 5–10%. Retains strength and toughness after heating in air for 2 hr at 370°C | 275, 563 |
| 34 | R = p-Bis(5-aminobenzoxazole-2-yl)benzene | A,1 | Endothermic transition temperature (DTA, $N_2$, 20°C/min) = 555°C; decomposition temperature (TGA, $N_2$, 15°C/min) = 560°C. Fiber had following properties: high crystallinity and orientation (X-ray); yellow; density = 1.46 g/cc; tenacity = 6.4 g/den.; elongation = 5.8%; initial modulus = 203 g/den. Performance of fibers at high temperatures, their heat aging characteristics and UV stability data are also described | 546 |
| 35 | R = m-Bis(5-aminobenzimidazole-2-yl)benzene | A,1 | Clear, brown, fairly flexible film | 275 |
| 36 | PAPI (Polyaniline polyisocyanate) | A,5 | Foam. Density = 3.97; K factor (open cell) = 0.26. Excellent thermal resistance: wt. loss after 14 days at 232°C is less than 2%. Has excellent solvent and chemical resistance, good fire resistance and good insulating properties | 223 |

## TABLE I.5
### Miscellaneous Polyimides

| No. | Structure | Method | Solubility | Molecular weight | Remarks and property data | References |
|-----|-----------|--------|------------|------------------|---------------------------|-----------|
| 1 | <br>4-Aminophthalic anhydride or derivative | A,1 | — | — | Useful in wire coating | 94, 152, 153 |
| 2 | <br>3-Aminophthalic anhydride or derivative | A,1 | conc. $H_2SO_4$ | $\eta_{inh}$ 0.05 $T_m > 350°C$ | | 153, 681, 682 |
| 3 | <br>4-(p-Aminobenzoyl)phthalic anhydride | A,1 | — | — | Coalescible powder prepared. It could be "molded" at 400°C and 15,000 psi pressure | 269, 270 |
| 4 | <br>Benzene-1,2,3,4-tetracarboxylic acid dianhydride or derivative and 4,4'-diaminodiphenyl ether | A,1 | — | — | Structure supported by IR | 519 |

## TABLE I.5—continued
### Miscellaneous Polyimides

Benzoyl pyromellitic dianhydride and

| No. | Structure | Method | Solubility | Molecular weight | Remarks and property data | References |
|-----|-----------|--------|------------|------------------|---------------------------|------------|
| 5 | $R=$ CH₂ <br> 4,4'-Diaminodiphenylmethane | A,1 | — | — | — | 646 |
| 6 | $R=$ O <br> 4,4'-Diaminodiphenyl ether | A,1 | — | — | — | 646 |

\* \* \*

| No. | Structure | Method | Solubility | Molecular weight | Remarks and property data | References |
|-----|-----------|--------|------------|------------------|---------------------------|------------|
| 7 | Thiophene-2,3,4,5-tetracarboxylic acid dianhydride or derivative and 2,11-diaminododecane | A,1 | — | — | — | 109 |

Pyridine-2,3,5,6-tetracarboxylic acid dianhydride or derivative and

| No. | R | | Conditions | Solubility | $\eta_{red}$ | Color | Ref. |
|---|---|---|---|---|---|---|---|
| 8 | 1,4-Diaminobenzene | | A,1 | Insoluble | — | Gray | 307 |
| 9 | 1,3-Diaminobenzene | | A,1 | — | — | Yellowish brown | 307 |
| 10 | 4,4'-Diaminobiphenyl | | A,1 | 95% $H_2SO_4$ | $\eta_{red}$ 0.34 | Gray-brown | 307 |
| 11 | 4,4'-Diamino-3,3'-dimethylbiphenyl | | A,1 | 95% $H_2SO_4$ | $\eta_{red}$ 0.12 | Gray-brown | 307 |
| 12 | 4,4'-Diamino-3,3'-dichlorobiphenyl | | A,1 | 95% $H_2SO_4$ | $\eta_{red}$ 0.50 | Gray | 307 |
| 13 | 4,4'-Diaminodiphenylmethane | | A,1 | 95% $H_2SO_4$ | $\eta_{red}$ 0.13 | Brown | 307 |

155

TABLE I.5—*continued*
Miscellaneous Polyimides

| No. | Structure | Method | Solubility | Molecular weight | Remarks and property data | References |
|-----|-----------|--------|-----------|------------------|---------------------------|------------|
| 14 | R =  4,4′-Diamino-3,3′-dichlorodiphenylmethane | A,1 | — | — | Light brown | 307 |
| 15 | R =  4,4′-Diaminodiphenyl ether | A,1 | 95% $H_2SO_4$  $\eta_{red}$ 0.52 | — | Brown | 307 |
| 16 | R =  4,4′-Diaminodiphenyl sulfone | A,1 | — | — | Yellow | 307 |
| 17 | R =  4,4′-Diaminodiphenylstilbene | A,1 | — | — | Gray-brown | 307 |
| 18 | R =  4,4′-Diaminodiphenyl disulfide | A,1 | — | — | Yellow | 307 |

* * *

Pyrazinetetracarboxylic acid dianhydride or derivative and

| 19 | R = <br> 2,5-Diamino-1,3,4-thiadiazole | A,1 | — | Transparent, smooth, flexible, red film. Zero strength temperature = 579° ($N_2$) and 592°C (air). Shows excellent thermal stability (no hydrogen atoms present in polymer backbone). No change observed after 25 hr at 400°C in air | 324 |
| 20 | R = <br> 4,4'-Diaminodiphenyl ether | A,1 | — | Tensile strength = 590–820 kg/cm²; TGA (air, 4.5°C/min): weight loss begins at 350°C and is equal to 50% at 500°C | 373, 460 |
| 21 | R = <br> Bis(4-aminophenyl)phenylphosphonate | A,1 | — | — | 283, 284 |

* * *

TABLE I.5—*continued*
Miscellaneous Polyimides

3,3',4,4'-Biphenyltetracarboxylic acid dianhydride or derivative and

| No. | Structure | Method | Solubility | Molecular weight | Remarks and property data | References |
|---|---|---|---|---|---|---|
| 22 | R = —CH=CHO$\{$Si(C$_6$H$_5$)(CH$_2$CH$_2$CH$_3$)O$\}_3$CH=CH—  α,ω-Bis(aminovinyl)phenyltrisiloxane | A,1 | — | — | — | *300* |
| 23 | R = (1,4-Diaminobenzene) | A,1 | — | — | Tough and thermally stable polymer. Structure supported by IR. Thermal imidation was studied | *7, 224* |
| 24 | R = 4,4'-Diaminobiphenyl | A,1 | — | — | Tough and thermally stable polymer. Structure supported by IR. Thermal imidation was studied | *7, 224* |
| 25 | R = 4,4'-Diaminodiphenyl ether | A,1 | — | — | Tough, thermally and irradiation stable polymer. Structure supported by IR. Thermal imidation was studied | *7, 224, 383* |
| 26 | R = P(C$_6$H$_5$)  4,4'-Diaminodiphenyl phenylphosphine oxide | A,1 | — | — | — | *283, 284* |

| No. | R = | Diamine / dianhydride | | | | Remarks | Ref. |
|---|---|---|---|---|---|---|---|
| 27 | Bis(4-aminophenyl) ether) of hydroquinone; 2,2′,6,6′-Biphenyltetracarboxylic acid dianhydride or derivative and | A,1 | — | — | — | Excellent thermal and irradiation resistance. Exhibits good mechanical properties within the temperature range of −200° to +400°C | 3, 383 |
| 28 | 1,3-Diaminobenzene | A,1; A,2 | — | — | — | Contains a 7-membered imide ring | 298 |
| 29 | 4,4′-Diaminodiphenyl ether; 2,3,5,6-Biphenyltetracarboxylic acid dianhydride or derivative and | A,1; A,2 | — | — | — | Contains a 7-membered imide ring. Excellent thermal stability up to 400°C in air (TGA) | 296, 298 |
| 30 | 4,4′-Diaminobiphenyl | A,1 | — | — | — | — | 361 |

TABLE I.5—*continued*

Miscellaneous Polyimides

| No. | Structure | Method | Solubility | Molecular weight | Remarks and property data | References |
|---|---|---|---|---|---|---|
| 31 | R = —CH₂—⬡ <br><br> **4,4′-Diaminodiphenylmethane** <br><br> * * * <br><br> (polyimide structure) <br><br> **3,3′,4,4′-Diphenylmethanetetracarboxylic acid dianhydride or derivative and** | A,1 | — | — | — | *361* |
| 32 |  R = ⬡ <br><br> **1,3-Diaminobenzene** <br> Copolymers of 32 | A,1 | — | — | Amorphous (X-ray). Coalescible powder prepared | 286 |
| 33 | 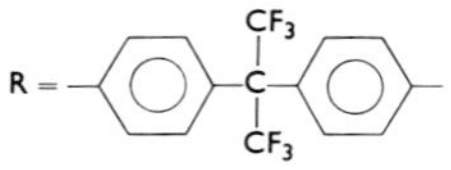 R = ⬡—C(CF₃)₂—⬡ <br><br> **2,2-Bis(4-aminophenyl))hexafluoropropane** | A,1; <br> A,2 | — <br> — | — <br> — | — <br> — | 286 <br><br> *564* |
| 34 | 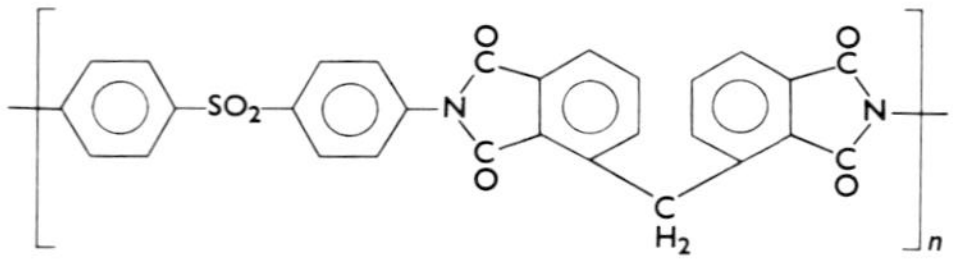 (polyimide structure with SO₂) <br><br> * * * <br><br> **2,2′,3,3′-Diphenylmethanetetracarboxylic acid dianhydride and 4,4′-diaminodiphenyl sulfone** | A,1 | — | — | — | *283, 284* |

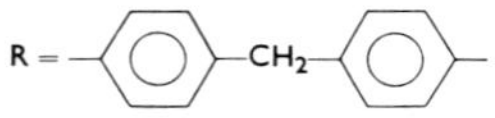

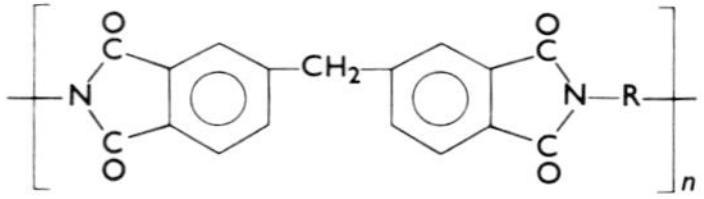

2,2-Bis(3,4-dicarboxyphenyl)propane dianhydride or derivative and

| No. | R | | | | Ref. |
|---|---|---|---|---|---|
| 35 | R = $-(CH_2)_4-$<br>1,4-Diaminobutane | A,1 | — | — | 301 |
| 36 | R = $-(CH_2)_6-$<br>1,6-Diaminohexane | A,1 | — | — | 301 |
| 37 | R =<br>1,4-Diaminobenzene | A,1 | — | Crystalline (X-ray), coalescible powder was prepared | 286 |
| 38 | R =<br>1,3-Diaminobenzene | A,1 | — | A molded film had following properties: density = 1.29 g/cm$^3$; tensile strength = 1500 kg/cm$^2$; elongation at break = 5.4%. A 500 hr treatment in boiling water resulted in 12% loss of strength. An amorphous (X-ray) coalescible powder was prepared | 243 286, 477 |
| | Copolymers containing the above dianhydride | A,1; A,2 | — | Some copolymers are compression-moldable | 243, 251, 484, 485 |
| 39 | R = $C(CF_3)_2$<br>2,2-Bis(4-aminophenyl)hexafluoropropane | A,1; A,2 | — | — | 564 |

TABLE I.5—*continued*

Miscellaneous Polyimides

| No. | Structure | Method | Solubility | Molecular weight | Remarks and property data | References |
|---|---|---|---|---|---|---|
| 40 | 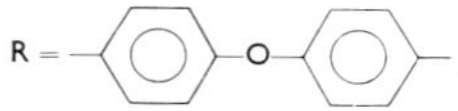 R = <br> 4,4'-Diaminodiphenyl ether | A,1 | — | — | Coalescible powder was prepared | 209, 237, 283, 284 |
| 41 | R = <br> α,ω-Bis(4-amino-2-chlorophenyl))dibutoxytrisiloxane <br><br> * * * * <br><br> <br> 2,2-Bis(3,4-dicarboxyphenyl)hexafluoropropane dianhydride or derivative and | A,1 | — | — | — | 300 |
| 42 | 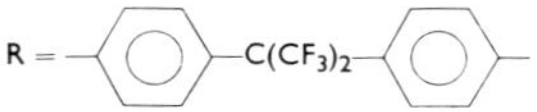 R = <br> 2,2-Bis(4-aminophenyl)hexafluoropropane | A,2 | CHCl₃, benzene, dioxane, acetone | — | $T_g = 340°C$; $T_m = >400°C$; useful in coatings. Thermal degradation in air begins at ~475°C | 564 |
|  | Copolymers of 42 <br><br> * * * | A,1 | — | — | Good adhesives | 27 |

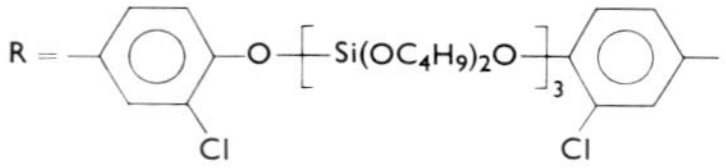

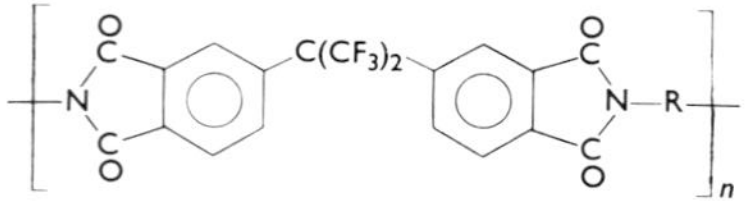

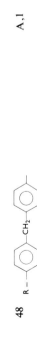

Bis(3,4-dicarboxyphenyl)ether dianhydride or derivative and

| No. | R | Method | Solubility | | Notes | References |
|---|---|---|---|---|---|---|
| 43 | R = $+(CH_2)_6+$  1,6-Diaminohexane | A,1 | — | — | — | 301 |
| 44 | R = 1,4-Diaminobenzene | A,1 | — | — | Thermal imidation was studied. A crystalline (X-ray) coalescible powder was prepared | 224, 286 |
| 45 | R = 1,3-Diaminobenzene | A,1 | — | — | Thermal imidation was studied. A crystalline (X-ray) coalescible powder was prepared | 224, 286 |
| 46 | R = 4,4'-Diaminobiphenyl | A,1 | — | — | Thermal imidation was studied | 224, 363 |
| 47 | R = $H_5C_6HN$ —— $NHC_6H_5$  3,3'-Diamino-4,4'-dianilinobiphenyl | A,1 | Sparingly soluble in boiling DMF and DMA and in strongly acidic solvents | — | IR supports structure. Begins to decompose in air at $\geqslant 400°C$ | 370 |
| 48 | R = $CH_2$  4,4'-Diaminodiphenylmethane | A,1 | — | — | — | 363 |

TABLE I.5—*continued*

Miscellaneous Polyimides

| No. | Structure | Method | Solubility | Molecular weight | Remarks and property data | References |
|-----|-----------|--------|------------|------------------|---------------------------|------------|
| 49 | R = —C(CF$_3$)$_2$— <br>2,2-Bis(4-aminophenyl)hexafluoropropane | A,1; A,2 | — | — | — | 564 |
| 50 | R = —O— <br>4,4'-Diaminodiphenyl ether | A,1 | H$_2$SO$_4$ | [$\eta$] 0.98 | Amorphous (X-ray); fusible. $T_g$ (via measurement of thermal expansion coefficient) = 250°C. At temperatures above 250°C cross-linking by transimidation takes place. Thermal imidation was studied. Mechanical properties: tensile strength = 16,500 psi; tensile modulus = 746,000 psi; elongation = 22% Other properties: stick temperature ~305°C; dielectric constant = 3.81; dissipation factor = 0.0012 | 209, 224, 237, 363, 375, 603 |
| | | A,1 | Fomal (10 parts by weight of phenol + 7 parts by weight of 2,4,6-trichlorophenol) | $\eta_{inh}$ 0.57 | Use was made of monofunctional anhydrides and amines as terminators to control polymer molecular weight. This was claimed to yield injection-moldable compositions, soluble in "fomal". The polymer $\eta_{inh}$ 0.57 was prepared with phthalic anhydride terminator; melt index (350°C, 11 lb) = 0.7 g/10 min. The molded product had flexural modulus of 350–450,000 psi (room temp.) and 200–250,000 psi (225°C). See text | 524 |

| No. | R = | A,1 | | | Comments | Refs. |
|---|---|---|---|---|---|---|
| 51 | Bis(4-aminophenyl)diethylsilane | A,1 | — | — | Coalescible powder was prepared | 283, 284 |
| 52 | 4,4'-diaminodiphenyl disulfide | A,1 | — | — | — | 5 |
| 53 | Bis(4-aminophenyl) ether of hydroquinone | A,1 | $H_2SO_4$ | $[\eta]$ 0.58 | IR supports structure. Crystallizes on annealing at 250°C | 5, 375, 380, 609 |
| 54 | 1,3-Bis(p-aminophenoxy)benzene | A,1 | — | — | Coalescible powder was prepared | 283, 284 |
| 55 | Bis(4-aminophenyl) ether of dithiohydroquinone | A,1 | — | — | IR supports structure | 5 |

* * *

Bis(3,4-dicarboxyphenyl) sulfone dianhydride or derivative and

| No. | R = | A,1 | | | Comments | Refs. |
|---|---|---|---|---|---|---|
| 56 | 1,4-Diaminobenzene | A,1 | — | — | Coalescible powder was prepared | 286 |

TABLE I.5—*continued*

Miscellaneous Polyimides

| No. | Structure | Method | Solubility | Molecular weight | Remarks and property data | References |
|---|---|---|---|---|---|---|
| 57 |  R = [benzene ring]<br>1,3-Diaminobenzene | A,1 | — | — | Coalescible powder was prepared | 286 |
| 58 | 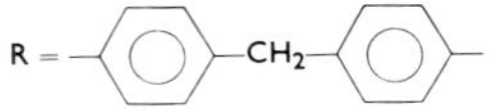 R = [phenyl–CH₂–phenyl]<br>4,4'-Diaminodiphenylmethane | A,2 | — | — | — | 719 |
| 59 | 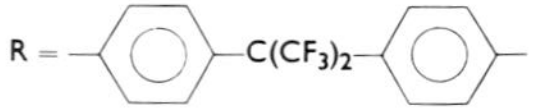 R = [phenyl–C(CF₃)₂–phenyl]<br>2,2-Bis(4-aminophenyl)hexafluoropropane | A,1;<br>A,2 | — | — | — | 564 |
| 60 | 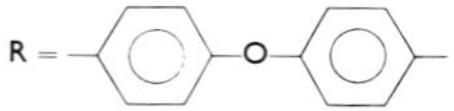 R = [phenyl–O–phenyl]<br>4,4'-Diaminodiphenyl ether | A,2 | — | — | — | 719 |
| 61 | 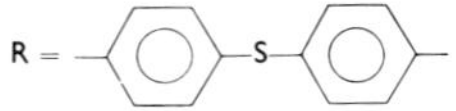 R = [phenyl–S–phenyl]<br>4,4'-Diaminodiphenyl sulfide | A,1 | — | — | Coalescible powder was prepared | 283, 284 |
| 62 | 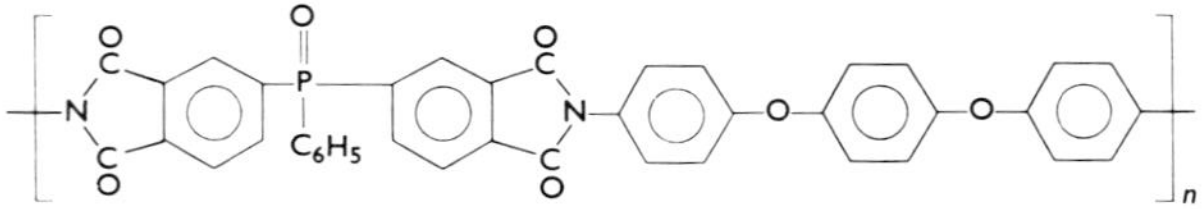 * * *<br><br>Bis(3,4-dicarboxyphenyl)phenylphosphine oxide dianhydride and bis(4-aminophenyl) ether of hydroquinone | A,1 | — | — | Structure supported by IR. Polymer is stable up to 380°C and is nonflammable | 8 |

| | Structure | Method | Properties | Ref. |
|---|---|---|---|---|
| 63 | Stilbene-3,3',4,4'-tetracarboxylic acid dianhydride and α,ω bis(4-aminophenyl)methylvinyltrisiloxane | A,1 | — | 300 |
| | Azobenzenetetracarboxylic acid dianhydride or derivative and | — | — | |
| 64 | R = $-(CH_2)_6-$ 1,6-Diaminohexane | A,1 | — | 64, 483 |
| 65 | R = 4,4'-Diaminodiphenylmethane | A,1 | Gives elastic films | 64, 483 |
| 66 | R = 4,4'-Diaminodicyclohexylmethane | A,1 | A wire coating had a heat distortion temperature >330°C | 64, 483 |
| 67 | R = 4,4'-Diaminodiphenyl ether | A,1 | Red, elastic film, insoluble in common organic solvents. A wire coated with this polymer was resistant to a thermal shock of 260°C; the coatings heat distortion temperature was >330°C and it displayed good shape stability at 200°C for 2 weeks | 64, 483 |

* * * *

TABLE I.5—*continued*
Miscellaneous Polyimides

| No. | Structure | Method | Solubility | Molecular weight | Remarks and property data | References |
|---|---|---|---|---|---|---|
| 68 | Azobenzene-3,3′,4,4′-tetracarboxylic acid d′anhydride and 1,3-diaminobenzene | A,1 | — | — | Useful in composites | 259 |
| 69 | Azoxybenzene-3,3′,4,4′-tetracarboxylic acid dianhydride and 1,3-diaminobenzene | A,1 | — | — | Flexible, transparent film possessing good thermal stability. Tensile strength ≃ 1090 kg/cm². Reduced by 18% only after 300 hr at 250°C. Polymer is useful in films, insulating coatings and glass fiber laminates | 258 |
| | Terephthaloyl diphthalic anhydride or derivative and | | | | | |
| 70 | R = 4,4′-Diaminodiphenylmethane | A,1 | — | — | — | 511 |
| 71 | R = 4,4′-Diaminodiphenyl ether | A,1 | — | — | Amorphous (X-ray). Can be hydrolyzed to the polyamic acid. Properties: volume resistivity ≃ $10^{16}$–$10^{17}$ ohm-cm; dielectric constant ≃ 6.1–7.8; dielectric loss ≃ 0.0045–0.0113; breakdown voltage ≃ 14–18 kV/0.1mm; thermal decomposition starts at >350°C | 511 |

* * *

| Structure | | | | Ref. |
|---|---|---|---|---|
| 2,3,6,7-Naphthalenetetracarboxylic acid dianhydride or derivative and | A,1 | — | — | 300 |
| 72  R = $-CH_2CHCH_2O[Si(CH_3)(C_6H_5)O]_3CH_2CHCH_2-$ with $CH_3$ substituents; α,ω-Bis(β-methyl-γ-aminopropyl)methyl phenyl trisiloxane | A,1 | — | — | |
| 73  R = $-OSi(C_2H_5)_2-O-$ ; Bis(4-aminophenoxy)diethylsilane | A,1 | — | Coalescible powder was prepared | 283, 284 |

* * *

| Structure | | | | Ref. |
|---|---|---|---|---|
| Naphthalene-1,4,5,8-tetracarboxylic acid dianhydride or derivative and | | | | |
| 74  R = $-(CH_2)_6-$ ; 1,6-Diaminohexane | A,1 | — | Contains a 6-membered imide ring. Structure supported by IR. TGA data given | 510 |
| 75  R = ; 1,4-Diaminobenzene | A,1 | — | Contains a 6-membered imide ring. Structure supported by IR. TGA data given | 510, 663 |

TABLE I.5— *continued*
Miscellaneous Polyimides

| No. | Structure | Method | Solubility | Molecular weight | Remarks and property data | References |
|---|---|---|---|---|---|---|
| 76 | R = <br> 1,3-Diaminobenzene | A,1 | conc. $H_2SO_4$, conc. $HNO_3$ | — | Contains a 6-membered imide ring. Structure supported by IR. TGA data given | 510, 540 |
| 77 | R = (CH₃) <br> 1,3-Diamino-4-methylbenzene | A,1 | m-cresol | — | Contains a 6-membered imide ring | 540 |
| 78 | R = <br> 4,4'-Diaminobiphenyl | A,1 | conc. $H_2SO_4$, conc. $HNO_3$ | — | Contains a 6-membered imide ring. Structure supported by IR. TGA data given | 510, 540, 663 |
| 79 | R = (CH₃, CH₃) <br> 4,4'-Diamino-3,3'-dimethylbiphenyl | A,1 | — | — | Contains a 6-membered imide ring. Structure supported by IR. TGA data given | 510 |
| 80 | R = (CH₂) <br> 4,4'-Diaminodiphenylmethane | A,1 | conc. $H_2SO_4$, conc. $HNO_3$ | — | Contains a 6-membered imide ring. Structure supported by IR. TGA data given | 510, 540 |
| 81 | R = (O) <br> 4,4'-Diaminodiphenyl ether | A,1 | conc. $H_2SO_4$, conc. $HNO_3$ | — | Contains a 6-membered imide ring. Structure supported by IR. TGA data given | 510, 540, 663 |

| No. | | Method | | | Remarks | Ref. |
|---|---|---|---|---|---|---|
| 82 | R = SO₂ structure; 4,4'-Diaminodiphenyl sulfone | A,1 | — | — | Contains a 6-membered imide ring. Structure supported by IR. TGA data given | 510, 663 |
| 83 | R = structure; N,N'-Diaminonaphthalene-1,4,5,8-tetracarboxydiimide | A,1 | — | — | Brittle, black, glassy solid of inferior thermal stability. Experimental evidence indicates that oxidative degradation proceeds via cleavage of the N—N bond | 199 |

* * *

| No. | | Method | | | Remarks | Ref. |
|---|---|---|---|---|---|---|
| 84 | Phenanthrene-10-carboxy-1,2,7,8-tetracarboxylic acid dianhydride and α,ω-Bis(7-aminoanthranyl-2)methyl benzyl polysiloxane | A,1 | — | — | — | 300 |
| | Various dianhydrides and | | | | | |
| 85 | R = structure; 2,5-Diaminothiadiazole | A,1 | — | — | Useful as fibers, films, and coatings | 88 |

TABLE I.5— *continued*
Miscellaneous Polyimides

| No. | Structure | Method | Solubility | Molecular weight | Remarks and property data | References |
|---|---|---|---|---|---|---|
| 86 | | A,1; A,2 | — | — | — | 499 |

Bis(R-substituted-4-aminophenyl)ethers of diphenols and bisphenols
Ar = variety of aryl groups; R = H, alkyl, alkoxy

| No. | Structure | Method | Solubility | Molecular weight | Remarks and property data | References |
|---|---|---|---|---|---|---|
| 87 | | A,1 | DMF, DMSO | — | Good thermal and oxidative stabilities claimed. Polymers are useful in films, fibers, coatings, and as molding materials | 92 |

Other similar structures. $\alpha,\omega$-Amino-terminated aromatic polyethers

TABLE I.6

Poly(amide-imides) from Diamino Amides

Pyromellitic anhydride or derivative and

| No. | Structure | Method | Remarks and property data | References |
|---|---|---|---|---|
| 1 | R = —HNCONH—<br>Carbohydrazide | A,7 | IR and thermal stability data | 658 |
| 2 | R = —HNCO$+$CH$_2$$+_4$CONH—<br>Adipic acid dihydrazide | A,7;<br>A,8 | Soluble in DMA prior to precipitation. Once precipitated is only swollen by DMA or $m$-cresol. Dissolves with degradation in conc. H$_2$SO$_4$. Slightly crystalline (X-ray). Structure is supported by spectral data (see text). TGA (air, 3°C/min): stable up to 360°C. $\eta_{inh}$ 0.38 | 278, 334,<br>367, 657 |
| 3 | Copolymers of 2 | A,7 | — | 367 |
| 3 | R = —HNCO$+$CH$_2$$+_7$CONH—<br>Azelaic acid dihydrazide | A,7 | — | 657 |
| 4 | R = —HNCO$+$CH$_2$$+_8$CONH—<br>Sebacic acid dihydrazide | A,7 | — | 367, 657 |
| 5 | R = ⟨ ⟩—CONH—<br>4-Aminobenzoylhydrazine | A,7 | IR indicates "imide" structure. TGA (argon or air, 1°C/min): rapid decomposition observed at 300°–350°C | 257, 574,<br>575 |

TABLE I.6—*continued*
Poly(amide-imides) from Diamino Amides

| No. | Structure | Method | Remarks and property data | References |
|---|---|---|---|---|
| 6 | R = —⟨⟩—CONH—<br>3-Aminobenzoylhydrazine | A,7 | — | 575 |
| 7 | R = —HNOC—⟨⟩—CONH—<br>Terephthalic acid dihydrazide | A,7;<br>A,8 | Soluble in DMA prior to precipitation. If once precipitated, only swollen by DMA or *m*-cresol. Dissolves with degradation in conc. $H_2SO_4$. "Imide" structure is supported spectrally. $\eta_{inh}$ 0.09 | 334, 367 |
| 8 | R = —HNOC—⟨⟩—CONH—<br>Isophthalic acid dihydrazide | A,7;<br>A,8 | Same solubility properties as 7. Slightly crystalline (X-ray). Brittle film. "Imide" structure supported spectrally. TGA (air, 3°C/min): 5% wt. loss observed up to 380°C. $\eta_{inh}$ 0.47 | 334, 367,<br>655 |
| 9 | R = —HNOC—⟨N⟩—CONH—<br>Pyridinedicarboxylic acid dihydrazides | A,7 | Thermal stability studied | 656 |
| 10 | R = —⟨⟩—CONH—⟨⟩—<br>4,4'-Diaminobenzanilide | A,7 | Brittle film. Wt. loss (400 hr, air, 325°C): 12.1% | 121, 122,<br>273 |

| | | | |
|---|---|---|---|
| **11** R= [structure] —CONH—[ring]<br>4,3'-Diaminobenzanilide | A,7 | Westinghouse designation: AI-8. Amorphous (X-ray). Develops crystallinity (X-ray) and simultaneously embrittles upon annealing at 400°C (vac., 4 min). $T_g$ (from modulus temperature measurements) = 360°C. It increases to 390°C upon annealing at 250°C (vac., 20 hr) due possibly to cross-linking or a more complete imidation. $T_g$ (from measurements of electrical dissipation factor) = 205°C. Mechanical properties: tensile strength = 11,000 psi; tensile modulus = 270,000 psi; elongation at break = 15%; wt. loss (400 hr, air, 325°C) = 11.9% | 121, 122, 182, 234, 238, 273, 274 |
| **12** R= [structure] —CONH—[ring]<br>3,4'-Diaminobenzanilide | A,7 | Flexible film. Excellent electrical properties and thermal stability claimed. Wt. loss (400 hr, air, 325°C) = 9.8%. The extrapolated thermal life of a wire enamel (based on thermal stability tests) was 100,000 hr at 210°C! Useful in laminates and electrical applications | 121, 122, 238, 273, 274, 581, 582 |
| Copolymers of 11 and 12 | A,7 | — | 238 |
| **13** R= [structure] —CONH—[ring]<br>3,3'-Diaminobenzanilide | A,7 | Flexible film. Weight loss (400 hr, air, 325°C) = 11.2% | 121, 122, 273 |
| **14** R= [structure]<br>3,5-Diaminobenzanilide | A,7 | Brittle | 121, 122, 273 |

TABLE I.6—*continued*
Poly(amide-imides) from Diamino Amides

| No. | Structure | Method | Remarks and property data | References |
|---|---|---|---|---|
| 15 | 3,5,2'-Triaminobenzanilide | A,7 | Under controlled conditions yields a benzimidazole-contain-ing polyimide. See text | 275 |
| 16 | R = —NHCONH— Diaminodiphenylurea | A,7 | — | 264 |
| 17 | R = —COO—...—NHCO— N,O-Bis(3-aminobenzoyl)-p-aminophenol | A,7 | Flexible film. Weight loss (400 hr, air, 325°C) = 27% | 121, 122, 273 |
| 18 | R = —NHCO—...—CONH— Isophthal(4-aminoanilide) | A,7 | Flexible film. Weight loss (400 hr, air, 325°C) = 20.4% | 121, 122, 273 |
| 19 | R = —CONH—...—NHCO— N,N'-m-Phenylene-bis(4-aminobenzamide) | A,7 | Flexible film. Weight loss (400 hr, air, 325°C) = 15.6% | 121, 122, 273 |

| No. | R = / Name | | Properties | Refs. |
|---|---|---|---|---|
| 20 | Isophthal(3-aminoanilide) | A,7 | Flexible film. Weight loss (400 hr, air, 325°C) = 13.2%. Useful in laminates | 121, 122, 238, 273, 274, 652, 653 |
| 21 | N,N'-m-Phenylene bis(3-aminobenzamide) | A,7 | Flexible film. Weight loss (400 hr, air, 325°C) = 20.3% | 121, 122, 238, 273, 274 |
| 22 | N,N'-Bis(3-aminobenzoyl)-2,4-diaminodiphenyl ether | A,7 | Flexible film. Weight loss (400 hr, air, 325°C) = 44.3% | 121, 122, 273 |
| 23 | N,N'-Bis(3-aminobenzoyl)-4,4'-diaminodiphenyl ether | A,7 | — | 238, 274 |
| 24 | R = residue of diamino-terminated polyamide from 4,4'-diaminodiphenylmethane and azelaic acid HOOC—(CH₂)₇—COOH | A,7 | Excellent thermal stability and mechanical and electrical properties claimed | 263 |
| 25 | R = residue of diamino-terminated polyamide from 1,3-diaminobenzene and terephthalic acid | A,7 | — | 238, 274 |

TABLE I.6—*continued*

Poly(amide-imides) from Diamino Amides

| No. | Structure | Method | Remarks and property data | References |
|---|---|---|---|---|
| 26 | R = residue of diamino-terminated polyamide from 1,3-diaminobenzene and terephthalic and isophthalic acids | A,7 | Excellent thermal stability and electrical properties claimed. Useful in laminates and electrical applications | 121, 122, 238, 273, 274, 582, 652, 653 |
|  | Copolymers incorporating the various amide structures above | — | Useful as wire coatings, laminating and molding resins | 129 |
| 27 | R = residue of diamino-terminated polyamide from 1,3-diaminobenzene and isophthalic acid | A,7 | Flexible, clear film | 121, 122, 238, 273, 274 |
|  | Copolymers of 27 | A,7 | — | 238 |
| 28 | R = residue of diamino-terminated polyamide from tolylenediamine and isophthalic acid | A,7 | Good thermal stability claimed | 346 |
| 29 | R = residue of diamino-terminated polyamide from 4,4'-diaminodiphenylmethane and isophthalic acid | A,7 | — | 238, 274 |
| 30 | R = residue of diamino-terminated polyamide from 4,4'-diaminodiphenyl ether and isophthalic acid | A,7 | Increase in amide content results in a decrease of the thermal and oxidative stability of the polymer | 518 |
| 31 | R = residue of diamino-terminated polyamide from p- or m-aminobenzoic acid hydrazide and various aromatic dibasic acids | A,7 | — | 393 |

* * *

3,3′,4,4′-Benzophenonetetracarboxylic acid dianhydride or derivative and

| No. | R | Conditions | Remarks | Ref. |
|---|---|---|---|---|
| 32 | R = —HNCO(CH₂)₄CONH— Adipic acid dihydrazide | A,7 | — | 444 |
| 33 | R = —HNCO(CH₂)₇CONH— Azelaic acid dihydrazide | A,7 | — | 444 |
| 34 | R = (benzene ring)CONH— 4-Aminobenzoic acid hydrazide | A,7 | — | 575 |
| 35 | R = (benzene ring)CONH— 3-Aminobenzoic acid hydrazide | A,7 | — | 575 |
| 36 | R = —HNCO(benzene ring)CONH— Terephthalic acid dihydrazide | A,7 | Apparently contains five-membered "imide" structure. Useful in coatings and electrical applications | 154, 444 |
| 37 | R = —HNOC(benzene ring)CONH— Isophthalic acid dihydrazide | A,7 | — | 444 |
| 38 | R = (benzene ring)CONH(benzene ring)— 3,4′-Diaminobenzanilide | A,7 | Stable at 300°C for "more" than 40 hr. Useful in wire coatings | 238, 274 |

TABLE I.6—*continued*

Poly(amide-imides) from Diamino Amides

| No. | Structure | Method | Remarks and property data | References |
|---|---|---|---|---|
| 39 | <br>R = <br>Isophthal(3-aminoanilide) | A,7 | Tough, flexible film | 238, 274 |
| | <br>Azobenzene-3,3′,4,4′-tetracarboxylic acid dianhydride or derivative and | | | |
| 40 | <br>p-Aminophenyl-N-methyl-N-(p-aminophenyl) carbamate | A,7 | $T_m = >400°C$ | 64, 483 |
| 41 | <br>N,N′-Dimethyl-N,N′-bis(p-aminophenyl)urea | A,7 | $T_m = >400°C$ | 64, 483 |
| 42 | <br>1,4,5,8-Naphthalenetetracarboxylic acid dianhydride and 3,4′-diaminobenzanilide | A,7 | Structure supported by IR | 238, 274 |
| 43 | Various poly(amide-imides) | — | — | 71 |

TABLE I.7

Poly(amide-imides) from Tricarboxylic Acid Derivatives

| No. | Structure | Method | Solubility | Molecular weight | $T_m$ (°C) | Remarks and property data | References |
|---|---|---|---|---|---|---|---|
| 1 | <br>1,2,3-Propanetricarboxylic acid and 1,10-diaminodecane | A,7 | — | — | 94 | Hard, tough, white solid | 271 |
| | <br>Maleopimaric acid or derivative and | | | | | | |
| 2 | $R = (CH_2)_6$<br>1,6-Diaminohexane | A,7 | $CHCl_3$, N-methylpyrrolidone | Mol. wt. = 3030 (VPO) | — | Good films obtained. IR supports structure | 586 |
| 3 | <br>1,4-Diaminobenzene | A,7 | — | — | — | — | 586 |
| 4 | <br>4,4'-Diaminodiphenylmethane | A,7 | DMF, $CHCl_3$, DMSO | $\eta_{inh}$ 0.1 Mol.wt. = 6070 (VPO) | 392 (capillary) | IR supports structure. TGA (air, 20°C/min): 11% wt. loss observed at 425°C. TGA ($N_2$, 20°C/min): 6% wt. loss observed at 400°C | 586 |

TABLE I.7—continued

Poly(amide-imides) from Tricarboxylic Acid Derivatives

| No. | Structure | Method | Solubility | Molecular weight | $T_m$ (°C) | Remarks and property data | References |
|---|---|---|---|---|---|---|---|
| 5 | R= (structure) 4,4'-Diaminodiphenyl ether | A,7 | — | — | — | — | 586 |
| | * * * 4-Carboxybenzylsuccinic anhydride or derivative and | | | | | | |
| 6 | R= (structure) 1,4-Diaminobenzene | A,7 | DMF, DMSO, N-methylpyrrolidone | — | — | Structure supported by IR | 365 |
| 7 | R= (structure) 1,3-Diaminobenzene | A,7 | DMF, DMSO, N-methylpyrrolidone | — | — | Structure supported by IR | 365 |
| 8 | R= (structure) 4,4'-Diaminobiphenyl | A,7 | DMF, DMSO, N-methylpyrrolidone | — | — | Structure supported by IR | 365 |
| 9 | R= (structure) 4,4'-Diaminodiphenylmethane | A,7 | DMF, DMSO, N-methylpyrrolidone | — | — | Structure supported by IR | 365 |

| No. | R = | | | Notes | Ref. |
|---|---|---|---|---|---|
| 10 | 4,4'-Diaminodiphenyl ether | A,7 | DMF, DMSO, N-methylpyrrolidone | — | Structure supported by IR | 365 |
| 11 | 4,4'-Diaminodiphenyl sulfone | A,7 | DMF, DMSO, N-methylpyrrolidone | — | Structure supported by IR | 365 |

* * *

3-Carboxybenzyl succinic anhydride or derivative and

| No. | R = | | | Notes | Ref. |
|---|---|---|---|---|---|
| 12 | 1,4-Diaminobenzene | A,7 | DMF, DMSO, N-methylpyrrolidone | — | TGA ($N_2$): decomposes rapidly at 450°C. Structure supported by IR | 365 |
| 13 | 1,3-Diaminobenzene | A,7 | DMF, DMSO, N-methylpyrrolidone | — | TGA ($N_2$): decomposes rapidly at 450°C. Structure supported by IR | 365 |
| 14 | 4,4'-Diaminobiphenyl | A,7 | DMF, DMSO, N-methylpyrrolidone | — | — | 365 |
| 15 | 4,4'-Diaminodiphenylmethane | A,7 | DMF, DMSO, N-methylpyrrolidone | — | — | 365 |

TABLE I.7—continued

Poly(amide-imides) from Tricarboxylic Acid Derivatives

| No. | Structure | Method | Solubility | Molecular weight | $T_m$ (°C) | Remarks and property data | References |
|---|---|---|---|---|---|---|---|
| 16 | 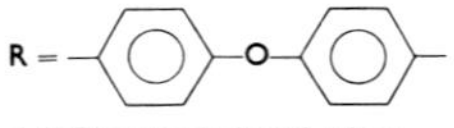 R = 4,4'-Diaminodiphenyl ether | A,7 | DMF, DMSO, N-methylpyrrolidone | — | — | — | 365 |
| 17 | 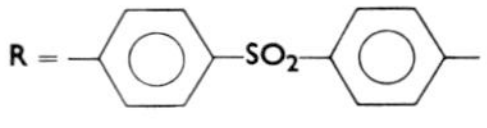 R = 4,4'-Diaminodiphenyl sulfone | A,7 | DMF, DMSO, N-methylpyrrolidone | — | — | — | 365 |

* * *

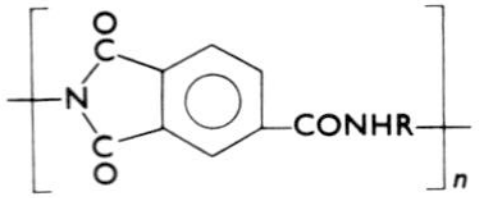

Trimellitic anhydride or derivative and

| No. | Structure | Method | Solubility | Molecular weight | $T_m$ (°C) | Remarks and property data | References |
|---|---|---|---|---|---|---|---|
| 18 | R = —CH₂— Methylene diisocyanate | A,9 | Solvesso 500 | — | — | Useful in laminates and electrical applications | 555 |
| 19 | 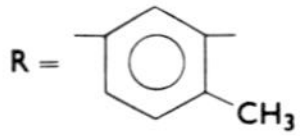 R = Toluene diisocyanate | A,9 | — | — | — | Useful in coatings | 252 |
| 20 | 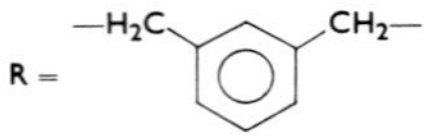 R = —H₂C— ... —CH₂— 1,3-Bis(isocyanatomethyl)benzene | A,9 | — | — | — | — | 252 |

| | | | | | | |
|---|---|---|---|---|---|---|
| 21 | R = [4,4'-Diaminodiphenylmethane structure, CH₂ linking two aminophenyl rings] **4,4'-Diaminodiphenylmethane** | A,7; A,9 | DMF, DMA, N-methylpyrrolidone | $[\eta]$ 0.92 | 320 | Tensile strength = 11,300 psi. Has excellent hardness and abrasion resistance. Useful in coatings, varnishes, enamels, and electrical insulation | 114, 140, 252, 256, 412, 624 |
| | | A,7 | m-Cresol | — | — | A carboxyl-terminated polymer. Can be condensed with polyesters, ex. poly(ethylene glycol/glycerol terephthalate) to give resins useful in coatings, lacquers, and as molding materials | 130 |
| | Copolymers of 21 | A,7 | — | — | — | — | 146 |
| 22 | R = [4,4'-Diaminodiphenyl ether structure, O linking two aminophenyl rings] **4,4'-Diaminodiphenyl ether** | A,7; A,9 | $H_2SO_4$ | $\eta_{inh}$ 0.55 | — | Structure supported by IR. Decomposition temperature = 480°C. Tenacity (at 22°C) = 10.7 kg/cm²; elongation (at 22°C) = 7%. Useful in coatings | 50, 252, 461, 578, 624 |
| | Copolymers of trimellitic anhydride containing a siloxane diamine | A,7 | — | — | — | Improved resistance to corona claimed | 331 |
| 23 | R = residue of diamino-terminated polyamide from 1,3-diaminobenzene and terephthalic and isophthalic acids. | A,7 | — | — | — | Tough, flexible films | 136, 242 |

* * *

TABLE I.7—*continued*

Poly(amide-imides) from Tricarboxylic Acid Derivatives

| No. | Structure | Method | Solubility | Molecular weight | $T_m$ (°C) | Remarks and property data | References |
|-----|-----------|--------|-----------|------------------|------------|---------------------------|------------|
| | 3,4,4'-Diphenyl ether tricarboxylic acid or derivative and | | | | | | |
| 24 | R = 4,4'-Diaminobiphenyl | A,7 | — | — | — | Good thermal stability up to 300°C (TGA) | 362 |
| 25 | R = 4,4'-Diaminodiphenylmethane | A,7 | — | — | — | TGA (air, 2°C/min); practically no weight loss observed up to 400°C | 362 |
| 26 | R = 4,4'-Diamino-3,3'-dimethyldiphenylmethane | A,7 | — | — | — | TGA (air, 2°C/min); weight loss at 300°C = 1.5% | 362 |

TABLE I.8

Poly(amide-imides) from Amide Anhydrides

| No. | Structure | Method | Solubility | Molecular weight | $T_m$ (°C) | Remarks and property data | References |
|---|---|---|---|---|---|---|---|
| 1 | Hydrochloride of the amide-amine prepared from maleopimaric acid monochloride and 1,6-diaminohexane | A,7 | CHCl$_3$ | — | — | — | 586 |
| | * * * | | | | | | |
| | Dianhydride from reaction of maleopimaric acid chloride with 1,6-diaminohexane and | | | | | | |
| 2 | R = —(CH$_2$)$_6$— 1,6-Diaminohexane | A,7 | CHCl$_3$ | — | — | — | 586 |
| 3 | Various diamines * * * | A,7 | — | — | — | — | 586 |

TABLE I.8—*continued*
Poly(amide-imides) from Amide Anhydrides

| No. | Structure | Method | Solubility | Molecular weight | $T_m$ (°C) | Remarks and property data | References |
|---|---|---|---|---|---|---|---|
| 4 | Dianhydride from reaction of trimellitic acid or derivative with 1,2-diaminoethane and 1,4-diaminobenzene | A,7 | — | — | >400 | — | 423 |
| 5 | Dianhydride obtained by the reaction of trimellitic acid or derivative with 1,4-diaminobenzene and 4,4'-diaminodiphenyl ether | A,7 | — | — | — | Good resistance to aqueous acid and alkali claimed. Useful in laminates and wire coatings | 155 |
| | * * * Dianhydride obtained by the reaction of trimellitic acid or derivative with N,N'-diphenyl-1,4-diaminobenzene and | | | | | | |
| 6 | $R = \text{(CH}_2\text{)}_6$ 1,6-Diaminohexane | A,7 | — | — | — | — | 282 |
| 7 | $R = \text{C}_6\text{H}_4$ 1,3-Diaminobenzene | A,7 | N-Methylpyrrolidone | — | ~300 | Hard flexible film. Useful in laminates and high-temperature coatings | 156, 329, 332 |
| | Copolymers of the above dianhydride | — | — | — | — | — | 156, 329, 332 |

TABLE I.9

Poly(ester-imides) from Ester Anhydrides

| No. | Structure | Method | $T_m$ (°C) | Remarks and property data | References |
|-----|-----------|--------|------------|---------------------------|------------|
| | 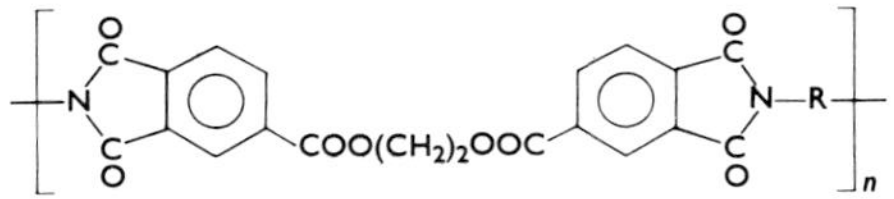 \nBis(trimellitic anhydride ester) of ethylene glycol and | | | | |
| 1 |  R = \n1,4-Diaminobenzene | A,10 | — | — | 423 |
| 2 | R = mixture of\n\nand\n\nMixture 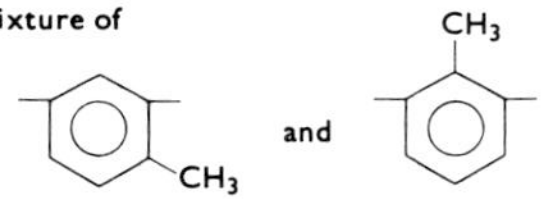 of 2,4- and 2,6-toluene diisocyanates | A,11 | — | Soluble in cresol. Ester-imide struc-ture supported by IR. Useful as a wire coating when used in admixture with a terephthalate polyester | 453 |
| 3 | 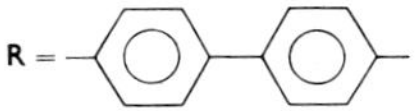 R = \n4,4′-Diisocyanatobiphenyl | A,11 | — | — | 453 |
| 4 | 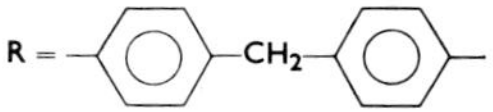 R = —CH₂—\n4,4′-Diaminodiphenylmethane | A,10; A,11 | — | — | 282, 453 |

* * *

TABLE I.9—*continued*

Poly(ester-imides) from Ester Anhydrides

| No. | Structure | Method | $T_m$ (°C) | Remarks and property data | References |
|---|---|---|---|---|---|
| |  Bis(trimellitic anhydride ester) of hydroquinone and | | | | |
| 5 | R = $+CH_2\frac{}{2}$  1,2-Diaminoethane | A,10 | — | — | 423 |
| 6 | R = —HNCOCONH—  Oxalyl dihydrazide | A,10 | >400 | Tough flexible film. Polymer has a 5-membered imide structure | 424 |
| 7 | R = $+CH_2\frac{}{6}$  1,6-Diaminohexane | A,10 | >360 | — | 423 |
| 8 | R = —HNCO$+CH_2\frac{}{8}$CONH—  Sebacic acid dihydrazide | A,10 | >350 | Tough, flexible film. Structure: See Remarks under entry No. 6 above | 424 |
| 9 | R = 1,4-Diaminobenzene | A,10 | >500 | Tensile strength = 16,800 psi (room temp.), 11,300 psi (200°C); elongation = 3% (room temp.); 3% (200°C); tensile modulus = 1,310,000 psi (room temp.); 549,000 psi (200°C) | 422, 423 |

| | R = | | | | |
|---|---|---|---|---|---|
| 10 | $H_3C$ $CH_3$ / $CH_3$ $CH_3$ <br> Durenediamine | A,10 | >500 | Brittle films | 422, 423 |
| 11 | 1,3-Diaminobenzene | A,10 | >500 | Tensile strength = 15,800 psi (room temp.), 6500 psi (200°C); elongation = 9% (room temp.), 15% (200°C); tensile modulus = 454,000 psi (room temp.), 255,000 psi (200°C); wt. loss (air, 100 hr, 300°C) 3.56%. Excellent wire insulator | 422, 423 |
| 12 | COOH <br> 3,5-Diaminobenzoic acid | A,10 | >500 | Flexible film | 422, 423 |
| | Various ester-imide copolymers | A,10 | — | — | 423 |
| 13 | —HNOC— —CONH— <br> Isophthaloyl dihydrazide | A,10 | >500 | TGA (air): decomposition of polymer starts at ~350°C. TGA (nitrogen): wt. loss (800°C, 3 hr) = 75% | 424, 426 |
| 14 | 4,4'-Diaminobiphenyl | A,10 | >540 | Flexible, heat-resistant films | 421–423 |

TABLE 1.9—*continued*

Poly(ester-imides) from Ester Anhydrides

| No. | Structure | Method | $T_m$ (°C) | Remarks and property data | References |
|---|---|---|---|---|---|
| 15 | 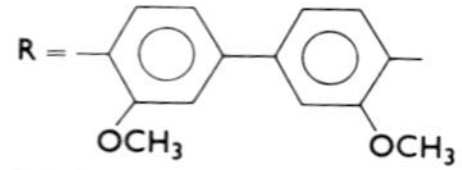 <br>$R = $ 3,3'-Dimethoxy-4,4'-diaminobiphenyl | A,10 | >500 | Flexible films | *422, 423* |
| | Copolymers | A,10 | — | — | *422, 423* |
| 16 | 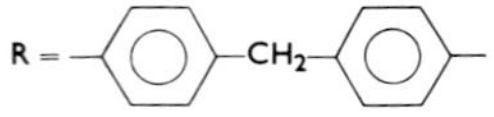 <br>$R = $ —CH$_2$—<br>4,4'-Diaminodiphenylmethane | A,10 | >500 | Tensile strength = 14,500 psi (room temp.), 6100 psi (200°C); elongation = 6% (room temp.), 21% (200°C); tensile modulus = 544,000 psi (room temp.), 191,000 psi (200°C) | *144, 262, 422, 423* |
| 17 | 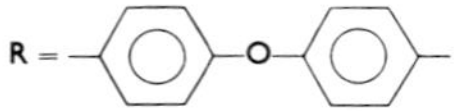 <br>$R = $ —O—<br>4,4'-Diaminodiphenyl ether | A,10 | >500 | Tensile strength = 15,000 psi (room temp.), 6300 psi (200°C); elongation = 14% (room temp.), 23% (200°C); tensile modulus = 440,000 psi (room temp.), 193,000 psi (200°C); wt. loss (air), 100 hr at 300°C = 3.42%, at 325°C = 7.95%, at 350°C = 29.0%. Claimed to have excellent wire insulating properties. | *422, 423* |
| 18 | 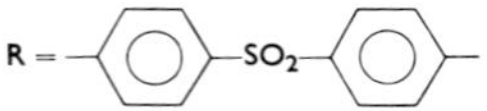 <br>$R = $ —SO$_2$—<br>4,4'-Diaminodiphenyl sulfone | A,10 | >500 | Brittle films | *422, 423* |

* * * *

19

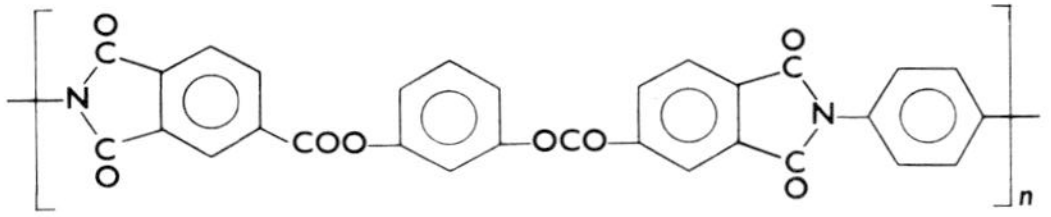

Bis(trimellitic anhydride ester) of resorcinol and 1,4-diaminobenzene

A,10    >540    *423*

20

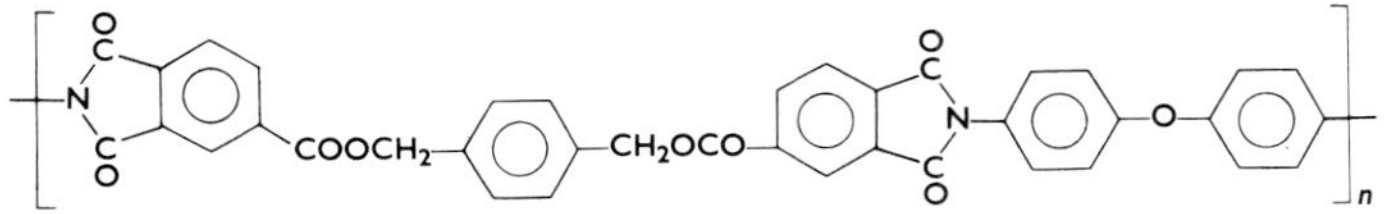

Bis(trimellitic anhydride ester) of 1,4-bis(hydroxymethyl)benzene and 4,4′-diaminodiphenyl ether

A,10    >350    —    *341*

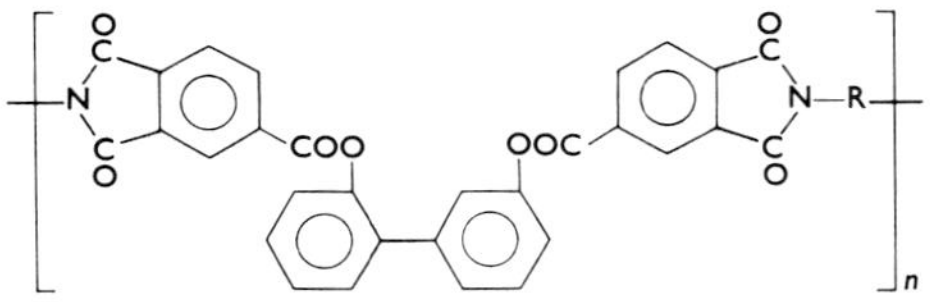

Bis(trimellitic anhydride ester) of 2,2′-dihydroxybiphenyl and

21

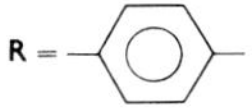

R = 1,4-Diaminobenzene

A,10    >500    Brittle films    *422*

22

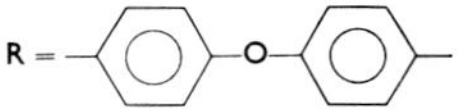

R = 4,4′-Diaminodiphenyl ether

A,10    235    Flexible films. Weight loss: air, 100 hr at 260°C = 8.54%, 100 hr at 300°C = 9.25%    *422, 423*

* * *

TABLE I.9—*continued*

Poly(ester-imides) from Ester Anhydrides

| No. | Structure | Method | $T_m$ (°C) | Remarks and property data | References |
|---|---|---|---|---|---|
| | Bis(trimellitic anhydride ester) of bisphenol-A and | | | | |
| 23 | 1,4-Diaminobenzene | A,10 | >500 | Flexible films | *422, 423* |
| 24 | $R =$ 4,4'-Diaminobiphenyl | A,10 | >500 | Tensile strength = 13,400 psi (room temp.), 6200 psi (200°C); elongation = 6% (room temp.), 10% (200°C); tensile modulus = 471,000 psi (room temp.), 208,000 psi (200°C); wt. loss (air), 100 hr at 260°C = 8.6%; 100 hr at 300°C = 12.00% | *422, 423* |
| 25 | $R =$ 4,4'-Diaminodiphenyl ether | A,10 | 390 | Tensile strength = 9900 psi (room temp.), 6300 psi (200°C); elongation = 4% (room temp.), 7% (200°C); tensile modulus = 360,000 psi (room temp.),152,000 psi (200°C); wt. loss (air),100 hr at 260°C = 7.46%,100 hr, at 300°C = 15.8% | *422, 423* |

*
*
*

Bis(trimellitic anhydride ester) of 4,4'-dihydroxydiphenyl sulfone and

| | | | | |
|---|---|---|---|---|
| 26  R = (1,3-Diaminobenzene) | A,10 | >500 | Brittle films | 422, 423 |
| 27  R = (CH$_2$) 4,4'-Diaminodiphenylmethane | A,10 | >500 | Brittle films | 422 |
| 28  R = (O) 4,4'-Diaminodiphenyl ether | A,10 | >500 | Brittle films | 422 |
| Copolymers of the above dianhydride | A,10 | — | — | 423–425 |

TABLE I.10

Poly(ester-imides) from Diamino Esters

| No. | Structure | Method | Remarks and property data | References |
|-----|-----------|--------|---------------------------|------------|
| | 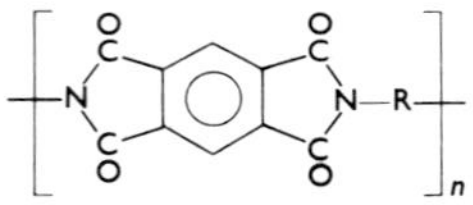 Pyromellitic anhydride and | | | |
| 1 | 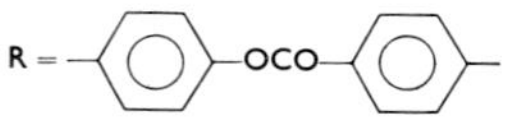 4,4'-Diaminophenyl benzoate | A,10 | Brittle films. Weight loss (air, 400 hr, 325°C) = 9.7% | 121, 122, 273 |
| 2 | 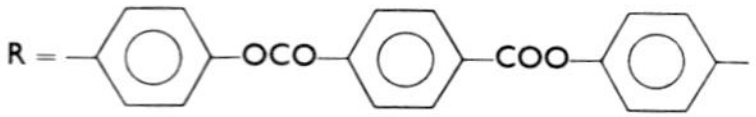 Bis(4-aminophenyl) terephthalate | A,10 | Good thermal stability claimed | 4 |
| 3 | 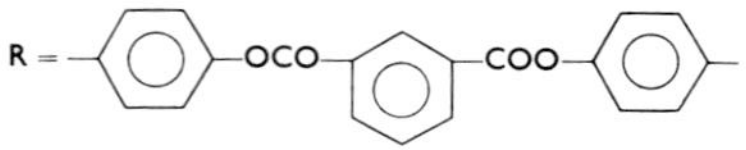 Bis(4-aminophenyl) isophthalate | A,10 | Flexible films. Weight loss (air, 400 hr, 325°C) = 15% | 4, 121 122, 273 |
| 4 | 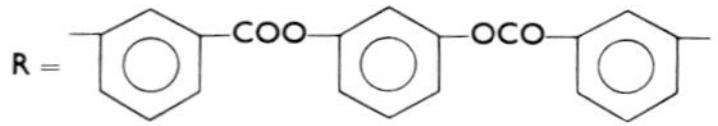 Resorcinol bis(3-aminobenzoate) | A,10 | Brittle films. Weight loss (air, 400 hr, 325°C) = 22.7% | 121, 122, 273 |

* * *

3,3',4,4'-Diphenyl ether tetracarboxylic acid dianhydride and

| | | | | |
|---|---|---|---|---|
| 5 | R = Bis(4-aminophenyl) terephthalate | A,10 | Good thermal stability claimed | 4 |
| 6 | R = Bis(4-aminophenyl) isophthalate | A,10 | Good thermal stability claimed | 4 |

TABLE I.11

Poly(ester-imides) from Dimethyl Bis(trimellitimidates)

Structure:

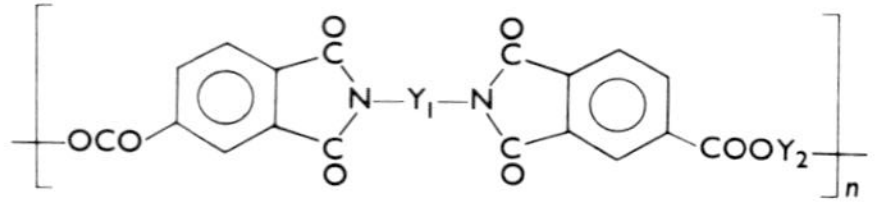

$Y_1 = (CH_2)_2$

Dimethyl $N,N'$-(1,2-ethylene)bis(trimellitimidate) and

| No. | Structure | Method | Solubility | Molecular weight | $T_m$ (°C) | Remarks and property data | References |
|-----|-----------|--------|-----------|------------------|-----------|---------------------------|------------|
| 1 | $Y_2 = (CH_2)_4$ <br> 1,4-Butanediol | A,12 | — | — | — | $T_g = 115°C$; tensile modulus = 320,000 psi; yield strength = 8400 psi | 356 |
| 2 | $Y_2 = (CH_2)_6$ <br> 1,6-Hexanediol | A,12 | — | — | — | $T_g = 110°C$ | 356 |

$Y_1 = (CH_2)_6$

Dimethyl $N,N'$-(hexamethylene)bis(trimellitimidate) and

| No. | Structure | Method | Solubility | Molecular weight | $T_m$ (°C) | Remarks and property data | References |
|-----|-----------|--------|-----------|------------------|-----------|---------------------------|------------|
| 3 | $Y_2 = (CH_2)_2$ <br> Ethylene glycol | A,12 | — | — | — | $T_g = 98°C$; tensile modulus = 340,000 psi; yield strength = 7700 psi | 356 |
| 4 | $Y_2 = (CH_2)_4$ <br> 1,4-Butanediol | A,12 | $p$-Chloro-phenol | $\eta_{red}$ 0.88 (61.4°C) | 240 (on fibers) | Yields "tough crystalline fibers" on drawing. $T_g = 80°C$; tensile modulus = 305,000 psi; yield strength = 6800 psi | 356 |

| No. | | | | | | Ref. |
|---|---|---|---|---|---|---|
| 5 | $Y_2 = \!+\!CH_2\!\rightarrow_5$ <br> 1,5-Pentanediol | A,12 | — | — | $T_g = 73°C$; tensile modulus = 283,000 psi; yield strength = 7000 psi | 356 |
| 6 | $Y_2 = \!+\!CH_2\!\rightarrow_2 O\!+\!CH_2\!\rightarrow_2$ <br> Diethylene glycol | A,12 | $p$-Chlorophenol | $\eta_{red}$ 0.55 (53°C) | — | Can be drawn into tough, amorphous fibers | 356 |
| 7 | $Y_2 = \!+\!CH_2\!\rightarrow_6$ <br> 1,6-Hexanediol | A,12 | — | — | $T_g = 67°C$; tensile modulus = 262,000 psi; yield strength = 7500 psi | 356 |
| 8 | $Y_2 = \!+\!CH_2\!\rightarrow_7$ <br> 1,7-Heptanediol | A,12 | — | — | $T_g = 45°C$; tensile modulus = 211,000 psi; yield strength = 4100 psi | 356 |

$Y_1 = \!+\!CH_2\!\rightarrow_3 O\!+\!CH_2\!\rightarrow_3$

Dimethyl $N,N'$-(3-oxa-1,7-heptylene) bis(trimellitimidate) and

| No. | | | | | | Ref. |
|---|---|---|---|---|---|---|
| 9 | $Y_2 = \!+\!CH_2\!\rightarrow_6$ <br> 1,6-Hexanediol | A,12 | $p$-Chlorophenol | $\eta_{red}$ 0.62 (61°C) | — | $T_g = 61°C$; tensile modulus = 225,000 psi | 356 |

TABLE I.12
Polyiminolactones

Pyromellitic anhydride and

| No. | Structure | Method | Solubility | Molecular weight | Remarks and property data | References |
|---|---|---|---|---|---|---|
| 1 | 1,4-Diaminobenzene | A,18 | $H_2SO_4$ | $\eta_{inh}$ 0.23 | Deep yellow. Properties: tensile modulus = 596,000 psi (23°C), 673,000 psi (105°C); tensile strength = 14,700 psi (23°C), 12,700 psi (105°C); elongation = 12.2 % (23°C), 2.8 % (105°C); impact = 1.38 kg cm/mil; density = 1.413 g/cm³; tear strength = 2.9 g/mil; zero strength temp. $\cong$ 760°C; volume resistivity = $2.1 \times 10^{11}$ ohm-cm (23°C); $1.6 \times 10^{11}$ ohm-cm (100°C) | 386 |
| 2 | 2,4-Diaminocumene | A,18 | DMA | $\eta_{inh}$ 1.0 | Structure supported by IR and UV. Density = 1.303 g/cm³; impact = 0.02 kg cm/mil; zero strength temperature $\cong$ 420°C; volume resistivity = $4.7 \times 10^{16}$ ohm-cm (23°C), $5.4 \times 10^{16}$ ohm-cm (100°C). Reactions with alcohols and thiols were described. Cross-linking with di-*tert*-butyl peroxide was reported | 19, 326, 386, 471, 642 |
| 3 | 4,4'-Diaminobiphenyl | A,18 | $H_2SO_4$ | $\eta_{inh}$ 0.57 | Deep orange. Properties: tensile modulus = 579,000 psi (23°C), 621,000 psi (105°C); tensile strength = 14,800 psi (23°C), 13,000 psi (105°C); elongation = 9.5 % (23°C), 3.3 % (105°C); density = 1.351 g/cm³; impact = 1.25 kg cm/mil; tear strength = 6.1 g/mil; zero strength temperature $\cong$ 840°C; volume resistivity = $1.4 \times 10^{12}$ ohm-cm (23°C), $1.4 \times 10^{12}$ ohm-cm (100°C) | 386 |

| | | | | |
|---|---|---|---|---|
| 4 <br> R = — ⟨⟩—CH₂—⟨⟩— <br> **4,4'-Diaminodiphenylmethane** | A,18 | DMF | — | Structure supported by IR. Properties: tensile modulus = 350,000 psi; tensile strength = 9800 psi; elongation = 3.4%. Thermal rearrangement to the corresponding polyimide was described | *19, 386* |
| 5 <br> R = — ⟨⟩—O—⟨⟩— <br> **4,4'-Diaminodiphenyl ether** | A,18 | H₂SO₄,THF η₁ₙₕ 0.31 | Structure supported by IR. Properties: tensile modulus = 276,000 psi (23°C), 257,000 psi (105°C); tensile strength = 7900 psi (23°C), 6400 psi (105°C); elongation = 3.7% (23°C), 2.7% (105°C); density = 1.353 g/cm³; impact = 0.19 kg cm/mil; tear strength = 8.7 g/mil; zero strength temperature ≅ 730°C; volume resistivity = 2.6 × 10¹² ohm-cm. <br> Reactions: <br> (a) Thermal isomerization to the polyimide; <br> (b) with ammonia, amines, alcohols and thiols; <br><br> (c) with hydrazoic acid; <br> (d) with (C₂H₅)₃NHOCOCH₃ (triethylamine acetate) to give the corresponding polyimide or poly(imide-iminolactone) copolymers | *57, 386, 469* <br><br><br><br><br><br><br> *19* <br> *29, 386, 471, 642* <br> *644* <br> *386* |
| 6 <br> R = — ⟨⟩—SO₂—⟨⟩— <br> **4,4'-Diaminodiphenyl sulfone** | A,18 | — | — | — | *386, 471* |

\* \* \*

TABLE I.12—*continued*

Polyiminolactones

| No. | Structure | Method | Solubility | Molecular weight | Remarks and property data | References |
|---|---|---|---|---|---|---|
| 7 | 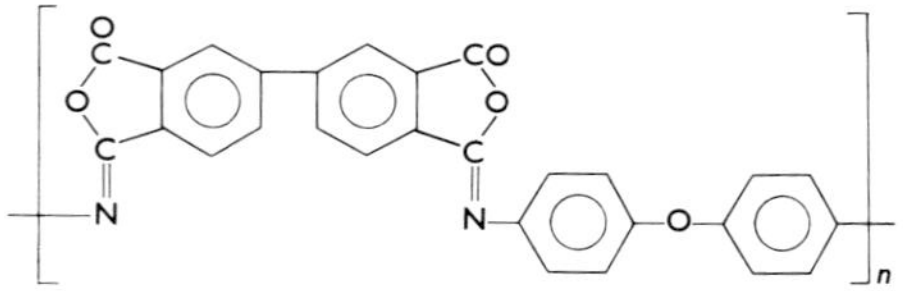 3,3',4,4'-Biphenyl tetracarboxylic acid dianhydride and 4,4'-diaminodiphenyl ether | A,18 | — | — | — | *386, 471* |
|  | 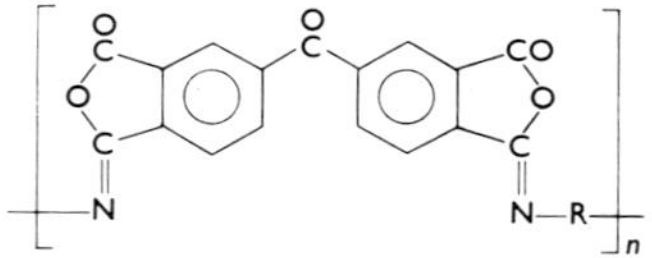 3,3',4,4'-Benzophenone tetracarboxylic acid dianhydride and |  |  |  |  |  |
| 8 |  R = 1,3-Diaminobenzene | A,18 | — | — | — | *386, 471* |
| 9 | 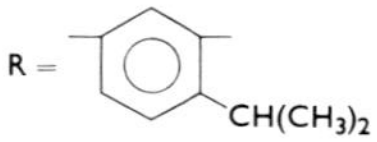 R = CH(CH₃)₂  2,4-Diaminocumene | A,18 | — | — | Cross-linking with di-*tert*-butyl peroxide was reported | *326, 471* |

\* \* \*

| | | | | | |
|---|---|---|---|---|---|
| 10 | <br>Bis(3,4-dicarboxyphenyl) ether dianhydride and 4,4'-diaminodiphenyl sulfide | A,18 | — | — | — | 386 |
| 11 | Poly(imide-iminolactone) copolymers | — | — | — | — | 386, 387 |

TABLE I.13
Poly(imide-imidazopyrrolones)

| No. | Structure | Method | Solubility | Molecular weight | Remarks and property data | References |
|---|---|---|---|---|---|---|
| 1 | Pyromellitic anhydride and 1,2,4-triaminobenzene | A,13; A,14 | 97% $H_2SO_4$ | [η] 0.631 | IR indicates presence of imide and imidazopyrrolone units. TGA (10°C/min) shows wt. loss of 32% at 1000°C and of 38% at 1176°C | 185 |
| 2 | Poly(imide-imidazopyrrolones) from pyromellitic anhydride, 3,3',4,4'-tetraaminobiphenyl and 4,4'-diaminodiphenyl ether | A,13 | — | — | Percent of imidazopyrrolone units in the copolymer was varied from 0 to 100. Chemical resistance and high-temperature stability were studied | 98 |
| 3 | 3,3',4,4'-Benzophenonetetracarboxylic acid dianhydride and 1,2,4-triaminobenzene | A,13 | — | — | — | 185 |
| 4 | 1,4,5,8-Naphthalenetetracarboxylic acid dianhydride and 1,2,4-triaminobenzene | A,13; A,14 | 97% $H_2SO_4$ | [η]0.882 | IR indicates presence of imide and imidazopyrrolone units. TGA (10°C/min) shows wt. loss of 26% at 1000°C and of 31% at 1176°C | 185 |
| 5 | Poly(imide-imidazopyrrolones) from 1,4,5,8-naphthalenetetracarboxylic acid dianhydride 3,3',4,4'-tetraaminobiphenyl and 4,4'-diaminodiphenyl ether | A,14 | — | — | Various compositions were prepared. Thermal and oxidative stabilities were studied | 100 |

TABLE I.14
Polymers containing Pyrrole and Related Rings

| No. | Structure | Method | Solubility | Remarks and property data | References |
|---|---|---|---|---|---|
| 1 |  Polytetracyanoethylene | B,1 | Insoluble | $T_m = 350°Cd$. Electrical conductivity $10^{-7}$–$10^{-9}$ ohm$^{-1}$ cm$^{-1}$. Thermal stability studied. **Black**. Copolymers with anthracene, napthacene, pentacene | 104, 105, 106 |
| 2 |  Polyphthalocyanine | B,2 | Insoluble | From pyromellitonitrile. From metal-containing polyphthalocyanines. Blue black powders. Electrical properties studied | 176, 177, 433, 654, 683 |

TABLE I.14—*continued*

Polymers containing Pyrrole and Related Rings

| No. | Structure | Method | Solubility | Remarks and property data | References |
|---|---|---|---|---|---|
| 3 | | B,3 | — | Tan. 10% wt. loss at 500° (air); 600° ($N_2$). Molded at 380° ± 20°C/10–20,000 psi | 534, 535 |
| 4 | | B,3 | — | Dark red. Stable to 600° (A). Molded at 350°–400°C to yield specimens with tensile strength of 13,000 psi | 533, 535 |

| | | | | Cross-linked structure | |
|---|---|---|---|---|---|
| 5 | B,3 | — | | | *434* |

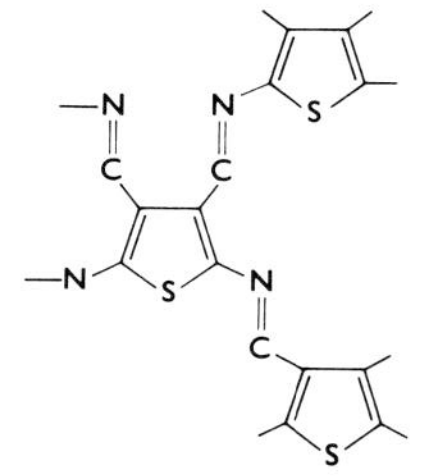

2,5-Diamino-3,4-dicyanothiophene

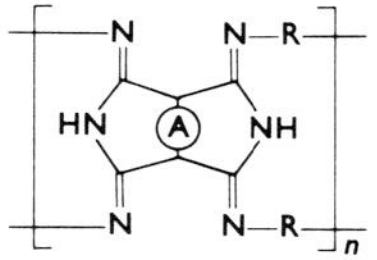

| 6 | B,3 | Insoluble | Black | | *434* |

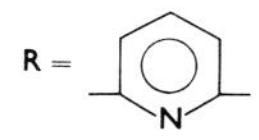

$A=$ >C=C<

$R=$ (pyridine)

| 7 | B,3 | Insoluble | Black | | *434* |

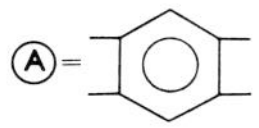

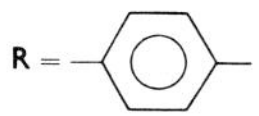

$A=$ (phenylene)

$R=$ (p-phenylene)

*p*-Phenylene

TABLE I.14—*continued*
Polymers containing Pyrrole and Related Rings

| No. | Structure | Method | Solubility | Remarks and property data | References |
|-----|-----------|--------|------------|---------------------------|------------|
| 8 | A = (benzene ring); R = (pyridine) | B,3 | Insoluble | Black | *434* |
| 9 | A = (thiophene); R = (p-Phenylene) | B,3 | Insoluble | Black | *434* |
| 10 | A = (thiophene); R = (pyridine) | B,3 | Insoluble | Black | *434* |
| 11 | A = (pyrrole, N−⁺N(CH₃)₄); R = (pyridine) | B,3 | Insoluble | Black | *434* |

| No. | Structure | Method | Solubility | Color | Properties | Ref. |
|---|---|---|---|---|---|---|
| 12 | Ⓐ= (structure), R= pyridine | B,3 | Insoluble | Black | | *434* |
| 13 | Ⓐ= Cu(S₄) complex, R= pyridine | B,3 | Insoluble | Black | | *434* |
| 14 | Ⓐ= Fe(S₄) complex, R= pyridine | B,3 | Insoluble | Black | | *434* |
| 15 | 1,2,4,5-Tetracyanobenzene and 3,3'-diaminobenzidine | B,3 | — | Cross-linked | | *505* |
| 16 | (polymer structure) | B,5 | $H_2SO_4$, DMF | $\eta_{inh}$ 1.25. $T_m$ = >400°C. Yellow-brown. Crystalline | | *693, 697, 698* |

TABLE I.15

Poly(1,3-oxazinones) and Poly(1,3-oxazindiones)

| No. | Structure | Method | Solubility | Molecular weight | $T_m$ (°C) | Remarks and property data | References |
|---|---|---|---|---|---|---|---|
| 1 | (from 2,5-Diaminoterephthalic acid) | C,1 | $H_2SO_4$ | $\eta_{inh}$ 0.06 | — | Self-condensation in polyphosphoric acid | 396, 398, 400 |
| 2 | | C,1 | Insoluble | — | — | Elongation = 3% | 576, 577 |
| 3 | | C,1 | Insoluble | — | — | Opaque, brittle. Co-polymers. Thermal stability studied. | 392, 576, 577 |
| 4 | | C,1 | — | — | — | — | 576, 577 |
| 5 | | C,1 | — | — | — | — | 576, 577 |

211

| No. | Structure | | Solvent | $\eta_{inh}$ | Temp. | Notes | Refs. |
|---|---|---|---|---|---|---|---|
| 6 | (from 4,4'-Diamino-3,3'-biphenyldicarboxylic acid) | C,1 | $H_2SO_4$ | $\eta_{inh}$ 0.23 | — | Self-condensation in polyphosphoric acid | 396, 398, 400 |
| 7 | | C,1 | $H_2SO_4$ | $\eta_{inh}$ 1.81 | — | — | 704 |
| 8 | "trans" | C,1 | $H_2SO_4$ | $\eta_{inh}$ 0.67 | — | — | 303, 347, 348 |
| 9 | "cis" | C,1 | $H_2SO_4$ | $\eta_{inh}$ 0.25 | — | — | 303, 347, 348 |
| 10 | "cis" | C,1 | $H_2SO_4$ | $\eta_{inh}$ 0.19 | — | — | 303, 347, 348 |
| 11 | | C,1 | $H_2SO_4$ | $\eta_{inh}$ 1.95 | 540–550 | Dark blue. Copolymers. Decomposes 500° (air) and 550° ($N_2$) | 401, 695, 699, 701–707, 709, 710, 712 |

TABLE I.15—continued

Poly(1,3-oxazinones) and Poly(1,3-oxazindiones)

| No. | Structure | Method | Solubility | Molecular weight | $T_m$ (°C) | Remarks and property data | References |
|---|---|---|---|---|---|---|---|
| 12 | | C,1 | $H_2SO_4$ | — | >500 | Copolymers also. Decomposes 500° (air) and 550° ($N_2$) | 401, 695, 704, 705, 710, 712 |
| 13 | | C,1 | $H_2SO_4$ | $\eta_{inh}$ 1.51 | — | — | 708 |
| 14 | | C,1 | — | — | >550 | Cyclized by air oxidation | 694 |
| 15 | | C,1 | $H_2SO_4$ | $\eta_{inh}$ 0.64 | — | Self-condensation in polyphosphoric acid | 396, 398, 400 |

(from 5,5'-Methylenedianthranilic acid)

| No. | Structure | | | | | | Ref. |
|---|---|---|---|---|---|---|---|
| 16 | | — | — | — | — | Cyclized chemically | 705 |
| 17 | | C,1 | $H_2SO_4$ | $\eta_{inh}$ 2.02 | — | — | 704 |
| 18 | | C,2 | — | $\eta_{inh}$ 0.23 | — | — | 16, 17 |
| 19 | | C,2 | — | — | — | Followed by IR studies | 16, 17 |
| 20 | (from Dimethyl tartrate) | C,2 | NMP | $\eta_{inh}$ 0.13 | — | Tan | 16, 17 |

**TABLE I.15—*continued***
Poly(1,3-oxazinones) and Poly(1,3-oxazindiones)

| No. | Structure | Method | Solubility | Molecular weight | $T_m$ (°C) | Remarks and property data | References |
|---|---|---|---|---|---|---|---|
| 21 | | C,2 | — | — | — | — | 16, 17 |
| 22 | | C,3 | — | — | — | Light yellow, flexible film | 21 |
| 23 | | C,3 | — | — | — | Tough, flexible film | 21 |
| 24 | | C,2 | — | — | — | — | 22 |
| 25 | | C,2 | — | — | — | Yellow, tough, flexible film | 22 |

| | | | | | | |
|---|---|---|---|---|---|---|
| 26 | HNCOO— (structure, Cl-substituted) | C,2 | — | — | — | 22 |
| 27 | HNCOO— (structure, biphenyl) | C,2 | — | — | — | 22 |
| 28 | HNCOO— (structure, OCH₃ / OCH₃) | C,2 | — | — | — | 22 |
| 29 | HNCO (structure, CH₂) | C,3 | — | — | — | 21 |
| 30 | HNCO (structure, O) | C,3 | — | — | — | 21 |
| 31 | HNCOO— (structure, O) | C,2 | — | — | — | 22 |

TABLE I.15—*continued*
Poly(1,3-oxazinones) and Poly(1,3-oxazindiones)

| No. | Structure | Method | Solubility | Molecular weight | $T_m$ (°C) | Remarks and property data | References |
|-----|-----------|--------|------------|------------------|------------|---------------------------|------------|
| 32 | | C,2 | — | — | — | — | 22 |
| 33 | | C,3 | Cresol | $\eta_{red}$ 2.01 | — | — | 115 |
| 34 | | C,2; C,3 | — | — | — | — | 16, 17 |
| 35 | | C,2 | — | — | — | Tough, flexible film | 16, 17 |
| 36 | | C,2 | — | — | — | Tough, flexible film | 16, 17, 260 |

| No. | Structure | | Solvent | $\eta$ | | Properties | Ref. |
|---|---|---|---|---|---|---|---|
| 37 | (CH₃-substituted) | C,2 | — | — | — | Cyclization followed by IR | 16, 17 |
| 38 | (Cl-substituted) | C,2 | — | — | — | Tough, flexible film | 16, 17 |
| 39 | (biphenyl) | C,2 | — | — | — | Tough, flexible film | 16, 17 |
| 40 | (OCH₃-substituted) | C,2 | — | — | — | Tough, flexible film | 16, 17 |
| 41 | (CH₂ bridge) | C,2; C,3 | — | — | — | — | 16, 17 |
| 42 | (O bridge) | C,2; C,3 | DMF | $\eta_{red}$ 2.9 | — | Tough, flexible, transparent, colorless film. Tensile strength 1000 kg/cm²; elongation 70% | 16, 17, 260 |

TABLE I.15—continued
Poly(1,3-oxazinones) and Poly(1,3-oxazindiones)

| No. | Structure | Method | Solubility | Molecular weight | $T_m$ (°C) | Remarks and property data | References |
|---|---|---|---|---|---|---|---|
| 43 | | C,2; C,3 | — | — | — | Tough, flexible films. Copolymers of 4,4'- and 2,4'-diaminodiphenyl ether | 16, 17, 76, 108 |
| 44 | | C,2 | — | — | — | — | 16, 17 |
| 45 | | C,3 | Cresol | $\eta_{red}$ 2.07 | — | — | 115 |
| 46 | | C,3 | — | — | — | Tough, flexible film | 21 |
| 47 | | C,2 | — | — | — | Tough, flexible film | 22 |

| No. | Structure | | Solvents | | Properties | Ref. |
|---|---|---|---|---|---|---|
| 48 | | C,2 | — | — | Tough, flexible film | 16, 17 |
| 49 | 4:1 ratio of 4,4′- to 2,4′-Diaminodiphenyl ether | C,2 | $H_2SO_4$, cresol DMSO, DMA, NMP | — | Colorless, transparent films. Retained elasticity after 1 month at 275°C (air) | 76, 108 |
| 50 | Copolymer | C,2 | — | — | Cast film stretched 3:1 at 285°C. Tensile strength = 40 kg/cm²; elongation = 40% | 557 |
| 51 | | C,3 | Cresol, TCE, pyridine | $\eta_{red}$ 1.49  230–240 | Cast films | 115 |

TABLE I.15—*continued*

Poly(1,3-oxazinones) and Poly(1,3-oxazindiones)

| No. | Structure | Method | Solubility | Molecular weight | $T_m$ (°C) | Remarks and Property data | References |
|---|---|---|---|---|---|---|---|
| 52 | (structure) | C,3 | Cresol, TCE, pyridine | $\eta_{red}$ 1.74 | 210–215 | Crystalline | 115 |
| 53 | (structure) | C,3 | Cresol, pyridine | $\eta_{red}$ 2.02 | 240–250 | — | 115 |
| 54 | (structure) | C,3 | Cresol, TCE, $CH_2Cl_2$ | $\eta_{red}$ 1.88 | 255–270 | Cast films | 115 |
| 55 | (structure) | C,3 | Cresol, TCE, pyridine | $\eta_{red}$ 1.45 | 270–285 | Cast films | 115 |

| | | | Cresol, TCE, pyridine | $\eta_{red}$ 1.63 | >380 | Crystalline | 115 |
|---|---|---|---|---|---|---|---|
| 56 | | C,3 | | | | | |
| 57 | | C,3 | — | — | — | — | 21 |
| 58 | | C,2 | — | — | — | — | 22 |
| 59 | | C,2 | — | — | — | — | 16, 17 |

TABLE I.16
Polydioxins

| No. | Structure | Method | Solubility | Molecular weight | Remarks and property data | References |
|-----|-----------|--------|------------|------------------|---------------------------|------------|
| 1 | Tetrahydroxy-$p$-benzoquinone and 2,3,7,8-tetrachloro-1,4,6,9-tetraazaanthracene | D,1 | $H_2SO_4$ | $\eta_{inh}$ 0.11 | Brown | 685 |
| 2 | 1,2,4,5-Tetrahydroxybenzene and 2,3,7,8-tetrachloro-1,4,6,9-tetraazaanthracene | D,1 | $H_2SO_4$ | $\eta_{inh}$ 0.03 | Black | 685 |
| 3 | Tetrahydroxy-$p$-benzoquinone and 2,2′,3,3′-tetrachloro-6,6′-bis(quinoxaline) | D,1 | $H_2SO_4$ | $\eta_{inh}$ 0.06 | Brown | 685 |
| 4 | Tetrahydroxy-$p$-benzoquinone and 2,2′,3,3′-tetrachloro-6,6′-bis(quinoxalyl) ether | D,1 | Insoluble | — | Black. TGA dec. 400°C (He) | 685 |

5

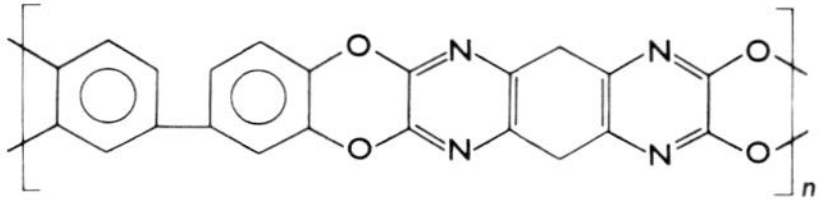

3,3´,4,4´-Tetrahydroxydiphenyl and
2,3,7,8-tetrachloro-1,4,6,9-tetraazaanthracene

D,1   H$_2$SO$_4$   $\eta_{inh}$ 0.25   Brown. TGA dec. 450°C (He)   *685*

6

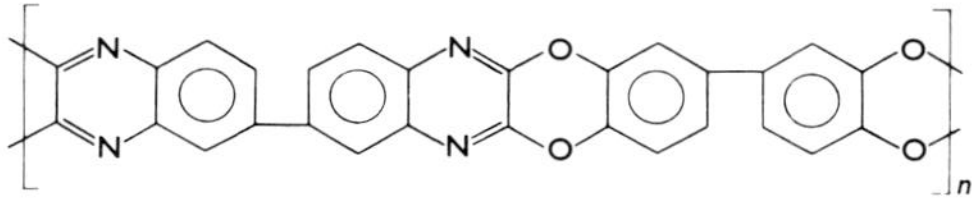

3,3´,4,4´-Tetrahydroxybiphenyl and
2,2´,3,3´-tetrachloro-6,6-bis(quinoxaline)

D,1   H$_2$SO$_4$   $\eta_{inh}$ 0.4   Brown. TGA dec. 450°C (He)   *685*

7

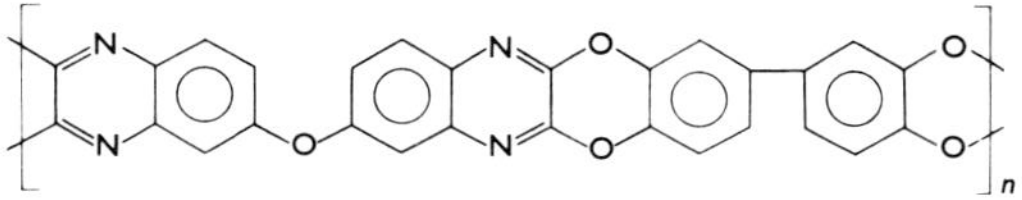

3,3´,4,4´-Tetrahydroxybiphenyl and
2,2´,3,3´-tetrachloro-6,6´-bis(quinoxalyl) ether

D,1   H$_2$SO$_4$   $\eta_{inh}$ 0.35   Light brown. TGA dec. 500°C (He)   *685*

8

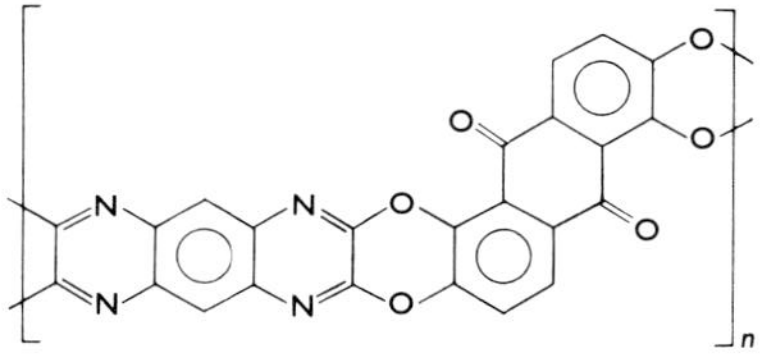

1,2,5,6-Tetrahydroxyanthraquinone and
2,3,7,8-tetrachloro-1,4,6,9-tetraazaanthracene

D,1   H$_2$SO$_4$   $\eta_{inh}$ 0.06   Brown   *685*

TABLE I.16—continued

Polydioxins

| No. | Structure | Method | Solubility | Molecular weight | Remarks and property data | References |
|---|---|---|---|---|---|---|
| 9 | 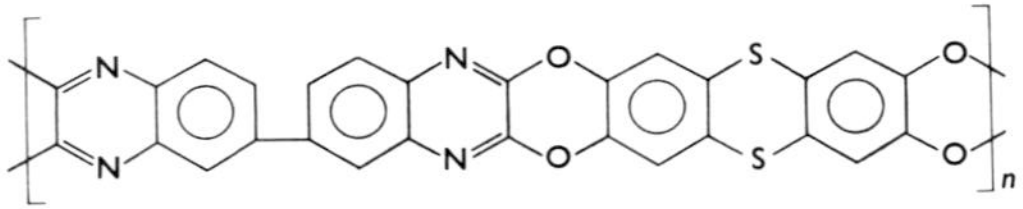 2,3,6,7-Tetrahydroxythianthrene and 2,2′,3,3′-tetrachloro-6,6′-bis(quinoxaline) | D,1 | Insoluble | — | Brown. TGA dec. 400°C (He) | 685 |
| 10 | 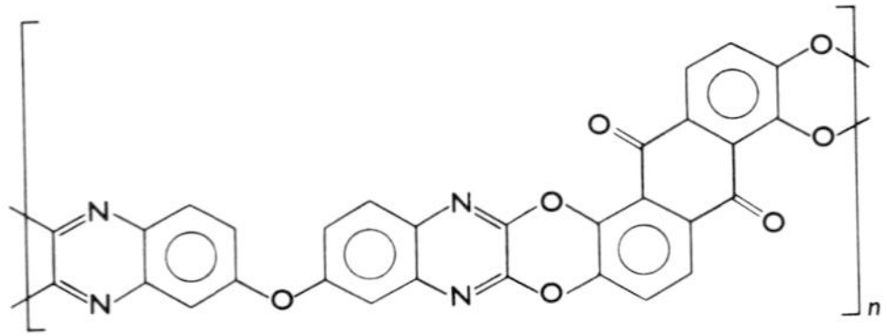 1,2,5,6-Tetrahydroxyanthraquinone and 2,2′,3,3′-tetrachloro-6,6′-bis(quinoxalyl) ether | D,1 | $H_2SO_4$ | $\eta_{inh}$ 0.09 | Dark brown | 685 |

TABLE I.17
Polyoxazines

| No. | Structure | Method | Solubility | Molecular weight | Remarks and property data | References |
|---|---|---|---|---|---|---|
| 1 | | D,3 | HMP | $\eta_{inh}$ 0.36 | TGA dec. 260°C | 628, 632 |
| 2 | <br>X = OH, Cl | D,3 | HMP, HCO$_2$H | $\eta_{inh}$ 0.18 | TGA dec. 225°C | 628, 632 |
| 3 | <br>2,5-Diamino-1,4-dihydroxybenzene and 2,3,6,7-tetrachloro-1,4,5,8-tetraazaanthracene, or 2,3,6,7-tetrahydroxy-1,4,5,8-tetraazaanthracene | D,2 | CH$_3$SO$_3$H, H$_2$SO$_4$ | $\eta_{inh}$ 0.85 | Black | 442, 521, 686 |
| 4 | <br>3,3'-Dihydroxybenzidine and 2,3,6,7-tetrachloro-1,4,5,8-tetraazaanthracene, or 2,3,6,7-tetrahydroxy-1,4,5,8-tetraazaanthracene | D,2 | CH$_3$SO$_3$H, H$_2$SO$_4$ | $\eta_{inh}$ 0.68 | Brown | 442, 521, 686 |

TABLE I.17—*continued*
Polyoxazines

| No. | Structure | Method | Solubility | Molecular weight | Remarks and property data | References |
|---|---|---|---|---|---|---|
| 5 | 2,5-Diamino-1,4-dihydroxybenzene and 2,2',3,3'-tetrahydroxy-6,6'-bis(quinoxaline) | D,2 | $CH_3SO_3H$, $H_3PO_4$, $H_2SO_4$ | $\eta_{inh}$ 0.48 | Black | 442, 521, 686 |
| 6 | 2,5-Diamino-1,4-dihydroxybenzene and 2,2',3,3'-tetrachloro-6,6'-diquinoxalyl ether | D,2 | $H_2SO_4$ | $\eta_{inh}$ 0.35 | Black | 442, 686 |
| 7 | 3,3'-Dihydroxybenzidine and 2,2',3,3'-tetrahydroxy-6,6'-bis(quinoxaline) | D,2 | $CH_3SO_3H$, $H_2SO_4$ | $\eta_{inh}$ 0.49 | Black | 442, 521, 686 |
| 8 | 3,3'-Dihydroxybenzidine and 2,2',3,3'-tetrachloro-6,6'-diquinoxalyl ether | D,2 | $H_2SO_4$ | $\eta_{inh}$ 0.40 | Black | 442, 686 |

TABLE I.18
Polythiazines

| No. | Structure | Method | Solubility | Molecular weight | Remarks and property data | References |
|---|---|---|---|---|---|---|
| 1 | 2,5-Dichloro-p-benzoquinone and 3,3'-dimercaptobenzidine | D,3 | $H_2SO_4$ | $\eta_{inh}$ 0.54 | Black; wt. loss of 3% at 500°C (He) | 442, 523 |
| 2 | 3,3'-Dimercaptobenzidine and 1,4-bis(α-bromoacetyl)benzene | D,4 | $p$-ClC$_6$H$_4$OH | 11,456 (Br. analysis) | — | 116, 117 |
| 3 | 3,3'-Dimercaptobenzidine and 1,3-bis(α-bromoacetyl)benzene | D,4 | $p$-ClC$_6$H$_4$OH | 13,000 (Br. analysis) | — | 116, 117 |
| 4 | 3,3'-Dimercaptobenzidine and 4,4'-bis(α-bromoacetyl)biphenyl | D,4 | $p$-ClC$_6$H$_4$OH | 25,000 (Br. analysis) | — | 116, 117 |

TABLE I.18—*continued*

Polythiazines

| No. | Structure | Method | Solubility | Molecular weight | Remarks and property data | References |
|---|---|---|---|---|---|---|
| 5 | 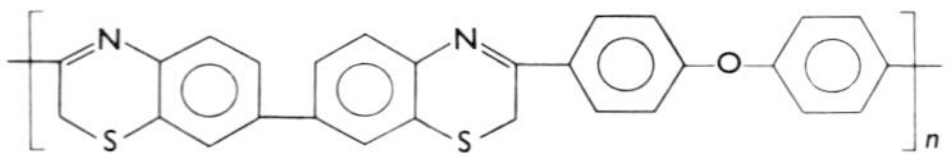 <br>3,3'-Dimercaptobenzidine and<br>4,4'-bis(α-bromoacetyl)diphenyl ether | D,4 | *m*-Cresol | 20,500 (Br. analysis), $\eta_{inh}$ 0.34 | Weight loss of 10% when heated to 438°C | *116, 117* |
| 6 | 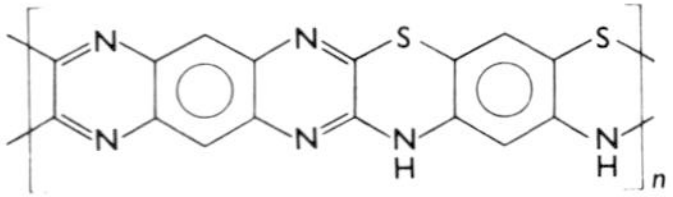 <br>2,5-Diamino-1,3-dithiophenol and<br>2,3,6,7-tetrachloro-1,4,5,8-tetraazaanthracene, or<br>2,3,6,7-tetrahydroxy-1,4,5,8-tetraazaanthracene | D,2 | CH₃SO₃H | $\eta_{inh}$ 1.19 | Black | *442, 521, 522* |
| 7 | 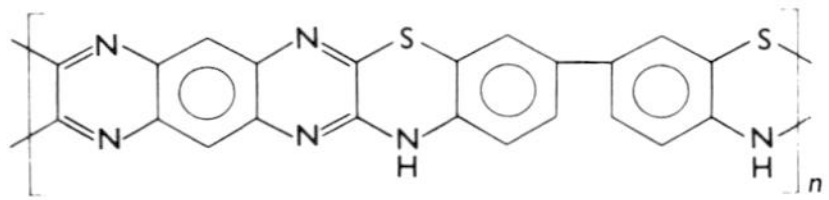 <br>3,3'-Dimercaptobenzidine and<br>2,3,6,7-tetrachloro-1,4,5,8-tetraazaanthracene or<br>2,3,6,7-tetrahydroxy-1,4,5,8-tetraazaanthracene | D,2 | CH₃SO₃H, H₂SO₄ | $\eta_{inh}$ 1.37 | Black | *442, 521, 522* |

| 8 | 4,6-Diamino-1,3-dithiophenol and 2,2',3,3'-tetrachloro-6,6'-bis(quinoxaline) | D,2 | CH$_3$SO$_3$H | $\eta_{inh}$ 1.47 | Brown | *442, 521, 522* |
| 9 | 4,6-Diamino-1,3-dithiophenol and 2,2',3,3'-tetrachloro-6,6'-diquinoxalyl ether | D,2 | CH$_3$SO$_3$H | $\eta_{inh}$ 1.25 | Dark red | *442, 521, 522* |
| 10 | 3,3'-Dimercaptobenzidine and 2,2',3,3'-tetrachloro-6,6'-bis(quinoxaline) | D,2 | H$_2$SO$_4$ | $\eta_{inh}$ 1.52 | Brown | *442, 521, 522* |
| 11 | 3,3'-Dimercaptobenzidine and 2,2',3,3'-tetrachlorodi(quinoxalyl) ether | D,2 | CH$_3$SO$_3$H | $\eta_{inh}$ 1.54 | Dark orange | *442, 521, 522* |

TABLE I.19
Polyquinoxalines and Related Polymers

| No. | Structure | Method | Solubility | Molecular weight | $T_m$ (°C) | Remarks and Property data | References |
|---|---|---|---|---|---|---|---|
| 1 | | E,1 | $H_2SO_4$ | Various D.P. | — | Black | 537, 691 |
| 2 | | E,2; E,3 | $H_2SO_4$ | $\eta_{inh}$ 0.42 | — | $10^{13}$–$10^{14}$ ohm-cm Blue-black. Loses 10% wt. to 600°C (He) | 44, 336 |
| 3 | | E,1 | — | — | — | | 335 |
| 4 | | E,2 | — | — | — | TGA dec. 500°C (air) | 632 |
| 5 | | E,2 | HMP | $\eta_{inh}$ 1.06 | — | TGA dec. 390°C (air); 400°C ($N_2$) | 231, 627 |

1,2,4,5-Tetraaminobenzene and
3,3,6,6-tetramethylcyclohexan-1,2,4,5-tetraone

| No. | Structure / Name | Method | Solvent | η | Temp | Properties | Ref. |
|---|---|---|---|---|---|---|---|
| 6 |  6,7-Diamino-1,4-dimethyl-2,3-dioxy-1,2,3,4-tetrahydroquinoxaline | E,3 | — | — | — | — | 342 |
| 7 | Condensation of 1,2,4,5-tetraaminobenzene with: | | | | | | |
| | (a) 2,3,6,7-Tetrahydroxy-1,4,5,8-tetraazaanthracene | E,3 | $CH_3SO_3H$ | $\eta_{inh}$ 0.8 | >350 | Black | 195 |
| | (b) 2,3,6,7-Tetrachloro-1,4,5,8-tetraazaanthracene | E,3 | $CH_3SO_3H$ | $\eta_{inh}$ 0.6 | >350 | Black | 195 |
| | (c) 2,3,6,7-Tetraphenoxy-1,4,5,8-tetraazaanthracene | E,3 | $CH_3SO_3H$ | $\eta_{inh}$ 0.6 | >350 | Black | 195 |
| | Self-condensation of: | | | | | | |
| | (a) 2,3-Dihydroxy-6,7-diaminoquinoxaline | E,3 | $CF_3CO_2H$, $H_2SO_4$, $CH_3SO_3H$ | $[\eta]$ 0.80; 60,000–70,000 | — | Black. TGA dec. 350°C | 195, 342 |
| | (b) 2,3-Diphenoxy-6,7-diaminoquinoxaline | E,3 | $CH_3SO_3H$ (80% sol.) | $\eta_{inh}$ 1.0 | — | Black | 195, 342 |
| 8 | 1,2,4,5-Tetraaminobenzene and 2,5-dihydroxybenzoquinone | E,2 | HMP | $\eta_{inh}$ 1.40 | — | TGA dec. 525° (air); 535°C ($N_2$) | 429, 570, 629, 631, 632 |

TABLE I.19—*continued*

Polyquinoxalines and Related Polymers

| No. | Structure | Method | Solubility | Molecular weight | $T_m$ (°C) | Remarks and property data | References |
|---|---|---|---|---|---|---|---|
| 9 | X = OH or Cl, F 3,6-Dichloro(difluoro)-2,5-dihydroxy-p-benzoquinone | E,2 | — | — | — | — | 631 |
| 10 | | E,2 | *m*-Cresol | $[\eta]$ 0.97 | — | $T_g$ 254°C; TGA dec. 520°C (vac.) | 687, 689 |
| 11 | | E,2 | *m*-Cresol | $[\eta]$ 0.58 | — | $T_g$ 218°C; TGA dec. 520°C (vac.). | 687, 689 |
| 12 | 1,2,4,5-Tetraaminobenzene and 1,2,6,7-tetraketopyrene | E,2 | HMP, hexafluoroiso-propanol | $\eta_{inh}$ 0.27 | — | TGA dec. 460° (air); 683°C ($N_2$) | 429, 630, 631, 632 |
| 13 | 1,2,4,5-Tetraaminobenzene and 1,2,5,6-tetraketoanthracene | E,2 | 1,3-Dichloro-1,1,3,3-tetra-fluoro-2,2-di-hydroxypropane | $\eta_{inh}$ 2.02 | — | — | 631 |

| No. | Structure / Name | | Solvent | Viscosity | | Properties | Ref. |
|---|---|---|---|---|---|---|---|
| 14 | | E,3 | $H_2SO_4$ | $\eta_{inh}$ 0.43 | — | Loses 20% wt. to 900°C | 44 |
| | 3,3'-Diaminobenzidine | | | | | | |
| 15 | | E,2 | $H_2SO_4$ | $\eta_{inh}$ 1.16 | — | Brown. Tough, flexible cast film. TGA dec. 445° (air); 450°C (He) | 315 |
| | 3,3'-Diaminobenzidine and 1,10-diphenyldecane-1,2,9,10-tetraone | | | | | | |
| 16 | | E,3 | $H_2SO_4$ | $\eta_{inh}$ 0.3 | — | Red | 343 |
| | 3,3'-Diaminobenzidine and 2,3,2',3'-tetraphenoxy-6,6'-bis(quinoxaline) | | | | | | |
| 17 | | E,2 | $H_2SO_4$ | $\eta_{inh}$ 0.51 | 380–400 | Crystalline. Brittle films. TGA dec. 500°C (air) | 191, 192, 570, 634–637, 684 |
| 18 | | E,2 | m-Cresol | $[\eta]$ 2.10 | 480 | Yellow crystalline. $T_g$ 317°C; TGA dec. 540°C (vac.); elongation = 17%; tensile tear strength = $6.6 \times 10^8$ dynes/cm²; dielectric constant = 4.6 | 687–689 |
| 19 | | E,2 | $H_2SO_4$, HMP | $\eta_{inh}$ 2.42 | — | TGA dec. 425°C | 637, 684 |

TABLE I.19—continued

Polyquinoxalines and Related Polymers

| No. | Structure | Method | Solubility | Molecular weight | $T_m$ (°C) | Remarks and property data | References |
|---|---|---|---|---|---|---|---|
| 20 | | E,2 | m-Cresol | [$\eta$] 0.90 | — | $T_g$ 320°C; TGA dec. 540°C (vac.) | 687, 689 |
| 21 | | E,2 | $H_2SO_4$ | $\eta_{inh}$ 1.42 | — | $T_g$ 386°C; TGA dec. 530° (air); 540°C (He) | 317, 318, 320 |
| 22 | | E,2 | $H_2SO_4$ | $\eta_{inh}$ 1.71 | — | $T_g$ 410°C; TGA dec. 530° (air); 550°C (He) | 317, 318, 320 |
| 23 | | E,2 | $H_2SO_4$, HMP, m-Cresol | $\eta_{inh}$ 2.06 | — | $T_g$ 388°C. Yellow. Crystalline. Thermal stability data. Prepolymers used for laminate preparation | 191, 192, 316–318, 320, 462, 570, 637 |
| 24 | | E,2 | $H_2SO_4$ | $\eta_{inh}$ 2.05 | — | $T_g$ 420°C; TGA dec. 520° (air); 550°C (He) | 317, 318, 320 |
| 25 | | E,3 | $H_2SO_4$ | $\eta_{inh}$ 0.3 | — | Copper color. TGA dec. 300°C | 343 |

3,3'-Diaminobenzidine and 2,3,6,7-tetrahydroxy-1,4,5,8-tetraazaanthracene (or the analogous tetraphenoxy compound)

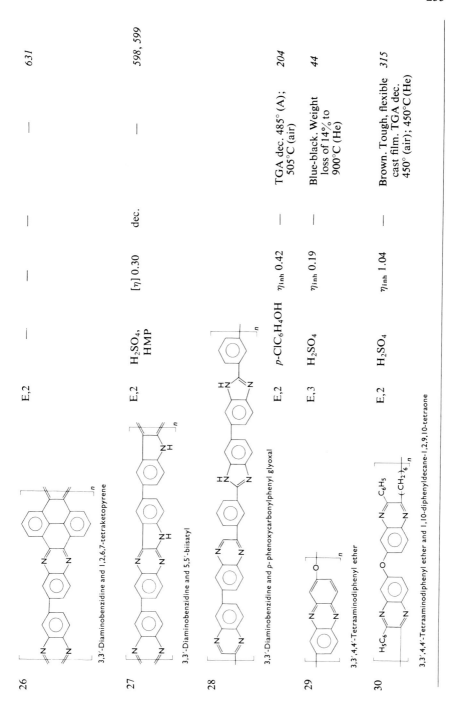

| No. | Structure | Method | Solvent | Viscosity | m.p. | Properties | Ref. |
|---|---|---|---|---|---|---|---|
| 26 | 3,3′-Diaminobenzidine and 1,2,6,7-tetraketopyrene | E,2 | — | — | — | — | 631 |
| 27 | 3,3′-Diaminobenzidine and 5,5′-biisatyl | E,2 | H₂SO₄, HMP | [η] 0.30 | dec. | — | 598, 599 |
| 28 | 3,3′-Diaminobenzidine and p-phenoxycarbonylphenyl glyoxal | E,2 | $p$-ClC₆H₄OH | $\eta_{inh}$ 0.42 | — | TGA dec. 485° (A); 505°C (air) | 204 |
| 29 | 3,3′,4,4′-Tetraaminodiphenyl ether | E,3 | H₂SO₄ | $\eta_{inh}$ 0.19 | — | Blue-black. Weight loss of 14% to 900°C (He) | 44 |
| 30 | 3,3′,4,4′-Tetraaminodiphenyl ether and 1,10-diphenyldecane-1,2,9,10-tetraone | E,2 | H₂SO₄ | $\eta_{inh}$ 1.04 | — | Brown. Tough, flexible cast film. TGA dec. 450° (air); 450°C (He) | 315 |

TABLE I.19—*continued*
Polyquinoxalines and Related Polymers

| No. | Structure | Method | Solubility | Molecular weight | $T_m$ (°C) | Remarks and property data | References |
|---|---|---|---|---|---|---|---|
| 31 | (structure) | E,2 | $H_2SO_4$ | $\eta_{inh}$ 0.35 | 325–330 (soft.) | TGA dec. 400°C | 192, 637, 684 |
| 32 | (structure) | E,2 | m-Cresol | $[\eta]$ 1.30 | — | $T_g$ 298°C; TGA dec. 515°C (vac.) | 687–689 |
| 33 | (structure) | E,2 | $H_2SO_4$, HMP | $\eta_{inh}$ 1.76 | — | — | 637, 684 |
| 34 | (structure) | E,2 | m-Cresol | $[\eta]$ 0.50 | — | $T_g$ 253°C; TGA dec. 520°C (vac.) | 687, 689 |
| 35 | (structure) | E,2 | $H_2SO_4$ | $\eta_{inh}$ 1.75 | — | $T_g$ 390°C; TGA dec. 520° (air); 540°C (He) | 316–318, 320 |
| 36 | (structure) | E,2 | $H_2SO_4$, m-Cresol, HMP | — | — | $T_g$ 345°, 382°C. Red, yellow cast films. Tough and flexible. Amorphous. TGA dec. 480° (air); 540°C (He) | 192, 316–318, 320, 570, 637, 684 |

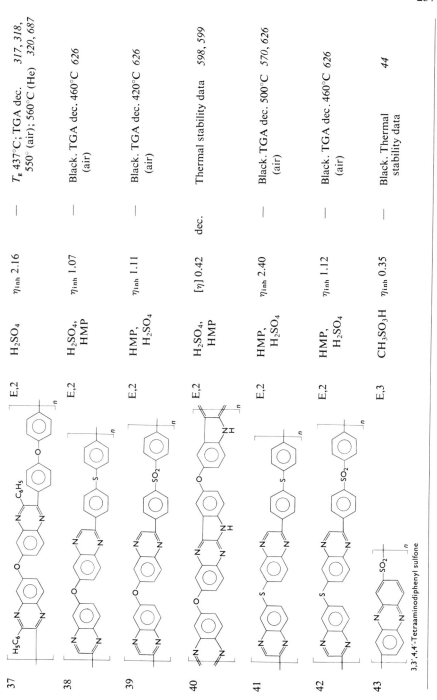

| | | | | | |
|---|---|---|---|---|---|
| 37 | | E,2 | H₂SO₄ | $\eta_{inh}$ 2.16 | — | $T_g$ 437°C; TGA dec.   *317, 318,* <br> 550° (air); 560°C (He)  *320, 687* |
| 38 | | E,2 | H₂SO₄, HMP | $\eta_{inh}$ 1.07 | — | Black. TGA dec. 460°C  *626* <br> (air) |
| 39 | | E,2 | HMP, H₂SO₄ | $\eta_{inh}$ 1.11 | — | Black. TGA dec. 420°C  *626* <br> (air) |
| 40 | | E,2 | H₂SO₄, HMP | $[\eta]$ 0.42 | dec. | Thermal stability data  *598, 599* |
| 41 | | E,2 | HMP, H₂SO₄ | $\eta_{inh}$ 2.40 | — | Black. TGA dec. 500°C  *570, 626* <br> (air) |
| 42 | | E,2 | HMP, H₂SO₄ | $\eta_{inh}$ 1.12 | — | Black. TGA dec. 460°C  *626* <br> (air) |
| 43 | | E,3 | CH₃SO₃H | $\eta_{inh}$ 0.35 | — | Black. Thermal <br> stability data  *44* |

3,3′,4,4′-Tetraaminodiphenyl sulfone

TABLE 1.19—*continued*
Polyquinoxalines and Related Polymers

| No. | Structure | Method | Solubility | Molecular weight | $T_m$ (°C) | Remarks and property data | References |
|---|---|---|---|---|---|---|---|
| 44 | | E,2 | HMP, $H_2SO_4$ | $\eta_{inh}$ 0.80 | — | Brown. TGA dec. 440°C (air) | *626* |
| 45 | | E,2 | HMP, $H_2SO_4$ | $\eta_{inh}$ 1.22 | — | Brown. TGA dec. 420°C (air) | *626* |
| 46 | | E,2 | — | — | — | — | *631* |
| 47 | | E,2 | HMP | $\eta_{inh}$ 0.55 | — | TGA dec. 200° (air); 400°C ($N_2$) | *627* |

2,3,6,7-Tetraaminodibenzo-p-dioxin and 2,5-dihydroxybenzoquinone

2,3,6,7-Tetraaminodibenzo-p-dioxin and 3,3,6,6-tetramethylcyclohexan-1,2,4,5-tetraone

TABLE I.20
Polypiperazines and Polydiketopiperazines

| No. | Structure | Method | Solubility | Molecular weight | $T_m$ (°C) | Remarks and property data | References |
|---|---|---|---|---|---|---|---|
| 1 | | F,1 | — | "Varied" | — | — | 149 |
| 2 | | F,1 | — | Low | — | — | 149 |
| 3 | <br>α,α'-Diaminosuberic acid | F,2 | — | — | — | — | 10, 228 |
| 4 | <br>α,α'-Diaminosebacic acid | F,2 | m-Cresol, $HCO_2H$ | — | — | Copolymers also | 10, 228 |

## TABLE 1.20—*continued*
### Polypiperazines and Polydiketopiperazines

| No. | Structure | Method | Solubility | Molecular weight | $T_m$ (°C) | Remarks and property data | References |
|---|---|---|---|---|---|---|---|
| 5 | *(structure: diketopiperazine ring with N—$(CH_2)_6$ linkage)* | F,2 | $m$-Cresol, $HCO_2H$, acetic acid | — | 235 | Tough. Fibers. Copolymers | *10, 228* |
| | Hexamethylene bis(iminoacetic acid) | | | | | | |
| 6 | *(structure: diketopiperazine with $CH_2CONH(CH_2)_6NHCOCH_2$ chain)* | F,2 | — | — | — | — | *357* |
| 7 | *(structure: diketopiperazine with $(CH_2)_2CONH(CH_2)_6NHCO(CH_2)_2$ chain)* | F,2 | $Cl_2CHCO_2H$ | $[\eta]$ 1.80 | 178–82 | Melt spun to fibers at 200°C. Drawn at 98°C. Tensile strength = 3.1 g/den.; elongation = 18% | *358* |
| 8 | *(structure: diketopiperazine with phenylene ring)* | F,3 | $Cl_2CHCO_2H$, $HCO_2H$ | $\eta_{inh}$ 0.21 | — | Crystalline. Yellow-brown | *339* |

| | | | | | | |
|---|---|---|---|---|---|---|
| 9 | | F,3 | $Cl_2CHCO_2H$, $HCO_2H$ | $\eta_{inh}$ 0.18 | — | Amorphous. Yellow-brown | 339 |
| 10 | | F,3 | $Cl_2CHCO_2H$, $HCO_2H$ | $\eta_{inh}$ 0.33 | — | Crystalline. Yellow-brown | 339 |
| 11 | | F,3 | $Cl_2CHCO_2H$, $HCO_2H$ | $\eta_{inh}$ 0.18 | — | Crystalline. Yellow-brown | 339 |

## TABLE I.21
### Polypyrimidines and Polytetraazopyrenes

| No. | Structure | Method | Solubility | Molecular weight | Remarks and property data | References |
|---|---|---|---|---|---|---|
| | Pyrimidines by nitrile polymerizations | | | | | |
| 1 | Suberonitrile | G,1 | — | — | Tacky, resinous polymer | 333 |
| 2 | Suberonitrile with acetonitrile | G,1 | $CH_2Cl_2$ | 700 (ebull.) | Glassy, solid | 333 |
| 3 | Sebaconitrile | G,1 | Insoluble | — | — | 333 |
| 4 | Sebaconitrile with acetonitrile | G,1 | dil. HCl | — | Viscous liquid to solid resin depending on polymerization conditions | 333 |
| 5 | | G,1 | $Cl_2CHCO_2H$ | $\eta_{inh}$ 0.15 | TGA dec. 320°C; 45% wt. loss to 618°C | 639 |
| 6 | | G,1 | $Cl_2CHCO_2H$ | $\eta_{inh}$ 0.16 | TGA dec. 317°C; 43% wt. loss to 618°C | 639 |
| 7 | | G,1 | $Cl_2CHCO_2H$ | $\eta_{inh}$ 0.21 | TGA dec. 305°C; 45% wt. loss to 618°C | 639 |

| No. | Structure | | Solvent | $\eta_{inh}$ | Properties | Ref. |
|---|---|---|---|---|---|---|
| 8 | | G,1 | $Cl_2CHCO_2H$ | $\eta_{inh}$ 0.15 | TGA dec. 282°C; 43% wt. loss to 618°C | 639 |
| 9 | | G,2 | $H_2SO_4$ | $\eta_{inh}$ 0.25 | Blue-black. TGA dec. 450° (air); 600°C ($N_2$) | 38 |
| 10 | | G,2 | DMSO | $\eta_{inh}$ 0.46 | TGA dec. 410° ($N_2$); 340°C (air). Black. Resistivity $5 \times 10^{13}$ ohm-cm; dielectric constant = 4 (100 megacycles) | 187, 435, 439 |
| 11 | | G,2 | DMSO, $H_2SO_4$ | $\eta_{inh}$ 0.52 | TGA dec. 410° ($N_2$); 340°C (air). Black. Resistivity $1 \times 10^{15}$ ohm-cm; dielectric constant = 3 (100 megacycles) | 187, 435, 439 |
| 12 | | G,2 | $H_2SO_4$ | $\eta_{inh}$ 0.34 | Black | 187, 435, 439 |
| 13 | | G,2 | Insoluble | — | TGA dec. 390° ($N_2$); 360°C (air). Black. Resistivity $2 \times 10^{14}$ ohm-cm; dielectric constant = 3 (100 megacycles) | 187, 435, 439 |

TABLE I.22
Polyquinazolones

| No. | Structure | Method | Solubility | Molecular weight | $T_m$ (°C) | Remarks and property data | References |
|-----|-----------|--------|------------|------------------|------------|---------------------------|------------|
| 1 | | H,3 | Insoluble | — | — | — | 268, 552 |
| 2 | | H,2 | $o$-ClC$_6$H$_4$OH | $\eta_{inh}$ 0.42 | — | Flexible film | 604 |
| 3 | | H,2 | HCO$_2$H | $\eta_{inh}$ 1.02 | — | — | 604 |
| 4 | | H,2 | $o$-ClC$_6$H$_4$OH | $\eta_{inh}$ 0.97 | — | Flexible film | 604 |
| 5 | | H,2 | Cresylic acid | $[\eta]$ 0.62 | — | Flexible film. Lost 50% wt. at 300°C in 100 hr | 592 |

| No. | Structure | | Solvent | Viscosity | | Properties | Ref. |
|---|---|---|---|---|---|---|---|
| 6 | | H,2 | Cresylic acid | $[\eta]$ 0.46 | — | Flexible film. Lost 50% wt. at 300°C in 100 hr | *592* |
| 7 | | H,2 | $o$-ClC$_6$H$_4$OH | $\eta_{inh}$ 0.67 | — | Flexible film. Elongation = 14%. Dec. at 400°C in air or argon | *604* |
| 8 | | H,2 | $m$-Cresol | $[\eta]$ 0.24 | — | — | *193* |
| 9 | | H,3 | DMSO | $[\eta]$ 0.82 | — | — | *595, 640* |
| 10 | | H,3 | H$_2$SO$_4$, HCO$_2$H, NMP, HMP | $\eta_{inh}$ 1.07 | 540–600 | Tough, cast films | *268, 552, 700* |
| 11 | | H,3 | — | — | — | — | *397* |

TABLE I.22—continued
Polyquinazolones

| No. | Structure | Method | Solubility | Molecular weight | $T_m$ (°C) | Remarks and property data | References |
|---|---|---|---|---|---|---|---|
| 12 | | H,3 | $HCO_2H$ | $[\eta]$ 0.95 | — | Brown. TGA dec. 490°C in air or argon | 268, 552, 595 |
| 13 | | H,2 | m-Cresol | — | — | — | 297 |
| 14 | | H,2 | $o\text{-}ClC_6H_4OH$ | $\eta_{\text{inh}}$ 0.4 | — | Dec. 425°C in air or argon | 604–606, 608 |
| 15 | | H,2 | m-Cresol | $[\eta]$ 0.48 | — | — | 193 |
| 16 | | H,2 | m-Cresol | $[\eta]$ 0.88 | — | Cast, yellow, trans-parent film. Dec. 425° (argon); 325°C (air) | 193, 464 |

| No. | Structure | Solubility | Solvent | Viscosity | Softening | Description | Refs. |
|---|---|---|---|---|---|---|---|
| 17 | | H,3 | — | — | >450 | Yellow-brown powder | 344 |
| 18 | | H,2 | $HCO_2H$ | $\eta_{inh}$ 3.68 | — | Dec. 425°C in air or argon | 604–606, 608 |
| 19 | | H,2 | m-Cresol | $[\eta]$ 0.97 | — | Cast, yellow, transparent film. Dec. 425° (argon), 325°C (air) | 193 |
| 20 | | H,3 | $H_2SO_4$ | $\eta_{inh}$ 0.65 | >540 | — | 344 |
| 21 | | H,3 | $o\text{-}ClC_6H_4OH$ | $\eta_{inh}$ 0.2 | — | Flexible film. Dec. 425°C in air or argon | 604–606, 608 |
| 22 | | H,3 | — | — | — | — | 268, 552 |

TABLE I.22—*continued*

Polyquinazolones

| No. | Structure | Method | Solubility | Molecular weight | $T_m$ (°C) | Remarks and property data | References |
|---|---|---|---|---|---|---|---|
| 23 | 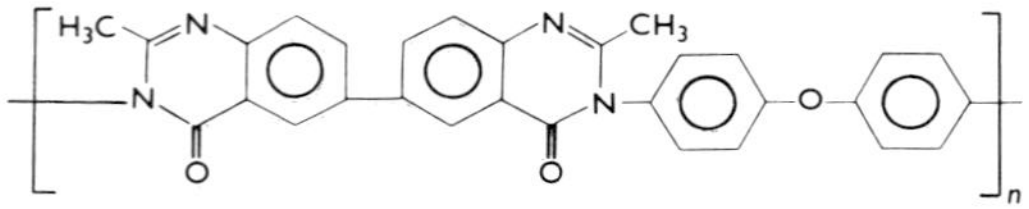 (polyquinazolone; $H_3C$–, $CH_3$, –O–) | H,2 | m-Cresol | $\eta_{inh}$ 2.34 | 325 soft. | Cast films. Elongation = 80%; tensile strength = 13,500 psi; dec. 425°C (argon) | 297, 406 |
| 24 | 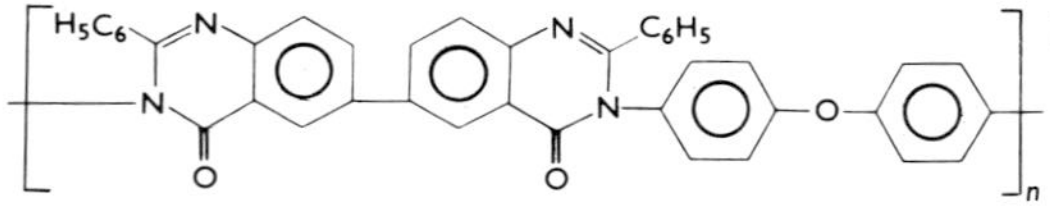 (polyquinazolone; $H_5C_6$–, $C_6H_5$, –O–) | H,2 | o-ClC$_6$H$_4$OH | $\eta_{inh}$ 1.12 | — | Flexible films. Elongation = 11%; tensile strength = 15,500 psi; dec. 425°C in air or argon | 604–606, 608 |
| 25 | (polyquinazolone; –(CH$_2$)$_8$–, CH$_2$) | H,3 | DMA | $\eta_{inh}$ 0.22 | — | Stable to 425° (N$_2$); 525°C (air) | 99 |
|  | 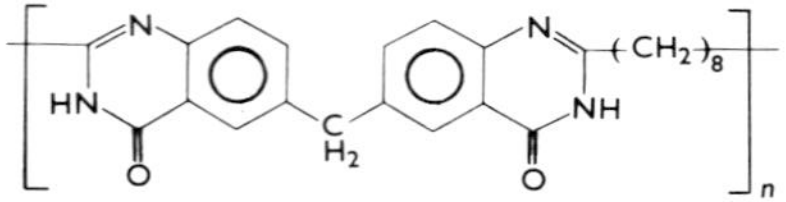  | H,3 | DMSO | $\eta_{inh}$ 0.14 | — | — | 640 |
| 26 | 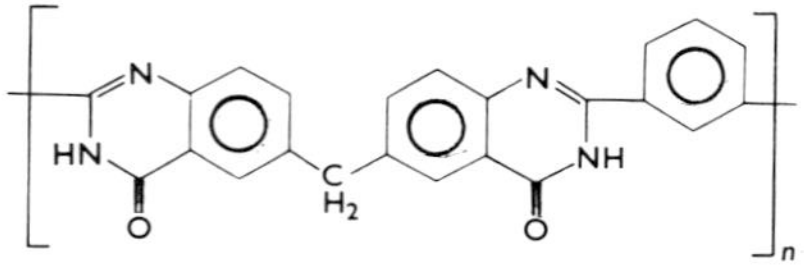 (polyquinazolone; phenylene, CH$_2$) | H,3 | DMSO | $[\eta]$ 0.75 | — | — | 595, 640 |

TABLE I.23

Polyquinazolinediones

| No. | Structure | Method | Solubility | Molecular weight | $T_m$ (°C) | Remarks and property data | References |
|---|---|---|---|---|---|---|---|
| 1 | | H,4 | — | — | — | "Good" high-temperature properties | 23 |
| 2 | | H,4 | $H_2SO_4$ | $\eta_{inh}$ 0.95 | >550 | Crystalline | 458 |
| 3 | | H,4 | — | — | — | "Excellent" thermal stability | 24 |
| 4 | | H,5 | $H_2SO_4$ | $\eta_{inh}$ 0.49 | — | 5% wt. loss at 520° (argon); 455°C (air) | 549–551 |

TABLE I.23—continued

Polyquinazolinediones

| No. | Structure | Method | Solubility | Molecular weight | $T_m$ (°C) | Remarks and property data | References |
|---|---|---|---|---|---|---|---|
| 5 | (structure) | H,4 | $H_2SO_4$, $HCO_2H$, DMSO | $\eta_{inh}$ 1.15 | — | Crystalline. Tough, flexible films. Dec. >550°C | 458 |
| 6 | (structure) | H,4 | $H_2SO_4$, DMA, DMSO, NMP, fuming $HNO_3$ | $\eta_{inh}$ 1.91 | >300 | Tough, cast films. Low crystallinity. Dec. >550°C ($N_2$) | 402, 458, 650, 696, 711, 713 |
| 7 | (structure) | H,4 | — | — | — | — | 650 |
| 8 | (structure) | H,4 | $H_2SO_4$, DMSO, NMP | $\eta_{inh}$ 1.05 | >500 | Tough, flexible, cast films. Dec. 600–800°C | 24, 458, 650, 713 |
| 9 | (structure) | H,4 | $H_2SO_4$ | $\eta_{inh}$ 0.96 | >450 | — | 458 |

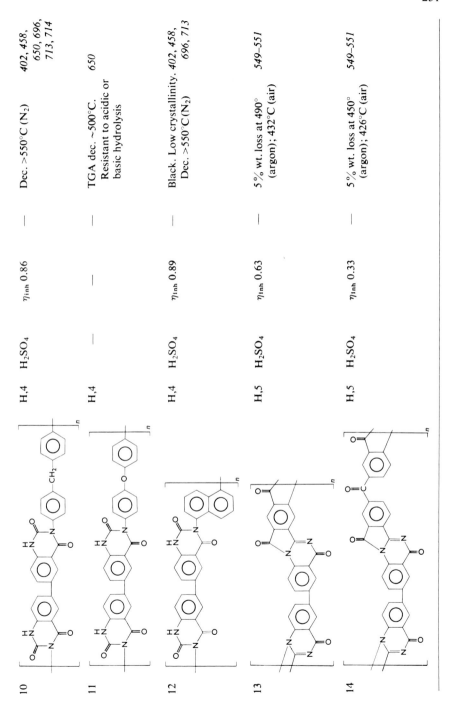

| No. | | Solvent | $\eta_{inh}$ | | Properties | References |
|---|---|---|---|---|---|---|
| 10 | H,4 | H₂SO₄ | $\eta_{inh}$ 0.86 | — | Dec. >550°C (N₂) | 402, 458, 650, 696, 713, 714 |
| 11 | H,4 | — | — | — | TGA dec. ~500°C. Resistant to acidic or basic hydrolysis | 650 |
| 12 | H,4 | H₂SO₄ | $\eta_{inh}$ 0.89 | — | Black. Low crystallinity. Dec. >550°C (N₂) | 402, 458, 696, 713 |
| 13 | H,5 | H₂SO₄ | $\eta_{inh}$ 0.63 | — | 5% wt. loss at 490° (argon); 432°C (air) | 549–551 |
| 14 | H,5 | H₂SO₄ | $\eta_{inh}$ 0.33 | — | 5% wt. loss at 450° (argon); 426°C (air) | 549–551 |

TABLE I.23—*continued*
Polyquinazolinediones

| No. | Structure | Method | Solubility | Molecular weight | $T_m$ (°C) | Remarks and property data | References |
|---|---|---|---|---|---|---|---|
| 15 | | H,5 | $H_2SO_4$ | $\eta_{inh}$ 0.42 | — | 5% wt. loss 455° (argon); 430°C (air) | 549–551 |
| 16 | | H,4 | — | — | — | — | 24 |
| 17 | | H,4 | $H_2SO_4$ | $\eta_{inh}$ 1.01 | >550 | — | 458 |
| 18 | | H,5 | $H_2SO_4$ | — | — | TGA dec. 400°C in air or argon | 404 |
| 19 | | H,4 | $H_2SO_4$ | $\eta_{inh}$ 0.40 | — | — | 696 |

| No. | Structure | | | | | Ref. |
|---|---|---|---|---|---|---|
| 20 | | H,4 | $H_2SO_4$ | $\eta_{inh}$ 0.92 | — | — | *458* |
| 21 | | H,4 | $H_2SO_4$ | $\eta_{inh}$ 1.40 | — | 5% wt. loss at 550°C | *458* |
| 22 | | H,4 | $H_2SO_4$ | $\eta_{inh}$ 1.20 | — | Tough, flexible, heat-resistant, cast films | *458* |
| 23 | | H,4 | $H_2SO_4$ | $[\eta]$ 0.95 | — | Dark yellow. Dec. >550°C ($N_2$) | *142, 458* |
| 24 | | H,4 | $H_2SO_4$ | $\eta_{inh}$ 0.36 | — | — | *696* |
| 25 | | H,4 | $H_2SO_4$ | $\eta_{inh}$ 0.86 | — | — | *458* |

TABLE I.24
Polyacetals and Polyketals

| No. | Structure | Method | Solubility | Molecular weight | $T_m$ (°C) | Remarks and property data | References |
|---|---|---|---|---|---|---|---|
| 1 | $\left[ \text{OCH}_2 - \text{C} < ^{\text{CH}_2-\text{O}}_{\text{CH}_2-\text{O}} > \text{CHCH} \right]_n$ | I,1 | — | — | — | Clear, viscous syrup cast and baked to insoluble film. White, amorphous, insoluble powder | 390, 554 |
| 2 | $\left[ \text{OCH}_2 - \text{C} < ^{\text{CH}_2\text{O}}_{\text{CH}_2\text{O}} > \text{CHCH}_2\text{CH} \right]_n$ | I,1 | — | ~3300 | >300 | Cream color | 179, 180, 385 |
| 3 | $\left[ \text{OCH}_2 - \text{C} < ^{\text{CH}_2\text{O}}_{\text{CH}_2\text{O}} > \text{CH}(\text{CH}_2)_2\text{CH} \right]_n$ | I,1 | Phenol: $m$-cresol (2:1) | $\eta_{sp}$ 0.53 | 235 | 73% Crystalline. Solution-spun, cold-drawable fibers | 171, 385 |
| 4 | $\left[ \text{OCH}_2 - \text{C} < ^{\text{CH}_2\text{O}}_{\text{CH}_2\text{O}} > \text{CH}(\text{CH}_2)_3\text{CH} \right]_n$ | I,1 | Phenol: trichlorophenol (2:1) | $\eta_{sp}$ 0.71 | 250 | 72% Crystalline. Solution-spun, cold-drawable fibers. Many other preparative variations gave lower molecular weights | 46, 171, 178, 179, 180, 292, 385, 560 |
| 5 | $\left[ \text{OCH}_2 - \text{C} < ^{\text{CH}_2\text{O}}_{\text{CH}_2\text{O}} > \text{CHCH}_2\text{CH}(\text{CH}_3)\text{CH}_2\text{CH} \right]_n$ | I,1 | — | — | 208–211 | — | 279 |
| 6 | $\left[ \text{OCH}_2 - \text{C} < ^{\text{CH}_2\text{O}}_{\text{CH}_2\text{O}} > \text{CH} - \text{C} < ^{\text{CH}_3}_{\text{CH}_2\text{OCH}_3} - \text{CH}_2 - \text{CHCH} < ^{\text{CH}_3} \right]_n$ | I,1 | — | — | — | Clear, hard resin. Plasticizer for cellulose acetate. Copolymers with ethylene glycol | 1 |

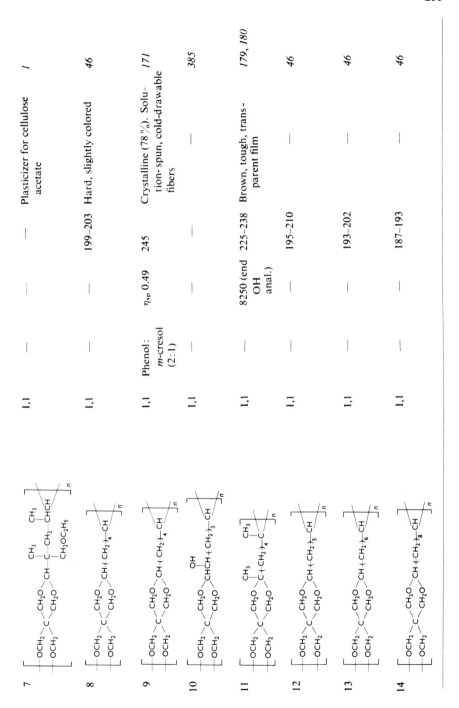

| No. | Structure | I,I | Solvent | η / mol. data | m.p. | Description | Ref. |
|---|---|---|---|---|---|---|---|
| 7 | | I,I | — | — | — | Plasticizer for cellulose acetate | 1 |
| 8 | | I,I | — | — | 199–203 | Hard, slightly colored | 46 |
| 9 | | I,I | Phenol: m-cresol (2:1) | $\eta_{sp}$ 0.49 | 245 | Crystalline (78%). Solution-spun, cold-drawable fibers | 171 |
| 10 | | I,I | — | — | — | — | 385 |
| 11 | | I,I | — | 8250 (end OH anal.) | 225–238 | Brown, tough, transparent film | 179, 180 |
| 12 | | I,I | — | — | 195–210 | — | 46 |
| 13 | | I,I | — | — | 193–202 | — | 46 |
| 14 | | I,I | — | — | 187–193 | — | 46 |

TABLE I.24—*continued*
Polyacetals and Polyketals

| No. | Structure | Method | Solubility | Molecular weight | $T_m$ (°C) | Remarks and property data | References |
|---|---|---|---|---|---|---|---|
| 15 | | 1,1 | — | — | 170–172 | Fiber properties: tensile strength = 1.7 g/den.; Young's modulus = 35.0 g/den.; elongation = 10.7%; Film properties: tensile strength = 4770 psi; Modulus = 185,000 psi; elect. resistivity $10^{16}$–$10^{17}$ ohm-cm; dielectric constant 2.4 | 46 |
| 16 | | 1,1 | Hexafluoro-isopropa-nol, sulfuric acid | [$\eta$] 0.049; 10,000 | 350 | White. Crystalline (95%). TGA$_{1/2}$ dec. at 450°C. Copolymers with cyclo-hexanone | 43, 672 |
| 17 | Mixed 1,3- and 1,4-dialdehydes | 1,1 | — | — | — | Brown, viscous, hard resin. Plasticizes cellulose acetate | 1 |
| 18 | | 1,1 | Hexafluoro-isopropa-nol, DMSO (hot) | [$\eta$] 0.092; ~30,000 | 300d | Highly crystalline. TGA$_{1/2}$ dec. at 365°C | 43, 672 |
| 19 | Copolymer of 16 and 18 | 1,1 | Hexafluoro-isopropa-nol | [$\eta$] 0.045 | — | Crystalline | 43, 672 |

| | | | | | | | |
|---|---|---|---|---|---|---|---|
| 20 | 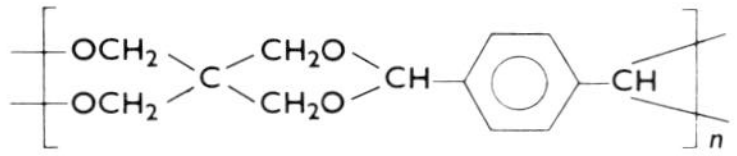 | I,1 | — | 2000 | >300 | — | *180* |
| 21 | 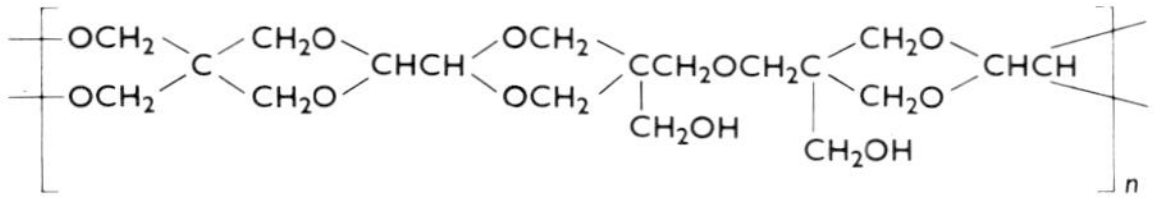 Glyoxal and **88**/12 pentaerythritol/dipentaerythritol | I,1 | — | 200 | — | — | *179*, *180*, *385* |
| 22 | 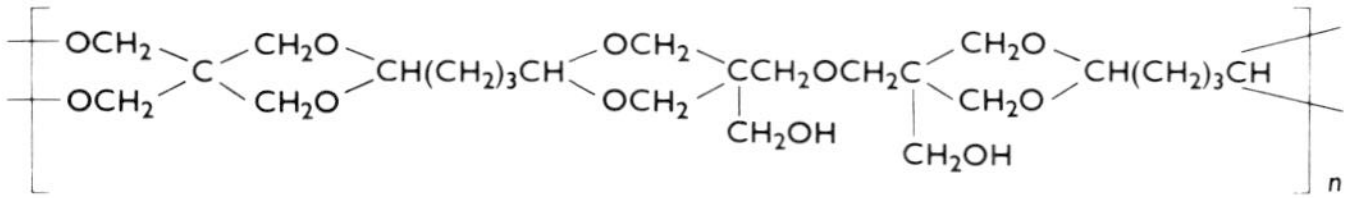 Glutarldehyde and **88**/12 pentaerythritol/dipentaerythritol | I,1 | Pyridine/ $Ac_2O$ | $[\eta]$ 0.28; ~10,000 | 233–238 | Crystalline (20–40%) | *178*, *179* |
| 23 | As in 22, but with 76/24 pentaerythritol/dipentaerythritol | I,1 | Pyridine/ $Ac_2O$ | $[\eta]$ 0.18; 1200 | 181–186 | Crystalline (20–40%) | *178* |
| 24 | As in 22, but with 3-methylglutaraldehyde | I,1 | Cresylic acid | ~9000 | 205–208 | — | *179*, *180*, *385* |
| 25 | 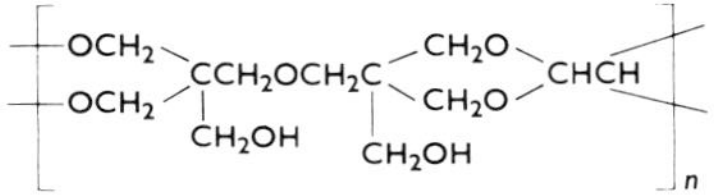 | I,1 | — | — | — | Syrupy product. Cast, baked films. Used for treating textiles | *390* |
| 26 | 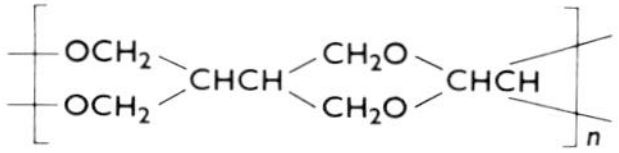 | I,1 | — | — | — | — | *585* |

TABLE I.24—continued
Polyacetals and Polyketals

| No. | Structure | Method | Solubility | Molecular weight | $T_m$ (°C) | Remarks and property data | References |
|---|---|---|---|---|---|---|---|
| 27 | $\left[OCH_2-CH(CH_2)_6-CH\big\langle{}^{CH_2O}_{CH_2O}\big\rangle CH(CH_2)_4-CH\right]_n$ | 1,1 | — | — | 200–204 | — | 46 |
| 28 | $\left[OCH_2-CH(CH_2)_6-CH\big\langle{}^{CH_2O}_{CH_2O}\big\rangle CH(CH_2)_6-CH\right]_n$ | 1,1 | — | — | 167–172 | Gray, transparent | 46 |
| 29 | $\left[OCH_2-CH(CH_2)_m-CH\big\langle{}^{CH_2O}_{CH_2O}\big\rangle CH(CH_2)_x-CH\right]_n$  $m = x = 6{-}10$ | 1,1 | Chlorinated hydrocarbons | — | 146–205 | Yellow-white solids. Film and fiber forming. Dielectric constant 2.1–2.4. Resistivity $10^{16}$–$10^{17}$ ohm-cm | 585 |

### Five-Membered Ring Polyacetals

| No. | Structure | Method | Solubility | Molecular weight | $T_m$ (°C) | Remarks and property data | References |
|---|---|---|---|---|---|---|---|
| 30 | $\left[{}^{OCH_2}_{OCHCH_2OC}\big\langle(CH_2)_7-CH\right]_n$  (O on the OC group) | 1,2 | — | 6000 | — | — | 449 |
| 31 | $\left[{}^{OCH_2}_{OCHCH_2OCHCCH_2CHCH}\big\langle\right]_n$ with $CH_3$, $CH_3$, $CH_2OCH_3$ | 1,1 | — | — | — | Dark-semisolid resin | 1 |
| 32 | $\left[OCH_2\,{}^{CH_2O}_{OCH(CH_2)_2CHO}CH(CH_2)_3-CH\right]_n$ | 1,1 | Xylene | — | — | Clear, cast coatings | 560 |

| No. | Structure | | Solvent | Mol wt | Temp (°C) | Uses | Ref |
|---|---|---|---|---|---|---|---|
| 33 | $\left[\begin{array}{c}OCH_2\\OCHCH_2OCH_2CHO\end{array}\; {}^{CH_2O}\diagdown CH(CH_2)_3\diagup {}^{CH}\right]_n$ | I,1 | — | — | — | Clear, cast coatings | *560* |
| | ***Polythioketals*** | | | | | | |
| 34 | $\left[\begin{array}{c}SCH_2\\SCH_2\end{array}C\diagdown{}^{CH_3}_{CH_2S}\diagup C(CH_2)_8\diagdown{}^{CH_3}_{C}\right]_n$ | I,1 | $C_6H_6$ | 1090 | 153–167 | — | *127, 229* |
| 35 | $\left[\begin{array}{c}SCH_2\\SCH_2\end{array}C\diagdown{}^{CH_2S}_{CH_2S}\diagup\bigcirc\!\!S\right]_n$ | I,1 | Naphthalene | 613 (cryos.) | 314–318 | — | *127, 229* |

TABLE I.25
Polyquinones

| No. | Structure | Method | Solubility | Molecular weight | $T_m$ (°C) | Remarks and property data | References |
|---|---|---|---|---|---|---|---|
| 1 | | J,1 | Insoluble | ~6000 | — | Black. Reported to be amorphous when prepared in solution. Crystalline from bulk prep. Sp. conduct. $1.5 \times 10^{-5}$ | 102, 302, 455 |
| 2 | | J,1 | Insoluble | — | — | Black. Prepared from precursor polyether by treatment with S and $I_2$ | 101 |
| 3 | | J,1 | — | 5500 | >390 | Dark brown. Semiconductor. Heat-stable pigments | 456 |
| 4 | | J,1 | DMA, DMSO, NMP, HMP, $H_2SO_4$ | $\eta_{red}$ 0.08 | — | Brown. Crystalline. TGA dec. 250–300°C | 417 |

| | | | | | |
|---|---|---|---|---|---|
| 5 | | J,1 | — | ~5500 | >395d | Red-brown. Molded at 300°C/250 atm. Conductivity $8.9 \times 10^{-4}$ mho/cm | 454 |
| 6 | | J,1 | Insoluble | ~2700 | >365d | Molded at 250°C/300 atm | 455 |
| 7 | | J,1 | Insoluble | ~3600 | >375d | Brown-black. Pigment | 455 |
| 8 | | J,1 | Insoluble | 3000–4000 | 405d | Red-brown. Sp. conduct. $= 1.6 \times 10^{-6}$ mho/cm. Molded at 280°C/300 atm | 457 |

## REFERENCES

1. Abbott, L. S., Faulkner, D., and Hollis, C. E., U.S. Pat. 2,739,972 (Distillers Co., Ltd.) (1956); *Chem. Abstr.* **50**, 13513 (1956).
2. Abramo, J. G., U S Pat. 3,342,897 (E. I. duPont de Nemours Co.) (1967); *Chem. Abstr.* **67**, 10312, 109185y (1967).
3. Adrova, N. A., Bessonov, M. I., Dubnova, A. M., Koton, M. M., Moskvina, E. M., and Rudakov, A. P., Russ. Pat. 189,574 (Institute of High Molecular Weight Compounds, Academy of Sciences, USSR) (1966); *Chem. Abstr.* **68**, 366, 3657j (1968).
4. Adrova, N. A., Koton, M. M , and Benina, I. N., *Vysokomol. Soedin.*, Ser. *B* **10**, 137 (1968); *Chem. Abstr.* **68**, 9301, 96238m (1968).
5. Adrova, N. A., Koton, M. M., and Dubnova, A. M., *Vysokomol. Soedin , Ser. B.* **10**, 354 (1968); *Chem. Abstr.* **69**, 2619, 27956c (1968).
6. Adrova, N. A., Koton, M. M., and Ivanova, L. P., *Vysokomol. Soedin.*, Ser. *B* **9**, 22 (1967); *Chem. Abstr.* **66**, 7183, 76364e (1967).
7. Adrova, N. A., Koton, M. M., and Moskvina, E. M., *Dokl. Chem. (English transl.)* **165**, 1171 (1965); *Chem. Abstr.* **64**, 12815b (1966).
8. Adrova, N. A., Koton, M. M., and Prokhorova, L. K., *Dokl. Chem. (English transl.)* **177**, 1058 (1967); *Chem. Abstr* **68**, 3923, 40176x (1968).
9. Agolini, F., and Gay, F. P., *Polym. Prepr., Amer. Chem. Soc., Div. Polym. Chem.* **10** (1), 175 (1969).
10. Allen, S. J., and Drewitt, J. G. N., Brit. Pat. 610,304 (Celanese Co) (1948); *Chem. Abstr.* **43**, 4519 (1949).
11. Amborski, L. E., *Ind. Eng. Chem., Prod. Res. Develop.* **2**, 189 (1963); *Chem. Abstr.* **59**, 7719 (1963).
12. Amborski, L. E., *Polym. Prepr., Amer. Chem. Soc., Div. Polym. Chem.* **4** (1), 175 (1963).
13. Amborski, L. E,, and Weisenberger, W. P., Belg. Pat. 638,688 (E. I. duPont de Nemours Co.) (1964); *Chem. Abstr.* **62**, 6641 (1965).
14. Amborski, L. E., and Weisenberger, W. P., U.S. Pat. 3,310,506 (E. I. duPont de Nemours Co.) (1967).
15. Angelo, R. J., U.S. Pat. 3,073,785 (E. I. duPont de Nemours Co.) (1963); *Chem. Abstr.* **58**, 7006 (1963).
16. Angelo, R. J., U.S. Pat. 3,244,675 (E. I. duPont de Nemours Co.) (1966); *Chem. Abstr.* **65**, 822 (1966).
17. Angelo, R. J., U.S. Pat. 3,278,493 (E. I. duPont de Nemours Co.) (1966); *Chem. Abstr.* **67**, 3158, 33147q (1967).
18. Angelo, R. J., U.S. Pat. 3,282,897 (E. I. duPont de Nemours Co.) (1966); *Chem. Abstr.* **66**, 2846, 29538e (1967).
19. Angelo, R. J., U.S. Pat. 3,282,898 (E. I. duPont de Nemours Co.) (1966).
20. Angelo, R. J., U.S. Pat. 3,316,211 (E. I. duPont de Nemours Co.) (1967); *Chem. Abstr.* **67**, 2152, 22408z (1967).
21. Angelo, R. J., U.S. Pat. 3,326,850 (E. I. duPont de Nemours Co.) (1967); *Chem. Abstr.* **67**, 4191, 44416c (1967).
22. Angelo, R. J., U.S. Pat. 3,328,351 (E. I. duPont de Nemours Co.) (1967); *Chem. Abstr.* **67**, 6140, 64992r (1967).
23. Angelo, R. J., U.S. Pat. 3,345,333 (E. I. duPont de Nemours Co.) (1967); *Chem. Abstr.* **67**, 11099, 117576s (1967).
24. Angelo, R. J., U.S. Pat. 3,345,334 (E. I. duPont de Nemours Co.) (1967); *Chem. Abstr.* **67**, 1099, 117577t (1967).

25. Angelo, R. J., U.S. Pat. 3,345,342 (E. I. duPont de Nemours Co.) (1967).
26. Angelo, R. J., U.S. Pat. 3,420,795 (E. I. duPont de Nemours Co.) (1969).
27. Angelo, R. J., U.S. Pat. 3,424,718 (E. I. duPont de Nemours Co.) (1969).
28. Angelo, R. J., and Kreuz, J. A., U.S. Pat. 3,336,258 (E. I. duPont de Nemours Co.) (1967); *Chem. Abstr.* **67**, 7814, 82692x (1967).
29. Angelo, R. J., and Tatum, W. E., U.S. Pat. 3,316,212 (E. I. duPont de Nemours Co.) (1967).
30. Anonymous, *Chem. Eng.* **70**, 68 and 70 (1963).
31. Anonymous, *Mater. Des. Eng.* **57**, 145 (1963); *Chem. Abstr.* **58**, 6984 (1963).
32. Anonymous, *WADC Tech Rep.* TR-58-387, Part VII (1963).
33. Anonymous, *Mod. Plast.*, **42**, 117–118, 120, 122, 126, 128, 130, 132, 135–136, 159–162, 164–165, 171–173, 175, 177, and 179–180 (1965); *Chem. Abstr.* **62**, 10597 (1965).
34. Anonymous, *Mod. Plast.* **43**, 123 (1966).
35. Anonymous, *Reinf. Plast. (Boston)* **6**, 33 (1967).
36. Ardashnikov, A. Ya., Kardash, I. E., Kotov, B. V., and Pravednikov, A. N., *Dokl. Chem. (English transl.)* **164**, 1006 (1965).
37. Arnold, F. E., *Diss. Abstr.* **26**, 5713 (1966); *Chem. Abstr.* **65**, 3965 (1966).
38. Arnold, F. E., and Van Deusen, R. L., *J. Polym. Sci., Part B* **6**, 815 (1968).
39. Asahara, T., and Fukui, M., *Kogyo Kagaku Zasshi* **70**, 2388 (1967); *Chem. Abstr.* **69**, 1862, 19643h (1968).
40. Asahara, T., and Fukui, M., *Kogyo Kagaku Zasshi* **71**, 918 (1968); *Chem. Abstr.* **69**, 6341, 67788f (1968).
41. Australian Patent 58424 (E. I. duPont de Nemours Co.) (1960).
42. Australian Patent 19,579/67 (I.C.I. Ltd.) (1968).
43. Bailey, W. J., and Volpe, A. A., *Polym. Prepr., Amer. Chem. Soc., Div. Polym. Chem.* **8** (1), 292 (1967).
44. Banihashemi, A., Fabbro, D., and Marvel, C. S., *J. Polym. Sci., Part A-1* **7**, 2293 (1969); *Chem. Abstr.* **71**, 102308q (1969).
45. Barito, R. W., U.S. Pat. 3,364,166 (General Electric Co.) (1968); *Chem. Abstr.* **68**, 51062 (1968).
46. Belgian Patent 568,181 (European Research Associates) (1958); *Chem. Abstr.* **53**, 13660 (1959).
47. Belgian Patent 649,335 (E. I. duPont de Nemours Co.) (1964).
48. Belgian Patent 649,336 (E. I. duPont de Nemours Co.) (1964); *Chem. Abstr.* **64**, 12838 (1966).
49. Belgian Patent 650,774 (Monsanto Co.) (1964).
50. Belgian Patent 650,979 (Standard Oil Co.) (1964).
51. Belgian Patent 651,380 (Cella Lackfabrik, Dr. C. Schleussner G.m.b.H.) (1964); *Chem. Abstr.* **64**, 17865 (1966).
52. Belgian Patent 654,420 (Cella Lackfabrik, Dr. C. Schleussner G.m.b.H.) (1965); *Chem. Abstr.* **64**, 17865 (1966).
53. Belgian Patent 654,849 (E. I. duPont de Nemours Co.) (1965).
54. Belgian Patent 654,850 (E. I. duPont de Nemours Co.) (1965).
55. Belgian Patent 655,562 (E. I. duPont de Nemours Co.) (1965).
56. Belgian Patent 655,654 (E. I. duPont de Nemours Co.) (1965); *Chem. Abstr.* **65**, 2376 (1966).
57. Belgian Patent 656,047 (E. I. duPont de Nemours Co.) (1965).
58. Belgian Patent 656,048 (E. I. duPont de Nemours Co.) (1965).
59. Belgian Patent 656,049 (E. I. duPont de Nemours Co.) (1965).
60. Belgian Patent 656,050 (E. I. duPont de Nemours Co.) (1965).

61. Belgian Patent 657, 246 (E. I. duPont de Nemours Co.) (1965).
62. Belgian Patent 672,175 (I.C.I. Ltd.) (1966); *Chem. Abstr.* **65**, 13961 (1966).
63. Belgian Patent 672,858 (Imperial Metal Ind. Ltd.) (1966); *Chem. Abstr.* **66**, 9878, 105436n (1967).
64. Belgian Patent 672,985 (Farbenfabriken Bayer) (1966).
65. Belgian Patent 677,856 (American Cyanamid Co.) (1966).
66. Belgian Patent 677,907 (American Cyanamid Co.) (1966).
67. Belgian Patent 678,689 (E. I. duPont de Nemours Co.) (1966).
68. Belgian Patent 680,335 (Gelsenberg Benzin A.G.) (1966).
69. Belgian Patent 682,875 (Rhone Poulenc) (1966).
70. Belgian Patent 683,284 (E. I. duPont de Nemours Co.) (1966).
71. Belgian Patent 684,154 (Monsanto Co.) (1967).
72. Belgian Patent 685,241 (Rhone Poulenc) (1967).
73. Belgian Patent 688,028 (American Cyanamid Co.) (1967).
74. Belgian Patent 689,400 (Gelsenberg Benzin A.G.) (1967).
75. Belgian Patent 690,754 (E. I. duPont de Nemours Co.) (1967).
76. Belgian Patent 691,900 (Farbenfabriken Bayer) (1967).
77. Belgian Patent 692,208 (E. I. duPont de Nemours Co.) (1967).
78. Belgian Patent 694,121 (Philips N.V.) (1967).
79. Belgian Patent 695,165 (E. I. duPont de Nemours Co.) (1967).
80. Belgian Patent 695,166 (E. I. duPont de Nemours Co.) (1967).
81. Belgian Patent 695,454 (E. I. duPont de Nemours Co.) (1967).
82. Belgian Patent 695,459 (E. I. duPont de Nemours Co.) (1967).
83. Belgian Patent 696,713 (Nat. Res. Dev. Corp.) (1967).
84. Belgian Patent 702,772 (General Mills, Inc.) (1968).
85. Belgian Patent 705,977 (E. I. duPont de Nemours Co.) (1968).
86. Belgian Patent 706,049 (Nat. Res. Dev. Corp.) (1968).
87. Belgian Patent 707,978 (E. I. duPont de Nemours Co.) (1968).
88. Belgian Patent 708,341 (Monsanto Co.) (1968).
89. Belgian Patent 709,375 (Farbwerke Hoechst) (1968).
90. Belgian Patent 710,906 (Nat. Res. Dev. Corp.) (1968).
91. Belgian Patent 711,240 (Soc. Rhodiaceta) (1968).
92. Belgian Patent 711,729 (Farbenfabriken Bayer) (1968).
93. Belgian Patent 711,891 (E. I. duPont de Nemours Co.) (1968).
94. Belgian Patent 713,683 (1968).
95. Belgian Patent 714,069 (Rhone Poulenc) (1968).
96. Belgian Patent 715,661 (Soc. Rhodiaceta) (1968).
97. Bell, V. L., *J. Polym. Sci., Part B* **5**, 941 (1967); *Chem. Abstr.* **67**, 6469, 100459h (1967).
98. Bell, V. L., *Stabil. Plast., Soc. Plast. Eng., Reg. Tech. Conf., Washington, D.C.* B1-B9 (1967); *Chem. Abstr.* **69**, 6359, 67978t (1968).
99. Belohlav, L. R., and Costanza, J. R., U.S. Pat. 3,444,136 (Celanese Corp.) (1969); *Chem. Abstr.* **71**, 22469c (1969); equivalent to Ger. Pat. 1,806,295 (1969); *Chem. Abstr.* **72**, 22156u (1970).
100. Berlin, A. A., Belova, G. V., Liogon'kii, B. I., and Shamraev, G. M., *Vysokomol. Soedin., Ser. A* **10**, 1561 (1968); *Chem. Abstr.* **69**, 7305, 77924m (1968).
101. Berlin, A. A., Liogon'kii, B. I., and Gurov, A. A., *Vysokomol. Soedin., Ser. A* **10**, 1590 (1968); *Polym. Sci. USSR* **10**, 1841 (1968).
102. Berlin, A. A., Liogon'kii, B. I., Gurov, A. A., and Razvadovskii, E. F., *Vysokomol. Soedin., Ser. A* **9**, 532 (1967); *Polym. Sci. USSR* **9**, 596 (1967); *Chem. Abstr.* **67**, 3147, 33030w (1967).

103. Berlin, A. A., Liogon'kii, B. I., and Zapadinskii, B. I., *Vysokomol. Soedin., Ser. B* **10**, 315 (1968); *Chem. Abstr.* **69**, 2624, 28013e (1968).
104. Berlin, A. A., and Matveeva, N. G., *Dokl. Akad. Nauk SSSR* **140**, 368 (1961); *Dokl. Chem. (English transl.)* **140**, 899 (1961); *Resins, Rubbers, Plastics* p. 781 (1962)
105. Berlin, A. A., and Matveeva, N. G., *Dokl. Chem. (English transl.)* **167**, 283 (1966).
106. Berlin, A. A., Sherle, A. I., Kuzina, V. V., and Fomin, G. V., *Vysokomol. Soedin., Ser. A* **10**, 1013 (1968).
107. Bernier, G. A., and Kline, D. E., *J. Appl. Polym. Sci.* **12**, 593 (1968); *Chem. Abstr.* **68**, 10218, 105700k (1968).
108. Binsack, R., Bottenbruch, L., and Schnell, H., S. Afr. Pat. 68/08,037 (Farbenfabriken Bayer) (1969); *Chem. Abstr.* **72**, 32454d (1970).
109. Blomstrom, D. C., U.S. Pat. 3,207,728 (E. I. duPont de Nemours Co.) (1965); *Chem. Abstr.* **64**, 701 (1966).
110. Bogert, M. T., and Renshaw, R. R., *J. Amer. Chem. Soc.* **30**, 1135 (1908); *Chem. Abstr.* **2**, 2681 (1908).
111. Boldebuck, E. M., and Klebe, J. F., Fr. Pat. 1,437,751 (Thomson-Houston) (1966); *Chem. Abstr.* **66**, 1863, 19155p (1967).
112. Boldyrev, A. G., Adrova, N. A., Bessonov, M. I., Koton, M. M., Kuvshinskii, E. V., Rudakov, A. P., and Florinskii, F. S., *Dokl. Chem. (English transl.)* **163**, 753 (1965); *Chem. Abstr.* **64**, 4900 (1966).
113. Bolton, B. A., and Dieterle, R. J., *Plast. Des. Process* **7**, 13 (1967); *Chem. Abstr.* **66**, 4454, 46868c (1967).
114. Bolton, B. A., and Stephens, J. R., U.S. Pat. 3,320,202 (Standard Oil Co. Indiana) (1967).
115. Bottenbruch, L., and Schnell, H., U.S. Pat. 3,379,686 (Farbenfabriken Bayer) (1968); equivalent to Br. Pat. 1,129,617 and Belg. Pat. 670,676; *Chem. Abstr.* **65**, 10696 (1966).
116. Bottex, P., Sillion, B., and de Gaudemaris, G., Fr. Pat. 1,506,282 (Institut Français du Petrole, des Carburants et Lubrifiants) (1967); *Chem. Abstr.* **70**, 469, 4809p (1969).
117. Bottex, P., Sillion, B., and de Gaudemaris, G., *C. R. Acad. Sci., Ser. C* **267**, 711 (1968); *Chem. Abstr.* **70**, 12031u (1969).
118. Bottger, H., U.S. Pat. 3,415,903 (Cella Lackfabrik, Dr. C. Schleussner G.m.b.H.) (1968).
119. Bower, G. M., Fr. Pat. 1,461,155 (Westinghouse Electric Corp.) (1966); *Chem. Abstr.* **67**, 1156, 11980d (1967).
120. Bower, G. M., Freeman, J. H., Traynor, E. J., Jr., Frost, L. W., Burgman, H. A., and Ruffing, C. R. *J. Polym. Sci., Part A-1* **6**, 877 (1968); *Chem. Abstr.* **68**, 9318, 96439c (1968).
121. Bower, G. M., and Frost, L. W., *Polym. Prepr., Amer. Chem. Soc., Div. Polym. Chem.* **4** (1), 357 (1963); *Chem. Abstr.* **62**, 641 (1965).
122. Bower, G. M., and Frost, L. W., *J. Polym. Sci., Part A-1*, 3135 (1963); *Resins, Rubbers, Plastics* p. 1597 (1964).
123. Bradshaw, J. S., and Stevens, M. P., *J. Appl. Polym. Sci.* **10**, 1809 (1966).
124. Brebner, D. L., Edwards, W. M., Robinson, I. M., Squire, E. N., and Starkweather, H. W., U.S. Pat. 2,944,993 (E. I. duPont de Nemours Co.) (1960); *Chem. Abstr.* **54**, 25975 (1960).
125. Breed, L. W., Haggerty, W. J., and Harvey, J., *J. Org. Chem.* **25**, 1804 (1960); *Resins, Rubbers, Plastics* p. 1245 (1961).
126. British Patent 570,858 (E. I. duPont de Nemours Co.) (1945); *Chem. Abstr.* **40**, 7650 (1946).
127. British Patent 577,205 (Imperial Chemical Industries) (1946); *Chem. Abstr.* **42**, 5718 (1948).

128. British Patent 903,272 (E. I. duPont de Nemours Co.) (1962); *Chem. Abstr.* **58**, 9256 (1963).

129. British Patent 935,388 (Westinghouse Electric Corp.) (1963); *Chem. Abstr.* **59**, 14176 (1963).

130. British Patent 973,377 (Dr. Beck & Co.) (1964); *Chem. Abstr.* **62**, 5415 (1965).

131. British Patent 980,274 (E. I. duPont de Nemours Co.) (1965); *Chem. Abstr.* **62**, 16480 (1965).

132. British Patent 982,914 (E. I. duPont de Nemours Co.) (1965); *Chem. Abstr.* **63**, 1966 (1965).

133. British Patent 996,649 (Dr. Beck & Co.) (1965).

134. British Patent 1,028,887 (Dr. Beck & Co.) (1966).

135. British Patent 1,030,026 (Dr. Beck & Co.) (1966); *Chem. Abstr.* **65**, 10807b (1966).

136. British Patent 1,032,649 (Westinghouse Electric Corp.) (1966); *Chem. Abstr.* **65**, 13904 (1966).

137. British Patent 1,035,428 (American Cyanamid Co.) (1966); *Chem. Abstr.* **65**, 13913 (1966).

138. British Patent 1,044,813 (E. I. duPont de Nemours Co.) (1966); *Chem. Abstr.* **65**, 15621 (1966).

139. British Patent 1,059,929 (3M Co.) (1967).

140. British Patent 1,063,038 (Monsanto Co.) (1967).

141. British Patent 1,067,541 (Dr. K. Herberts & Co.) (1967).

142. British Patent 1,060,576 (Toyo Rayon) (1967); *Chem. Abstr.* **67**, 327, 3331c (1967).

143. British Patent 1,067,542 (Dr. K. Herberts & Co.) (1967).

144. British Patent 1,070,364 (Dr. Beck & Co.) (1967); *Chem. Abstr.* **67**, 65487s (1967).

145. British Patent 1,072,334 (E. I. duPont de Nemours Co.) (1967).

146. British Patent 1,075,284 (Dr. Beck & Co.) (1967); *Chem. Abstr.* **68**, 51058m (1968).

147. British Patent 1,083,168 (E. I. duPont de Nemours Co.) (1967).

148. British Patent 1,090,726 (General Electric Co.) (1967).

149. British Patent 1,099,065 (Soc. Rhodiaceta) (1968); *Chem. Abstr.*, **66**, 56000e (1967); equivalent to Fr. Pat. 1,448,914 and Neth. Pat. 66/08398.

150. British Patent 1,103,218 (General Electric Co.) (1968).

151. British Patent 1,105,437 (Soc. Rhodiaceta) (1968).

152. British Patent 1,109,079 (E. I. duPont de Nemours Co.) (1968); *Chem. Abstr.* **68**, 11120, 115225y (1968).

153. British Patent 1,109,080 (E. I. duPont de Nemours Co.) (1968); *Chem. Abstr.* **69**, 1056, 10964f (1968).

154. British Patent 1,117,070 (American Machine & Foundry Co.) (1968).

155. British Patent 1,119,791 (Sumitomo Electric Ind., Ltd.) (1968); *Chem. Abstr.* **69**, 6363, 68017r (1968).

156. British Patent 1,123,798 (General Electric Co.) (1968); *Chem. Abstr.* **69**, 9140, 97359j (1968).

157. Bruck, S. D., *Polymer* **5**, 435 (1964); *Chem. Abstr.* **62**, 5354 (1965); *Resins, Rubbers, Plastics* p. 1135 (1965).

158. Bruck, S. D., *Polym. Prepr., Amer. Chem. Soc., Div. Polym. Chem.* **5** (1), 148 (1964).

159. Bruck, S. D., *Polym. Prepr., Amer. Chem. Soc., Div. Polym. Chem.* **5** (2), 466 (1964); *Chem. Abstr.* **64**, 12825a (1966).

160. Bruck, S. D., *Polym. Prepr., Amer. Chem. Soc., Div. Polym. Chem.* **6** (1), 28 (1965).

161. Bruck, S. D., *Polymer* **6**, 49 (1965); *Chem. Abstr.* **62**, 10535 (1965); *Resins, Rubbers, Plastics* p. 2065 (1965).

162. Bruck, S. D., *J. Chem. Educ.* **42**, 18 (1965); *Chem. Abstr.* **62**, 9303 (1965).

163. Bruck, S. D., *J. Polym. Sci. Part C* **17**, 169 (1967); *Chem. Abstr.* **67**, 324, 3298x (1967).
164. Bruck, S. D., U.S. Pat. 3,362,917 (Research Corp.) (1968); *Chem. Abstr.* **68**, 3950, 40456p (1968).
165. Buckley, D. H., NASA Accession No. N66-16588, Rep. No. NASA-TN-D-3261; *Chem. Abstr.* **66**, 8084, 86158e (1967).
166. Burgman, H. A., Freeman, J. H., Frost, L. W., Bower, G. M., Traynor, E. J., Jr., and Ruffing, C. R., *J. Appl. Polym. Sci.* **12**, 805 (1968); *Chem. Abstr.* **68**, 11141, 115466c (1968).
167. Burks, R. E., Jr., Covington, E. R., and Ray, T. W., NASA, N66-32252, Southern Research Institute, pp. 86 and 88 (1966).
168. Canadian Patent 771,126 (Schenectady Chemicals, Inc.) (1967).
169. Canadian Patent 789,620 (E. I. duPont de Nemours Co.) (1968).
170. Canadian Patent 797,912 (Allied Chem. Corp.) (1968).
171. Capps, D. B., U.S. Pat. 2,889,290 (Chemstrand Corp.) (1959); *Chem. Abstr.* **53**, 16552 (1959).
172. Chalmers, J. R., Belg. Pat. 663,711 (E. I. duPont de Nemours Co.) (1965); *Chem. Abstr.* **64**, 17816 (1966).
173. Chalmers, J. R., U.S. Pat. 3,242,128 (E. I. duPont de Nemours Co.) (1966).
174. Chalmers, J. R., U.S. Pat. 3,342,768 (E. I. duPont de Nemours Co.) (1967).
175. Chalmers, J. R., and Victorius, C., U.S. Pat. 3,416,994 (E. I. duPont de Nemours Co.) (1968); *Chem. Abstr.* **70**, 38485v (1969).
176. Cherkashina, L. G., and Berlin, A. A., *Vysokomol. Soedin.* **8**, 627 (1966); *Polym. Sci. USSR* **8**, 687 (1966).
177. Cherkashina, L. G., and Berlin, A. A., *Vysokomol. Soedin.*, Ser. *B* **9**, 336 (1967); *Chem. Abstr.* **67**, 11083, 117424r (1967).
178. Cohen, S. M., Hunt, C. F., Kass, R. E., and Markhart, A. H., *J. Appl. Polym. Sci.* **6**, 508 (1962), *Resins, Rubbers, Plastics* p. 2933 (1962).
179. Cohen, S. M., and Lavin, E., U.S. Pat. 2,963,464 (Shawinigan Resins Corp.) (1960); *Chem. Abstr.* **55**, 6047 (1961); equivalent to Austral. Pat. 225,640 (1960).
180. Cohen, S. M., and Lavin, E., *J. Appl. Polym. Sci.* **6**, 503 (1962); *Resins, Rubbers, Plastics* p. 2931 (1962).
181. Collins, W. E., and Mass, K. A., U.S. Pat. 3,376,150 (E. I. duPont de Nemours Co.) (1968).
182. Cooper, S. L., Mair, A. D., and Tobolsky, A. V., *Text. Res. J.* **35**, 1110 (1965); *Chem. Abstr.* **64**, 9886h (1966); *Resins, Rubbers, Plastics* p. 895 (1966).
183. Courtright, J. R., *Plast. Des. Process.* **7**, 10 (1967); *Chem. Abstr.* **66**, 4455, 46877g (1967).
184. Cummings, W., and Lynch, E. R., Br. Pat. 1,077,243 (Monsanto Chem., Ltd.) (1967); *Chem. Abstr.* **68**, 41104j (1968).
185. D'Alelio, G. F., and Kieffer, H. E., *J. Macromol. Sci., Chem.* **2**, 1275 (1968).
186. Davis, A. C., and Tyler, G. J., Br. Pat. 1,037,374 (I.C.I. Fibers, Ltd.) (1966); *Chem. Abstr.* **65**, 12391b (1966).
187. Dawans, F., Reichel, B., and Marvel, C. S., *J. Polym. Sci. Part A* **2**, 5005 (1964); *Chem. Abstr.* **62**, 6572 (1965); *Resins, Rubbers, Plastics* p. 1619 (1965).
188. De Brunner, R. E., *Proc. Annu. Tech. Conf., SPI Reinf. Plast. Div.* p. 9-B (1967); *Chem. Abstr.* **67**, 5162, 54706m (1967).
189. De Brunner, R. E., *Mod. Plast.* **45**, 153–154, 158, 160, 162 and 164–165 (1968); *Chem. Abstr.* **69**, 7329, 78190f (1968).
190. De Brunner, R. E., and Fincke, J. K., U.S. Pat. 3,423,366 (Monsanto Research Corp.) (1969).

191. de Gaudemaris, G., and Sillion, B., *J. Polym. Sci., Part B* **2**, 203 (1964).
192. de Gaudemaris, G., Sillion, B., and Prévé, J., *Bull. Soc. Chim. Fr.* [5] p. 1793 (1964); *Chem. Abstr.* **62**, 1757 (1965); *Resins, Rubbers, Plastics* p. 1033 (1965).
193. de Gaudemaris, G., Sillion, B., and Prévé, J., *Bull. Soc. Chim. Fr.* [5] p. 171 (1965); *Chem. Abstr.* **62**, 13253 (1965).
194. Delman, A. D., *J. Macromol. Sci., Rev. Macromol. Chem.* **2**, 154 (1968).
195. De Schryver, F., and Marvel, C. S., *J. Polym. Sci., Part A-1* **5**, 545 (1967).
196. De Winter, W., *Rev. Macromol. Chem.* **1**, 329 (1966); *Chem. Abstr.* **66**, 5285, 55758w (1967).
197. DiLeone, R. R., Lucas, H. R., and French, J. C., *Soc. Plast. Eng., Newark Sect.* p. 98 (1967); *Chem. Abstr.* **67**, 3167, 33238v (1967).
198. Dine-Hart, R. A., U.S. Dep. Commun. AD 482,530 (1965); *Chem. Abstr.* **66**, 8991, 95736u (1967).
199. Dine-Hart, R. A., *J. Polym. Sci., Part A-1* **6**, 2755 (1968); *Chem. Abstr.* **69**, 7298, 77839n (1968).
200. Dine-Hart, R. A., and Wright, W. W., *J. Appl. Polym. Sci.* **11**, 609 (1967); *Chem. Abstr.* **67**, 3167, 33242s (1967).
201. Dine-Hart, R. A., and Wright, W. W., *Chem. Ind. (London)* p. 1565 (1967); *Chem. Abstr.* **67**, 9415, 100088m (1967).
202. Dunphy, J. F., and Parish, D. J., Fr. Pat. 1,446,178 (E. I. duPont de Nemours Co.) (1966); *Chem. Abstr.* **66**, 3708, 38586w (1967).
203. Dunphy, J. F., and Parish, D. J., Can. Pat. 767,073 (E. I. duPont de Nemours Co.) (1967).
204. Durif-Varambon, B., Sillion, B., and de Gaudemaris, G., *C. R. Acad. Sci. Ser. C* **267**, 471 (1968); *Chem. Abstr.* **69**, 10054, 107151b (1968).
205. Dyer, E., and Hartzler, J., *J. Polym. Sci., Part A-1*, **7**, 833 (1969).
206. Edwards, W. M., Belg. Pat. 614,941 (E. I. duPont de Nemours Co.) (1962); *Chem. Abstr.* **58**, 9333 (1963).
207. Edwards, W. M., Br. Pat. 898,651 (E. I. duPont de Nemours Co.) (1962); *Chem. Abstr.* **58**, 14147 (1963).
208. Edwards, W. M., U.S. Pat. 3,179,614 (E. I. duPont de Nemours Co.) (1965).
209. Edwards, W. M., U.S. Pat. 3,179,634 (E. I. duPont de Nemours Co.) (1965).
210. Edwards, W. M., U.S. Pat. 3,287,311 (E. I. duPont de Nemours Co.) (1966); *Chem. Abstr.* **66**, 1866, 19176w (1967).
211. Edwards, W. M., and Endrey, A. L., Br. Pat. 903,271 (E. I. duPont de Nemours Co.) (1962); *Chem. Abstr.* **58**, 3520 (1963).
212. Edwards, W. M., and Robinson, I. M., Br. Pat. 762,152 (E. I. duPont de Nemours Co.) (1956); *Chem. Abstr.* **51**, 8138 (1957).
213. Edwards, W. M., and Robinson, I. M., U.S. Pat. 2,710,853 (E. I. duPont de Nemours Co.) (1955); *Chem. Abstr.* **50**, 5753 (1956).
214. Edwards, W. M., and Robinson, I. M., U.S. Pat. 2,880,230 (E. I. duPont de Nemours Co.) (1959); *Chem. Abstr.* **53**, 16080 (1959).
215. Edwards, W. M., and Robinson, I. M., U.S. Pat. 2,900,369 (E. I. duPont de Nemours Co.) (1959); *Chem. Abstr.* **53**, 23076 (1959).
216. Edwards, W. M., Robinson, I. M., and Squire, E. N., U.S. Pat. 2,867,609 (E. I. duPont de Nemours Co.) (1959); *Chem. Abstr.* **53**, 8708 (1959).
217. Endrey, A. L., U.S., Pat. 3,073,784 (E. I. duPont de Nemours Co.) (1963); *Chem. Abstr.* **58**, 7005 (1963).
218. Endrey, A. L., U.S. Pat. 3,179,630 (E. I. duPont de Nemours Co.) (1965); *Chem. Abstr.* **63**, 18403 (1965).

219. Endrey, A. L., U.S. Pat. 3,179,631 (E. I. duPont de Nemours Co.) (1965); *Chem. Abstr.* **63**, 16503 (1965).
220. Endrey, A. L., U.S. Pat. 3,179,633 (E. I. duPont de Nemours Co.) (1965).
221. Endrey, A. L., U.S. Pat. 3,242,136 (E. I. duPont de Nemours Co.) (1966); *Chem. Abstr.* **64**, 19830 (1966).
222. Endrey, A. L., U.S. Pat. 3,410,826 (E. I. duPont de Nemours Co.) (1968); *Chem. Abstr.* **70**, 38453h (1969).
223. Farrissey, W. J., Rose, J. S., and Carleton, P. S., *Polym. Prepr., Amer. Chem. Soc. Div. Polym. Chem.* **9** (2), 1581 (1968).
224. Fedorova, E. F., Adrova, N. A., Kudryavtsev, V. V., Pokrovskii, E. I., and Koton, M. M., *Vysokomol. Soedin., Ser. B* **10**, 273 (1968); *Chem. Abstr.* **69**, 1059, 11002c (1968).
225. Fedorova, E. F., Pokrovskii, E. I., Kudryavtsev, V. V., and Koton, M. M., *Vysokomol. Soedin., Ser. B* **10**, 119 (1968); *Chem. Abstr.* **68**, 9301, 96243j (1968).
226. Fick, H. J., U.S. Pat. 3,395,057 (G.T. Schjeldahl Co.) (1968); *Chem. Abstr.* **69**, 5611, 59890t (1968).
227. Fincke, J. K., AFML-TR-66-188 (Monsanto Research Corp.) (1966).
228. Fisher, J. W., *Chem. Ind. (London)*, p. 244 (1952).
229. Fisher, N. G., and Wiley, R. H., U.S. Pat. 2,389,662 (E. I. duPont de Nemours Co.) (1945); *Chem. Abstr.* **40**, 1062 (1946).
230. Flavell, W., and Yarsley, V. E., *Chem. Brit.* **3**, 375 (1967); *Chem. Abstr.* **68**, 4867, 50051s (1968).
231. Freeburger, M. E., *Diss. Abstr.* **29**, 2012 (1968); *Chem. Abstr.* **70**, 68808v (1969).
232. Freeman, J. H., AFML-TR-65-188 (Westinghouse Research Labs.) (1965).
233. Freeman, J. H., *Mater. Des. Eng.* **64**, 80 (1966); *Chem. Abstr.* **66**, 367, 3232x (1967).
234. Freeman, J. H., Frost, L. W., Bower, G. M., and Traynor, E. J., Jr., *Conf. Struct. Plast., Adhesives Filament Wound Composites* (1962).
235. Freeman, J. H., Frost, L. W., Bower, G. M., Traynor, E. J., Burgman, H. A., and Ruffing, C. R., *Soc. Plast. Eng., Reg. Tech. Conf., Washington, D.C.* pp. D1–D42; (1967); *Chem. Abstr.* **69**, 7329, 78190f (1968).
236. Freeman, R. R., NASA Accession No. N66-16684, Rep. No. RIA-65-1904, Avail. CFSTI (1965); *Chem. Abstr.* **66**, 2811, 29210s (1967).
237. French Patent 83,564/1,256,203 (E. I. duPont de Nemours Co.) (1964).
238. French Patent 86,124/1,283,378 (Westinghouse Electric Corp.) (1964).
239. French Patent 86,080/1,368,741 (Dr. Beck & Co., G.m.b.H.) (1966).
240. French Patent 1,373,383 (Dr. Beck & Co., G.m.b.H.) (1964); *Chem. Abstr.* **62**, 10636c (1965).
241. French Patent 1,399,331 (Dorken, A.G.) (1965).
242. French Patent 1,421,681 (Westinghouse Electric Corp.) (1965).
243. French Patent 1,424,046 (E. I. duPont de Nemours Co.) (1966); *Chem. Abstr.* **65**, 17086d (1966).
244. French Patent 1,437,746 (Dr. Beck & Co., G.m.b.H.) (1966); *Chem. Abstr.* **66**, 20010n (1967).
245. French Patent 1,448,373 (American Cyanamid Co.) (1966).
246. French Patent 1,449,464 (Monsanto Co.) (1966).
247. French Patent 1,450,606 (Rhone-Poulenc) (1966).
248. French Patent 1,454,878 (I.C.I., Ltd.) (1966); *Chem. Abstr.* **67**, 7820, 82766z (1967).
249. French Patent 1,456,995 (Dr. K. Herberts & Co.) (1966).
250. French Patent 1,459,735 (Westinghouse Electric Corp.) (1966).
251. French Patent 1,462,801 (Republique Federale D'Allemagne) (1966).

252. French Patent 1,473,600 (Hitachi Chemical Co., Ltd.) (1967); *Chem. Abstr.* **68**, 14141n (1968).
253. French Patent 1,478,938 (Schenectady Chemicals, Inc.) (1967).
254. French Patent 1,491,016 (Dr. K. Herberts & Co.) (1967); *Chem. Abstr.* **68**, 96865p (1968).
255. French Patent 1,497,684 (Toyo Rayon) (1967).
256. French Patent 1,501,198 (Mobil Oil Corp.) (1967).
257. French Patent 1,506,279 (Institut Français du Petrole, des Carburants et Lubrifiants) (1967).
258. French Patent 1,506,518 (Rhone Poulenc) (1967); *Chem. Abstr.* **70**, 485, 4978t (1969).
259. French Patent 1,506,519 (Rhone Poulenc) (1967); *Chem. Abstr.* **70**, 482, 4956j (1969).
260. French Patent 1,507,149 (Farbenfabriken Bayer) (1967); *Chem. Abstr.* **70**, 469, 4810g (1969).
261. French Patent 1,509,730 (General Electric Co.) (1968).
262. French Patent 1,509,731 (General Electric Co.) (1968).
263. French Patent 1,511,317 (General Electric Co.) (1968).
264. French Patent 1,511,318 (General Electric Co.) (1968).
265. French Patent 1,511,961 (Schnectady Chemicals, Inc.) (1968).
266. French Patent 1,532,382 (General Electric Co.) (1968).
267. French Patent 1,533,925 (General Electric Co.) (1968).
268. French Patent 1,553,691 (Institut Français du Petrole, des Carburants et Lubrifiants) (1969); *Chem. Abstr.* **71**, 62358x (1969).
269. Fritz, C. G., U.S. Pat. 3,376,260 (E. I. duPont de Nemours Co.) (1968); *Chem. Abstr.* **68**, 10222, 105741z (1968).
270. Fritz, C. G., U.S. Pat. 3,391,120 (E. I. duPont de Nemours Co.) (1968); *Chem. Abstr.* **69**, 4170, 44359e (1968).
271. Frosch, C. J., U.S. Pat. 2,421,024 (Bell Telephone Labs.) (1947); *Chem. Abstr.* **41**, 5342 (1947).
272. Frost, L. W., Fr. Pat. 1,459,375 (Westinghouse Electric Corp.) (1966); *Chem. Abstr.* **67**, 2145, 22341x (1967).
273. Frost, L. W., and Bower, G. M., Fr. Pat. 1,283,378 (Westinghouse Electric Corp.) (1962).
274. Frost, L. W., and Bower, G. M., U.S. Pat. 3,179,635 (Westinghouse Electric Corp.) (1965).
275. Frost, L. W., Bower, G. M., Freeman, J. H., Burgman, H. A., Traynor, E. J., and Ruffing, C. R., *J. Polym. Sci., Part A-1* **6**, 215 (1968); *Chem. Abstr.* **68**, 7601, 78654q (1968).
276. Frost, L. W., and Kesse, I., *Polym. Prepr., Amer. Chem. Soc., Div. Polym. Chem.* **4** (1), 369 (1963).
277. Frost, L. W., and Kesse, I., *J. Appl. Polym. Sci.* **8**, 1039 (1964).
278. Fujita, S., and Itoga, M., Jap. Pat. 68/00,636 (Toyo Rayon) (1968); *Chem. Abstr.* **69**, 8227, 87769d (1968).
279. Fujiwara, T., Matsumoto, S., and Morishima, A., Jap. Pat. 65/20,989 (Chisso Corp.) (1965); *Chem. Abstr.* **64**, 885 (1966).
280. Fukuda, A., *Kogaku To Kogyo (Osaka)* **42**, 109 (1968); *Chem. Abstr.* **69**, 5594, 59713n (1968).
281. Gaertner, R. F., U.S. Pat. 3,356,691 (General Electric Co.) (1967); *Chem. Abstr.* **68**, 2975, 30425p (1968).
282. Gaertner, R. F., and Holub, F. F., U.S. Pat. 3,417,042 (General Electric Co.) (1968).
283. Gall, W. G., Fr. Pat. 1,365,545 (E. I. duPont de Nemours Co.) (1964); *Chem. Abstr.* **62**, 7898 (1965).

284. Gall, W. G., U.S. Pat. 3,249,588 (E. I. duPont de Nemours Co.) (1966).

285. Gall, W. G., U.S. Pat. 3,264,250 (E. I. duPont de Nemours Co.) (1966).

286. Gall, W. G., U.S. Pat. 3,422,061 (E. I. duPont de Nemours Co.) (1969).

287. Gall, W. G., U.S. Pat. 3,422,064 (E. I. duPont de Nemours Co.) (1969).

288. Gallard, J., Salle, R., Traynard, P., and Teyssie, P., Fr. Pat. 1,397,159 (Institut Français du Petrole, des Carburants et Lubrifiants) (1965); *Chem. Abstr.* **63**, 9836 (1965).

289. Gallard-Nechtschein, J., Pecher-Reboul, A., and Traynard, P., *Bull. Soc. Chim. Fr.* [5] p. 960 (1967); *Chem. Abstr.* **67**, 1149, 11903f (1967).

290. Gay, F. P., and Berr, C. E., *J. Polym. Sci., Part A-1* **6**, 1935 (1968); *Chem. Abstr.* **69**, 1871, 19731k (1968).

291. George, N. J., U.S. Pat. 3,428,486 (P. D. George Co.) (1969).

292. German Patent Application 1,210,964 (Monsanto Co.) (1966); *Chem. Abstr.* **64**, 16135 (1966).

293. German Patent 1,273,186 (Standard Oil Co.) (1968).

294. Gerow, C. W., Belg. Pat. 641,568 (E. I. duPont de Nemours Co.) (1964); *Chem. Abstr.* **63**, 3076 (1965).

295. Gerow, C. W., U.S. Pat. 3,356,759 (E. I. duPont de Nemours Co.) (1967); *Chem. Abstr.* **68**, 2998, 30662p (1968).

296. Gibbs, W. E., *J. Macromol. Sci. Chem.* **2**, 1291 (1968).

297. Giuliani, P., and Cohen, C., Fr. Pat. Addn. 91,854 (Institut Français du Petrole, des Carburants et Lubrifiants) (1968); *Chem. Abstr.* **71**, 13752n (1969).

298. Goins, O. K., and Van Deusen, R. L., *J. Polym. Sci., Part B* **6**, 821 (1968).

299. Green, D. E., U.S. Pat. 3,288,754 (Dow Corning Corp.) (1966).

300. Green, D. E., U.S. Pat. 3,338,859 (Dow Corning Corp.) (1967); *Chem. Abstr.* **67**, 8612, 91256k (1967).

301. Gresham, W. F., and Naylor, M. A., U.S. Pat. 2,731,447 (E. I. duPont de Nemours Co.) (1956); *Chem. Abstr.* **50**, 6096 (1956).

302. Gurov, A. A., Liogon'kii, B. I., and Berlin, A. A., *Vysokomol. Soedin., Ser. A* **9**, 2259 (1967); *Polym. Sci. USSR* **9**, 2555 (1967); *Chem. Abstr.* **68**, 324, 3241u (1968).

303. Hagiwara, Y., Ikeda, K., Kurihara, M., and Yoda, N., *Makromol. Chem.* **126**, 48 (1969); *Chem. Abstr.* **71**, 71018p (1969).

304. Hait, P. W., *Vacuum* **17**, 547 (1967); *Chem. Abstr.* **67**, 11123, 117803p (1967).

305. Hall, R. W., and Jeffreys, K. D., Br. Pat. 1,084,902 (Distillers Co., Ltd.) (1967); *Chem. Abstr.* **67**, 10321, 109274b (1967).

306. Haller, J. R., U.S. Pat. 3,428,602 (3M Co.) (1969).

307. Hashimoto, S., and Nagasuna, Y., *Kobunshi Kagaku* **24**, 633 (1967); *Chem. Abstr.* **68**, 11103, 115038q (1968).

308. Hayashi, K., *Nippon Secchaku Kyokai Shi* **2**, 167 (1966); *Chem. Abstr.* **66**, 10791, 115959z (1967).

309. Heacock, J. F., and Berr, C. E., *SPE Trans.* **5**, 105 (1965); *Chem. Abstr.* **63**, 742 (1965).

310. Heine, E., Du Pont Mag. March, p. 26 (1963).

311. Hendrix, W. R., Belg. Pat. 627,623 (E. I. duPont deNemours Co.) (1963); *Chem. Abstr.* **60**, 12136 (1964).

312. Hendrix, W. R., Belg. Pat. 638,689 (E. I. duPont de Nemours Co.) (1964); *Chem. Abstr.* **62**, 6642 (1965).

313. Hendrix, W. R., U.S. Pat. 3,179,632 (E. I. duPont de Nemours Co.) (1965).

314. Hendrix, W. R., U.S. Pat. 3,249,561 (E. I. duPont de Nemours Co.) (1966).

315. Hergenrother, P. M., *J. Polym. Sci., Part A-1* **6**, 3170 (1968); *Chem. Abstr.* **70**, 456, 4681r (1969).

316. Hergenrother, P. M., Kerkmeyer, J. L., and Levine, H. H., NASA Accession No. N65-34211, Rep. No. AD 466,574; *Chem. Abstr.* **68**, 2181, 22427a (1968).
317. Hergenrother, P. M., and Levine, H. H., *Polym. Prepr., Amer. Chem. Soc., Div. Polym. Chem.* **8**, 501 (1967).
318. Hergenrother, P. M., and Levine, H. H., *J. Polym. Sci., Part A-1* **5**, 1453 (1967); *Chem. Abstr.* **67**, 1161, 12031g (1967).
319. Hergenrother, P. M., and Levine, H. H., AD 642,043; *Chem. Abstr.* **67**, 3167, 33236t (1967).
320. Hergenrother, P. M., and Levine, H. H., *U.S. Clearinghouse Fed. Sci. Tech. Inform.* AD 678,975 (1968); *Chem. Abstr.* **71**, 4040b (1969).
321. Hergenrother, P. M., Wrasidlo, W. J., and Levine, H. H., *U.S. Govt. Res. Rep.* AD 602,679 (1964).
322. Hermans, P. H., and Streef, J. W., *Makromol. Chem.* **74**, 133 (1964).
323. Heslinga, H., *Plastica* **18**, 550 (1965); *Chem. Abstr.* **67**, 7000, 74072e (1967).
324. Hirsch, S. S., *Polym. Prepr., Amer. Chem. Soc., Div. Polym. Chem.* **8** (2), 1155 (1967).
325. Hoegger, E. F., U.S. Pat. 3,342,774 (E. I. duPont de Nemours Co.) (1967).
326. Hoegger, E. F., U.S. Pat. 3,423,365 (E. I. duPont de Nemours Co.) (1969).
327. Hofbauer, E. I., and Nesterova, E. I., Russ. Pat. 219,784 (1968); *Chem. Abstr.* **70**, 12136g (1969).
328. Holub, F. F., U.S. Pat. 3,277,043 (General Electric Co.) (1966).
329. Holub, F. F., Fr. Pat. 1,471,562 (General Electric Co.) (1967); *Chem. Abstr.* **67**, 11094, 117536d (1967).
330. Holub, F. F., U.S. Pat. 3,325,450 (General Electric Co.) (1967).
331. Holub, F. F., U.S. Pat. 3,392,144 (General Electric Co.) (1968); *Chem. Abstr.* **69**, 4165, 44299h (1968).
332. Holub, F. F., U.S. Pat. 3,410,875 (General Electric Co.) (1968).
333. Howard, E. G., Jr., U.S. Pat. 3,166,521 (E. I. duPont de Nemours Co.) (1965); *Chem. Abstr.* **62**, 10544e (1965).
334. Imai, Y., Uno, K., and Iwakura, Y., *Makromol. Chem.* **94**, 114 (1966).
335. Inoue, H., Adachi, M., Fukui, M., and Imoto, E., *Kogyo Kagaku Zasshi*, **63**, 1014 (1960); *Resins, Rubbers, Plastics., Ser. B* p. 91 (1961).
336. Inoue, H., and Imoto, E., *Bull. Univ. Osaka Prefect., Ser. B* **10**, 61 (1961); *Chem. Abstr.* **57**, 195 (1962).
337. Irwin, R. S., and Smullen, R. S., U.S. Pat. 3,415,782 (E. I. duPont de Nemours Co.) (1968).
338. Irwin, R. S., and Sweeny, W., *J. Polym. Sci., Part C* **19**, 41 (1967).
339. Iwakura, Y., Izawa, S., and Hayano, F., *J. Polym. Sci., Part A-1* **6**, 1097 (1968); *Chem. Abstr.* **69**, 2624, 28011c (1968).
340. Iwakura, Y., Uno, K., Imai, Y., and Fukui, M., Jap. Pat. 65/29,034 (Teijin, Ltd.) (1965); *Chem. Abstr.* **64**, 12844 (1966).
341. Izumi, M., Matsumura, S., and Asano, N., Jap. Pat. 68/5911 (Sumitomo Electric Ind., Ltd.) (1968); *Chem. Abstr.* **69**, 19762w (1968).
342. Jackson, W. G., and Schroeder, W., U.S. Pat. 3,326,915 (Burdick and Jackson Laboratories) (1967); *Chem. Abstr.* **67**, 5152, 54606d (1967); equivalent to Br. Pat. 1,113,325.
343. Jadamus, H., De Schryver, F., De Winter, W., and Marvel, C. S., *J. Polym. Sci., Part A-1* **4**, 2831 (1966).
344. Japanese Patent 68/2476 (Toyo Rayon) (1968); *Chem. Abstr.* **69**, 28122q (1968).
345. Japanese Patent 68/13473 (Sumitomo Electric Ind., Ltd.) (1968).
346. Japanese Patent 68/22998 (Sumitomo Electric Ind., Ltd.) (1968).
347. Japanese Patent 69/8236 (Toyo Rayon) (1969).

348. Japanese Patent 69/8239 (Toyo Rayon) (1969).
349. Jones, J. I., *Prepr., Int. Symp. Macromol. Chem., 1967* p. 1 (4/30) (1967).
350. Jones, J. I., *J. Macromol. Sci., Rev. Macromol. Chem.* **2**, 303 (1968); *Chem. Abstr.* **69**, 3419, 36435p (1968).
351. Jones, J. I., Ochynski, W. F., and Rackley, F. A., *Chem. Ind. (London)* p. 1686 (1962); *Chem. Abstr.* **57**, 15336 (1963).
352. Jones, J. W., U.S. Pat. 3,427,188 (E. I. duPont de Nemours Co.) (1969).
353. Kachi, H., and Sekiguchi, H., Jap. Pat. 68/12,837 (Furukawa Electric Co., Ltd.) (1968); *Chem. Abstr.* **69**, 7310, 77987j (1968).
354. Kardash, I. E., and Pravednikov, A. N., *Vysokomol Soedin., Ser. B* **9**, 873 (1967); *Chem. Abstr.* **68**, 5806, 59963y (1968).
355. Kinosaki, T., and Young, R. R., *J. Polym. Sci., Part C* **23**, Part 1, 57 (1968); *Chem. Abstr.* **69**, 4914, 52486g (1968).
356. Kluiber, R. W., U.S. Pat. 3,274,159 (Union Carbide Corp.) (1966).
357. Kobayashi, H., Yamaguchi, K., and Yamashita, T., Jap. Pat. 68/15,833 (Asahi) (1968); *Chem. Abstr.* **70**, 20516w (1959).
358. Kobayashi, H., Yamaguchi, K., and Yamashita, T., Jap. Pat. 69/7,953 (Asahi) (1969); *Chem. Abstr.* **71**, 125848e (1969).
359. Koehler, A. M., Measday, D. F., and Morrill, D. H., *Nucl. Instrum. & Methods* **33**, 341 (1965); *Chem. Abstr.* **63**, 737 (1965).
360. Kolesnikov, G. S., Fedotova, O. Ya, and Hofbauer, E. I., *Vysokomol. Soedin., Ser. A* **10**, 1511 (1968).
361. Kolesnikov, G. S., Fedotova, O. Ya., Hofbauer, E. I., and Al-Sufi, H. H. M. A., *Vysokomol. Soedin.* **8**, 1440 (1966).
362. Kolesnikov, G. S., Fedotova, O. Ya., Hofbauer, E. I., and Nesterova, E. I., *Vysokomol. Soedin., Ser. A* **10**, 1845 (1968); *Chem. Abstr.* **70**, 12035y (1969).
363. Kolesnikov, G. S., Fedotova, O. Ya., Hofbauer, E. I., and Shelgaeva, V. G., *Vysokomol. Soedin., Ser. A* **9**, 612 (1967); *Chem. Abstr.* **67**, 2150, 22389u (1967).
364. Kolesnikov, G. S., Fedotova, O. Ya., Hofbauer, E. I., and Shelgaeva, V. G., *Vysokomol. Soedin., Ser. B* **9**, 201 (1967); *Chem. Abstr.* **67**, 320, 3254c (1967).
365. Konsaka, T., Shono, T., and Oda, R., *Kogyo Kagaku Zasshi* **71**, 1738 (1968).
366. Korshak, V. V., Fedorova, L. S., and Mozgova, K. K., *Vysokomol. Soedin., Ser. B* **9**, 736 (1967); *Chem. Abstr.* **68**, 2166, 22280x (1968).
367. Korshak, V. V., Fedorova, L. S., and Mozgova, K. K., *Khim. Geterotsikl. Soedin.* p. 993 (1967); *Chem. Abstr.* **68**, 6720, 69433q (1968).
368. Korshak, V. V., Fedorova, L. S., Mozgova, K. K., and Shatskaya, N. A., *Vysokomol. Soedin., Ser. A* **11**, 43 (1969).
369. Korshak, V. V., and Krongauz, E. S., *Usp. Khim.* **33**, 1409 (1964); *Russ. Chem. Rev.* **33**, 609 (1964); *Chem. Abstr.* **62**, 9237 (1965).
370. Korshak, V. V., Rusanov, A. L., Katsarova, R. D., and Tugushi, D. S., *Izv. Akad. Nauk SSSR, Ser. Khim.* p. 1654 (1968); *Bull. Acad. Sci. USSR, Div. Chem. Sci.* p. 1570 (1968); *Chem. Abstr.* **69**, 9130, 97258a (1968).
371. Korshak, V. V., Tseitlin, G. M., Azarov, V. I., and Pavlov, A. I., *Izv. Akad. Nauk SSSR, Ser. Khim.* p. 226 (1968); *Bull. Acad. Sci. USSR, Div. Chem. Sci.* p. 230 (1968); *Chem. Abstr.* **68**, 6720, 69434r (1968).
372. Korshak, V. V., Tseitlin, G. M., Azarov, V. I., and Pavlov, A. I., Russ. Pat. 218,424 (D. I. Mendeleev Chem. Tech. Inst. Moscow) (1968); *Chem. Abstr.* **69**, 7308, 77966b (1968).
373. Korshak, V. V., Tseitlin, G. M., Pavlov, A. I., Pogorelova, T. G., and Smirnova, V. A., *Izv. Akad. Nauk SSSR, Ser. Khim.* p. 1900 (1968); *Bull. Acad. Sci. USSR, Div. Chem. Sci.* p. 1811 (1968); *Chem. Abstr.* **69**, 9130, 97260v (1968).

374. Korshak, V. V., Vinogradova, S. V., Vygodskii, Ya. S., Pavlova, S. A., and Boiko, L. V., *Izv. Akad. Nauk SSSR, Ser. Khim.* p. 2267 (1967); *Bull. Acad. Sci. USSR, Div. Chem. Sci.* p. 2172 (1967); *Chem. Abstr.* **68**, 327, 3274g (1968).
375. Korshak, V. V., Vinogradova, S. V., Vygodskii, Ya. S., and Yudin, B. N., *Izv. Akad. Nauk SSSR, Ser. Khim.* p. 1405 (1968); *Bull. Acad. Sci. USSR, Div. Chem. Sci.* p. 1331 (1968); *Chem. Abstr.* **69**, 9929, 106129p (1968).
376. Koton, M. M., *Russ. Chem. Rev.* **31**, 81 (1962).
377. Koton, M. M., Adrova, N. A., Bessonov, M. I., Layus, L. A., Rudakov, A. P., and Florinskii, F. S., *Plaste Kaut.* **14**, 730 (1967); *Chem. Abstr.* **68**, 5819, 60080q (1968).
378. Koton, M. M., Adrova, N. A., Dubnova, A. M., Bessonov, M. I., and Rudakov, A. P., Russ. Pat. 188,005 (1966); *Chem. Abstr.* **68**, 366, 3658k (1968).
379. Koton, M. M., Adrova, N. A., Dubnova, A. M., Rudakov, A. P., Korzhavin, L. N., Frenkel, S. Ya., and Bessonov, M. I., Russ. Pat. 220,425 (Institute of High Molecular Weight Compounds, Academy of Sciences, USSR) (1968); *Chem. Abstr.* **69**, 7349, 78390w (1968).
380. Koton, M. M., Florinskii, F. S., Adrova, N. A., Bessonov, M. I., Dubnova, A. M., and Rudakov, A. P., Russ. Pat. 192,400 (Institute of High Molecular Weight Compounds, Academy of Sciences, USSR) (1967); *Chem. Abstr.* **69**, 4164, 44297f (1968).
381. Koton, M. M., Kudryavtsev, V. V., Rudakov, V. V., and Bessonov, M. I., Russ. Pat. 173,931 (Institute of High Molecular Weight Compounds, Academy of Sciences, USSR) (1965); *Chem. Abstr.* **64**, 3723 (1966).
382. Koton, M. M., Kudryavtsev, V. V., Rudakov, A. P., Korzhavin, L. N., Frenkel, S. Ya., Florinskii, F. S., and Bessonov, M. I., Russ. Pat. 205,208 (Institute of High Molecular Weight Compounds, Academy of Sciences, USSR) (1967); *Chem. Abstr.* **69**, 1091, 11347u (1968).
383. Koton, M. M., Rudakov, A. P., and Frenkel, S. Ya., *Vestn. Akad. Nauk SSSR* **36**, 56 (1966); *Chem. Abstr.* **65**, 18777b (1966).
384. Koton, M. M., Yakovlev, B. I., Rudakov, A. P., Knyazeva, T. S., Florinskii, F. S., Bessonov, M. I., Kuleva, M. M., Tolparova, G. A., and Laius, L. A., *Zh. Prikl. Khim.* (*Leningrad*) **38**, 2728 (1965); *Chem. Abstr.* **64**, 12808 (1966).
385. Kress, B. H., U.S. Pat. 2,785,996 (Quaker Chem. Prod.) (1957); *Chem. Abstr.* **51**, 10085 (1957).
386. Kreuz, J. A., U.S. Pat. 3,271,366 (E. I. duPont de Nemours Co.) (1966).
387. Kreuz, J. A., U.S. Pat. 3,413,267 (E. I. duPont de Nemours Co.) (1968).
388. Kreuz, J. A., Endrey, A. L., Gay, F. P., and Sroog, C. E., *J. Polym. Sci., Part A-1* **4**, 2607 (1966).
389. Krieger, R. B., Jr., Politi, R. E., *Nat. Symp. Joining Mater. Aerosp. Syst,* [*Proc.*], *9th, 1965*, Pap. No. V-4 (1965); *Chem. Abstr.* **67**, 11123, 117800k (1967).
390. Kropa, E. L., and Thomas, W. M., U.S. Pat. 2,643,236 (American Cyanamid Co.) (1953); *Chem. Abstr.* **47**, 9625d (1953).
391. Kuckertz, H., *Makromol. Chem.* **98**, 101 (1966).
392. Kudryavtsev, G. I., Odnoralova, V. N., Bogomolova, T. B., and Shablygin, M. V., *Vysokomol. Soedin., Ser. B* **10**, 295 (1968); *Chem. Abstr.* **69**, 1044, 10829r (1968).
393. Kudryavtsev, G. I., Tokarev, A. V., and Chikenina, L. V., Russ. Pat. 194,229 (Sci. Res. Inst. of Synthetic Fibers) (1967); *Chem. Abstr.* **68**, 4947, 50907a (1968).
394. Kudryavtsev, V. V., Rudakov, A. P., and Koton, M. M., *Vysokomol. Soedin., Ser. A* **9**, 1985 (1967); *Chem. Abstr.* **67**, 11080, 117397j (1967).
395. Kudryavtsev, V. V., Rudakov, A. P., Koton, M. M., Moldovskii, B. L., Batalin, O. E. and Rubinshtein, E. I., *Zh. Prikl. Khim.* (*Leningrad*) **41**, 1825 (1968); *Chem. Abstr.* **69**, 10054, 107149g (1968).

396. Kurihara, M., *Makromol. Chem.* **105**, 84 (1967); *Chem. Abstr.* **67**, 7792, 82459b (1967).
397. Kurihara, M., and Ida, N., Jap. Pat. 68/29958 (Toyo Rayon) (1968); *Chem. Abstr.* **70**, 97610e (1969).
398. Kurihara, M., Saito, H., Nukada, K., and Yoda, N., *J. Polym. Sci., Part A-1* **7**, 2897 (1969); *Chem. Abstr.* **72**, 3823t (1970).
399. Kurihara, M., Toyama, S., Kobayashi, A., and Yoda, N., Jap. Pat. 69/19,880 (Toyo Rayon) (1969); *Chem. Abstr.* **72**, 4025w (1970).
400. Kurihara, M., and Yoda, N., *Makromol. Chem.* **107**, 112 (1967); *Chem. Abstr.* **68**, 1307, 13471h (1968).
401. Kurihara, M., and Yoda, N., *J. Macromol. Sci., Part A* **1**, 1069 (1967); *Chem. Abstr.* **68**, 11105, 115060r (1968).
402. Kurihara, M., and Yoda, N., *J. Polym. Sci., Part A-1* **5**, 1765 (1967); *Chem. Abstr.* **67**, 7793, 82460v (1967).
403. Kurihara, M., and Yoda, N., Jap. Pat. 67/26,877 (Toyo Rayon) (1967); *Chem. Abstr.* **68**, 7612, 78762y (1968).
404. Kurihara, M., and Yoda, N., *J. Polym. Sci., Part B* **6**, 875 (1968); *Chem. Abstr.* **70**, 47951w (1969).
405. Laius, L. A., Bessonov, M. I., Kallistova, E. V., Adrova, N. A., and Florinskii, F. S., *Vysokomol. Soedin., Ser. A* **9**, 2185 (1967); *Chem. Abstr.* **68**, 319, 3196h (1968).
406. Lateltin, P., and Mallet, P., Fr. Pat. 1,457,460 (Rhone Poulenc) (1966); *Chem. Abstr.* **67**, 1174, 12150v (1967); equivalent to Br. Pat. 1,123,467.
407. Lavin, E., Markhart, A. H., and Hunt, C. F., Belg. Pat. 615,937 (Shawinigan Resins Corp.) (1962); *Chem. Abstr.* **59**, 5367 (1963).
408. Lavin, E., Markhart, A. H., and Hunt, C. F., Belg. Pat. 616,815 (Shawinigan Resins Corp.) (1962); *Chem. Abstr.* **58**, 10406 (1963).
409. Lavin, E., Markhart, A. H., and Hunt, C. F., Belg. Pat. 617,901 (Shawinigan Resins Corp.) (1962); *Chem. Abstr.* **58**, 9331 (1963).
410. Lavin, E., Markhart, A. H., and Kass, R. E., Fr. Pat. 1,325,462 (Shawinigan Resins Corp.) (1963).
411. Lavin, E., Markhart, A. H., and Kass, R. E., U.S. Pat. 3,190,856 (Shawinigan Resins Corp.) (1965).
412. Lavin, E., Markhart, A. H., and Sauter, J. O., U.S. Pat. 3,260,691 (Monsanto Co.) (1966); *Chem. Abstr.* **65**, 7412 (1966).
413. Lee, H., Stoffey, D., and Neville, K., "New Linear Polymers". McGraw-Hill, New York (1967).
414. Lee, H., Stoffey, D., and Neville, K., "New Linear Polymers", pp. 205–264, McGraw-Hill, New York (1967).
415. Lindsey, W. B., U.S. Pat. 3,361,586 (E. I. duPont de Nemours Co.) (1968).
416. Lindsey, W. B., U.S. Pat. 3,361,589 (E. I. duPont de Nemours Co.) (1968).
417. Liogon'kii, B. I., Gurov, A. A., and Berlin, A. A., *Vysokomol. Soedin., Ser. A* **10**, 1890 (1968); *Polym. Sci. USSR* **10**, 2192 (1968); *Chem. Abstr.* **69**, 9129, 97255x (1968).
418. Loncrini, D. F., Fr. Pat. 1,412,896 (Thomson-Houston) (1965); *Chem. Abstr.* **65**, 5556e (1966).
419. Loncrini, D. F., U.S. Pat. 3,182,073 (General Electric Co.) (1965).
420. Loncrini, D. F., U.S. Pat. 3,182,074 (General Electric Co.) (1965); *Chem. Abstr.* **63**, 8269 (1965).
421. Loncrini, D. F., Fr. Pat. 1,422,925 (Compagnie Française Thomson-Houston) (1966); *Chem. Abstr.* **66**, 2847, 29545e (1967).
422. Loncrini, D. F., *J. Polym. Sci., Part A-1* **4**, 1531 (1966).
423. Loncrini, D. F., U.S. Pat. 3,355,427 (General Electric Co.) (1967).

424. Loncrini, D. F., U.S. Pat. 3,360,502 (General Electric Co.) (1967).
425. Loncrini, D. F., U.S. Pat. 3,407,176 (General Electric Co.) (1968); *Chem. Abstr.* **70**, 12154m (1969).
426. Loncrini, D. F., Walton, W. L., and Hughes, R. B., *J. Polym. Sci.*; *Part A-1* **4**, 440 (1966).
427. Lopez, A. H., Span. Pat. 284,662 (1963); *Chem. Abstr.* **59**, 4169 (1963).
428. McKeown, J. J., and Wright, C. D., U.S. Pat. 3,389,111 (3M Co.) (1968); *Chem. Abstr.* **69**, 3442, 36706y (1968).
429. Mainen, E. L., *Diss. Abstr.* **29**, 547 (1968).
430. Mair, A. D., Shen, M. C., and Tobolsky, A. V., AD 604,010; *U.S. Govt. Res. Rep.* **39**, 67 (1964); *Chem. Abstr.* **62**, 4165 (1965).
431. Mair, A. D., Shen, M. C., and Tobolsky, A. V., "High Temperature Polymers: H-Film and SP-Polymer." ONR Tech. Rep. (1964).
432. Manaka, K., *Shinku Kagaku* **14**, 34 (1966); *Chem. Abstr.* **68**, 2939, 30077b (1968).
433. Manecke, G., and Wöhrle, D., *Makromol. Chem.* **120**, 176 (1969).
434. Manecke, G., and Wöhrle, D., *Makromol. Chem.* **120**, 192 (1969).
435. Marvel, C. S., *Polym. Prepr., Amer. Chem. Soc., Div. Polym. Chem.* **5** (1), 167 (1964).
436. Marvel, C. S., *Polym. Prepr., Amer. Chem. Soc., Div. Polym. Chem.* **6** (1), 15 (1965).
437. Marvel, C. S., *Proc. Robert A. Welch Found. Conf. Chem. Res.* **10**, 59–75 (1966); *Chem. Abstr.* **67**, 11101, 117589y (1967).
438. Marvel, C. S., *J. Macromol. Sci., Part A* **1**, 7 (1967); *Chem. Abstr.* **67**, 2140, 22297n (1967).
439. Marvel, C. S., U.S. Pat. 3,301,828 (E. I. duPont de Nemours Co.) (1967); *Chem. Abstr.* **66**, 6241, 66052n (1967).
440. Marvel, C. S., *Stabil. Plast., Soc. Plast. Eng., Reg. Tech. Conf., Washington, D.C.* J1–J23 (1967); *Chem. Abstr.* **69**, 6335, 67718h (1968).
441. Marvel, C. S., *Pure Appl. Chem.* **16**, 351 (1968); *Chem. Abstr.* **69**, 7289, 77740y (1968).
442. Marvel, C. S., Okada, M., and De Schryver, F., U.S. Pat. 3,475,374 (Research Corp.) (1969); *Chem. Abstr.* **72**, 13751g (1970).
443. Matray, A., U.S. Pat. 3,356,760 (E. I. duPont de Nemours Co.) (1967); *Chem. Abstr.* **68**, 2185, 22470j (1968).
444. Maxwell, V. W., Fr. Pat. 1,507,461 (American Machine & Foundry Co.); *Chem. Abstr.* **70**, 12260t (1969).
445. Meyer, J. F., Sheffer, H. E., and Zalewski, E. J., U.S. Pat. 3,425,866 (Schenectady Chemicals, Inc.) (1969).
446. Meyer, J. F., Sheffer, H. E., and Zalewski, E. J., U.S. Pat. 3,426,098 (Schenectady Chemicals, Inc.) (1969).
447. Meyers, R. A., *Polym. Prepr., Amer. Chem. Soc., Div. Polym. Chem.* **10**, 186 (1969).
448. Milek, J. T., AD 475,505 (Hughes Aircraft Co.) (1965).
449. Miller, W. R., Cowan, J. C., and Pryde, E. H., U.S. Pat. 3,285,880 (United States of America) (1966).
450. Mita, T., *Enka Biniiru To Porima* **6**, 14 (1966); *Chem. Abstr.* **66**, 8982, 95639q (1967).
451. Mortillaro, L., *Mater. Plast. Elastomeri* **33**, 14 (1967); *Chem. Abstr.* **67**, 319, 3238c (1967).
452. Mueller, G., and Merten, R., U.S. Pat. 3,314,923 (Farbenfabriken Bayer, A. G.) (1967); *Chem. Abstr.* **67**, 2147, 22362e (1967).
453. Mueller, G., Merten, R., Duenwald, W., and Mielke, K. H., Belg. Pat. 666,934 (Farbenfabriken Bayer, A. G.) (1965); *Chem. Abstr.* **65**, 3997c (1966).
454. Naarman, H., Ger. Pat. 1,179,715 (Badische Anilin-und-Soda Fabrik) (1964); *Chem. Abstr.* **62**, 662 (1965).

455. Naarman, H., Ger. Pat. 1,179,716 (Badische Anilin-und-Soda Fabrik) (1964); *Chem. Abstr.* **62**, 1773 (1965).
456. Naarman, H., Ger. Pat. 1,197,228 (Badische Anilin-und-Soda Fabrik) (1965); *Chem. Abstr.* **63**, 11808 (1965).
457. Naarman, H., Ger. Pat. 1,198,553 (Badische Anilin-und-Soda Fabrik) (1965); *Chem. Abstr.* **63**, 11809 (1965).
458. Nakanishi, R., Yoda, N., and Kurihara, M., U.S. Pat. 3,424,728 (Toyo Rayon) (1969).
459. Netherlands Patent 298,294 (E. I. duPont de Nemours Co.) (1965); *Chem. Abstr.* **64**, 14386 (1966).
460. Netherlands Patent 298,295 (E. I. duPont de Nemours Co.) (1965); *Chem. Abstr.* **64**, 14386 (1966).
461. Netherlands Patent Application 64/00422 (Standard Oil Co. Ind.) (1964); *Chem. Abstr.* **62**, 1817 (1965).
462. Netherlands Patent Application 64/05435 (Institut Français du Petrole, des Carburants et Lubrifiants) (1964); *Chem. Abstr.* **62**, 10545 (1965); equivalent to Ger. Pat. 1,217,614; Belg. Pat. 648,040; Fr. Pat. 1,388,650.
463. Netherlands Patent Application 64/06900 (E. I. duPont de Nemours Co.) (1964); *Chem. Abstr.* **62**, 14857 (1965).
464. Netherlands Patent Application 64/08197 (Institut Français du Petrole, des Carburants et Lubrifiants) (1965); *Chem. Abstr.* **63**, 752 (1965).
465. Netherlands Patent Application 64/08298 (Monsanto Co.) (1965); *Chem. Abstr.* **63**, 3129 (1965).
466. Netherlands Patent Application 64/10211 (Badische Anilin-und-Soda Fabrik) (1965); *Chem. Abstr.* **63**, 5855 (1965).
467. Netherlands Patent Application 64/12661 (E. I. duPont de Nemours Co.) (1965); *Chem. Abstr.* **63**, 13503 (1965).
468. Netherlands Patent Application 64/12662 (E. I. duPont de Nemours Co.) (1965); *Chem. Abstr.* **63**, 13512 (1965).
469. Netherlands Patent Application 64/13549 (E. I. duPont de Nemours Co.) (1965); *Chem. Abstr.* **63**, 15057 (1965).
470. Netherlands Patent Application 64/13550 (E. I. duPont de Nemours Co.) (1965); *Chem. Abstr.* **63**, 15010 (1965).
471. Netherlands Patent Application 64/13551 (E. I. duPont de Nemours Co.) (1965); *Chem. Abstr.* **63**, 15010 (1965).
472. Netherlands Patent Application 64/13552 (E. I. duPont de Nemours Co.) (1965); *Chem. Abstr.* **63**, 15010 (1965).
473. Netherlands Patent Application 64/14419 (Dow Corning Corp.) (1965); *Chem. Abstr.* **64**, 6787 (1966).
474. Netherlands Patent Application 64/14424 (E. I. duPont de Nemours Co.) (1965); *Chem. Abstr.* **64**, 839 (1966).
475. Netherlands Patent Application 64/14467 (Dow Corning Corp.) (1965); *Chem. Abstr.* **64**, 838 (1966).
476. Netherlands Patent Application 65/00135 (E. I. duPont de Nemours Co.) (1965); *Chem. Abstr.* **64**, 5010 (1966).
477. Netherlands Patent Application 65/00212 (E. I. duPont de Nemours Co.) (1966); *Chem. Abstr.* **66**, 348, 3025e (1967).
478. Netherlands Patent Application 65/00641 (E. I. duPont de Nemours Co.) (1965); *Chem. Abstr.* **64**, 883 (1966).
479. Netherlands Patent Application 65/04004 (E. I. duPont de Nemours Co.) (1965); *Chem. Abstr.* **64**, 11370 (1966).

480. Netherlands Patent Application 65/10178 (E. I. duPont de Nemours Co.) (1966); *Chem. Abstr.* **65**, 828 (1966).
481. Netherlands Patent Application 65/12671 (E. I. duPont de Nemours Co.) (1966).
482. Netherlands Patent Application 65/12886 (E. I. duPont de Nemours Co.) (1966); *Chem. Abstr.* **67**, 10334, 109405v (1967).
483. Netherlands Patent Application 65/15693 (Farbenfabriken Bayer) (1966); *Chem. Abstr.* **65**, 17085 (1966).
484. Netherlands Patent Application 65/16980 (German Republic) (1966).
485. Netherlands Patent Application 65/16986 (German Republic) (1966); *Chem. Abstr.* **67**, 1166, 12079 (1967).
486. Netherlands Patent Application 66/00689 (E. I. duPont de Nemours Co.) (1966); *Chem. Abstr.* **66**, 1863, 19153m (1967).
487. Netherlands Patent Application 66/01893 (N.V. Philips Gloeilampenfabricken) (1967); *Chem. Abstr.* **67**, 10321, 109275c (1967).
488. Netherlands Patent Application 66/03420 (American Cyanamid Co.) (1967); *Chem. Abstr.* **68**, 10216, 105685j (1968).
489. Netherlands Patent Application 66/04263 (E. I. duPont de Nemours Co.) (1966); *Chem. Abstr.* **66**, 2868, 29779j (1967).
490. Netherlands Patent Application 66/05964 (E. I. duPont de Nemours Co.) (1966); *Chem. Abstr.* **67**, 333, 3384x (1967).
491. Netherlands Patent Application 66/08731 (Dr. K. Herberts & Co.) (1967).
492. Netherlands Patent Application 66/09025 (E. I. duPont de Nemours Co.) (1966); *Chem. Abstr.* **67**, 3189, 33535q (1967).
493. Netherlands Patent Application 66/09214 (Soc. Rhodiaceta) (1967); *Chem. Abstr.* **67**, 2146, 22355e (1967).
494. Netherlands Patent Application 66/14346 (American Cyanamid Co.) (1967); *Chem. Abstr.* **68**, 6748, 69725m (1968).
495. Netherlands Patent Application 66/16621 (E. I. duPont de Nemours Co.) (1967).
496. Netherlands Patent Application 66/17078 (E. I. duPont de Nemours Co.) (1967).
497. Netherlands Patent Application 67/04836 (Toyo Rayon) (1967).
498. Netherlands Patent Application 67/12020 (E. I. duPont de Nemours Co.) (1967).
499. Netherlands Patent Application 67/13214 (Farbenfabriken Bayer, A. G.) (1968).
500. Netherlands Patent Application 68/02157 (Nat. Res. Dev. Corp.) (1968).
501. Netherlands Patent Application 68/02533 (Corning Glass Works) (1969).
502. Netherlands Patent Application 68/03527 (Stamicarbon) (1968).
503. Netherlands Patent Application 68/04027 (E. I. duPont de Nemours Co.) (1968).
504. Netherlands Patent Application 68/05343 (Nat. Res. Dev. Corp.) (1968).
505. Netherlands Patent Application 68/06299 (Nat. Res. Dev. Corp.) (1968).
506. Netherlands Patent Application 68/09244 (TRW, Inc.) (1968).
507. Netherlands Patent Application 68/09665 (TRW, Inc.) (1968).
508. Netherlands Patent Application 69/08835 (Farbenfabriken Bayer) (1969).
509. Nishizaki, S., *Kogyo Kagaku Zasshi* **68**, 574 (1965); *Chem. Abstr.* **63**, 7129 (1965).
510. Nishizaki, S., *Kogyo Kagaku Zasshi* **68**, 1756 (1965); *Resins, Rubbers, Plastics* p. 545 (1966).
511. Nishizaki, S., *Kogyo Kagaku Zasshi* **69**, 1069 (1966); *Chem. Abstr.* **68**, 6720, 69435s (1968).
512. Nishizaki, S., *J. Chem. Soc. Jap., Ind. Chem. Sect.* **69**, 1393 (1966); *Chem. Abstr.* **66**, 4453, 46857a (1967).
513. Nishizaki, S., and Fukami, A., *J. Chem. Soc. Jap., Ind. Chem. Sect.* **66**, 382 (1963); *Chem. Abstr.* **59**, 11673 (1963).

514. Nishizaki, S., and Fukami, A., *Mitsubishi Denki Lab. Rep.* **4**, 521 (1963); *Chem. Abstr.* **66**, 1113, 11227t (1967).
515. Nishizaki, S., and Fukami, A., *Kogyo Kagaku Zasshi* **67**, 474 (1964); *Chem. Abstr.* **61**, 16181 (1964); *Resins, Rubbers, Plastics* p. 443 (1965).
516. Nishizaki, S., and Fukami, A., *Mitsubishi Denki Lab. Rep.* **5**, 83 (1964); *Chem. Abstr.* **64**, 17729c (1966).
517. Nishizaki, S., and Fukami, A., *Mitsubishi Denki Lab. Rep.* **5**, 257 (1964); *Chem. Abstr.* **64**, 17729 (1966).
518. Nishizaki, S., and Fukami, A., *Kogyo Kagaku Zasshi* **71**, 1565 (1968); *Chem. Abstr.* **70**, 38185x (1969).
519. Nishizaki, S., and Moriwaki, T., *Kogyo Kagaku Zasshi* **71**, 1559 (1968).
520. Nyrkov, E. S., *Elektrotekhnika* **37**, 39 (1966); *Chem. Abstr.* **66**, 366, 3213f (1967).
521. Okada, M., and Marvel, C. S., *Polym. Prepr., Amer. Chem. Soc., Div. Polym. Chem.* **8** (1), 229 (1967).
522. Okada, M., and Marvel, C. S., *J. Polym. Sci., Part A-1* **6**, 1259 (1968); *Chem. Abstr.* **69**, 319, 3212f (1968).
523. Okada, M., and Marvel, C. S., *J. Polym. Sci., Part A-1* **6**, 1774 (1968); *Chem. Abstr.* **69**, 1042, 10808h (1968).
524. Olivier, K. L., U.S. Pat. 3,234,181 (E. I. duPont de Nemours Co.) (1966).
525. Oromi, J. C., Sp. Pat. 304,410 (1965); *Chem. Abstr.* **62**, 10634g (1965).
526. Oromi, J. C., Sp. Pat. 304,411 (1965); *Chem. Abstr.* **62**, 10553f (1965).
527. Oromi, J. C., Sp. Pat. 304,412 (1965); *Chem. Abstr.* **62**, 11939 (1965).
528. Orth, H., *Kunststoffe* **41**, 454 (1951).
529. Overberger, C. G., and Mukamal, H., ML-TDR-64-80, Part II (1965).
530. Overberger, C. G., Ozaki, S., and Mukamal, H., ML-TDR-64-80, Part I (1963).
531. Overberger, C. G., Ozaki, S., and Mukamal, H., *Proc. Battelle Symp. Thermal Stabil. Polym., 1963* pp. K1–K2 (1964); *Chem. Abstr.* **60**, 10796a (1964).
532. Overberger, C. G., Ozaki, S., and Mukamal, H., *J. Polym. Sci., Part B* **2**, 627 (1964).
533. Packham, D. I., and Rackley, F. A., *Chem. Ind. (London)* p. 1254 (1967); *Chem. Abstr.* **67**, 7795, 82489m (1967).
534. Packham, D. I., and Rackley, F. A., *Chem. Ind. (London)* p. 1566 (1967); *Chem. Abstr.* **67**, 9473, 100497u (1967).
535. Packham, D. I., and Rackley, F. A., *Polymer* **10**, 559 (1969); *Chem. Abstr.* **72**, 44190c (1970).
536. Paris, A., *Ind. Chim. (Paris)* **54**, 91 (1967); *Chem. Abstr.* **68**, 1301, 13404p (1968).
537. Paushkin, Ya. M., Vishnyakova, T. P., Nisova, S. A., Lunin, A. F., Omarov, O. Yu., Markov, Yu. Ya., Machus, F. F., Golubeva, I. A., Polak, L. S., Patalakh, I. I., Stychenko, V. A., and Sokolinskaya, T. A., *J. Polym. Sci., Part A-1* **5**, 1203 (1967).
538. Penczek, P., *Polimery* **12**, 214 (1967); *Chem. Abstr.* **68**, 6600, 68523g (1968).
539. Petrov, K. A., Nifantev, E. E., Goltsova, R. G., and Solintseva, L. M., *Vysokomol. Soedin.* **4**, 1219 (1962); *Resins, Rubbers, Plastics* p. 441 (1963).
540. Plonka, Z. J., and Albrecht, W. M., *Vysokomol. Soedin.* **7**, 2177 (1965); *Polym. Sci. USSR* **7**, 2387 (1965).
541. Preston, J., and Black, W. B., *Polym. Prepr., Amer. Chem. Soc., Div. Polym. Chem.* **6** (2), 757 (1965).
542. Preston, J., and Black, W. B., *J. Polym. Sci., Part B* **3**, 845 (1965).
543. Preston, J., and Black, W. B., *J. Polym. Sci., Part A-1* **5**, 2429 (1967).
544. Preston, J., and Black, W. B., *Man-Made Fibers* **2**, 365 (1968); *Chem. Abstr.* **69**, 313, 3144k (1968).

545. Preston, J., and Black, W. B., *Polym. Prepr., Amer. Chem. Soc., Div. Polym. Chem.* **9** (2), 1143 (1968).

546. Preston, J., Black, W. B., and De Winter, W., *Polym. Prepr., Amer. Chem. Soc., Div. Polym. Chem.* **9** (2), 1165 (1968).

547. Prince, M. I., and Hornyak, J., *J. Polym. Sci., Part B* **4**, 601 (1966); *Chem. Abstr.* **65**, 15535f (1966).

548. Pruckmayr, G., U.S. Pat. 3,349,061 (E. I. duPont de Nemours Co.) (1967).

549. Rabilloud, G., Sillion, B., and de Gaudemaris, G., *C. R. Acad. Sci., Ser. C* **263**, 862 (1966); *Chem. Abstr.* **66**, 1113, 11226s (1967).

550. Rabilloud, G., Sillion, B., and de Gaudemaris, G., *Makromol. Chem.* **108**, 18 (1967); *Chem. Abstr.* **67**, 11076, 117374z (1967).

551. Rabilloud, G., Sillion, B., and de Gaudemaris, G., Fr. Pat. 1,504,425 (Institut Français du Petrole, des Carburants et Lubrifiants) (1967); *Chem. Abstr.* **70**, 482, 4960f (1969).

552. Rabilloud, G., Sillion, B., and de Gaudemaris, G., *Makromol. Chem.* **125**, 264 (1969).

553. Rafikov, S. R., Derevyanchenko, V. P., and Zhubanov, B. A., *Izv. Akad. Nauk Kaz. SSR, Ser. Khim.* **16**, 101 (1966); *Chem. Abstr.* **66**, 2810, 29205u (1967).

554. Read, J., *J. Chem. Soc. (London)* **101**, 2090 (1912).

555. Redman, E. G., and Skinner, J. S., Fr. Pat. 1,501,198 (Mobil Oil Corp.) (1967); *Chem. Abstr.* **69**, 8231, 87799p (1968).

557. Reese, E., Roehm, W., Hoffmeir, H., Bottenbruch, L., and Schnell, H., S. Afr. Pat. 68/00829 (Farbenfabriken Bayer) (1968); *Chem. Abstr.* **70**, 48341j (1969).

558. Reichel, B., Fr. Pat. 1,493,350 (Badische Anilin-und-Soda Fabrik. A. G.) (1967); *Chem. Abstr.* **68**, 106146w (1968).

559. Reimschuessel, H. K., *J. Polym. Sci., Part B* **4**, 953 (1966).

560. Reinhardt, H. F., U.S. Pat. 3,232,907 (E. I. duPont de Nemours Co.) (1966); *Chem. Abstr.* **64**, 11234a (1966).

561. Reynolds, R. J. W., and Seddon, J. D., *J. Polym. Sci., Part C-1* **23**, 45–56 (1968); *Chem. Abstr.* **69**, 4917, 52526v (1968).

562. Rodia, J. S., Fr. Pat. 1,404,741 (Minnesota Mining and Manufacturing Co.) (1965); *Chem. Abstr.* **65**, 4070h (1966).

563. Rodia, J. S., U.S. Pat. 3,247,165 (Minnesota Mining and Manufacturing Co.) (1966).

564. Rogers, F. E., U.S. Pat. 3,356,648 (E. I. duPont de Nemours Co.) (1967); *Chem. Abstr.* **68**, 2974, 30419g (1968).

565. Rudakov, A. P., Adrova, N. A., Bessonov, M. I., and Koton, M. M., *Dokl. Akad. Nauk SSSR* **172**, 899 (1967); *Chem. Abstr.* **66**, 7197, 76505b (1967).

566. Rudakov, A. P., Bessonov, M. I., Koton, M. M., and Florinskii, F. S., *Khim. Volokna* **20**, (1966); *Chem. Abstr.* **66**, 1156, 11694t (1967).

567. Rudakov, A. P., Bessonov, M. I., Koton, M. M., Pokrovskii, E. I., and Fedorova, E. F., *Dokl. Akad. Nauk SSSR* **161**, 617 (1965); *Chem. Abstr.* **63**, 740d (1965); *Dokl. Chem. (English transl.)* **161**, 318 (1965).

568. Russo, M., *Mater. Plast. Elastomeri* **32**, 949 (1966); *Chem. Abstr.* **66**, 2834, 29429v (1967).

569. Russo, M., *Mater. Plast. Elastomeri* **33**, 810,851 (1967); *Chem. Abstr.* **67**, 10295, 109009u (1967).

570. Russo, M., *Mater. Plast. Elastomeri* **34**, 515 (1968); *Chem. Abstr.* **69**, 4918, 52527w (1968).

571. Russo, M., *Mater. Plast. Elastomeri* **34**, 630 (1968); *Chem. Abstr.* **69**, 7329, 78193j (1968).

572. Saga, M., Shono, T., and Shiura, K., *Bull. Chem. Soc. Jap.* **39**, 1795 (1966).

573. Salle, R., Sillion B., and de Gaudemaris, G., Fr. Pat. 1,488,977 (Institut Français du Petrole, des Carburants et Lubrifiants (1967); *Chem. Abstr.* **68**, 5817, 60075s (1968).

574. Salle, R., Sillion, B., and de Gaudemaris, G., *Bull. Soc. Chim. Fr.* [5] p. 4296 (1967); *Chem. Abstr.* **68**, 6723, 69470z (1968).

575. Salle, R., Sillion, B., and de Gaudemaris, G., Fr. Pat. 1,506,279 (Institut Français du Petrole, des Carburants et Lubrifiants (1967); *Chem. Abstr.* **69**, 10068, 107284x (1968).

576. Salle, R., Sillion, B., and de Gaudemaris, G., *C. R. Acad. Sci., Ser. C* **267**, 1213 (1968); *Chem. Abstr.* **70**, 29423w (1969).

577. Salle, R., Sillion, B., and de Gaudemaris, G., Ger. Pat. 1,908,609 (Institut Français du Petrole, des Carburants et Lubrifiants) (1969); *Chem. Abstr.* **71**, 125434s (1969).

578. Sambeth, J., Fr. Pat. 1,498,015 (Société Rhodiaceta) (1967); *Chem. Abstr.* **69**, 6355, 67926z (1968).

579. Santangelo, J. G., Can. Pat. 764,523 (Celanese Corp.) (1967).

580. Saunders, F. C., and Caldwell, D. S., Br. Pat. 1,002,460 (Midland Silicones) (1965); *Chem. Abstr.* **63**, 13500f (1965).

581. Scala, L. C., and Hickam, W. M., *J. Appl. Polym. Sci.* **9**, 245 (1965); *Chem. Abstr.* **62**, 10608 (1965).

582. Scala, L. C., Hickam, W. M., and Marschik, I., *J. Appl. Polym. Sci.* **12**, 2339 (1968); *Chem. Abstr.* **69**, 10071, 107318m (1968).

583. Schmidt, K., Hansch, F., and Rombrecht, H.-M., U.S. Pat. 3,306,771 (Dr. Beck and Co., G.m.b.H.) (1967).

584. Schmidt, P. G., U.S. Pat. 3,375,131 (E. I. duPont de Nemours Co.) (1968); *Chem. Abstr.* **68**, 9326, 96508z (1968).

585. Schors, A., and Lohnizen, O., *Khim. Tekhnol. Polimerov* p. 140 (1960).

586. Schuller, W. H., Lawrence, R. V., and Culbertson, B. M., *J. Polym. Sci., Part A-1* **5**, 2204 (1967); *Chem. Abstr.* **67**, 8603, 91155b (1967).

587. Schulz, H., and Wagner, H., *Angew. Chem.* **62**, 105 (1950).

588. Schulz, R. C., *Angew. Makromol. Chem.* **4/5**, 1 (1968).

589. Sekiguchi, H., Jap. Pat. 67/677 (Furukawa Electric Co., Ltd.) (1967); *Chem. Abstr.* **67**, 3409j (1967).

590. Sekiguchi, H., and Konishi, T., Jap. Pat. 67/5354 (Furukawa Electric Co., Ltd.) (1967); *Chem. Abstr.* **67**, 5167, 54756c (1967).

591. Serlin, I., Lavin, E., and Markhart, A. H., Belg. Pat. 670,611 (Monsanto Co.) (1966); *Chem. Abstr.* **65**, 10810a (1966).

592. Serlin, I., and Markhart, A. H., *J. Polym. Sci.* **60**, S59 (1962); *Resins, Rubbers, Plastics* p. 2613 (1962).

593. Sheffer, H. E., Fr. Pat. 1,472,442 (Schenectady Chemicals, Inc.) (1967); *Chem. Abstr.* **67**, 83046b (1967).

594. Shelton, C. F., Jr., U.S. Pat. 3,408,453 (Cerro Corp.) (1968); *Chem. Abstr.* **70**, 12368j (1969).

595. Shono, T., Izumi, M., Matsumura, S., and Asano, N., Jap. Pat. 68/15,993 (Sumitomo Electric Industries, Ltd.) (1968); *Chem. Abstr.* **70**, 12135f (1969).

596. Shono, T., Saga, M., Obora, M., and Hachihama, Y., *Technol. Rep. Osaka Univ.* **15**, (638–657), 1–7 (1965); *Chem. Abstr.* **64**, 3701g (1966).

597. Shono, T., Saga, M., Obora, M., and Shiura, K., *J. Chem. Soc. Jap., Ind. Chem. Sect.* **70**, 1250 (1967).

598. Shopov, I., and Popov, N., *Vysokomol. Soedin., Ser. B* **9**, 415 (1967); *Chem. Abstr.* **67**, 7792, 82457z (1967).

599. Shopov, I., and Popov, N., *J. Polym. Sci., Part A-1* **7**, 1803 (1969); *Chem. Abstr.* **71**, 71023m (1969).

600. Shulman, G. P., and Tusing, C. R., *Amer. Chem. Soc., Div. Org. Coatings Plast. Chem., Preprints.* **27** (1), 228 (1967); *Chem. Abstr.* **66**, 9866, 105321w (1967).

601. Sibilia, J. P., Reimschuessel, H. K., and Babbitt, G. E., *Polym. Prepr., Amer. Chem. Soc., Div. Polym. Chem.* **8** (2), 1109 (1967).
602. Sidorovich, A. V., Bessonov, M. I., Rudakov, A. P., and Koton, M. M., *Dokl. Akad. Nauk SSSR* **165**, 848 (1965); *Chem. Abstr.* **64**, 9883f (1966); *Dokl. Chem. (English transl.)* **165**, 1168 (1965).
603. Sidorovich, A. V., and Kuvshinskii, E. V., *Vysokomol. Soedin., Ser. A* **10**, 1401 (1968); *Chem. Abstr.* **69**, 5590, 59653t (1968).
604. Sillion, B., and de Gaudemaris, G., *J. Polym. Sci., Part C* **22**, 827 (1967); *Chem. Abstr.* **71**, 102354b (1969).
605. Sillion, B., and de Gaudemaris, G., *C. R. Acad. Sci., Ser. C* **265**, 1234 (1967); *Chem. Abstr.* **68**, 5808, 59982d (1968).
606. Sillion, B. and de Gaudemaris, G., *Chem. Abstr.* **68**, 2958, 30265m (1968).
607. Sillion, B., and de Gaudemaris, G., *Entropie* No. 21, 67–77 (1968); *Chem. Abstr.* **69**, 4166, 44304f (1968).
608. Sillion, B., and de Gaudemaris, G., Fr. Pat. 1,523,133 (Institut Français du Petrole, des Carburants et Lubrifiants) (1968); *Chem. Abstr.* **71**, 81928z (1969).
609. Smirnova, V. E., Rudakov, A. P., Adrova, N. A., Bessonov, M. I., Koton, M. M., and Florinskii, F. S., *Dokl. Akad. Nauk SSSR* **174**, 1352 (1967); *Chem. Abstr.* **68**, 4881, 50204u (1968); *Dokl. Chem. (English transl.)* **174**, 586 (1967).
610. Smith, R. E., Jr., and Gardner, J. M., U.S. Pat. 3,168,417 (Haveg Industries, Inc.) (1965).
611. Sorenson, W. R., Fr. Pat. 1,399,077 (E. I. duPont de Nemours Co.) (1965); *Chem. Abstr.* **64**, 3775d (1966).
612. Sorenson, W. R., U.S. Pat. 3,312,663 (E. I. duPont de Nemours Co.) (1967).
613. South African Patent 66/7015 (Schenectady Chemicals, Inc.) (1967).
614. South African Patent 67/1914 (General Electric Co.) (1967).
615. South African Patent 67/2020 (General Electric Co.) (1967).
616. South African Patent 67/2064 (General Electric Co.) (1967).
617. South African Patent 67/2065 (General Electric Co.) (1967).
618. South African Patent 67/2066 (General Electric Co.) (1967).
619. Spain, R. G., and Ray, J. D., *Stabil. Plast., Soc. Plast. Eng., Tech. Conf., Washington, D.C.* pp. E1–E14 (1967); *Chem. Abstr.* **69**, 6356, 67941a (1968).
620. Sroog, C. E., *J. Polym. Sci., Part C* No. 16, Part 2, 1191–1209 (1967); *Chem. Abstr.* **67**, 6113, 64712z (1967).
621. Sroog, C. E., Abramo, S. V., Berr, C. E., Edwards, W. M., Endrey, A. L., and Olivier, K. L., *Polym. Prepr., Amer. Chem. Soc., Div. Polym. Chem.* **5** (1), 132 (1964).
622. Sroog, C. E., Endrey, A. L., Abramo, S. V., Berr, C. E., Edwards, W. M., and Olivier, K. L., *J. Polym. Sci., Part A* **3**, 1373 (1965).
623. Standage, A. E., and Turner, W. N., *J. Mater. Sci.* **2**, 103 (1967); *Chem. Abstr.* **67**, 1163, 12051p (1967).
624. Stephens, J. R., and Van Strien, R. E., U.S. Pat. 3,347,828 (Standard Oil Co.) (1967).
625. Stille, J. K., *Encycl. Polym. Sci. Technol.* **11**, 389 (1969); *Chem. Abstr.* **72**, 44150q (1970).
626. Stille, J. K., and Arnold, F. E., *J. Polym. Sci., Part A-1* **4**, 551 (1966).
627. Stille, J. K., and Freeburger, M. E., *J. Polym. Sci., Part B* **5**, 989 (1967); *Chem. Abstr.* **67**, 11076, 117373y (1967).
628. Stille, J. K., and Freeburger, M. E., *J. Polym. Sci., Part A-1* **6**, 161 (1968); *Chem. Abstr.* **68**, 6720, 69438v (1968).
629. Stille, J. K., and Mainen, E. L., *J. Polym. Sci., Part B* **4**, 39 (1966); *Chem. Abstr.* **64**, 9827g (1966).
630. Stille, J. K., and Mainen, E. L., *J. Polym. Sci., Part B* **4**, 665 (1966).

631. Stille, J. K., and Mainen, E. L., *Macromolecules* **1**, 36 (1968); *Chem. Abstr.* **68**, 11106, 115064v (1968).
632. Stille, J. K., Mainen, E. L., Freeburger, M. E., and Harris, F. W., *Polym. Prepr.*, *Amer. Chem. Soc.*, *Div. Polym. Chem.* **8** (1), 244 (1967).
633. Stille, J. K., and Morgan, R. A., *J. Polym. Sci.*, *Part A* **3**, 2397 (1965).
634. Stille, J. K., and Williamson, J. R., *Polym. Prepr.*, *Amer. Chem. Soc.*, *Div. Polym. Chem.* **5** (1), 185 (1964).
635. Stille, J. K., and Williamson, J. R., *J. Polym. Sci.*, *Part A* **2**, 3867 (1964).
636. Stille, J. K., and Williamson, J. R., *J. Polym. Sci.*, *Part B* **2**, 209 (1964).
637. Stille, J. K., Williamson, J. R., and Arnold, F. E., *J. Polym. Sci.*, *Part A* **3**, 1013 (1965); *Chem. Abstr.* **62**, 13252c (1965); *Resins, Rubbers, Plastics* p. 2197 (1965).
638. Strauss, E. L., *SPE J.* **22**, 45 (1966); *Chem. Abstr.* **66**, 8098, 86310y (1967).
639. Strul, M., and Zugravescu, I., *Rev. Roum. Chim.* **12**, 867 (1967); *Chem. Abstr.* **69**, 1043, 10821g (1968).
640. Suga, M., Shono, T., and Shiura, K., *Kogyo Kagaku Zasshi* **69**, 1529 (1966); *J. Chem. Soc. Jap.*, *Ind. Chem. Sect.* **69**, 1529 (1966); *Chem. Abstr.* **66**, 4435, 46644d (1967).
641. Tabushi, I., Tanimura, N., and Oda, R., *Kogyo Kagaku Zasshi* **67**, 1084 (1964); *Chem. Abstr.* **61**, 16167g (1964).
642. Tatum, W. E., U.S. Pat. 3,261,811 (E. I. duPont de Nemours Co.) (1966).
643. Tatum, W. E., Amborski, L. E., Gerow, C. W., Heathcock, J., and Mallouck, R. S., *Elec. Insul. Conf.*, *1963* AIEE Doc. T-153-3, Minutes, p. 1.
644. Tatum, W. E., and Thornton, R. L., U.S. Pat. 3,326,863 (E. I. duPont de Nemours Co.) (1967); *Chem. Abstr.* **67**, 6159, 65171j (1967).
645. Ting, M.-H., *Hua Hsueh Tung Pao* **7**, 385 (1965); *Chem. Abstr.* **64**, 6769d (1966).
646. Tocker, S., U.S. Pat. 3,299,101 (E. I. duPont de Nemours Co.) (1967); *Chem. Abstr.* **68**, 11119, 115217x (1968).
647. Tocker, S., U.S. Pat. 3,326,851 (E. I. duPont de Nemours Co.) (1967); *Chem. Abstr.* **67**, 6132, 64908t (1967).
648. Todd, N. W., *Nat. SAMPE* (*Soc. Aerosp. Mater. Process Eng.*) *Symp.*, *Prepr. 8th 1965* 9 pp. (1965); *Chem. Abstr.* **68**, 4902, 50441u (1968).
649. Todd, N. W., Wolff, F. A., Mallouck, R. S., and Schweitzer, F. E., *Mach. Des.* **36**, 228 (1964).
650. Tohyama, S., Kurihara, M., Ikeda, K., and Yoda, N., *J. Polym. Sci.*, *Part A-1* **5**, 2523 (1967); *Chem. Abstr.* **68**, 1308, 13472j (1968).
651. Tokarev, A. V., Lyubova, T. A., and Kudryavtsev, G. I., *Vysokomol. Soedin.*, *Ser. B* **9**, 634 (1967); *Chem. Abstr.* **67**, 10300, 109061e (1967).
652. Traynor, E. J., Jr., and Luck, R. M., Ger. Pat. 1,221,797 (Westinghouse Electric Corp.) (1966); *Chem. Abstr.* **66**, 360, 3157v (1967).
653. Traynor, E. J., Jr., and Luck, R. M., U.S. Pat. 3,371,009 (Westinghouse Electric Corp.) (1968).
654. Tuemmler, W. B., U.S. Pat. 3,245,965 (Monsanto Co.) (1966).
655. Unishi, T., *J. Polym. Sci.*, *Part B* **3**, 679 (1965).
656. Unishi, T., and Hasegawa, M., *Kogyo Kagaku Zasshi* **70**, 2392 (1967); *Chem. Abstr.* **69**, 1861, 19623b (1968).
657. Unishi, T., and Tsujimura, T., *Yuki Gosei Kagaku Kyokai Shi* **23**, 1028 (1965); *Chem. Abstr.* **64**, 5223d (1966).
658. Unishi, T., and Tsujimura, T., *Kogyo Kagaku Zasshi* **68**, 2275 (1965); *Chem. Abstr.* **64**, 12811h (1966).
659. Vayson de Pradenne, H., Fr. Pat. 1,430,968 (Société Générale de Constructions Electriques et Mécaniques) (1966); *Chem. Abstr.* **66**, 349, 3036z (1967).

660. Veitch, J., and Yeomans, B., Br. Pat. 1,094,251 (Distillers Co. Ltd.) (1967); *Chem. Abstr.* **68**, 2974, 30418p (1968).
661. Vinogradova, S. V., Korshak, V. V., and Vygodskii, Ya. S., Russ. Pat. 171,552 (1965); *Chem. Abstr.* **63**, 16503f (1965).
662. Vinogradova, S. V., Korshak, V. V., and Vygodskii, Ya. S., *Vysokomol. Soedin.* **8**, 809 (1966).
663. Vinogradova, S. V., Korshak, V. V., and Vygodskii, Ya. S., Russ. Pat. 183,383 (1966); *Chem. Abstr.* **66**, 348, 3022g (1967).
664. Vinogradova, S. V., Korshak, V. V., and Vygodskii, Ya. S., Russ. Pat. 215,493 (Institute of High Molecular-Weight Compounds, Academy of Sciences, USSR) (1968); *Chem. Abstr.* **69**, 1877, 19809s (1968).
665. Vinogradova, S. V., Korshak, V. V., Vygodskii, Ya. S., and Lokshin, B. V., *Dokl. Akad. Nauk SSSR* **171**, 1329 (1966); *Chem. Abstr.* **66**, 5300, 55928b (1967); *Dokl. Chem. (English transl.)* **171**, 1194 (1966).
666. Vinogradova, S. V., Korshak, V. V., Vygodskii, Ya. S., and Lokshin, B. V., *Vysokomol. Soedin., Ser. A* **9**, 1091 (1967); *Chem. Abstr.* **67**, 6991, 73984s (1967).
667. Vinogradova, S. V., Pavlova, S. A., Korshak, V. V., Vygodskii, Ya. S., Boiko, L. V., and Golubeva, N. A., *Vysokomol Soedin., Ser. B* **10**, 398 (1968); *Chem. Abstr.* **69**, 4919, 52546b (1968).
668. Vlasova, K. N., and Chernova, A. G., *Plast. Massy* No. 10, p. 16 (1967); *Chem. Abstr.* **68**, 1322, 13630j (1968); *Sov. Plast.* No. 10, p. 19 (1967).
669. Vlasova, K. N., Chernova, A. G., Dobrokhotova, M. L., Tanunina, P. M., Gershkokhen, S. L., Pilyaeva, V. F., Emel'yanova, L. N., Bublik, L. S., Shcherba, N. S., and Chechik, A. I., *Plast. Massy* No. 5, p. 15 (1968); *Chem. Abstr.* **69**, 1901, 20076g (1968).
670. Vollmert, B., *Kunststoffe* **56**, 680 (1966); *Chem. Abstr.* **66**, 1109, 11181y (1967).
671. Vollmert, B., and Reichel, B., Fr. Pat. 1,488,924 (Badische Anilin-und-Soda Fabrik. A.G.) (1967); *Chem. Abstr.* **68**, 7612, 78763z (1968).
672. Volpe, A. A., and Bailey, W. J., AD 641,873; *Chem. Abstr.* **68**, 2953, 30212s (1968).
673. Vygodskii, Ya. S., Vinogradova, S. V., and Korshak, V. V., *Vysokomol. Soedin., Ser. B* **9**, 587 (1967); *Chem. Abstr.* **67**, 9473, 100498v (1967).
674. Wallach, M. L., *Polym. Prepr., Amer. Chem. Soc., Div. Polym. Chem.* **6** (1), 53 (1965); *Chem. Abstr.* **65**, 20227g (1966).
675. Wallach, M. L., *Polym. Prepr., Amer. Chem. Soc., Div. Polym. Chem.* **8** (1), 656 (1967).
676. Wallach, M. L., *Polym. Prepr., Amer. Chem. Soc., Div. Polym. Chem.* **8** (2), 1170 (1967).
677. Wallach, M. L., *J. Polym. Sci., Part A-2* **5**, 653 (1967); *Chem. Abstr.* **67**, 4168, 44176z (1967).
678. Wallach, M. L., *J. Polym. Sci., Part A-2* **6**, 953 (1968); *Chem. Abstr.* **69**, 323, 3259b (1968).
679. Wallenberger, F. T., *Polym. Eng. Sci.* **6**, 369 (1966); *Chem. Abstr.* **65**, 20290b (1966).
680. Wang, P.-J., *Ko Fen Tzu T'ung Hsun* **8**, 41 (1966); *Chem. Abstr.* **66**, 325, 2782j (1967).
681. Werntz, J. H., Fr. Pat. 1,446,620 (E. I. duPont de Nemours Co.) (1966); *Chem. Abstr.* **66**, 4448, 46788d (1967).
682. Werntz, J. H., U.S. Pat. 3,414,546 (E. I. duPont de Nemours Co.) (1968).
683. Wildi, B. S., and Katon, J. E., *J. Polym. Sci., Part A* **2**, 4709 (1964).
684. Williamson, J. R., *Diss. Abstr.* **26**, 107 (1965); *Chem. Abstr.* **63**, 16540c (1965).
685. Wolf, R., and Marvel, C. S., *J. Polym. Sci., Part A-1* **7**, 2481 (1969); *Chem. Abstr.* **71**, 125052r (1969).
686. Wolf, R., Okada, M., and Marvel, C. S., *J. Polym. Sci., Part A-1* **6**, 1503 (1968); *Chem. Abstr.* **69**, 319, 3213g (1968).

687. Wrasidlo, W. J., and Augl, J. M., *Polym. Prepr., Amer. Chem. Soc., Div. Polym. Chem.* **10**, 1353 (1969).

688. Wrasidlo, W. J., and Augl, J. M., *J. Polym. Sci., Part B* **7**, 281 (1969); *Chem. Abstr.* **71**, 39526s (1969).

689. Wrasidlo, W. J., and Augl, J. M., *J. Polym. Sci., Part A-1* **7**, 3393 (1969); *Chem. Abstr.* **72**, 44176c (1970).

690. Wrasidlo, W. J., Hergenrother, P. M., and Levine, H. H., *Polym. Prepr., Amer. Chem. Soc., Div. Polym. Chem.* **5** (1), 141 (1964); *Chem. Abstr.* **64**, 3699g (1966).

691. Wu, H.-C., Chu, C.-C., and Ho, T.-C., *K'o Hsueh Ch'u Pan She* pp. 280–285 (1963); *Chem. Abstr.* **64**, 2186c (1966).

692. Yoda, N., *Kagaku Kogyo* **18**, No. 5, 49 (1967); *Chem. Abstr.* **67**, 7787, 82393a (1967).

693. Yoda, N., and Baba, Y., Jap. Pat. 67/14,469 (Toyo Rayon Co.) (1967); *Chem. Abstr.* **68**, 2968, 30365u (1968).

694. Yoda, N., Ikeda, K., and Kurihara, M., Jap. Pat. 69/20,638 (1969); *Chem. Abstr.* **71**, 125252f (1969).

695. Yoda, N., Ikeda, K., Kurihara, M., Tohyama, S., and Nakanishi, R., *J. Polym. Sci., Part A-1* **5**, 2359 (1967); *Chem. Abstr.* **68**, 5804, 59933p (1968).

696. Yoda, N., and Kurihara, M., Jap. Pat. 67/10,629 (Toyo Rayon Co.) (1967); *Chem. Abstr.* **67**, 6133, 64915t (1967).

697. Yoda, N., and Kurihara, M., Jap. Pat. 67/14,466 (Toyo Rayon Co.) (1967); *Chem. Abstr.* **68**, 2974, 30422k (1968).

698. Yoda, N., and Kurihara, M., Jap. Pat. 67/14,467 (Toyo Rayon Co.) (1967); *Chem. Abstr.* **68**, 2975, 30424n (1968).

699. Yoda, N., and Kurihara, M., Jap. Pat. 68/06,075 (Toyo Rayon Co.) (1968); *Chem. Abstr.* **69**, 1877, 19808r (1968).

700. Yoda, N., and Kurihara, M., Jap. Pat. 68/11,239 (Toyo Rayon Co.) (1968); *Chem. Abstr.* **70**, 470, 4813k (1969).

701. Yoda, N., and Kurihara, M., Jap. Pat. 68/15,837 (Toyo Rayon Co.) (1968); *Chem. Abstr.* **70**, 12137h (1969).

702. Yoda, N., and Kurihara, M., Jap Pat. 68/15,838 (Toyo Rayon Co.) (1968); *Chem. Abstr.* **70**, 12138j (1969).

703. Yoda, N., Kurihara, M., and Ikeda, K., Jap. Pat. 68/15,839 (Toyo Rayon Co.) (1968); *Chem. Abstr.* **70**, 20509w (1969).

704. Yoda, N., Kurihara, M., and Ikeda, K., Jap. Pat. 68/15,997 (Toyo Rayon Co.) (1968); *Chem. Abstr.* **70**, 68892t (1969).

705. Yoda, N., Kurihara, M., and Ikeda, K., Jap Pat. 68/15,998 (Toyo Rayon Co.) (1968); *Chem. Abstr.* **70**, 12134e (1969).

706. Yoda, N., Kurihara, M., and Ikeda, K., Jap. Pat. 68/27,633 (Toyo Rayon Co.) (1968); *Chem. Abstr.* **70**, 88451b (1969).

707. Yoda, N., Kurihara, M., and Ikeda, K., Jap. Pat. 68/28,480 (Toyo Rayon Co.) (1968); *Chem. Abstr.* **70**, 78559d (1969).

708. Yoda, N., Kurihara, M., and Ikeda, K., Jap. Pat. 69/08,239 (Toyo Rayon Co.) (1969); *Chem. Abstr.* **71**, 71226e (1969).

709. Yoda, N., Kurihara, M., and Ikeda, K., Jap. Pat. 69/20,635 (Toyo Rayon Co.) (1969); *Chem. Abstr.* **72**, 13378r (1970).

710. Yoda, N., Kurihara, M., Ikeda, K., Toyama, S., and Nakanishi, R., *J. Polym. Sci., Part B* **4**, 551 (1966).

711. Yoda, N., Kurihara, M., Toyama, S., Dogoshi, N., and Ikeda, K., Jap. Pat. 69/19555 (Toyo Rayon Co.) (1969); *Chem. Abstr.* **71**, 113490u (1969).

712. Yoda, N., Nakanishi, R., Kubota, T., Kurihara, M., and Ikeda, K., Fr. Pat. 1,448,897 (Toyo Rayon Co.) (1966); *Chem. Abstr.* **66**, 6242, 66058u (1967).
713. Yoda, N., Nakanishi, R., Kurihara, M., Baruba, Y., Shunzoku, T., and Ikeda, K., *J. Polym. Sci., Part B* **4**, 11 (1966).
714. Yoda, N., Toyama, S., Kurihara, M., and Ikeda, K., Jap. Pat. 69/19,556 (Toyo Rayon Co.) (1969); *Chem. Abstr.* **71**, 113497b (1969).
715. Yokoyama, M., and Konya, S., *Kogyo Kagaku Zasshi* **71**, 599 (1968); *Chem. Abstr.* **69**, 2619, 27952y (1968).
716. Zakoshchikov, S. A., Pomerantseva, K. P., and Emel'yanova, L. N., *Vysokomol. Soedin., Ser. B* **9**, 757 (1967); *Chem. Abstr.* **68**, 1308, 13477q (1968).
717. Zakoshchikov, S. A., Rozhkov, V. S., Ruzhentseva, G. A., and Zubareva, G. M., Russ. Pat. 173,930 (1965); *Chem. Abstr.* **64**, 3723a (1966).
718. Zakoshchikov, S. A., and Ruzhentseva, G. A., *Polym. Sci. USSR* **8**, 1355 (1966).
719. Zakoshchikov, S. A., Vlasova, K. N., and Zolotareva, G. M., *Vysokomol. Soedin., Ser. B* **9**, 234 (1967); *Chem. Abstr.* **67**, 1140, 11807c (1967).
720. Zakoshchikov, S. A., Vlasova, K. N., Zubareva, G. M., Krasnova, N. M., and Ruzhentseva, G. A., *Plast. Massy* No. 1, p. 14 (1966); *Chem. Abstr.* **64**, 15994f (1966); *Sov. Plast.* No. 1 p. 16 (1966).
721. Zakoshchikov, S. A., Zubareva, G. M., and Zolotareva, G. M., *Sov. Plast.* No. 4, p. 13 (1967).

## CHAPTER II

# Intra–Intermolecular Polymerizations Leading to Heterocyclic Rings

Intra–intermolecular polymerization or cyclopolymerization, is a widely used method for the synthesis of ring-containing polymers. The generalized reaction is shown in Eq. (II-1). The formation of polymers containing carbo-

$$(\text{II-1})$$

cyclic rings by this route was discussed in Chapter II of Part A. The reaction is also applicable to the synthesis of heterocyclic ring-containing polymers. In fact, the first well-defined polymer prepared by intra–intermolecular polymerization was a polymeric piperidinium salt (*63, 72, 76, 77*) [Eq. (II-2)].

$$(\text{II-2})$$

287

Since this initial work, many other classes of heterocyclic polymers have been prepared by this method. They will be discussed in subsequent sections of this chapter.

The tendency of certain classes of monomers to yield linear ring-containing polymers is indeed remarkable. Staudinger's hypothesis (*260*) predicts that monomers containing two or more double bonds should yield cross-linked polymers. The generality of cyclopolymerization raises an obvious question. What makes a particular monomer yield a linear ring-containing polymer instead of a cross-linked gel? Probability calculations (*82*) indicate that intra–intermolecular polymerization should not occur to a significant extent, especially when the reaction is carried out at high monomer concentrations (bulk). Butler postulates (*64, 70*) that spatial interaction of the vinyl groups in the monomer is responsible for the observed tendency to cyclopolymerize. Spectral studies (*70, 74, 86*) do in fact indicate that such an interaction exists. Interestingly, a similar interaction was demonstrated in *o*-phthalaldehyde (*29*). The latter yields cyclopolymers when polymerized by a variety of initiators (*29*).

[7]        [8]        (II-3)

Steric factors in the monomer molecule are also important (*135*). For example, the dialdehydes, [9] and [10], do not cyclopolymerize (*29*).

[9]        [10]        (II-4)

The cyclopolymerization of methacrylic anhydride was explained by assuming that ring conformations [11] or [12] are important in the monomer (*139, 140*). Based on the specific gravity of poly(methacrylic anhydride), the molar shrinkage value for one double bond in this polymerization was calculated to be 13.7 ml/mole (*139*). A comparison with the value of 22–23 ml/mole for acrylic esters supports the hypothesis that the monomer exists

$$(II-5)$$

[11]        [12]

in a cyclic conformation which probably provides the driving force for cyclopolymerization.

Many studies of intra–intermolecular polymerization have resulted in the following generalizations:

(1) Cyclization is favored when the reactivities of the two double bonds are as similar as possible (32, 33, 93, 117, 164, 229). Dissimilar reactivities lead to the preferential reaction of one double bond with itself (99). The relative rate of the cyclization step becomes slow compared with the self-propagation rate of the more reactive double bond. When the double bonds are identical, cyclization will occur not only in difunctional systems, but in tri- and tetrafunctional molecules as well. This is shown in Eq. (II-6) (272). Similarly, tetraallyl-ammonium bromide yields a linear, cyclic polymer (272).

$$(II-6)$$

[13]        [14]        1. Cyclization  2. Monomer        [15]

The relative reactivities of two different double bonds can change with the type of initiator used. Thus, starting with the same monomer different polymers may be obtained, as was elegantly shown for the case of S-vinyl-N-vinylthio-carbamate (229). The S-vinyl group in [16] is more reactive than the N-vinyl group with radical initiation. The N-vinyl group is more reactive than the

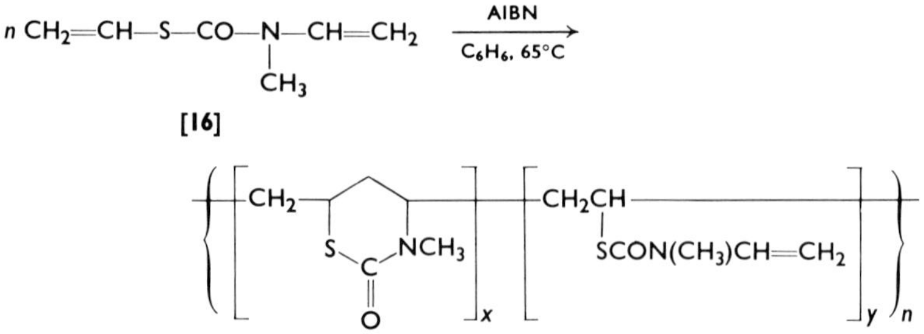

$$n\ CH_2{=}CH{-}S{-}CO{-}N{-}CH{=}CH_2 \quad \xrightarrow[\text{C}_6\text{H}_6,\ 65°\text{C}]{\text{AIBN}}$$

with $CH_3$ on the N.

[16]

(II-7)

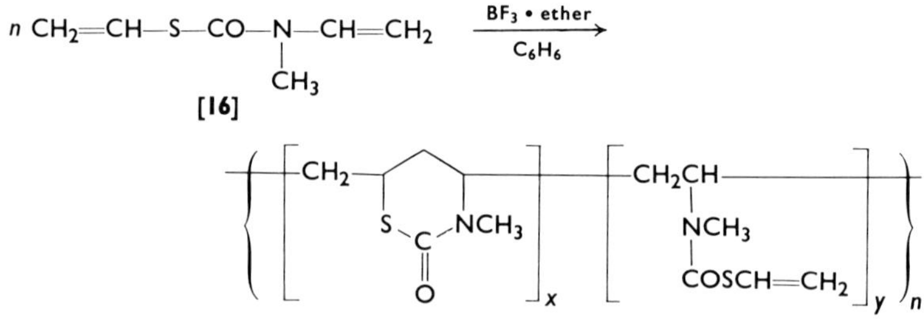

[17]

$$n\ CH_2{=}CH{-}S{-}CO{-}N{-}CH{=}CH_2 \quad \xrightarrow[\text{C}_6\text{H}_6]{\text{BF}_3 \cdot \text{ether}}$$

with $CH_3$ on the N.

[16]

(II-8)

[18]

*S*-vinyl group with cationic initiation. Interestingly, the free-radical polymerization of [16] gave less cyclization when the *N*-methyl group was replaced by larger aliphatic radicals. It was concluded that steric inhibition to ring closure was responsible for this phenomenon (*229*).

(2) Cyclization is more favored when polymerization is carried out at low monomer concentrations (*117, 229, 273–275, 277–280*). Bulk polymerizations can often lead to insoluble, cross-linked polymers.

(3) High conversions favor formation of insoluble cross-linked materials (*152, 229, 236, 237, 273–275, 277–280*).

Intra–intermolecular polymerizations have been the subject of several kinetic studies and reviews (*15, 36, 64, 67, 98, 158, 165, 169, 195, 246*).

## METHODS

The methods that have been used to effect intra–intermolecular polymerizations are grouped into the following four categories. They are shown in the order in which they appear in the discussion.

1. Free-Radical Initiation. 2. Ziegler Initiation. 3. Cationic Initiation. 4. Anionic Initiation. For example, cationic initiation is always *Method 3*.

If a class of monomers was cyclopolymerized by Ziegler and anionic initiations only, the text will start with *Method 2* followed by *Method 4*.

## A. Polymerization of Dialdehydes

Intra–intermolecular polymerization of a dialdehyde is shown in Eq. (II-9). The following methods have been reported for this monomer class.

$$(II-9)$$

[19]                [20]

METHOD 1. FREE-RADICAL INITIATION

Heating glutaraldehyde in bulk or toluene solution yields [22] (*228*, *281*).

$$(II-10)$$

[21]                [22]

Similarly, irradiation of *o*-phthalaldehyde with $\gamma$-rays in methylene chloride solution gives a low molecular weight material that possesses a linear, cyclic

$$(II-11)$$

[23]                [24]

structure (*28*, *29*) [24]. The structure of the phthalan ring in [24] was investigated by nuclear resonance spectroscopy (*29*). Eighty-seven percent of the five-membered rings possessed cis substituents. The possibility that this polymerization actually proceeded by a cationic mechanism was also considered (*29*).

Polymerization of glyoxal, the first member of the homologous dialdehyde series, was studied. Glyoxal polymerizes to a ring-containing polymer. However, the ring does not form during polymerization. Rather, polyaddition through one carbonyl group takes place. The pendant carbonyl functions then form a ring via hydrogen bonding of a partially enolized structure. This is illustrated by Eq. (II-12). Glyoxal polymerizes at room temperature and 100 mm pressure in 48 hours to a waxy, colorless polymer [26] (*134*). The polymer

[25]                                              [26]

(II-12)

decomposes at 150°C. Its structure [26] was supported by infrared spectroscopy. This tool also indicated the presence of lactone rings, presumably a product of a secondary Tischenko-type reaction in the polymer chain (*134*). Irradiation of glyoxal with ultraviolet light yields a white, insoluble polymer (*208, 221*). Infrared spectroscopy of this material indicated the presence of acetal linkages and the absence of carbonyl groups.

### METHOD 2. ZIEGLER INITIATION

*o*-Phthalaldehyde was cyclopolymerized by triethylaluminum/titanium tetrachloride or triethylaluminum/bis(cyclopentadienyl)titanium dichloride. The reactions were performed in toluene solution. Infrared and nuclear magnetic resonance spectroscopy and the solubility properties of the polymer indicated that it possessed a cyclic, linear structure (*29*). However, important differences were noted on comparing the present polymer with the one prepared by Method 1. The "free-radical" polymer softened at 130°–132°C; its phthalan ring substituents had a predominantly cis configuration (87%) (*29*). On the other hand, the softening point of the present material was 139°–154°C. Nuclear magnetic resonance spectroscopy indicated that the substituents on the phthalan ring were only 37–40% cis (*29*).

### METHOD 3. CATIONIC INITIATION

As far back as 1902, Harries observed (*129, 130*) that low molecular weight, glassy materials resulted when succinaldehyde or glutaraldehyde were left at 0°C or room temperature. This spontaneous polymerization was studied recently with 3-methyl- and 3-phenylglutaraldehydes [27] (*18*). It was established that low molecular weight polymers (mol.wt. 700–800) are obtained that possess structure [28] (*18*). The spontaneous polymerization can be catalyzed by the addition of small amounts (5-10%) of materials that possess an active hydrogen atom (*18*). Water, ethanol, and benzyl alcohol are suitable catalysts.

$$\text{(II-13)}$$

[27]

R = CH₃ or C₆H₅

[28]

This finding suggests that the polymerization may be cationic, although no definite proof is available at the present time.

Other studies on the spontaneous polymerization of glutaraldehyde and its 3-methyl and 3-phenyl derivatives at 0°C support structure [29] for the polymers

$$\text{(II-14)}$$

[29]

R = H, CH₃, C₆H₅

(198, 199). These polymers were thermally unstable. However, acetylation of the hydroxyl end groups significantly improved their thermal stability (199).

In an attempt to prepare polymers of glutaraldehyde by the action of aluminum alkoxides via a Tishchenko-type polymerization, polymers containing some cyclic tetrahydropyran units along with the expected ester groups were obtained (285). It was assumed that spontaneous cyclopolymerization, possibly catalyzed by traces of water, was responsible for the presence of these cyclic units. Interestingly, under these conditions, cyclopolymerization was favored over the Tishchenko polycondensation by the use of lower temperatures (286).

The polymerization of dialdehydes in the presence of weak Lewis acids was studied in several investigations. Examples of the catalysts used included equimolar mixtures of triethylaluminum and water (16–18), diisobutyl-aluminum chloride (226), diethylzinc (228), and tri-tert-butylaluminum (228). The polymerizations are preferably run in toluene at −76° to −78°C. Incomplete cyclization was usually observed and the polymers contained pendant aldehyde groups in variable amounts. Linear and cyclic dialdehydes could be polymerized by this method. Depending on the experimental conditions used, the reactions were accompanied by variable amounts of insoluble, cross-linked polymer. An example is shown in Eq. (II-15). The polymerization

[30]                [31]    (II-15)

yielded a material [31] that was partly soluble (17.9%) in benzene. The intrinsic viscosity of the soluble fraction was 0.26. The ratio $x/y$ varied with experimental conditions over the range of 6 to 16 (16). A similar polymerization is observed with *trans*-cyclohexane-1,2-dicarboxaldehyde (226).

[32]                [33]    (II-16)

A stronger Lewis acid, widely used for the ring-forming polymerization of dialdehydes, is boron trifluoride etherate. Glutaraldehyde, treated with this reagent in tetrahydrofuran at −70°C, yielded a partially cyclized polymer that was soluble in methylene chloride (209), chloroform, benzene, and acetone (228). At 0°C, a cross-linked insoluble product was obtained (228). The choice of solvent appears to be very important in these polymerizations. For instance, when glutaraldehyde was treated with boron trifluoride etherate in toluene or petroleum ether at −50° to −78°C, gelation occurred (199). The polymers were insoluble in the usual solvents, but could be dissolved, although with degradation, in *m*-cresol.

A cyclic, soluble polymer was obtained from *o*-phthalaldehyde by treatment with triphenylmethyl fluoroborate in methylene chloride (28, 29). It was soluble in methylene chloride ([$\eta$] 0.30), tetrahydrofuran, and toluene and softened at 125°–127°C. The cis content of the phthalan ring was 77% (nuclear magnetic resonance spectroscopy).

The ring-forming polymerization of succinaldehyde by boron trifluoride etherate gave low molecular weight polymer with a reduced viscosity of 0.05 (24). However, the use of an equimolar mixture of triethylaluminum and water yielded high molecular weight material ([$\eta$] 0.59) (24). These results were rationalized in terms of the transition state [34], which was believed to favor cyclization (24).

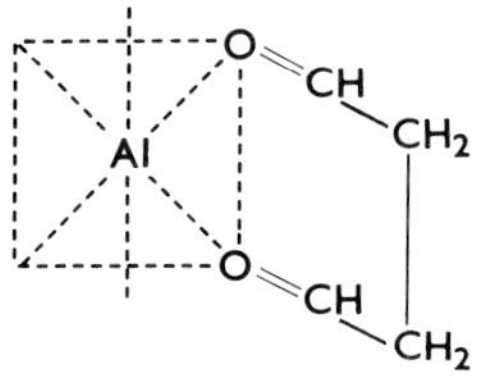

(II-17)

[34]

The effect of the size of the incipient ring on the degree of cyclization was shown for the polymerization of adipinaldehyde and suberaldehyde by Lewis acids. While the degree of cyclization was high with glutaraldehyde [six-membered rings formed (*16*)], it was lower for adipinaldehyde [seven-membered rings formed (*20, 21*)]. Suberaldehyde (*160, 161*) gave an insoluble polymer. Infrared spectroscopy suggested the presence of some cyclic units in the material. Presumably, the result reflected the reluctance to form the energetically unfavorable nine-membered rings. An interesting observation was made by studying the copolymerization of adipinaldehyde with chloral (*20, 21*). The high degree of cyclization observed under these conditions was influenced by the trichloromethyl group of the chloral (*20, 21*).

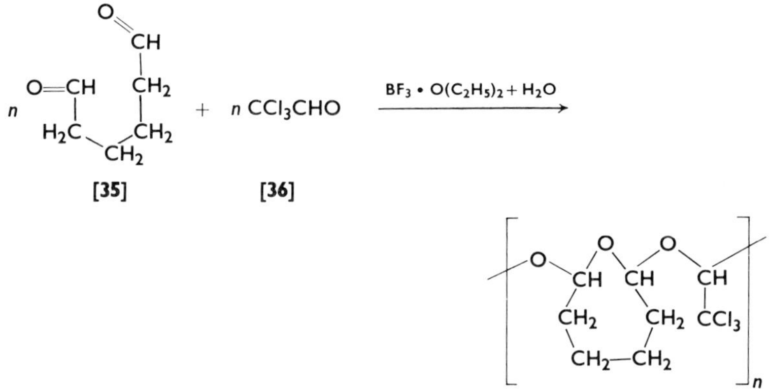

[35]   [36]

(II-18)

[37]

## METHOD 4. ANIONIC INITIATION

Treatment of *o*-phthalaldehyde with lithium *tert*-butoxide in tetrahydrofuran gave a cyclopolymer ([η] 0.13) that was soluble in methylene chloride, tetrahydrofuran, and toluene (*29*). Infrared and nuclear magnetic resonance spectroscopic studies indicated that the geometry of the phthalan ring of this polymer corresponded to 60% cis isomer. The polymer softened at 117°–120°C. Comparing the properties of the polymer made from *o*-phthalaldehyde by the four initiation methods shows that in each case a different material was obtained. Similar variations in polymer properties were observed with succinaldehyde (23). This behavior indicates that there are important mechanistic differences when a dialdehyde is polymerized by these different catalysts.

Treatment of glyoxal with sodium naphthalene in tetrahydrofuran at $-78°C$ yielded polymer ($\eta_{inh}$ 0.3) (54, 55). Its structure was presumed to be the same as the one shown earlier (Method 1, [26]). X-rays showed the polymer to be amorphous. Infrared spectroscopy supported the assigned structure [26] (54, 55).

The cyclopolymerization of acrolein and methacrolein dimers by Lewis acids is related to the reactions that are discussed in this section. In the case of these monomers, ring formation takes place by the interaction of an aldehyde group with a carbon-to-carbon double bond (26, 27). An example is shown in Eq. (II-19). The catalysts useful for this polymerization include

$$\text{[38]} \qquad\qquad\qquad\qquad \text{[39]} \qquad\qquad (II-19)$$

tin tetrachloride and boron trifluoride etherate. The use of lower temperatures results in a lower degree of cyclization and yields polymers containing unsaturation. The proposed structure [39] is supported by infrared and nuclear magnetic resonance data.

## B. Polymerization of Diepoxides

The cyclopolymerization of a diepoxide is shown in Eq. (II-20).

$$\text{[40]} \qquad\qquad\qquad\qquad \text{[41]} \qquad\qquad (II-20)$$

METHOD 1. FREE-RADICAL INITIATION

The irradiation of solid or liquid o-di(epoxyethyl)benzene with γ-rays or $^{60}Co$ rays at temperatures from $-78°$ to $+40°C$ gave polymer (19, 25). It was partially soluble in tetrahydrofuran. The soluble portion was assigned structure [43], which results from cyclopolymerization.

$$[42] \qquad\qquad [43] \qquad\qquad\qquad \text{(II-21)}$$

## METHOD 3. CATIONIC INITIATION

Cationic catalysts have been widely used for the polymerization of diepoxides. Useful catalysts include boron trifluoride etherate (*19, 25*), phosphorus pentafluoride (*261–263*), partially hydrolyzed phosphorus pentafluoride (*261–263*), mixtures of aluminum alkoxides with zinc chloride (*19, 25*), and aluminum or zinc alkyls with or without water or alumina as cocatalysts (*19, 25, 261–263*). The reactions were carried out in bulk or with solvents. Benzene, toluene, nitrobenzene, methylene chloride, 1,2-dichloroethane, and *n*-heptane were used as solvents. The polymerization of 1,2,5,6-diepoxyhexane [44], shown in Eq. (II-22) is an example (*261–263*).

$$[44] \qquad\qquad\qquad\qquad [45] \qquad\qquad \text{(II-22)}$$

The diepoxide [46] gave only cross-linked material when polymerized by several cationic catalysts (*263*). The difference in behavior between [44] and [46] was

$$\text{(II-23)}$$

$$[46]$$

explained on steric grounds. The diepoxide [46] should give a polymer containing five-membered rings while six-membered rings are formed from monomer [44]. Apparently, differences in the stability of the incipient ring can again determine the course of polymerization.

Recently the polymerization of the *dl* and *meso* forms of *o*-di(epoxyethyl)-benzene [42] was investigated (*264*). Phosphorus pentafluoride in methylene chloride or aluminum triisobutyl in benzene were used as catalysts. The polymer obtained from the *dl* form had a different melting point (85°C by DTA) than the product from the *meso* isomer ($T_m = 125$°C). The authors concluded that both the intra and inter propagation steps were stereospecific and proceeded with inversion of configuration (*264*).

An interesting example involving the copolymerization of an oxirane ring with a double bond was described by Nikolaev and co-workers (*216*). Polymerization of isoprene oxide [47] with diethylaluminum chloride in a variety of solvents yielded an oligomeric product. Analysis of the product [48] indicated the presence of oxetane rings in its structure [Eq. (II-24)].

$$(n + 2)\ CH_2-C-CH=CH_2 \xrightarrow[\substack{Solvent,\\0-80°C}]{(C_2H_5)_2AlCl}$$

[47]

$$HOCH_2-C \underset{CH_3}{\overset{CH=CH_2}{\underset{\displaystyle |}{\overset{\displaystyle |}{\Big|}}}} \left[ CH_2-CH \underset{\displaystyle |}{\overset{CH_3}{\Big|}} \right]_n OCH_2CHCH=CH_2 \quad (II\text{-}24)$$

[48]

## METHOD 4. ANIONIC INITIATION

Diepoxides were polymerized to ring-containing polymers by several anionic catalysts. Triethylamine in bulk (*19, 25*) and potassium *tert*-butoxide in benzene or tetrahydrofuran (*38*) were useful. With *o*-di(epoxyethyl)benzene (*19, 25*), triethylamine-catalyzed polymerization proceeded as shown in Eq. (II-21) and a polymer with structure [43] was obtained. However, polymerization of 1,2,5,6-diepoxyhexane [44] with potassium *tert*-butoxide took a different course from that in Eq. (II-22) (*38*). A five-membered ring-containing polymer was obtained instead [Eq. (II-25)]. Infrared and nuclear

$$n\ CH_2-CH-(CH_2)_2-CH-CH_2 \xrightarrow[50°C]{tert-C_4H_9OK}$$

[44]

$$\left[ CH_2-\underset{O}{\overset{}{\diagdown}}-CH_2O \right]_n \quad (II\text{-}25)$$

[49]

magnetic resonance data supported structure [49]. Similarly, $N,N$-diglycidyl-aniline gave polymer [51], not the expected [52] (*38*).

**C. Polymerization of Divinyl Acetals and Ketals**

The ring-forming polymerization of divinyl acetals and ketals is shown in Eq. (II-27).

METHOD 1. FREE-RADICAL INITIATION

Most of the polymerization research with divinyl acetals and ketals was performed with free-radical initiators. Benzoyl peroxide and $\alpha,\alpha'$-azodiiso-butyronitrile in bulk (*1, 7, 12, 20, 168, 177, 205*) or solution (*1, 180, 207*) have been used as catalysts. Benzene (*1, 207*), toluene (*180*), and methyl ethyl ketone (*207*) were preferred solvents. Polymerization temperatures were generally within the 50°–100°C range. This class of monomers has also been polymerized by ultraviolet light either alone (*11, 207, 231*) or in combina-tion with a free-radical source such as $\alpha,\alpha'$-azodiisobutyronitrile (*12, 177*).

The kinetics of the polymerization of divinyl formal were studied (*205*). Contradictory reports about the products obtainable from [55] have appeared.

$$
\begin{array}{c}
\underset{\substack{\text{CH}_2 \quad \text{CH}_2 \\ \| \qquad \| \\ n \ \text{CH} \quad \text{CH} \\ | \qquad | \\ \text{O} \qquad \text{O} \\ \diagdown \text{C} \diagup \\ \text{H}_2}}{[55]}
\end{array}
\quad \longrightarrow \quad
\begin{array}{c}
\left[\text{CH}_2 - \underset{\text{O} \quad \text{O} \\ \diagdown \text{CH}_2 \diagup}{\bigcirc}\right]_n \\ [56]
\end{array}
\qquad \text{(II-28)}
$$

The formation of a soluble polymer [56] is reported in most instances (*11, 12, 34, 35, 157, 168, 174, 177, 190*). However in one case, solution polymerization of [55] was claimed to yield an insoluble material (*207*). Probably some slight variation in the experimental conditions was responsible for the differences in the product.

It is interesting to note that although a variety of divinyl acetals and ketals of structure [57] readily give polymer,

$$
\begin{array}{cc}
\underset{\substack{\text{CH}_2 \quad \text{CH}_2 \\ \| \qquad \| \\ \text{CH} \quad \text{CH} \\ | \qquad | \\ \text{O} \qquad \text{O} \\ \diagdown \text{C} \diagup \\ \text{R} \quad \text{R}}}{[57]}
&
\underset{\substack{\text{CH}_2 \quad \text{CH}_2 \\ \| \qquad \| \\ \text{CH}_3 - \text{C} \qquad \text{C} - \text{CH}_3 \\ | \qquad | \\ \text{O} \qquad \text{O} \\ \diagdown \text{C} \diagup \\ \text{H} \quad \text{R}}}{[58]}
\end{array}
\qquad \text{(II-29)}
$$

$$
\text{R} = \text{H, alkyl, aryl} \qquad \text{R} = \text{H}, n\text{—}\text{C}_3\text{H}_7
$$

substituted derivatives possessing structure [58] do not polymerize (*175*). These results were explained as being due to steric hindrance between the reactive centers (*175*). Matsoyan and co-workers (*185*) studied the polymerizations of the series of divinyl acetals represented by [59]. Polymers

$$
\begin{array}{c}
\underset{\substack{\text{CH}_2 \quad \text{CH}_2 \\ \| \qquad \| \\ \text{CH} \quad \text{CH} \\ | \qquad | \\ \text{O} \qquad \text{O} \\ \diagdown \text{CH} \diagup \\ | \\ \bigcirc\text{—OR}}}{[59]}
\end{array}
\qquad \text{(II-30)}
$$

that were soluble in chloroform, benzene, and dioxane were obtained. The R group had a definite influence on the rate of polymerization. Polymerization rates decreased in the order shown in the sequence

$$
\text{R} = \text{CH}_3 > \text{C}_2\text{H}_5 > n\text{-}\text{C}_3\text{H}_7 > n\text{-}\text{C}_4\text{H}_9 > n\text{-}\text{C}_5\text{H}_{11}
$$

Branching of R also decreased the reaction rate. Interestingly, increasing the chain length of R gave polymers with increasing melting points. Branching had the same effect (*185*).

It is generally assumed that poly(divinyl acetals or ketals) contain only six-membered rings as shown in structure [60]. However, in one instance a

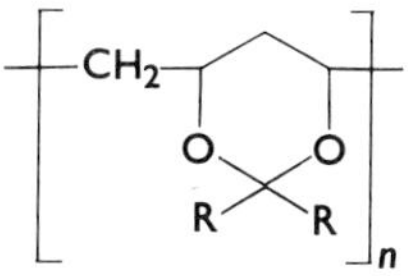

(II-31)

[60]

different structure [62] was reported for these polymers [Eq. (II-32)] (*267*). Structure [62] was determined by hydrolyzing the polymers and then oxidizing the hydrolysis products (*267*). A mixture of 1,2- and 1,3-glycols was obtained.

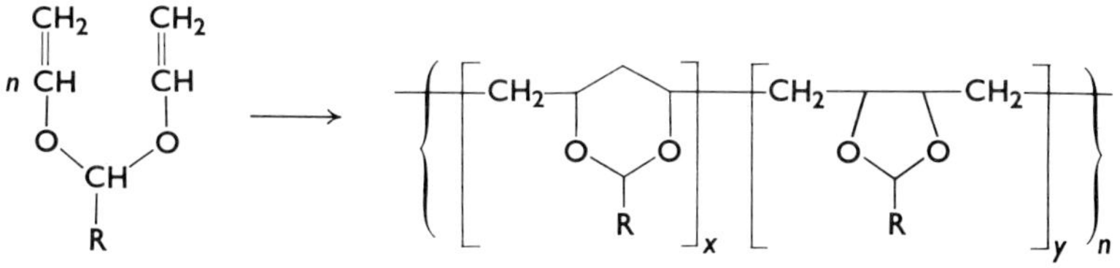

[61]    [62]

(II-32)

The ratio $x/y$ was 77/23 and remained the same for a variety of R groups. However, because most publications describe poly(divinyl acetals and ketals) in terms of structure [60] it will be used in this book.

Retention of stereochemistry was observed in the polymerization of the divinyl acetal of (R)(+)-3,7-dimethyloctanal [63] (*1*) [Eq. (II-33)]. The polymer

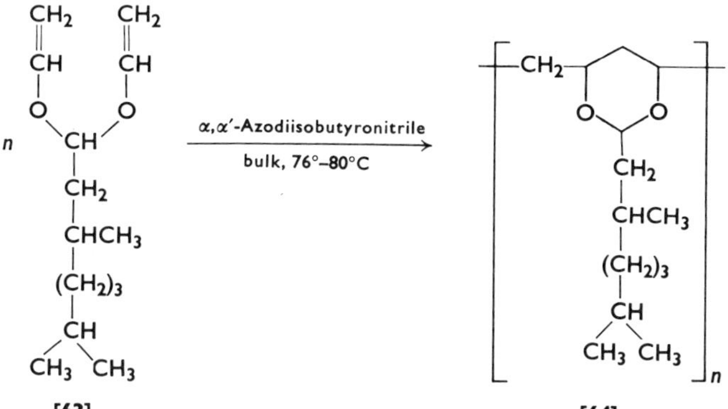

[63]    [64]

(II-33)

[64] was a viscous oil, soluble in chloroform and benzene (*1*). Its optical rotation was positive. On the other hand, polymerization of [63] with boron trifluoride etherate gave a solid polymer that exhibited negative rotations. A difference in stereoregularity was suggested to explain these results (*1*).

The cyclopolymerization of the divinyl acetal of furfural [65] was studied by several groups (*170, 171*). Initially, structure [66] was assigned to the polymer

$$\text{[65]} \xrightarrow{\alpha,\alpha'\text{-Azodiisobutyronitrile}} \text{[66]} \qquad \text{(II-34)}$$

(*170*). A more detailed study has shown that different initiators can give different polymers (*231, 232*). In the presence of α,α′-azodiisobutyronitrile, polymer [66] is obtained. However, with benzoyl peroxide or in the presence of ultraviolet light and oxygen, oxidation takes place and the reaction yields [69].

$$\text{[65]} \longrightarrow \text{[67]} \longrightarrow$$

$$\text{[68]}$$

$$\downarrow$$

$$\text{[69]} \qquad \text{(II-35)}$$

Copolymers of divinyl acetals and ketals with olefins were prepared (*168, 177*). They contain cyclic structures derived from the acetal or ketal components. Copolymerizations of two acetals or ketals yielded copolymers with two types of ring structures in their chains (*168, 177*). Cross-linkable polymers were also prepared (*180*). For example, monomer [70] was cyclopolymerized by α,α′-azodiisobutyronitrile to polymer [71] (*180*). Infrared

[70]                              [71]                              (II-36)

evidence supports structure [71]. It agrees with the known reluctance of vinyl ethers toward radical polymerization. Heating polymer [71] at 100°C yielded a hard, cross-linked product.

The polymerizations that follow do not use a divinyl acetal or ketal as a starting monomer. However, they are discussed here because of their structural resemblance to the examples in this section.

### a. Polymerization of Trivinyl Orthoformate [72]

Free-radical initiation yields a polymer that contains a bicyclic ring system

[72]                    [73]

[74]                              [75]                              (II-37)

(*183*). Appreciable cyclization to a [4.4.2] bicyclic ring structure also took place during the similar radical-initiated polymerization of triallyl orthoformate (*272*).

## b. Copolymerization of Divinyl Ether

Polymer is formed as shown in Eq. (II-38). The reactions were initiated by

(II-38)

potassium persulfate in water, or benzoyl peroxide and $\alpha,\alpha'$-azodiisobutyro-nitrile in benzene, xylene, or dimethylformamide. Maleic anhydride, N-phenylmaleimide, acrylonitrile, fumaronitrile, tetracyanoethylene, and 4-vinylpyridine were also described as comonomers. In some instances (maleic anhydride, N-phenylmaleimide, fumaronitrile) the idealized copolymer [82] resulted regardless of the composition of the monomer feed. Compositional differences were observed in other cases and the factors responsible were discussed (37, 71, 85).

## c. Homopolymerization of Divinyl Ether [76]

Divinyl ether yields a polymer when treated with a free-radical source (30, 31). Infrared, nuclear magnetic resonance, and chemical degradation data indicate that the following reactions occur during the polymerization (30, 31).

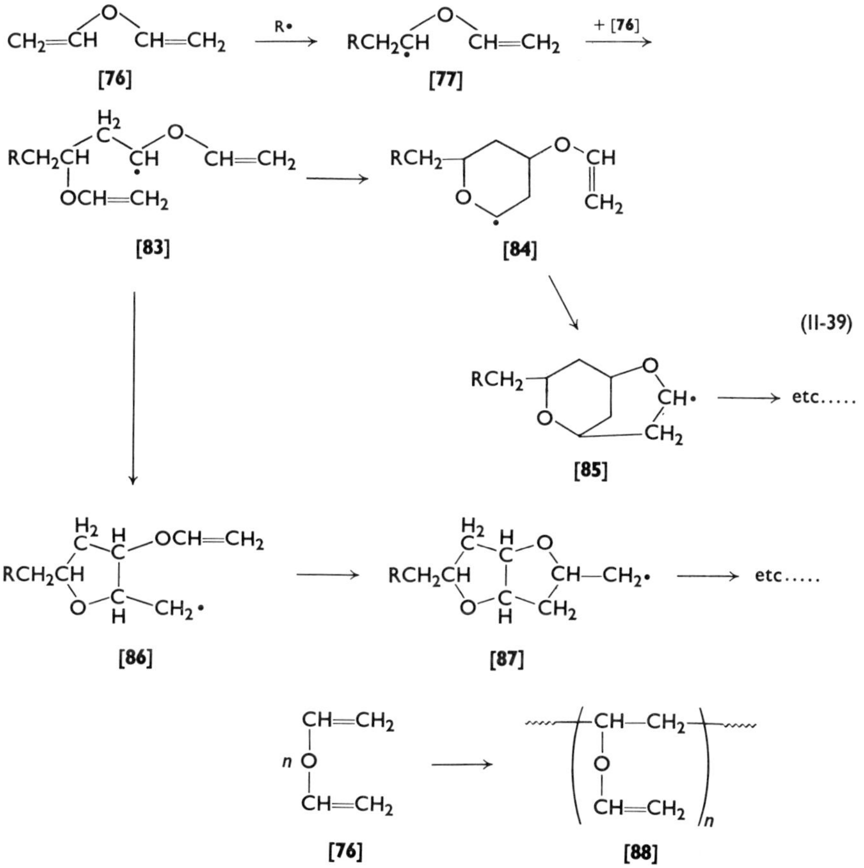

$$\text{(II-39)}$$

The data indicated that the structural units [85], [87], and [88] were present in the polymer in approximately equivalent amounts. Thus, the final polymer structure is [89].

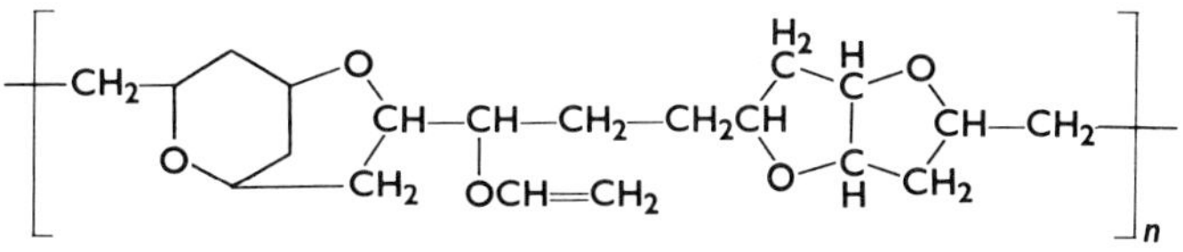

$$\text{(II-40)}$$

[89]

METHOD 2. ZIEGLER INITIATION

The polymerization of divinyl formal [55] by triethylaluminum and vanadium oxytrichloride was reported (207). An insoluble polymer was obtained.

METHOD 3. CATIONIC INITIATION

Divinyl acetals can be polymerized to linear ring-containing polymers by the use of Lewis acids. Boron trifluoride etherate, tin tetrachloride, ferric chloride, and zinc chloride (1, 193) were used. The reactions were performed at low temperatures in diethyl ether or methylene chloride. Linear and cross-linked polymers were obtained from aliphatic divinyl acetals (193). On the other hand, almost exclusively linear polymers resulted in the aromatic series (193). Reaction temperature and the use of a solvent appear of utmost importance in these reactions. Thus, Lewis acid-catalyzed bulk polymerization of divinyl formal, divinyl ethanal, and divinyl butyral was reported to only yield cross-linked, tridimensional materials (168, 177). In one instance, the boron trifluoride etherate-catalyzed low-temperature solution polymerization of divinyl formal yielded an infusible polymer as the sole product (207).

## D. Polymerization of Unsaturated Esters

Two types of unsaturated esters undergo ring-forming polymerization reactions. They are unsaturated glycidyl esters and diunsaturated esters. The glycidyl derivatives will be discussed first.

## 1. POLYMERIZATION OF UNSATURATED GLYCIDYL ESTERS

An example is shown in Eq. (II-41).

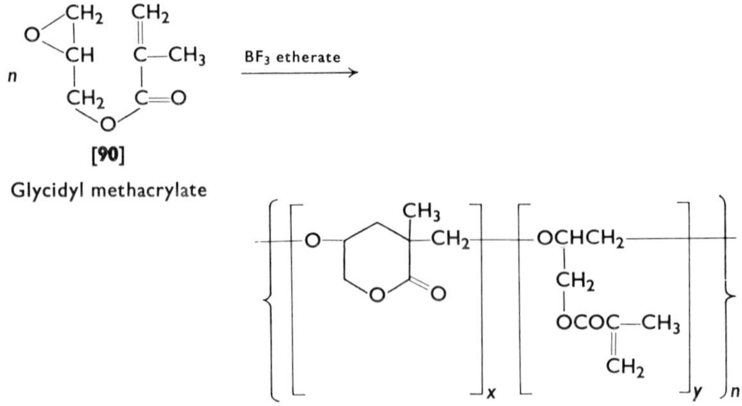

[90]

Glycidyl methacrylate

(II-41)

[91]

## METHOD 3. CATIONIC INITIATION

The functional groups that polymerize in a glycidyl ester containing olefinic unsaturation are the epoxide group and the carbon–carbon double bond. They possess different reactivities, and, as a result, polymerization is accompanied by an appreciable amount of homopolymerization. The functions that homopolymerize vary with the type of initiation. Cationic initiation gives ring-containing polymers that contain olefinic unsaturation (*4, 5*). No epoxide groups are present. This indicates that cationic polymerization of these esters proceeds predominantly via the epoxide groups. A variety of cationic initiators have been used. They include boron trifluoride, boron trifluoride etherate, aluminum chloride, tin tetrachloride, sulfuric and phosphoric acids (*4, 5, 101, 240*). Reactions are performed in bulk at $-15°$ to $+70°C$. Solvents such as tetrahydrofuran or benzene led to epoxide polymerization only (*224*).

The particular catalyst that is used strongly influences the degree of cyclization. This is illustrated for glycidyl methacrylate [90] [Eq. (II-41)]. Boron trifluoride etherate gave a polymer that was 84% cyclized. With aluminum chloride, the degree of cyclization was 94% (*4, 5*).

These cyclopolymers cross-link on heating in bulk at temperatures above $70°C$ due to the pendant double bonds of the polymer (*4, 5, 240*).

## METHOD 4. ANIONIC INITIATION

Butyl lithium-catalyzed polymerization of glycidyl methacrylate was studied and a polymer soluble in acetone and dioxane was obtained (*5*). Analysis indicated the virtual absence of olefinic unsaturation, but the polymer contained 80% of epoxide groups. Thus, the polymerization proceeded predominantly via the carbon-to-carbon double bond (*5*).

## 2. POLYMERIZATION OF DIUNSATURATED ESTERS

Diunsaturated esters polymerize as shown in Eq. (II-42) to yield [93]. Structure [94] can also be formed.

(II-42)

[92]            [93]            [94]

METHOD 1. FREE-RADICAL INITIATION

All the polymerizations of diunsaturated esters used free-radical initiation. The catalysts included benzoyl peroxide (*6, 9, 10, 32, 151, 210*), azodiisobutyro-nitrile (*32, 210, 239*), azodiisobutyronitrile in the presence of ultraviolet light (*210*), ultraviolet light (*51*), cyclohexyl percarbonate (*6*), and di-*tert*-butyl peroxide (*6*). The reactions were carried out in bulk or solution. Benzene, toluene, carbon tetrachloride, acetonitrile, dioxane, and dimethylformamide were useful solvents.

Polymerization of diunsaturated esters was studied extensively in the past. Various unusual phenomena were noted in these studies that led to the belief that some cyclization accompanied polymerization. In the case of ethylene glycol dimethacrylate, the double-bond content of the polymer was lower than expected (*13*). In the divinyl adipate–vinyl acetate system, as well as in other related copolymerizations (*120, 121, 131, 283*), a "delay" in reaching the gel point was observed; in other words, gelation occurred at conversions that were significantly higher than those predicted by theory. The low cross-linking efficiency of ethylene diacrylate in the polymerization of methyl methacrylate (*136*) was also explained by cyclization. Similarly, cyclizations were postulated in the polymerizations of diallyl esters (*119, 133, 135, 162, 218, 251, 252*) and in many other cases (*58, 88, 116, 217, 239, 245*). In fact, the beginnings of intra–intermolecular polymerization can be traced back to these studies.

Brief discussions of some interesting examples of cyclopolymerization of diunsaturated esters follow.

*a. Mono- and Diallyl Maleates and Citraconates*

The polymerization of diallyl maleate yields a ring-containing polymer (*9*) [Eq. (II-43)]. Analysis of the polymer indicated that it contained structures

(II-43)

that resulted from both allylic–maleic and allylic–allylic interactions. The former leads to a six-membered ring while the latter gives a ten-membered ring. Allylic–maleic interaction also takes place in the free-radical polymerization of monoallyl maleate and citraconate (10). Homopolymerization at the two double bonds accompany the ring-forming reaction. A detailed study of

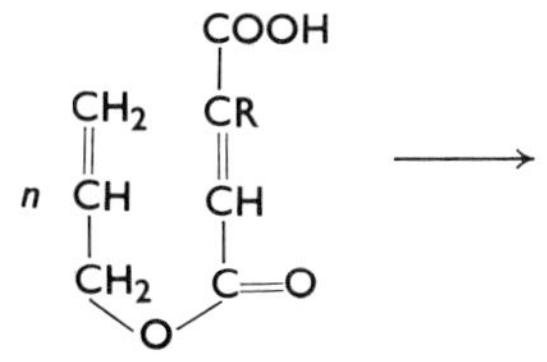

[97]

R = H (maleate)
R = CH₃ (citraconate)

(II-44)

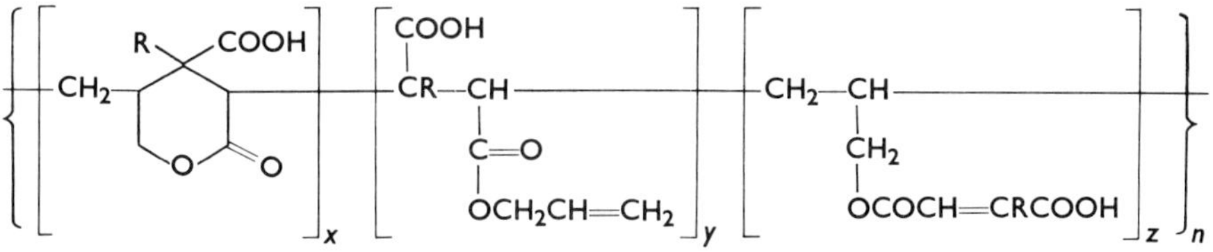

[98]

this cyclopolymerization revealed that five-membered rings are also formed (6). It was also shown that increasing the polymerization temperature gave more five-membered rings in the polymer. No explanation for this phenomenon was offered.

### b. Divinyl Carbonate

The polymerization of divinyl carbonate was studied in bulk and solution (210). Azodiisobutyronitrile alone, or with ultraviolet light, was used for the initiator. The solvents that were used included benzene, toluene, carbon tetrachloride, and acetonitrile. The polymer had structure [99]. The values

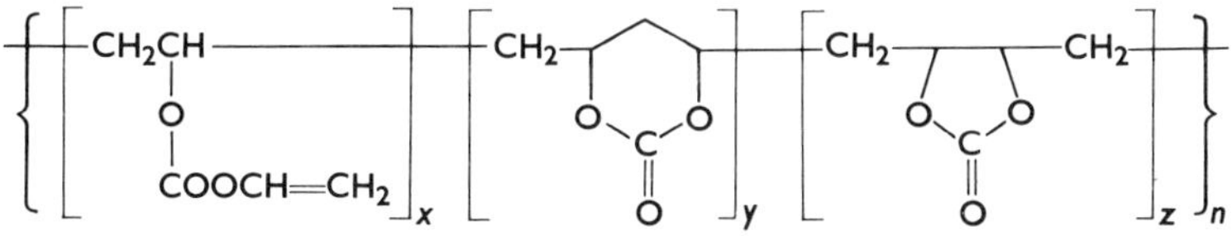

(II-45)

[99]

of $x$, $y$, and $z$ were dependent on the experimental conditions. Insoluble product was also formed in amounts that increased with increasing conversions.

### c. Allyl and Methallyl Crotonates

Five-membered rings are formed in the polymerization of allyl and methallyl crotonates (*32*). It is interesting that radical [102] is formed in preference to radical [104], which is favored by resonance. The preference was explained

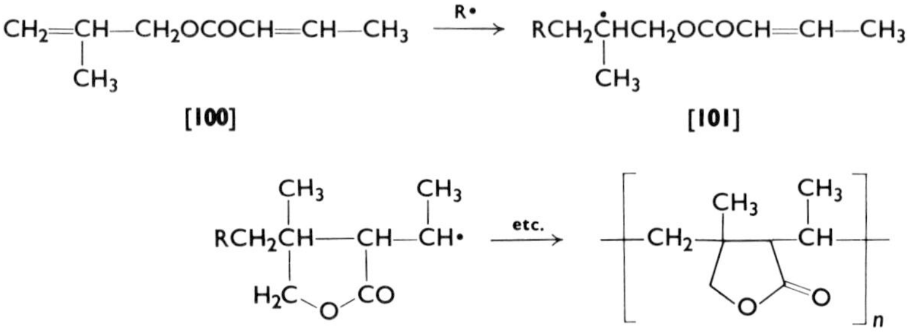

$$CH_2=CH-CH_2OCOCH=CH-CH_3 \xrightarrow{\text{R}\cdot} RCH_2\overset{\cdot}{C}HCH_2OCOCH=CH-CH_3 \longrightarrow$$
$$\underset{CH_3}{|} \qquad\qquad\qquad\qquad \underset{CH_3}{|}$$
$$[100] \qquad\qquad\qquad\qquad\qquad [101]$$

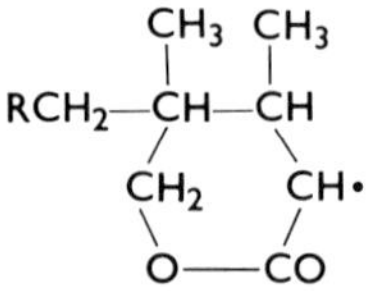

(II-46)

[102]    [103]

(II-47)

[104]

on steric grounds (*32*). The degree of cyclization in these polymers was low, with the difference in the reactivity of the two double bonds being responsible (*32*). Similar reasons were given for the relatively low degree of cyclization that was observed in the free-radical polymerizations of a series of methyl allyl (or butenyl) maleates and fumarates (*33, 93*). Along related lines, practically no cyclopolymerization took place in the case of vinyl atropate [105] (*164*). Again, the dissimilarity of the double bonds was considered to be the determining factor for homo- versus cyclopolymerization (*164*).

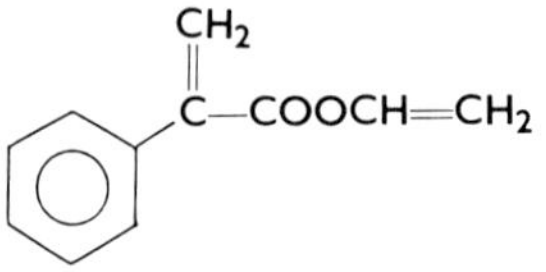

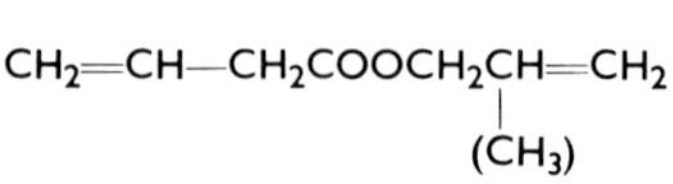

$$CH_2=CH-CH_2COOCH_2CH=CH_2 \qquad (II-48)$$
$$\underset{(CH_3)}{|}$$

[105]    [106]

In contrast to the allyl and methallyl crotonates, allyl and methallyl 3-butenoates [106] gave only viscous oils when initiated with benzoyl peroxide or azodiisobutyronitrile (*32*). The reluctance of [106] for free-radical polymerization was not explained.

### d. Allyl Acrylate and Methacrylate

The free-radical polymerization of these monomers was studied extensively. The polymerization of allyl acrylate with azodiisobutyronitrile in benzene solution yields a polymer that is soluble in benzene and in pyridine (247, 276). Bromination experiments established that its structure was [108] (247).

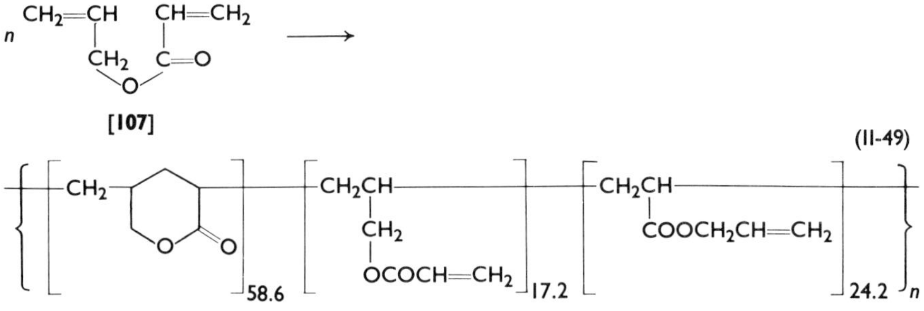

[107]

[108]

(II-49)

Poly(allyl methacrylate) has been reported several times. Bulk polymerization with ultraviolet light with or without benzoyl peroxide (51, 91) gave a soluble polymer at conversions below 27%. A somewhat conflicting result was obtained by others (152). In this latter study (152) the polymerization was carried out in toluene using azodiisobutyronitrile as catalyst. No soluble polymer resulted. Degradative chain transfer via the allyl group [Eq. (II-50)]

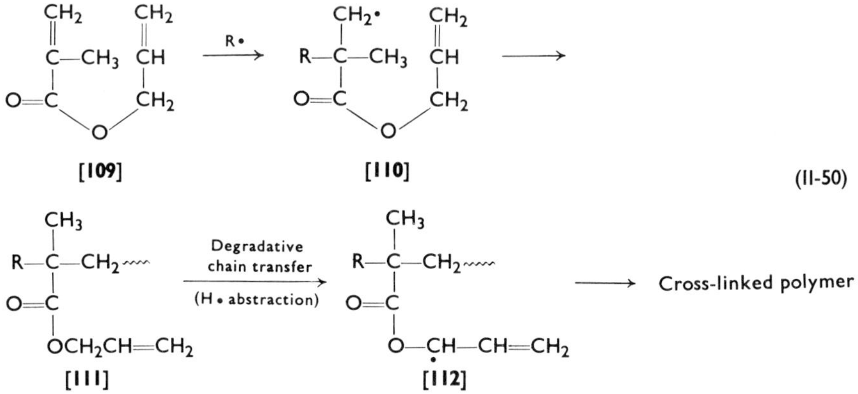

[109]          [110]          (II-50)

[111]          [112]          Cross-linked polymer

was believed to be responsible for cross-linking (152). In contrast, no chain transfer of this type can take place when the allyl group is replaced by a vinyl group. A soluble polymer was in fact obtained from vinyl methacrylate at conversion below 25% (*152*). The degree of cyclization of [**114**] was estimated to be 50–60% by infrared analysis (*152*).

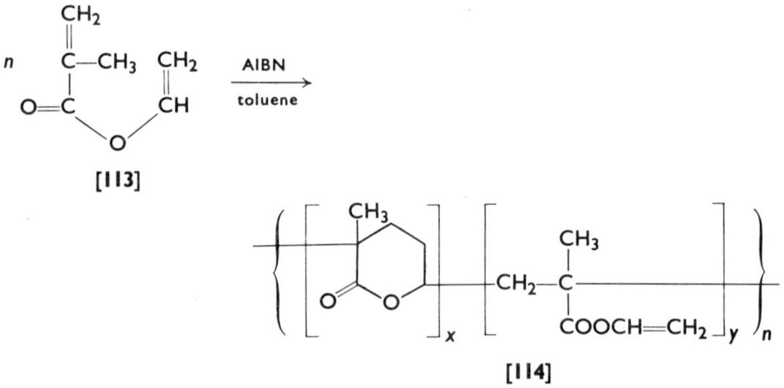

[**113**]

(II-51)

[**114**]

The degradative chain transfer discussed above cannot be the only factor responsible for cross-linking during the polymerization of allyl methacrylate. Recent studies have shown that a soluble polymer can be obtained at low conversions in benzene with azodiisobutyronitrile as initiator (*275, 277, 278*). Structure [**115**] was assigned to this polymer.

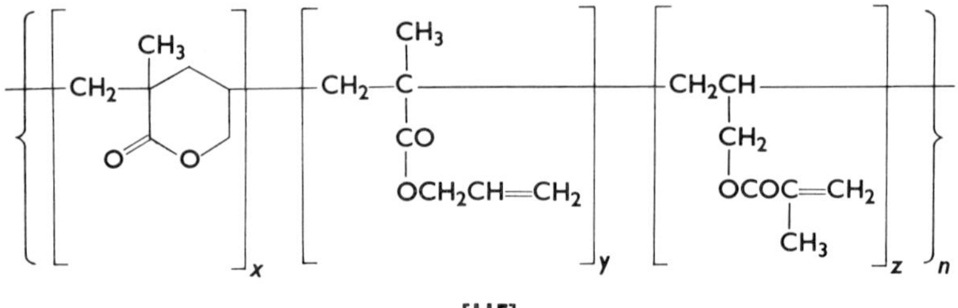

(II-52)

[**115**]

The values of $x$, $y$, and $z$ depended strongly on experimental conditions. At monomer conversions less than 10%, the polymer contained approximately 20% of $\delta$-lactone rings. It was established that low conversions and low monomer concentrations were essential in order to avoid cross-linking and obtain a cyclized structure (*278*). Consequently, the result of Kawai (*152*) was explained by the fact that the monomer concentration that was utilized was high and led to cross-linking rather than cyclopolymerization. The fact that ultraviolet light in bulk gave a soluble polymer (*51, 91*) at conversions of up to 27% could be due to a difference in the mechanism of the photoinitiated polymerization. It is known that allyl groups are reluctant to polymerize under these conditions (*51*). Experimental evidence indicated that the photopolymerization of allyl methacrylate proceeded preferentially at the methacrylate double bond (*51*).

### e. Vinyl Cinnamates

The polymerization of *trans*-vinyl cinnamate proceeded differently in bulk than it did in solution (230). Azodiisobutyronitrile in bulk gave insoluble

$$C_6H_5\overset{\displaystyle H}{C}=\overset{\displaystyle}{C}COOCH=CH_2 \qquad (II\text{-}53)$$

$$\text{[116]}$$

polymer. Polymerization in benzene or dioxane gave a soluble material when the monomer concentration and conversion were lower than 30% and 50%, respectively. Chemical and infrared examination of the polymer indicated the presence of γ-lactone rings. The degree of cyclization was 80–85%. It increased with decreasing monomer concentration. The noncyclized portion of the polymer was derived almost exclusively from the polymerization of the vinyl double bond.

The formation of a γ-lactone ring can take place as shown in Eq. (II-54). Two γ-lactone structures, [118] and [121] are possible. Structure [118] could be formed by two distinct mechanisms. Lactone [118] was preferred, although

$$(II\text{-}54)$$

[121] could not be ruled out on the basis of infrared evidence (*230, 238*). The exact mechanism of the different propagation steps was not established.

The copolymerization of *trans*-vinyl cinnamate has been studied (*238*). Vinylpyrrolidone, vinyl acetate, methacrylonitrile, and styrene were used as co-monomers. The copolymerization parameters were determined. A very interesting influence of the comonomer on the copolymer structure was found. With methacrylonitrile, vinylpyrrolidone, and vinyl acetate, the noncyclized part of the cinnamate was formed, as in the case of the parent homopolymer, predominantly via vinyl polymerization. In contrast, the copolymers obtained with styrene contained significant amounts of vinyl unsaturation (*238*).

Solution polymerization of a series of cinnamates [122] was also studied (*151*). The reactions were effected at 75°C in toluene or dimethylformamide.

$$CH{=}CHCOOCH{=}CH_2 \qquad \text{(II-55)}$$

[122]

R = H, CH₃, CH₃O, Cl, Br, CN

R = H, $CH_3$, $CH_3O$, Cl, Br, CN

Benzoyl peroxide was the initiator. Infrared studies gave evidence for cyclopolymerization with formation of $\gamma$-lactone rings. The dimethylformamide polymerizations were homogeneous; those carried out in toluene were heterogeneous. It was found that the nature of R had no effect on the degree of cyclization in the toluene experiments; in dimethylformamide, cyclization was favored when R was electron-withdrawing.

### f. Diallyl Tartrate

Strictly speaking, the polymerization of diallyl tartrate is not of the intra–intermolecular type. It is described here because diallyl tartrate is a diunsatur-

[123]   [124]   (II-56)

ated ester that yields a ring-containing polymer when treated with a free-radical initiator. Initiation by benzoyl peroxide or azodiisobutyronitrile yields a mixture that is partly soluble in methanol (*250*). Structures [123] and [124] were assigned to the soluble and insoluble fractions, respectively (*250*).

### g. Diallyl Muconate

Diallyl muconate was polymerized in toluene by di-*tert*-butyl peroxide (*8*). A cyclopolymerization involving the interaction of four double bonds takes place under these conditions. The softening point of polymer [126] was above 230°C.

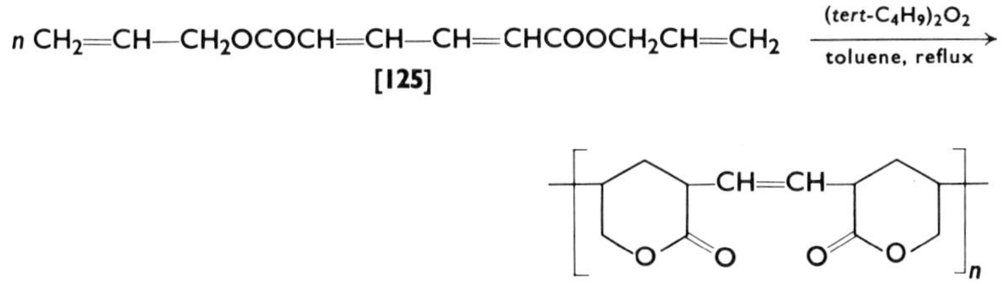

$$n \; CH_2\!\!=\!\!CH\!-\!CH_2OCOCH\!\!=\!\!CH\!-\!CH\!\!=\!\!CHCOOCH_2CH\!\!=\!\!CH_2 \xrightarrow[\text{toluene, reflux}]{(\textit{tert-}C_4H_9)_2O_2}$$

[125]

(II-57)

[126]

### h. Propargylic Esters

The ring-forming polymerization of propargylic esters was investigated by Arbuzova and co-workers (*194*). Dipropargyl maleate [127], propargyl crotonate [128], and propargyl methacrylate [129] were studied. The polymerizations were initiated by benzoyl peroxide or di-*tert*-butyl peroxide in bulk or dioxane solution. Chemical and spectroscopic evidence supported

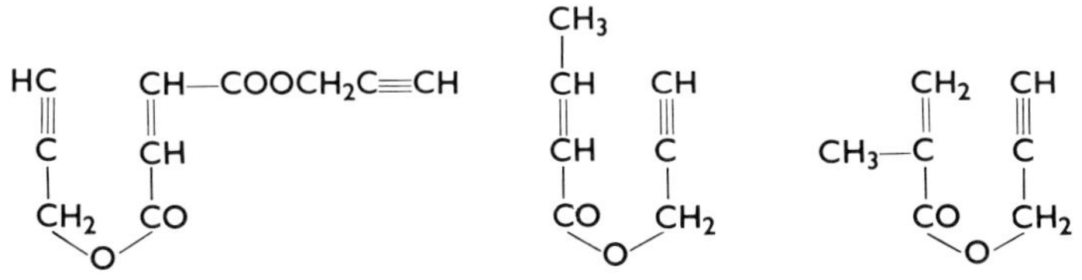

(II-58)

[127]                    [128]                    [129]

structures [130] and [131] for the polymers that were obtained from esters [127] and [128], respectively. Monomer [129] yielded a linear polymer derived from methacrylate polymerization only. The lack of cyclic structure was explained by the appreciable difference in the reactivity of the two multiple bonds toward radicals (*194*).

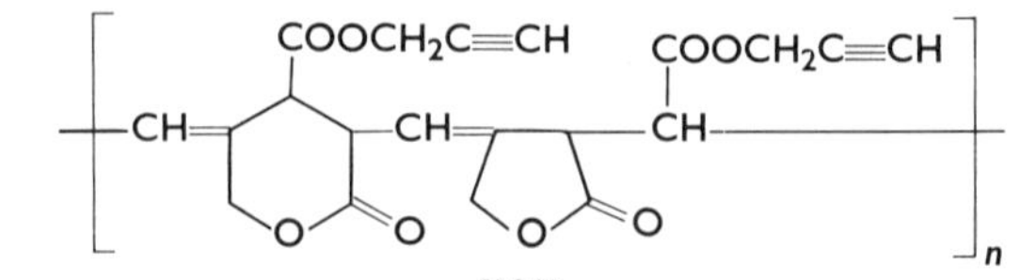

[130]

(II-59)

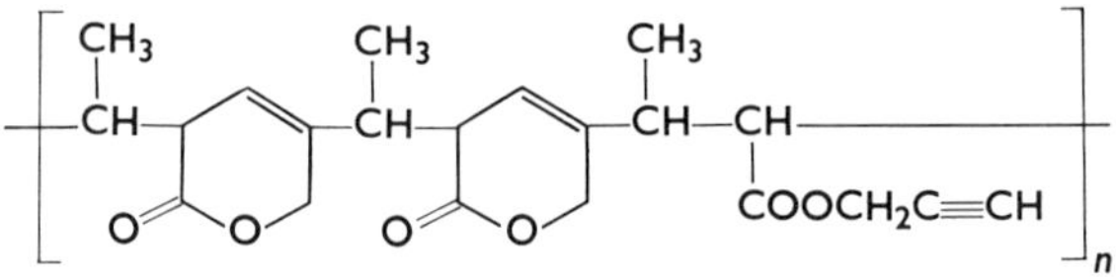

[131]

*i. Miscellaneous*

An interesting case of cyclization was observed in the radical-initiated emulsion polymerization of styrene, ethyl acrylate, and methyl methacrylate with 1–3% of vinyl acrylate [132] (97). Without vinyl acrylate, very high molecular weight polymer ($\eta_{red}$ 15–18) is obtained. In its presence, lower

$$CH_2{=}CH{-}COOCH{=}CH_2 \qquad\qquad\text{(II-60)}$$

[132]

molecular weight, soluble polymer ($\eta_{red} < 1$) results. There is no apparent decrease in the reaction rate. Furthermore, in the presence or absence of [132] conversions were high (>90%). These observations were rationalized in terms of the chain-transfer phenomenon shown in Eq. (II-61). Infrared

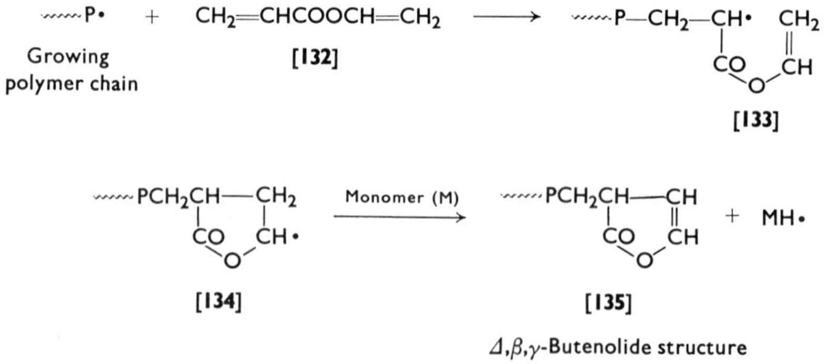

Growing polymer chain     [132]     [133]

[134]     [135]     (II-61)

$\Delta,\beta,\gamma$-Butenolide structure

spectroscopy has confirmed the presence of butenolide structures in these polymers.

## E. Polymerization of Diunsaturated Anhydrides

The polymerization of this class of monomers is shown in Eq. (II-62).

$$n \quad \underset{\substack{\text{[136]}\\ \text{Acrylic anhydride}}}{\text{CH}_2{=}\text{CH} \quad \text{CH}_2 \text{CH}} \quad \xrightarrow{\text{Initiation}} \quad \underset{\substack{\text{[137]}\\ \text{Poly(acrylic anhydride)}}}{} \tag{II-62}$$

METHOD 1. FREE-RADICAL INITIATION

Diunsaturated anhydrides readily yield soluble linear cyclopolymers with free-radical initiators. Peroxides (*57, 94, 95, 138, 139, 140, 148, 153, 206, 270*), α,α'-azodiisobutyronitrile (*112, 113, 138, 139, 153, 196, 254*), UV light (*137, 139*), and γ-irradiation (*219, 220, 270*) were used. They were run in bulk (*95, 138–140, 196*) and solution (*57, 94, 95, 112, 113, 137–140, 148, 153, 196, 206, 254, 270*). Acetone, dioxane, cyclohexanone, benzene, toluene, dimethylformamide, and dimethyl sulfoxide were used as solvents. Copolymers of acrylic and methacrylic anhydrides with a variety of unsaturated comonomers were prepared by similar techniques (*140, 253*).

A large number of studies on the intra–intermolecular polymerization of acrylic and methacrylic anhydrides have been published. The results are summarized in the following discussion.

### a. Acrylic Anhydride

Depending on the polymerization conditions, very high or low molecular weight polymer is obtained (*94*). For example, free-radical initiation in benzene was reported to yield a polymer with a DP of up to 750. On the other hand, use of dioxane as solvent resulted in only low polymer (*94*). A higher concentration of monomer as well as polymerization in bulk (*94, 95*) led to increased molecular weights. The solubility of the polymer (*94, 95, 202*) and the virtual absence of residual double bonds as demonstrated by infrared and bromination techniques, support structure [**138**]. Hydrolysis of [**138**] yields poly(acrylic acid)

$$\tag{II-63}$$

[**138**]

whose molecular weight was determined by viscosity measurements on aqueous sodium hydroxide solutions. This confirmed that [**138**] was a very high molecular weight polymer (95). The kinetics of this polymerization were studied in bulk and in solution (196). Reaction was initiated with α,α'-azodi-isobutyronitrile; dimethylformamide and cyclohexanone were solvents.

The question of whether this polymerization proceeds stereospecifically has been considered (94, 95). It was reported (95) that X-ray examination of poly(acrylic acid) obtained by hydrolysis of [138] showed that it possessed a more regular structure than the polyacid prepared by the direct polymerization. Recently, Mercier and Smets studied the structure of poly(acrylic anhydride) as a function of the experimental conditions used for its preparation (197). Infrared examination of polymer prepared in cyclohexanone at 35°C in the presence of α,α'-azodiisobutyronitrile indicated that it possessed the syndiotactic structure. Heating, or traces of acid, isomerized this material into its isotactic form. It was also shown that at higher polymerization temperatures (xylene, 115°C) a polymer that contained an appreciable amount of five-membered rings [139] was formed (197). The residual unsaturation in these polymers decreased with an increase in reaction temperature (197).

$$-CH_2-CH-\!\!-\!\!-CH-CH_2- \tag{II-64}$$

[139]

Since a cyclopolymer can be easily obtained from acrylic anhydride, the question "Why is this anhydride a cross-linking agent in the polymerization of vinyl monomers?" was asked (148). It was speculated (148) that in the case of the anhydride itself cyclopolymerization is favored because the two double bonds are identical (148). In the presence of another monomer, two reactions can take place, as shown in Eq. (II-65). Structure [141] does not lead to cross-linking, while structure [143] does (148). Obviously, the use of a reactive

[140]     + $CH_2$=CHX     $\longrightarrow$     [141]

[142]     + $CH_2$=CHX     $\longrightarrow$     [143]     (II-65)

vinyl monomer will favor formation of [143] and give a more highly cross-linked material (*140*). Copolymerization studies of acrylic anhydride with vinyl monomers did in fact indicate that the degree of cyclization of the copolymer varied with the comonomer that was used (*253*).

### b. Methacrylic Anhydride

Methacrylic anhydride [144] yields linear, soluble cyclopolymers under a variety of radical-initiated polymerization conditions (*57, 112, 113, 137, 139*).

$$CH_2{=}\underset{\underset{O}{\|}}{\overset{\overset{CH_3}{|}}{C}}{-}C{-}O{-}\underset{\underset{O}{\|}}{C}{-}\overset{\overset{CH_3}{|}}{C}{=}CH_2 \qquad\qquad (II\text{-}66)$$

[144]

These include the use of peroxides and α,α'-azodiisobutyronitrile in bulk or solution (*57, 112, 113, 139*), as well as UV radiation at low temperature (*137*). The ease with which cyclopolymerization of this monomer takes place is remarkable. It was suggested that this phenomenon is due to the interaction of the double bonds of the monomer, which results in conformations [145] or [146] (*140*). Based on the specific gravity of poly(methacrylic anhydride), the

[145]      or      [146]                    (II-67)

mole shrinkage value for one double bond in the polymerization was calculated to be 13.7 ml/mole (*139*). A comparison with the value of 22–23 ml/mole for acrylic esters supports the hypothesis that the monomer exists in a cyclic conformation which provides the driving force for cyclopolymerization. This theory is similar to Butler's postulate on the spatial interaction of the double bonds in nonconjugated dienes. According to Butler, this spatial interaction is responsible for the tendency of these monomers to undergo intra–intermolecular polymerizations (*74*).

The kinetics of the α,α'-azodiisobutyronitrile-initiated polymerization were studied in dimethylformamide solution (*102, 112, 113*). It was shown that high monomer concentration leads to partial gelation. Furthermore, this results also in an increased residual unsaturation content in the polymer (*112, 113*). Interestingly, only in one instance (*270*) was the free-radical-initiated polymerization of methacrylic anhydride reported to yield a cross-linked, insoluble material. Probably, impurities that were present in the monomer were responsible for this result.

It is generally agreed that poly(methacrylic anhydride) consists mainly of units possessing structure [**147**]. This agrees with the low residual unsaturation

[**147**]

(II-68)

of the polymer and with its solubility (*202*). Furthermore, it was shown that the cyclopolymer is very similar to the pyrolysis product of poly(methacrylic acid) (*57*). The observed differences between the two products were believed to be due to secondary reactions that occur during the pyrolysis. The stereo-chemistry of poly(methacrylic anhydride) was the object of several studies (*66, 90, 112, 113, 137, 204, 235, 243*). It was shown that the steric structure of the polymer depends on the experimental conditions that are used for its preparation.

Copolymers of methacrylic anhydride with various vinyl monomers were prepared (*140, 253*). The anhydride moiety of the copolymer is cyclized. However, the degree of cyclization depends on the co-monomer that is used (*253*). Generally, a low reactivity of the co-monomer in free-radical polymer-izations favors formation of soluble copolymers (*140*). The latter are also formed when the concentration of the co-reactants and the degree of conversion are low (*140*). One other important factor that determines the solubility of the copolymer is the mole ratio of the co-reactants. Ratios markedly different from one lead to soluble products (*140*).

METHOD 4. ANIONIC INITIATION

Linear polymers are obtained from propargylic anhydrides and anionic initiators (*100, 284*). The reactions were performed at 70°–100°C in the absence of air. Potassium salts, such as the thiocyanate, cyanide, iodide, bromide, and chloride, were preferred catalysts. The polymer obtained from propargylic anhydride [**148**] (R = H) was soluble in dimethylformamide, concentrated sulfuric acid, and aqueous sodium hydroxide. In the latter medium, hydrolysis

$$(\text{II-69})$$

[148]                    [149]

to poly(propiolic acid) took place. Structure [149] was also supported by infra-red data (100, 284). End-group analysis of polymer [149] (R = H) gave molecu-lar weights of up to 25,000 (284). The polymer gives an ESR signal, and its electroconductivity is in the semiconductor range ($\sim 10^{-9}$ ohm$^{-1}$cm$^{-1}$) (284).

## F. Polymerization of Diunsaturated Germanium Derivatives

The intra–intermolecular polymerization of diunsaturated heteroatomic compounds leads to several types of heterocyclic polymers. These polymeriza-tions are discussed in this and the following sections (F–N).

$$(\text{II-70})$$

[150]                    [151]

X = Ge, N, P, S, Si, Se, etc.

METHOD 2.  ZIEGLER INITIATION

A low molecular weight polymer was obtained from diallyldiethylgermanium by treatment with triethylaluminum and titanium tetrachloride. The reaction

$$(\text{II-71})$$

[152]

was performed at 60°C in petroleum ether. The polymer presumably possesses a cyclic structure (56, 159).

## G. Polymerization of Diunsaturated Ammonium Salts, Amine Oxides, and Amines

### METHOD 1. FREE-RADICAL INITIATION

The free-radical polymerization of allyl ammonium compounds was studied in depth by Butler and his students (72, 73, 76–79). It was as a result of this study that the intra–intermolecular polymerization mechanism was first recognized. Butler found that ammonium compounds which contained one or two allyl groups led to soluble polymers (78–80). Those derivatives that possessed three or four allyl functions gave insoluble polymer. In order to rationalize these results, Butler and Angelo (72) proposed the following mechanism for the polymerization of diallyldiethyl ammonium bromide [Eq. (II-72)].

The structure of the product was proven by degradative studies (76, 77). This mechanism occurs with all intra–intermolecular polymerizations and suggests the number of new polymer structures that can be prepared by this method.

Polymerization of diunsaturated ammonium compounds can be effected by a variety of free-radical initiators. Peroxides, hydroperoxides, azo derivatives, peracids, diacyl peroxides, hydrogen peroxide, and inorganic derivatives such as barium peroxide are useful (39, 41, 59, 60, 69, 72, 73, 75, 77, 108, 125–127, 146, 212, 215, 257). The reactions are effected in polar solvents, including water, methanol, acetone, dimethylformamide, and dimethyl sulfoxide. Redox initiation systems were also used successfully (214, 242, 244). Copolymers of diunsaturated ammonium salts with various monounsaturated compounds were prepared by similar techniques (42, 43, 108, 214, 242, 244).

A study of the polymerization of diallylamine hydrochlorides has shown that high monomer concentrations lead to cross-linked structures (73). Linear polymer is favored at low monomer concentration. Other studies have indicated that the anion can also have considerable influence on the polymerization (*69, 72, 212*). Generally, chlorides give higher molecular weight polymers than bromides (*212*). Ammonium persulfate in dimethyl sulfoxide and *tert*-butyl hydroperoxide in water were the initiators used in this latter work. The results were explained by assuming oxidation of the bromide anion to free bromine by the initiator. The bromine then acts as a polymerization inhibitor (*212*). However, it should be noted that high molecular weight bromides and other salts of the cyclopolymers can be obtained by ion exchange of initially formed chloride polymers (*60, 69*).

Polymerization of certain ammonium salts can lead to bi- and tricyclic systems. For example, triallylammonium hydrobromide [157] yields a bicyclic polymer [158] (*183*).

$$
n \text{ HBr} \cdot \text{N} \begin{array}{l} \text{CH}_2-\text{CH}=\text{CH}_2 \\ -\text{CH}_2-\text{CH}=\text{CH}_2 \\ \text{CH}_2-\text{CH}=\text{CH}_2 \end{array} \xrightarrow[\substack{\text{butyronitrile,} \\ \text{DMF, or} \\ \text{alcohol, 80°C}}]{\alpha,\alpha'\text{-Azodiiso-}} \quad \text{[158 structure]} \qquad \text{(II-73)}
$$

[157]                                                        [158]

Reaction probably proceeds via the following mechanism:

$$
\text{[157]} \xrightarrow{\text{R}\cdot} \text{[159]} \longrightarrow \qquad \text{(II-74)}
$$

[157]                    [159]

$$
\text{[160]} \xrightarrow[\text{etc.}]{\text{Monomer}} \text{Polymer}
$$

[160]

Similarly, polymer [162] was obtained from triallylethylammonium bromide [161] (*272*). A tricyclic polymer, soluble in water and methanol, was obtained

$$
n \text{ [161 structure]} \xrightarrow[\text{H}_2\text{O, 60°C}]{\text{tert-C}_4\text{H}_9\text{OOH}} \text{[162 structure]} \qquad \text{(II-75)}
$$

[161]                                                        [162]

by polymerization of tetraallylammonium bromide (272). Low monomer concentrations favor a high degree of cyclization in these systems (272).

Copolymerization of ammonium salts with sulfur dioxide is the subject of several studies (39, 125–128, 146). When a series of salts represented by formula [163] was investigated, some interesting results regarding copolymer structure were obtained (125, 126). When $R_1$ is hydrogen and $R_2$ is methyl, ethyl, n-propyl,

$$\begin{array}{c} R_1 \\ R_2 \end{array} \overset{+}{N} \begin{array}{c} CH_2-CH=CH_2 \\ CH_2-CH=CH_2 \end{array} \quad + SO_2 \quad \longrightarrow \quad Copolymer \qquad (II\text{-}76)$$
$$Cl^-$$

[163]

isopropyl, n-butyl, tert-butyl, benzyl, cyclohexyl, or β-cyanoethyl, only 1:1 copolymers are obtained regardless of feed compositions. They were soluble in dilute acid. Infrared data indicated that two types of cyclic units [164] and [165] were present in the polymer. A similar result was obtained when $R_1$ and $R_2$ were both ethyl groups, except that this polymer was soluble in water at

[164]                [165]                                    (II-77)

any pH. However, for monomers [166] and [167], the composition of the copolymer was dependent on the monomer feed. At higher sulfur dioxide contents, units [168] were formed in addition to [164] and [165]. The solubility of these latter copolymers was similar to the polymer from the N,N-diethyl compound (125, 126).

[166]                [167]                          [168]                  (II-78)

A variety of uses have been described for the polymers from diunsaturated ammonium salts and their copolymers with monounsaturated compounds. These include flocculating agents (41, 108, 214, 215, 257), strengthening agents for paper (*244*), agents for the preparation of electrically conductive paper (*42*), and as deemulsifiers for oil-in-water emulsions (*242*).

The copolymerization of a series of diallyl alkylamine *N*-oxides [**169**] with monomers such as acrylamide, acrylonitrile, styrene, methyl acrylate,

$$(CH_2{=}CH{-}CH_2)_2N{\overset{\displaystyle O}{\underset{\displaystyle R}{<}}} \qquad\qquad (II\text{-}79)$$

[**169**]

$$R = CH_3, C_2H_5, C_6H_5CH_2$$

and vinyl acetate was reported in a patent (*233*). Ammonium persulfate in water and $\alpha,\alpha'$-azodiisobutyronitrile in benzene were used as initiators. The copolymers were soluble and probably possessed the cyclic linear structure.

The cyclopolymerization of diunsaturated amines has been the subject of several reports. Free-radical initiators, including benzoyl peroxide, $\alpha,\alpha'$-azodiisobutyronitrile, and $^{60}$Co irradiation are useful (*48, 86, 96, 114, 115, 182, 282*). Reactions were carried out in bulk or solution. Dioxane, benzene, dimethylformamide, dimethyl sulfoxide, and *m*-cresol are useful solvents. A variation of this type of polymerization was reported for a series of diallyl amines (*150*). First the amines were reacted with zinc chloride. The complex that was obtained was then polymerized in toluene using $\alpha,\alpha'$-azodiisobutyronitrile as initiator. Soluble polymers and copolymers were obtained. The reaction rates were lower in the absence of zinc chloride (*150*).

Recent studies of the free radical-initiated cyclopolymerization of the series of amines, $[(CH_2{=}CHCH_2)_2N{-}R]$, uncovered an interesting effect that the R group has on the ease of polymerization (*182, 223*). Electron-withdrawing groups favor polymerization. Polymers formed rapidly where R was $CH_3CO$, $C_2H_5CO$, $ClCH_2CO$, $C_6H_5CO$, $CH_3OCO$, $C_6H_5SO_2$, $CH_3\text{-}p\text{-}C_6H_4SO_2$, and $CN$. The effectiveness of R in promoting polymerization was directly related to its electron-withdrawing properties. The rate sequence of $ClCH_2CO > CH_3CO > C_2H_5CO$ was observed. On the other hand, only low yields (1–7%) of low molecular weight polymer were obtained with *N*-benzyl, *N*-cyanomethyl, and *N*-($\beta$-cyanoethyl) derivatives (*223*).

The kinetics and mechanism of polymerization of *N*,*N*-diallylcyanamide [**170**] were studied (*282*). The structure of the polymer was supported by

[**170**]                      [**171**]                      (II-80)

infrared data. It was shown (282) that intramolecular abstraction of hydrogen is responsible for termination in this reaction.

The effect of the incipient ring size was shown in a study of the copolymerization of allyl-substituted ureas with lauryl methacrylate (48). The reaction of N,N-diallylurea [172] with lauryl methacrylate, initiated by benzoyl

$$(CH_2\!\!=\!\!CH\!-\!CH_2)_2NCONH_2 \qquad\qquad (II\text{-}81)$$

[172]

peroxide in dioxane gave copolymer that contained structure [173]. However, no cyclic structures were formed when N,N'-diallylurea [174] was used in the same reaction (48).

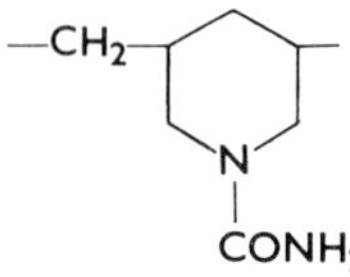

[173]

$$(CH_2\!\!=\!\!CH\!-\!CH_2NH)_2CO \qquad\qquad (II\text{-}82)$$

[174]

A very interesting case of cyclopolymerization was described by Chang and Price (86) for N,N-divinylaniline [175]. The monomer gave different products when the α,α'-azodiisobutyronitrile-initiated polymerization was performed in bulk or in benzene. A partly soluble product was obtained from the bulk reaction. The soluble portion of the product has structure [179]. The following mechanism was proposed:

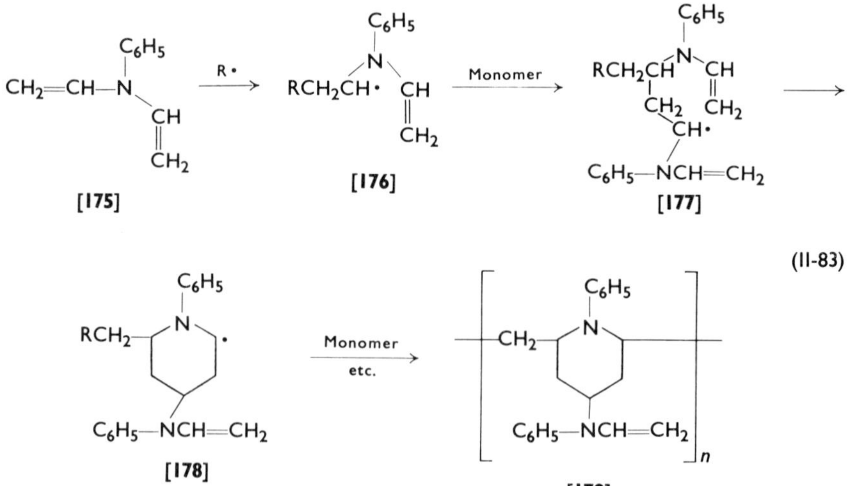

(II-83)

In benzene solution at conversions below 30%, a different, soluble polymer was obtained (86). Structure [181] was assigned to it. The copolymerization of

$$\text{(II-84)}$$

N,N-divinylaniline [175] with vinyl monomers was also studied (86). The copolymers obtained by bulk reaction were insoluble and cross-linked. However, in benzene or dimethylformamide at monomer concentrations below 30%, soluble materials resulted. Based on analytical data, structure [183] was assigned to these products. Interestingly, with diethyl fumarate as the

$$\text{(II-85)}$$

(II-86)

[184]

co-monomer, polymer [184] was obtained (86). Polymers of different properties were obtained from the polymerization of N,N-diallylmelamine [185].
The polymer obtained from $\alpha,\alpha'$-azodiisobutyronitrile initiation was shown

(II-87)

[185]                           [186]

R = 3,5-diaminotriazinyl

by infrared and chemical evidence to possess structure [186] (114, 115). Its
intrinsic viscosity was 0.1 and it was soluble in acids and dimethyl sulfoxide.
Solid-state polymerization of [185] (114, 115) with $^{60}$Co irradiation gave a
polymer with an intrinsic viscosity of 0.2 (acetic acid). Analysis again indicated
structure [186]. The material was soluble in acids but not in dimethyl sulfoxide.
It was postulated that this property difference is due to greater stereoregularity
of the polymer from the $^{60}$Co-catalyzed reaction. However, X-ray analysis
did not reveal any significant level of crystallinity in the less soluble material.
Molecular weight differences could also be the cause of the observed solubility
differences.

## H. Polymerization of Diunsaturated Amides

### Method 1. Free-Radical Initiation

The cyclopolymerization of diunsaturated amides can be initiated by typical
free-radical initiators. These include benzoyl peroxide, tert-butylperoxide,
$\alpha,\alpha'$-azodiisobutyronitrile, and ultraviolet irradiation (103, 122, 141, 152, 227,
247–249, 255, 256). The reactions are effected in bulk or solution. Methanol,
ethanol, acetone, benzene, toluene, dimethylformamide, and pyridine are
useful solvents. The intra–intermolecular polymerization of N,N'-divinylurea

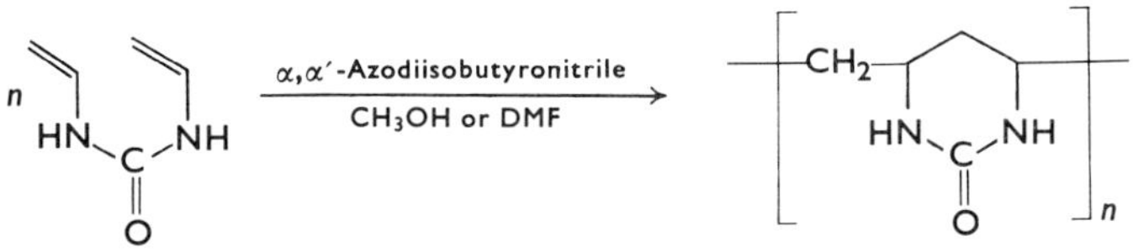

$$\text{[187]} \qquad \text{[188]} \qquad \text{(II-88)}$$

[187] is an illustration (141, 227). The structure of polymer [188] was supported by infrared data. Its insolubility raises the possibility that it might be partly cross-linked (141, 227). Its melting point was greater than 300°C.

Polymerization of dimethacrylamides [189] was the subject of several reports (122, 241, 255, 256). The structure of the polymer depends on whether the R

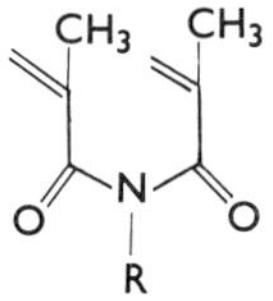

$$\text{(II-89)}$$

[189]

$$R = H, CH_3, C_2H_5, n—C_3H_7, C_6H_5$$

group was hydrogen, alkyl, or aryl. A copolymer [190] containing both five- and six-membered rings is formed from the unsubstituted derivative ([189]; R = H). On the other hand, polymers [191] containing only five-membered rings are obtained when the R group is methyl, ethyl, n-propyl, and phenyl

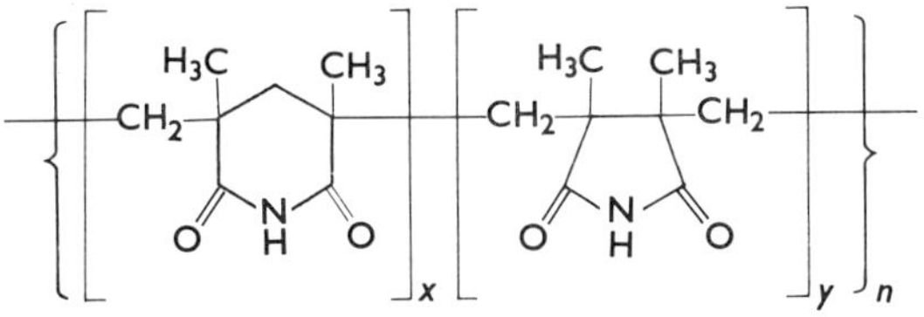

$$\text{(II-90)}$$

[190]

(241, 256). These observations were explained by steric and conformational factors.

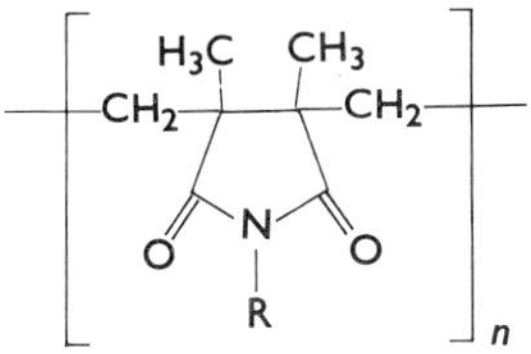

$$\text{(II-91)}$$

[191]

$$R = CH_3, C_2H_5, n—C_3H_7, C_6H_5$$

Contradictory results were published for the polymerization of N-allyl-acrylamide [192] and N-allylmethacrylamide [193] (152, 181, 249). Kawai

$$CH_2{=}CH{-}CH_2NHCOCH{=}CH_2 \quad CH_2{=}CH{-}CH_2NHCO\underset{\underset{CH_3}{|}}{C}{=}CH_2$$

$$[192] \qquad\qquad\qquad [193] \qquad\qquad (II\text{-}92)$$

(152) reported that the polymerization of these monomers by $\alpha,\alpha'$-azodiiso-butyronitrile in bulk or solution gave only cross-linked, insoluble materials. He concluded that allyl resonance stabilization leading to chain-transfer reactions was responsible for the formation of cross-links. Other investigators pointed out the importance of experimental conditions and showed that cyclopolymers as well as cross-linked materials can be obtained (181, 249). This was confirmed by Trosarelli and co-workers. It was shown that cross-linked products result in benzene or toluene solutions with $\alpha,\alpha'$-azodiisobuty-ronitrile as the catalyst (273–275, 278–280). However, in methanol at low mono-mer concentration and at conversions below 50%, the same initiator gives soluble polymers that contain six-membered lactam rings. Increased monomer concentration leads to a decrease in the degree of cyclization. The pendant double bonds of the soluble polymer were both allylic and acrylic types.

Ring-containing polymers containing large rings were studied by Schulz and his students (247, 248). The monomers were prepared by the reactions of Eq. (II-93).

$$CH_2{=}CHNCO + HOCH_2CH{=}CH_2 \longrightarrow CH_2{=}CHNHCOOCH_2CH{=}CH_2$$

$$\quad[194] \qquad\qquad [195] \qquad\qquad\qquad\qquad [196]$$

$$(II\text{-}93)$$

$$2\,CH_2{=}CHNCO \;+\; \left\{\begin{array}{c} HOROH \\ \text{or} \\ HORNH_2 \end{array}\right. \longrightarrow CH_2{=}CHNHCOXR'XCONHCH{=}CH_2$$

$$\qquad[194] \qquad\qquad\qquad\qquad\qquad\qquad [197]$$

Monomer [196] polymerizes to a soluble ring-containing material on treat-ment with $\alpha,\alpha'$-azodiisobutyronitrile in pyridine. Hydrogenation indicated that 68% of the units were cyclized. Since the diene [196] is asymmetric, two types of pendant double bonds (vinyl and/or allyl) can be present in the non-cyclized portion of the polymer. The nature of the residual unsaturation was not established.

The polymerization of the series of monomers [197] was studied in dimethyl-formamide solution at 50°C (248). Reaction was initiated by $\alpha,\alpha'$-azodiiso-butyronitrile. Ring-containing polymers were obtained. The type and size of the incipient ring and the degree of cyclization are tabulated in the accompanying table. Examination of the data in the table shows that high degrees of cycliza-tion were observed except for the piperazine derivative. The results should be contrasted with the work of Marvel and Garrison on the cyclopolymerization of $\alpha,\omega$-diolefins (166). In their study, all-carbon rings of comparable sizes were formed to much lesser extents.

Cyclopolymers from $CH_2{=}CHNHCOXR'XCONHCH{=}CH_2{}^a$

| $-XR'X-$ | Size of incipient ring | Degree of cyclization at 40% conversion[b] |
|---|---|---|
| $-O(CH_2)_2O-$ | 11 | 86.8 |
| $-O(CH_2)_4O-$ | 13 | 90.3 |
| $-O(CH_2)_6O-$ | 15 | 89.3 |
| $-O(CH_2)_8O-$ | 17 | 87.2 |
| $-O-\langle S \rangle-O-$ <br> cis | 13 | 68.7 |
| $-O-\langle S \rangle-O-$ <br> trans | 13 | 60.7 |
| $o-OC_6H_4O-$ | 11 | 74.3 |
| $m-OC_6H_4O-$ | 12 | 67.7 |
| $p-OC_6H_4O-$ | 13 | 66.3 |
| $-N \bigcirc N-$ | 11 | 32.5 |

[a] Reference (248).
[b] Determined by hydrolysis of the noncyclized units to acetaldehyde followed by a colorimetric determination of the latter.

## METHOD 4. ANIONIC INITIATION

The anionic polymerization of $N$-methyldimethacrylamide is shown in Eq. (II-94) (122). The polymer [199] was identical with the product obtained

$$\text{[198]} \xrightarrow[\text{or toluene}]{\underset{\text{Tetrahydrofuran}}{C_6H_5MgBr}} \text{[199]} \tag{II-94}$$

[198]                                    [199]

from poly(methyl methacrylate) and methylamine. Consequently, it contains six-membered rings. Monomer [**198**] gave only five-membered rings on free radical-initiated cyclopolymerization.

## I. Polymerization of Diisocyanates

The polymerization of a diisocyanate can give products possessing the following three structures. Units [**201**] and [**202**] result from ring-forming polymerizations.

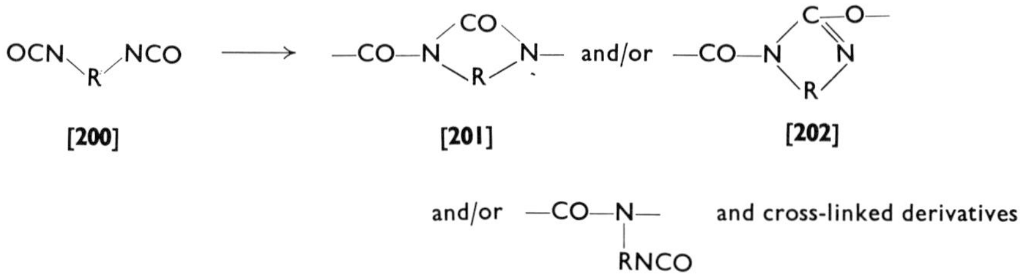
OCN
NCO
R
CO
—CO—N
N—
and/or
—CO—N
C—O—
N
and/or —CO—N—
RNCO
and cross-linked derivatives

[200] [201] [202] [203] (II-95)

METHOD 1. FREE-RADICAL INITIATION

Diisocyanate [**204**] gave a soluble polymer when subjected to $\gamma$-rays (*142*).

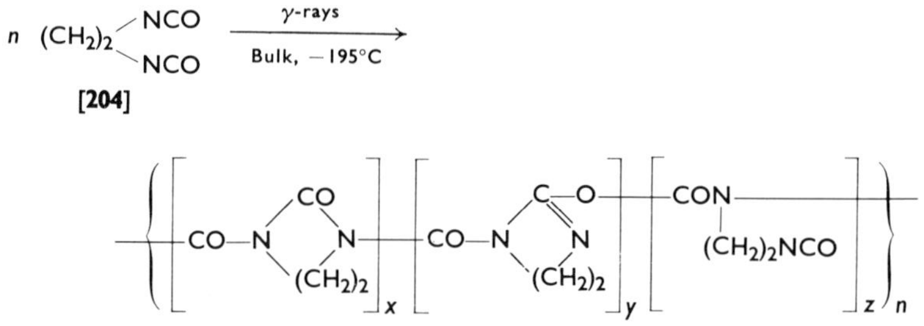
n (CH2)2
NCO
NCO
γ-rays
Bulk, −195°C
CO—N
CO
N
(CH2)2
C—O
CON
(CH2)2NCO

[204] [205] (II-96)

The material possessed structure [**205**] and was identical to the polymer obtained from [**204**] by anionic initiation. Tetramethylene diisocyanate gave an insoluble polymer when treated with $\gamma$-rays (*142*).

METHOD 4. ANIONIC INITIATION

The majority of cyclopolymerizations of diisocyanates have been initiated anionically. Sodium cyanide in dimethylformamide is the most commonly used catalyst/solvent combination (*49*, *92*, *142*, *144*, *154*). The use of sodium cyanide/*N*-methylpyrrolidone (*222*) and of butyllithium/triethylamine (*203*) was also disclosed. The reactions are carried out at temperatures in the range of $-65°$ to $-10°C$. Starting with appropriate monomers, monocyclic and bi-cyclic structures can be obtained. An example is shown in Eq. (II-97) (154).

[206]                                                [207]                    (II-97)

This method can also be applied to the synthesis of copolymers from di- and monoisocyanates (222). It is noteworthy that the tendency of some diisocyanates to cyclopolymerize is so high that ring-containing, linear materials are obtained on simple standing at $0°C$ without the use of any catalyst. Relatively high molecular weight polymers were obtained in this fashion from 1,2-ethylene, 1,2-propylene, and 1,2-cyclohexylene diisocyanates (154). Iwakura et al. (142) studied the polymerization of the series of diisocyanates [208]. For $m$ equal

[208]

$m = 1, 2, 3, 4$

(II-98)

[209]

to 2, 3, and 4, dimethylformamide was the solvent. Toluene was used for the polymerization of the methylene derivative ([208], $m = 1$). The latter reacted violently, yielding a product of undetermined structure on contact with dimethylformamide. Reaction in toluene resulted in a higher yield of polymer when performed at room temperature rather than at $-65°C$. It was also found that 1,2-ethylene diisocyanate could be polymerized to a soluble product by pyridine (142). Temperatures in the range of $23°$ to $80°C$ were adequate. Excess pyridine or epichlorohydrin were good reaction solvents (142). The structure of polymer [209] was supported by solubility, infrared, and chemical evidence. The following mechanism was proposed to explain formation of the cyclic units (142). The values of $x$, $y$, and $z$ in polymer [209] are a function of

$$\underset{[208]}{(CH_2)_m \overset{NCO}{\underset{NCO}{<}}} \xrightarrow{R^-} \underset{[210]}{RCO\underset{}{N} \overset{(CH_2)_m}{<} NCO} \longrightarrow$$

$$RCO\underset{}{N} \overset{(CH_2)_m}{\underset{\underset{O}{\overset{\shortmid}{C}}}{<}} N \underset{}{\overset{\nearrow}{\searrow}} \begin{array}{c} -CO-N \overset{CO}{\underset{(CH_2)_m}{<}} N- \\ [212] \\ -CO-N \overset{C-O-}{\underset{(CH_2)_m}{<}} N \\ [213] \end{array} \qquad \text{(II-99)}$$

[211]

the particular monomer. For example, when $m$ was 1 in formula [208], $x$ was 66% and ($y + z$) was 34%. On the other hand, in the case of $m = 2$, $x$ and $z$ were 0 and $y$ was 100% (142). The same polymer [214] was also obtained when 1,2-ethylene diisocyanate [204] was treated with sodium cyanate as shown in Eq. (II-100) (143).

$$n \underset{[204]}{(CH_2)_2 \overset{NCO}{\underset{NCO}{<}}} \xrightarrow[\text{DMF, } -54° \text{ to } -46°C]{N_2OCN} \underset{[214]}{\left[ \overset{CO}{\underset{N}{\overset{\mid}{\diagdown}}} \underset{N}{\overset{C}{<}} O \right]_n} \qquad \text{(II-100)}$$

Iwakura's research gave no indication that the cyclopolymerization of diisocyanates takes place only when rings of a particular size are formed (142). The polymers obtained from monomers [208] were all soluble, regardless of the value of $m$. Polymerization of 1,3-cyclohexylene diisocyanate was studied by

$$n \underset{[215]}{\overset{OCN \quad NCO}{\underset{H}{\bigcirc}}} \xrightarrow[\text{DMF, } -60°C]{NaCN}$$

$$\left\{ \left[ -CO-N \overset{CO}{\underset{\underset{H}{\bigcirc}}{<}} N \right]_x \left[ -CO-N \overset{C-O-}{\underset{\underset{H}{\bigcirc}}{<}} N \right]_y \right\}_n \qquad \text{(II-101)}$$

[216]

Corfield and Crawshaw (92). The formation of a soluble copolymer [216] was reported by these investigators. However, different conclusions were reached by King (154). According to this study, only those diisocyanates that can give five-membered rings will cyclopolymerize (154). All others ([217] and [218]) yield insoluble, cross-linked materials containing few cyclized structures.

$$
\begin{array}{ccc}
& {}^{\displaystyle \nearrow \text{NCO}} & \text{CH}_2\text{---NCO} \\
(\text{CH}_2)_3 & & | \\
& {}^{\displaystyle \searrow \text{NCO}} & \text{O} \\
& & | \\
& & \text{CH}_2\text{---NCO} \\
& \textbf{[217]} & \textbf{[218]}
\end{array}
\qquad (\text{II-102})
$$

The experimental conditions that were used by all of these workers are practically identical (sodium cyanide/dimethylformamide, low temperature). Possibly, the discrepancy between results may be due to monomer purity. Structure [209] due to Iwakura was discussed above (142). According to King (154), the polymers are all best represented by formula [209] in which $y$ and $z$ are 0. In other words, while Iwakura (142) favors two possible cyclizations leading to a urea and isourea, King favors only one (154).

In one instance (144) the use of a sodium metal/1,2-diethoxyethane combination was disclosed. It effected the preparation of a cyclic polymer from 1,3-propylene diisocyanate. The same report (144) describes the preparation of a material containing thiourea rings from 1,4-butylene diisothiocyanate by the action of sodium cyanide in dimethylformamide. It was also found that the thermal stability of diisocyanate- and diisothiocyanate-derived cyclopolymers was improved by post treatment with alcohols or amines (144).

## J. Polymerization of Dinitriles

The cyclopolymerization of dinitriles has already been discussed in Volume B—Part 1, Chapter I in conjunction with polymers containing multiple carbon–nitrogen bonds. It is also briefly considered here to provide a complete picture of the types of cyclopolymerization reactions that occur with nitrogen-containing compounds.

### METHOD 1. FREE-RADICAL INITIATION

Treatment of maleonitrile [219] or fumaronitrile [220] with *tert*-butyl-peroxide in bulk at 160°C gave cyclopolymer [221] (163). Under similar conditions succinonitrile [222] yields the copolymer [223] (163). Other free-radical

$$(II\text{-}103)$$

$$(II\text{-}104)$$

initiators, including benzoyl peroxide, *tert*-butyl hydroperoxide, and cumene hydroperoxide were also used successfully (*163*). The structures of these polymers were supported by infrared, ultraviolet, and nuclear magnetic resonance data. Several copolymers from monomers [219], [220], and [222] were also prepared (*163*).

## METHOD 2. ZIEGLER INITIATION, MISCELLANY

Treatment of acrylonitrile with a triethylaluminum/vanadium oxytrichloride catalyst yields a low molecular weight (~2355) polymer (*3*). It is 90–95% soluble in hot dimethylformamide. Infrared studies indicate that structures of type [224] were present in the polymer. The reaction path that

$$(II\text{-}105)$$

leads to structure [224] is an interesting example of a unique ring-forming polymerization.

## K. Polymerization of Diunsaturated Phosphorus-Containing Compounds

Diallylphosphine oxides and related phosphorus compounds cyclopoly-merize as shown in Eq. (II-106).

$$ \text{[225]} \qquad\qquad \text{[226]} \qquad\qquad\qquad \text{(II-106)}$$

### METHOD 1. FREE-RADICAL INITIATION

The polymerization of diunsaturated phosphines, phosphine oxides, phosphonates, and quaternary phosphonium salts is initiated by typical free-radical catalysts. Peroxides, hydroperoxides, peracids, and $\alpha,\alpha'$-azodiiso-butyronitrile are useful (*44, 45, 47, 65, 149, 183, 258, 259*). The reactions are performed in bulk or solution. Diethyl ether, dioxane, diethylene glycol, dimethylformamide, benzene, and chlorobenzene have been used as reaction media. In one instance, $\beta,\beta'$-dichlorodiethyl ether was a very useful solvent in conjunction with *tert*-butyl peroxide (*258, 259*). Thus, allylphenylallyl phosphonate [227] gave polymer [228] possessing a molecular weight of

$$ \text{[227]} \qquad\qquad\qquad\qquad\qquad \text{[228]} \qquad\qquad \text{(II-107)}$$

26,000 (*258*). On the other hand, polymerization in benzene or diethyl ether with *tert*-butyl peroxide as catalyst gave products that had molecular weights of only about 1000 (*258*). The temperatures at which phosphorus derivatives polymerize vary depending on the particular monomer, catalyst, and solvent. Temperatures ranging from 65° to 194°C have been employed. Copolymers from diunsaturated phosphorus derivatives with monoolefins were also prepared by this method (*53, 258, 259*).

Berlin and Butler (*44, 45*) studied the cyclopolymerization of a series of phosphine oxides. Benzoyl peroxide and $\alpha,\alpha'$-azodiisobutyronitrile were initiators. Reactions were performed in bulk. When R′ was alkyl, low yields

$$(II\text{-}108)$$

**[229]**

R = H or methyl
R′ = alkyl or aryl

of low molecular weight polymers were obtained with the AIBN initiator. No polymerization occurred with benzoyl peroxide (45). However, both catalysts gave low polymer when R′ was aryl (44). The results were explained by assuming that degradative chain-transfer takes place during the polymerization. The greater reluctance of the alkyl derivative to polymerize is due to the fact that there are more methylene groups capable of undergoing such chain-transfer processes in the latter than when R′ is aryl.

The polymerization of allylphenylallyl phosphonate [227] by benzoyl peroxide was first reported by Kamai and Kukhtin in 1955 (149). A low molecular weight material ($n = 2$–3) was obtained. The possibility of cyclopolymerization was not mentioned. Monomer [227] was also polymerized by Berlin and Butler (44). Benzoyl peroxide was used for the initiator. A polymer possessing an intrinsic viscosity of 0.09 was obtained with negligible residual unsaturation. A cyclic structure was assumed, but no decision was made between the two possibilities (44). Structure [230] may be preferred geometrically, but on the other hand, structure [228] will result if radical stability is the governing factor. The polymerization of [227] was described in later patents

$$(II\text{-}109)$$

**[230]**

(258, 259). As mentioned [see Eq. (II-107)] a high molecular weight material [228] was obtained by the use of the *tert*-butyl peroxide/$\beta,\beta'$-dichlorodiethyl ether system (258, 259).

The polymerization of a series of phosphonium salts capable of forming five-, six-, seven-, and eight-membered rings was studied (53). The first three groups of monomers yield saturated, soluble polymers indicative of cyclopolymerization. The monomers leading to eight membered ring-containing materials gave polymers that showed appreciable unsaturation (53). The results indicate that ring size and degree of cyclization are intimately related in these reactions.

The use of appropriate monomers leads to products that contain recurring bicyclic units. An example is shown in Eq. (II-110) (*183*). Poly(allyl phenyl-

$$\text{[231]} \qquad \qquad \text{[232]} \qquad \qquad \text{(II-110)}$$

allyl phosphonate) [**228**], its copolymer with lauryl methacrylate, and poly-(diallyl phenyl phosphine oxide) [**233**] were claimed to impart excellent

$$\text{[233]} \qquad \qquad \text{(II-111)}$$

thermal stability and detergent properties to engine oils which are used at high temperature and pressure under conditions requiring frequent cold-starting (*258, 259*).

METHOD 4. ANIONIC INITIATION

Soluble and insoluble polymers are obtained from di-*n*-propenyl- and diisopropenylphosphine oxides upon treatment with Grignard reagents (*111*).

## L. Polymerization of Diunsaturated Sulfur-Containing Compounds

Most cyclopolymerizations of sulfur compounds were performed with free-radical initiators. The several classes of compounds that were polymerized are described.

METHOD 1. FREE-RADICAL INITIATION

*a. Sulfides, Sulfones, and Sulfonium Compounds*

Formation of linear polymers from the diunsaturated sulfide [**234**] and sulfone [**235**] has been reported (*181*). Benzoyl peroxide was the initiator.

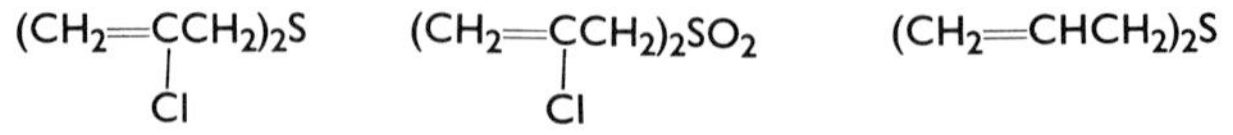

$$(CH_2\!=\!\underset{\underset{Cl}{|}}{C}CH_2)_2S \qquad (CH_2\!=\!\underset{\underset{Cl}{|}}{C}CH_2)_2SO_2 \qquad (CH_2\!=\!CHCH_2)_2S \tag{II-112}$$

**[234]**          **[235]**          **[236]**

Free-radical polymerization of diallyl sulfide **[236]** reportedly gave a high-boiling oil (*269*). On the other hand, treatment of a mixture of diallyl sulfide and acrylonitrile with α,α′-azodiisobutyronitrile in benzene yields an insoluble gel (*244*). A soluble copolymer was obtained from diallyl methyl sulfonium methosulfate **[237]** as shown in Eq. (II-113) (*244*). Cyclopolymerization of the diallyl moiety of **[237]** was postulated in this case (*244*).

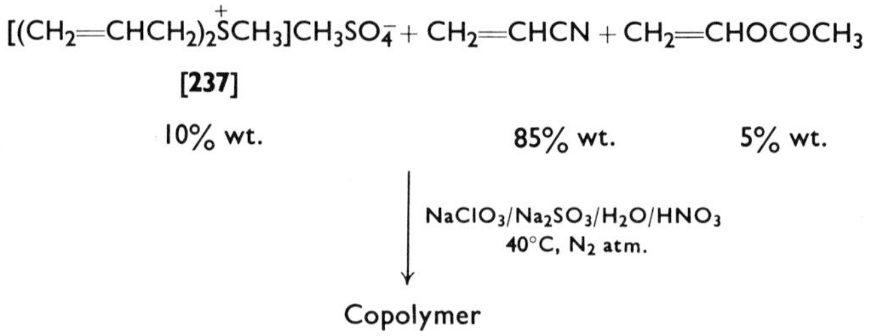

$$[(CH_2\!=\!CHCH_2)_2\overset{+}{S}CH_3]CH_3SO_4^- + CH_2\!=\!CHCN + CH_2\!=\!CHOCOCH_3$$

**[237]**

10% wt.                    85% wt.          5% wt.

$$\tag{II-113}$$

$$\Big\downarrow \begin{array}{l} NaClO_3/Na_2SO_3/H_2O/HNO_3 \\ 40°C, N_2 \text{ atm.} \end{array}$$

Copolymer

### b. Thioacetals and Thioketals

Divinyl thioacetals and ketals yield ring-containing polymers by free-radical polymerization (*188*). Their structure **[239]** was supported by infrared studies.

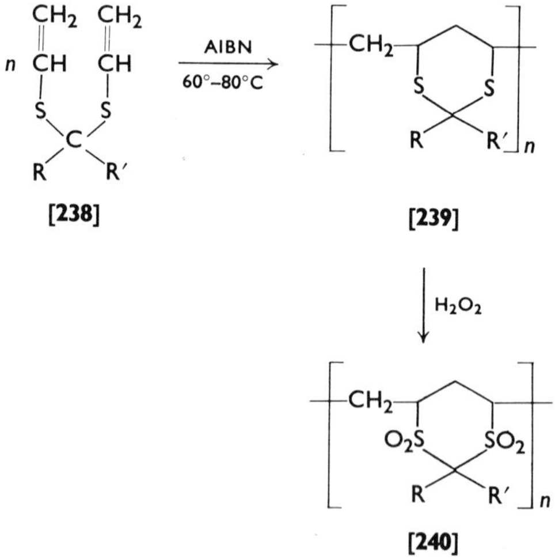

**[238]**                    **[239]**                    (II-114)

$$\Big\downarrow H_2O_2$$

**[240]**

Oxidation with hydrogen peroxide gave the cyclic sulfones **[240]**. Contradictory results were reported for polymerization of divinyl mercaptal (**[238]**;

R = R′ = H). The formation of a soluble polymer possessing structure [239] (R = R′ = H) was described (237). The experimental conditions were essentially those given in Eq. (II-114). The reaction was carried out in solution to low conversions (237). According to others (189), divinyl mercaptal does not polymerize under these conditions. However, polymerization was effected by exposure of the monomer to sunlight for 8 months (189).

### c. Sulfonates, Sulfonamides, Thiocarbonates, and Thiocarbamates

The polymerization of allyl vinyl sulfonate [241] gives a soluble material that contains six-membered rings (117). The reaction is run in benzene because insoluble polymer was obtained in bulk. Polymer [242] also contains some

(II-115)

**[241]**          **[242]**

residual unsaturation. Low monomer concentrations favor cyclization. A similar result was obtained with N-allyl vinyl sulfonamide [243] (118).

(II-116)

**[243]**          **[244]**

The polymerization of dithiolocarbonates [245] and [246] was described (225, 236). Monomer [245] gave mainly cross-linked material on treatment in

$$CH_2=CHSCH_2CH_2SCOSCH=CH_2 \qquad CH_2=CHSCOSCH=CH_2$$

**[245]**                              **[246]**

(II-117)

bulk with α,α′-azodiisobutyronitrile. In benzene, a soluble product was obtained even at conversions as high as 40%. Infrared investigation of this soluble material showed small amounts of vinyl unsaturation (236).

A series of S-vinyl-N-vinyl thiocarbamates [247] were polymerized with α,α′-azodiisobutyronitrile in benzene (229). A mixture of soluble and insoluble polymer was obtained. Based on chemical and infrared evidence structure

$$CH_2\!\!=\!\!CHSCONCH\!\!=\!\!CH_2$$

$$\underset{[247]}{\overset{R}{|}}$$

(II-118)

$$R = CH_3, C_2H_5, n\text{—}C_4H_9$$

[248] was assigned to the soluble material ($R = CH_3$). The percent of soluble product, as well as its degree of cyclization, increased with decreasing monomer concentration. Structure [248] is in agreement with the fact that the

(II-119)

[248]

S-vinyl group is more reactive under free-radical polymerization conditions than the N-vinyl group. It is of interest to note (see Method 3, "Cationic Initiation") that the reverse holds true for the Lewis acid-catalyzed polymerization of [247].

An interesting effect was observed by changing the R from methyl to n-butyl in monomer [247]. Lower conversions to polymer took place; the percent of insoluble material increased and the degree of cyclization of the soluble portion decreased. The phenomenon was explained as being due to steric inhibition of the ring-forming reaction (229).

### d. Dienes and Sulfur Dioxide

The copolymerization of 1,5-hexadiene with sulfur dioxide (64, 68, 265) is shown in Eq. (II-120). Hydrogen peroxide, benzoyl peroxide, and $\alpha,\alpha'$-azo-diisobutyronitrile are useful initiators. Reactions were carried out in bulk

(II-120)

[249]    [250]    [251]

or in n-heptane solution (265). The use of potassium persulfate/ammonium nitrate in water was also reported (68).

Similar cyclopolymerizations take place with other dienes. For example, cis,cis-1,5-cyclooctadiene [252] gave polymer [253] (104, 106). Its structure was supported by elemental analysis, solubility, infrared data, and absence

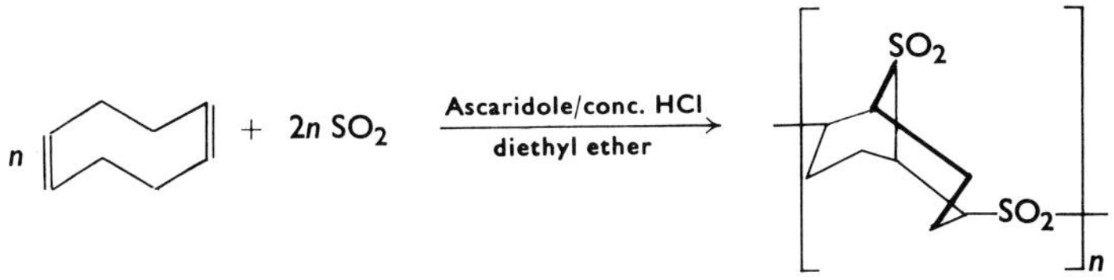

$$\text{(II-121)}$$

[252]     [250]                                    [253]

of residual unsaturation. The mechanism whereby these materials form is visualized to be the following (106):

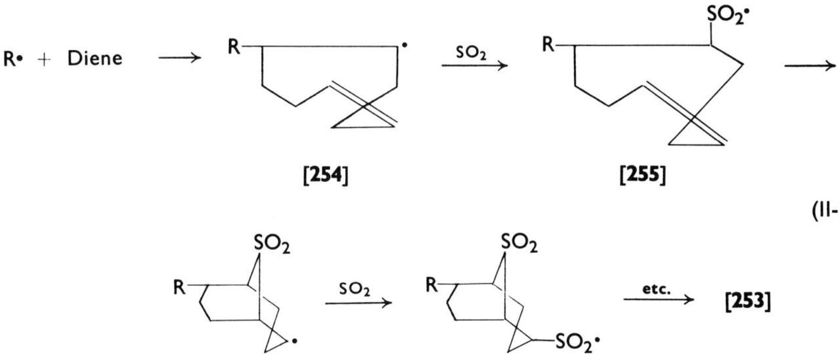

$$\text{(II-122)}$$

R• + Diene ⟶        [254]                    [255]

[256]                    [257]                    [253]

Copolymers [259] are obtained when cis,cis-1,5-cyclooctadiene reacts with sulfur dioxide and carbon monoxide simultaneously (105). Other examples

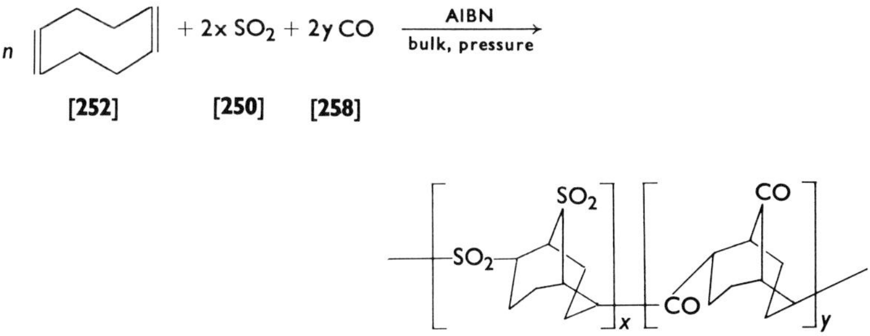

$$n \quad [252] + 2x\,SO_2 + 2y\,CO \xrightarrow[\text{bulk, pressure}]{\text{AIBN}}$$

[252]        [250]    [258]

$$\text{(II-123)}$$

[259]

of copolymerizations of dienes with sulfur dioxide are shown in Eq. (II-124) and (II-125) (200). A cross-linked material was obtained from the tetraunsaturated monomer [264]. No polymerization took place with the oxygenated monomers [265] and [266] (200).

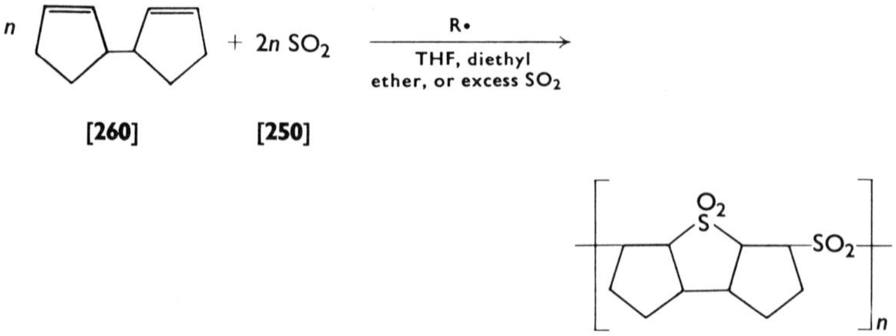

[260]       [250]

(II-124)

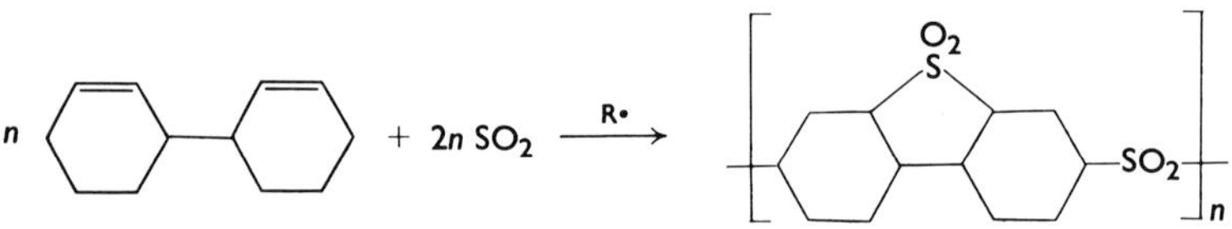

[261]

(II-125)

[262]      [250]          [263]

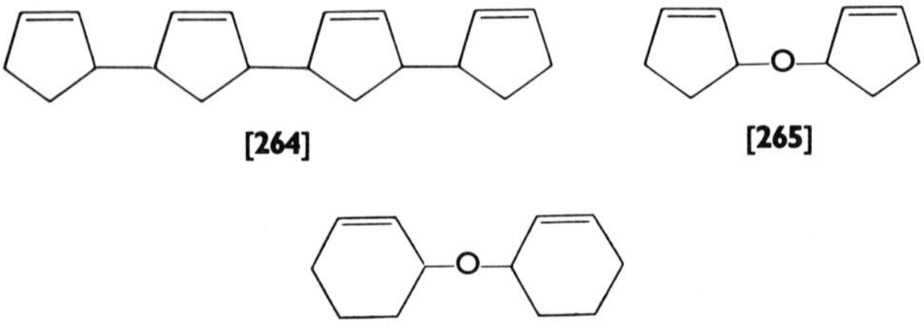

[264]          [265]

(II-126)

[266]

In all of the foregoing examples, two molecules of sulfur dioxide reacted with one molecule of the diene. A ring that contains the sulfone ($SO_2$) group is formed. However, if the two double bonds of the diene are spaced in a manner to allow the formation of a six-membered, all-carbon ring during the intramolecular cyclization step, a copolymer that contains only one molecule of sulfur dioxide per repeat unit forms (234). An example is shown in Eq. (II-127) (234).

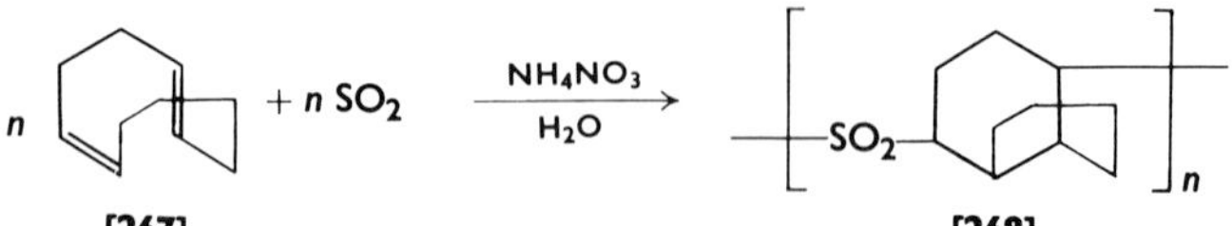

[267]            [268]

cis, trans-1,5-Cyclodecadiene

## e. Divinyl Sulfone and Olefins

The copolymerization of divinyl sulfone with maleic anhydride, dimethyl fumarate, or acrylonitrile yields ring-containing, linear polymers (64, 71, 81). The reactions proceed as shown in Eq. (II-128) for the case of maleic anhydride. Typical free-radical initiators are used. Benzene and dimethylformamide are

$$(II-128)$$

solvents. The ideal structure [273] is approached to various degrees depending on the monounsaturated comonomer and on the feed composition.

## METHOD 2. ZIEGLER INITIATION

The cyclopolymerization of diallyl sulfide to a high-boiling oil in the presence of coordination-type catalysts was reported (269).

## METHOD 3. CATIONIC INITIATION

The S-vinyl-N-vinyl thiocarbamates [274] are polymerized by boron trifluoride etherate in benzene solution (229). The N-methyl derivative gave

$$CH_2=CHSCONCH=CH_2 \tag{II-129}$$
$$\underset{R}{|}$$

[274]

$$R = CH_3, C_2H_5, n-C_4H_9$$

soluble polymer in an 85.7% conversion. It contained 42% of residual double bonds. Only $S$-vinyl unsaturation is present in the polymer [275] due to the

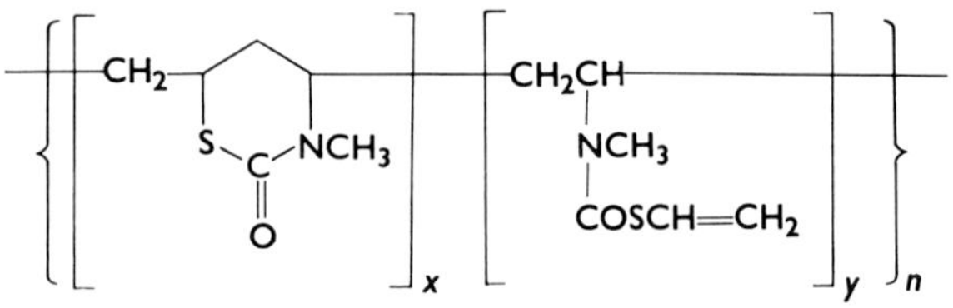

[275]

greater reactivity of $N$-vinyl groups under these conditions. The situation is the reverse of that which is encountered in the free-radical polymerization of [274]. In the latter case [see Eq. (II-119)] the cyclopolymer contains only pendant $N$-vinyl groups (229). Treatment of the $N$-ethyl and $N$-$n$-butyl derivatives [274] with boron trifluoride etherate yielded tacky polymers in low conversion (229).

The cationic polymerization of diallyl sulfide was mentioned in the literature (269). A high-boiling oil was obtained (269).

## M. Polymerization of Diunsaturated Silicon Derivatives

Several types of silicon-containing compounds were cyclopolymerized. The reactions and methods are discussed below.

### Method 1. Free-Radical Initiation

The free-radical intra–intermolecular polymerization of silicon-containing monomers is initiated by peroxides, hydroperoxides, peracids, and $\alpha,\alpha'$-azodiisobutyronitrile (65, 71, 84, 87, 109, 110, 201). Aromatic hydrocarbons are preferred solvents. Cobalt-60 irradiation in benzene solution was also reported (123, 124). An example is shown in Eq. (II-131) (84). Appropriate monomers lead to materials composed of bicyclic repeat units (124). The

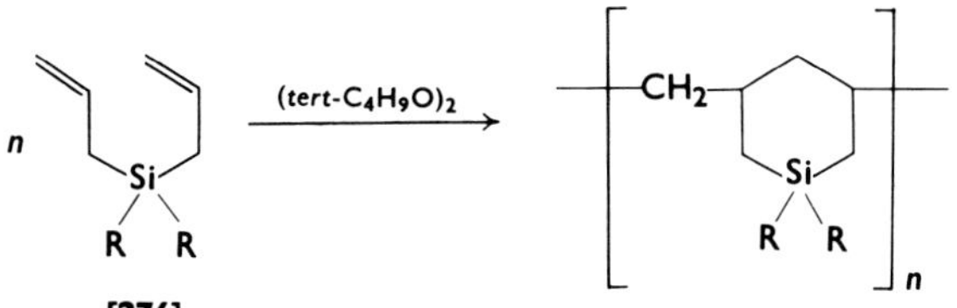

$$(II\text{-}131)$$

[276]

R = CH_3, C_6H_5

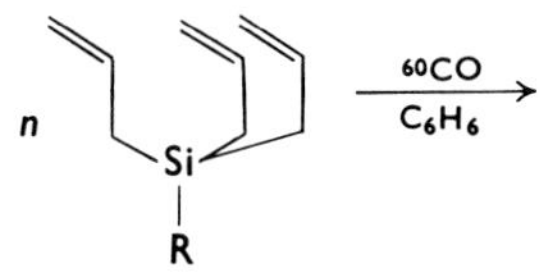

[278]

Triallylmethyl(or phenyl)silane

R = CH_3, C_6H_5

$$(II\text{-}132)$$

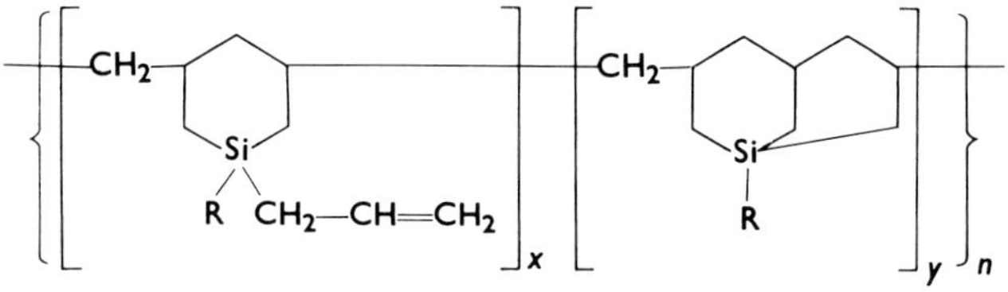

[279]

formation of polymer [279] proceeds via intermediates [280] and [281]. Structure [279] was supported by infrared data. The value of $x$ in [279] depends

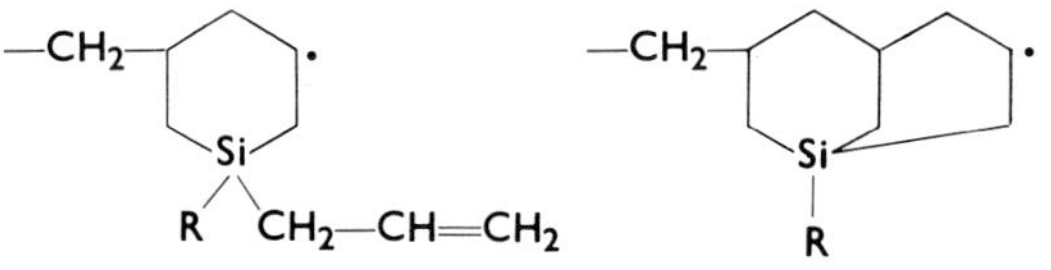

$$(II\text{-}133)$$

[280]        [281]

on the R group. The methyl-substituted polymer contained 18–20% of residual unsaturation. Lower levels (13–15%) were found when R was phenyl (*124*).

Linear polymers are prepared by the copolymerization of divinylsilanes with various monounsaturated comonomers (*64, 71*). The reaction is illustrated in Eq. (II-134). Maleic anhydride, fumaryl chloride, and vinyl acetate were used as co-monomers in these copolymerizations (*64, 71*).

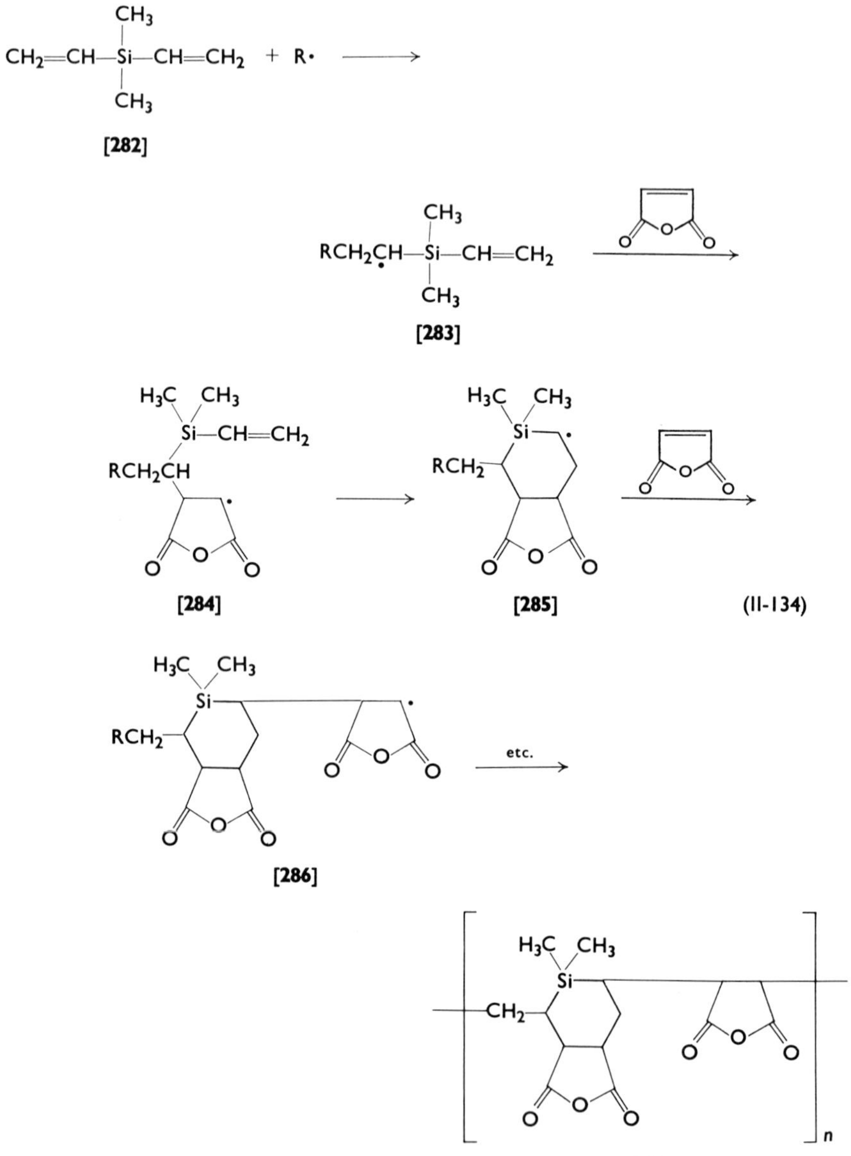

$$
\underset{[282]}{CH_2{=}CH{-}\underset{\underset{CH_3}{|}}{\overset{\overset{CH_3}{|}}{Si}}{-}CH{=}CH_2} \ + \ R\cdot \ \longrightarrow
$$

$$
\underset{[283]}{RCH_2\overset{\cdot}{CH}{-}\underset{\underset{CH_3}{|}}{\overset{\overset{CH_3}{|}}{Si}}{-}CH{=}CH_2}
$$

(II-134)

[284]                    [285]

etc.

[286]

[287]

The examples described above all deal with allyl- or vinyl-substituted silanes. Recent reports (*109*, *110*) describe the cyclopolymerization of divinyl dimethyl siloxane [**288**]. It is shown in Eq. (II-135). The structure of the polymer

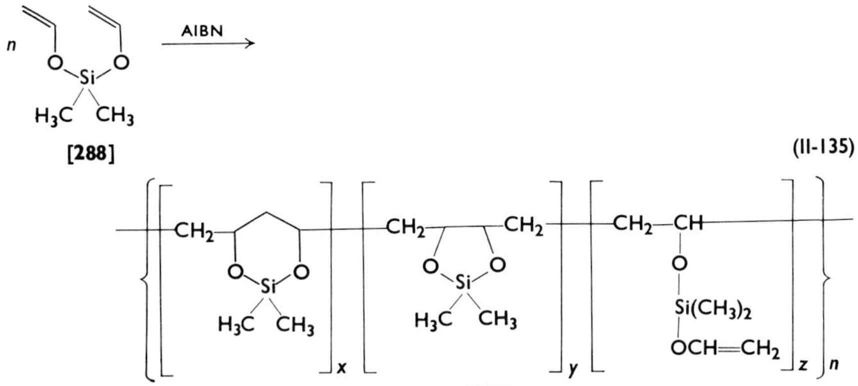

[288]                                                                    (II-135)

[289]

[289] is supported by infrared data. The presence of the five-membered rings was proven by hydrolysis to yield a poly(vinyl alcohol) in which infrared and nuclear magnetic resonance techniques detected "head-to-head" sequences (109).

METHOD 2.  ZIEGLER INITIATION

Diallylsilanes [290] have been cyclopolymerized by a catalyst composed of a trialkyl aluminum and titanium tetrachloride (52, 84, 159, 167). n-Heptane

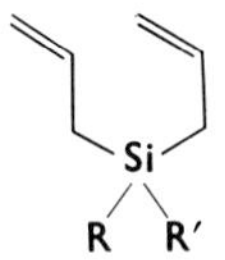

[290]                                                                    (II-136)

R and R' = CH$_3$ and/or C$_6$H$_5$

or petroleum ether were solvents. The triallyl derivative [291] gave polymer [292] (272). The residual unsaturation in this polymer (38%) was higher than when it was prepared by free-radical initiation [see Eq. (II-132)].

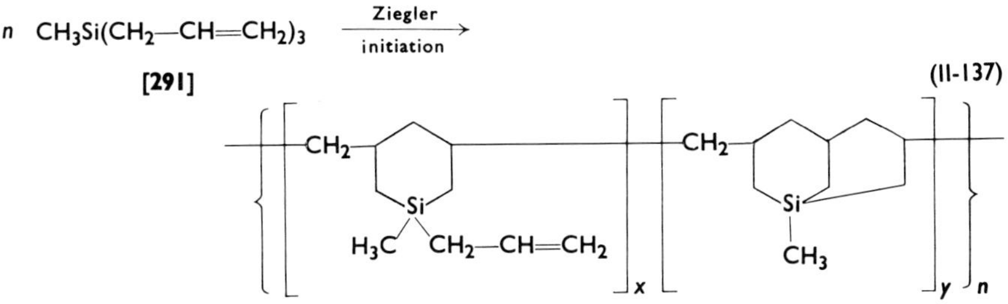

$n$  CH$_3$Si(CH$_2$—CH=CH$_2$)$_3$  $\xrightarrow[\text{initiation}]{\text{Ziegler}}$

[291]                                                                    (II-137)

[292]

METHOD 3. CATIONIC INITIATION

The cationic polymerization of divinyl dimethyl siloxane was studied (*268*). Proper choice of solvent was very important. There was no cyclization in toluene or nitroethane, but cyclic structures were obtained in methylene chloride (*268*).

METHOD 4. ANIONIC INITIATION

A soluble polymer [294] was obtained when *N*-methyl-*sym*-divinyltetra-methyl disilazane [293] was treated with butyllithium in a mixture of hexane and triethylamine (*266*). The formation of [294] was only favored at low

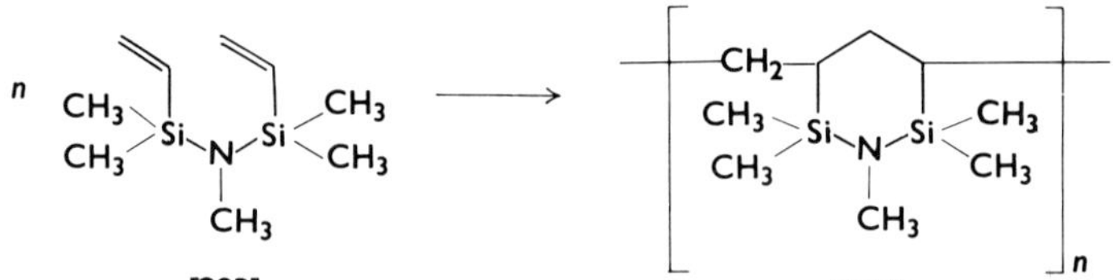

(II-138)

[293]                              [294]

monomer concentrations. High concentrations led to cross-linked materials (*266*).

## N. Miscellaneous Polymerizations

METHOD 1. FREE-RADICAL INITIATION

The copolymerization of 1,5-dienes with selenium dioxide has been de-scribed (*68*). Reaction proceeds as shown in Eq. (II-139). It was carried out in

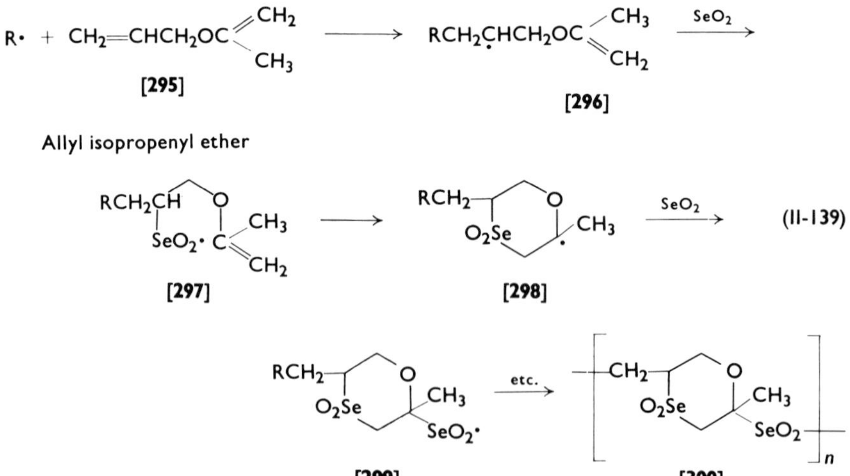

[295]

[296]

Allyl isopropenyl ether

[297]                    [298]                    (II-139)

[299]                    [300]

water with $\alpha,\alpha'$-azodiisobutyronitrile as initiator. Temperatures of 55°–60°C were preferred.

The formation of a cyclic polymer from diallyldiethyllead [301] was reported (65). Reaction was carried out with peracetic acid in benzene solution. Free-

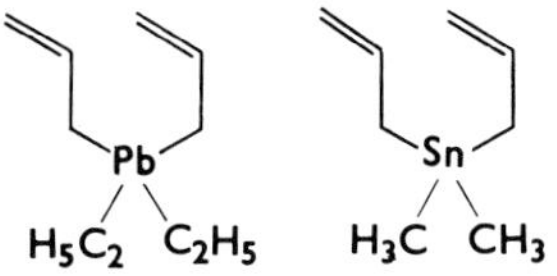

(II-140)

[301]     [302]

radical polymerization of the dimethyltin analog [302] was claimed to give a soluble polymer containing "some" residual unsaturation (63).

METHOD 2. ZIEGLER INITIATION

A ring-containing polymer was obtained when diallyldiethyllead [301] was treated with chromium(III) acetylacetonate/tributylaluminum (65).

TABLE II.1
Polydialdehydes

| No. | Structure | Method | Solubility | Molecular weight | $T_m$ (°C) | Remarks and property data | References |
|---|---|---|---|---|---|---|---|
| 1 | | A,1; A,4 | Sparingly in alcohol and hot water. In aqueous NaOH with strong degradation | $\eta_{inh}$ 0.3 | 144–150 d | — | 54,55,134 |
| 2 | | A,3 | $C_6H_6$, toluene | $\bar{M}_n =$ 2130 | 47 | Original polymer was 12% soluble in $C_6H_6$. All data on soluble fraction. Degree of cyclization $(x/y)$ is high | 26, 27 |
| 3 | | A,3 | $C_6H_6$, toluene | $[\eta]$ 0.59 | 100–105 | Original polymer was partly soluble in $C_6H_6$. All data on soluble fraction. % CHO $\cong$ 6–10% | 2, 22, 24 |
| 4 | Poly(3,4-dibenzylglutaconaldehyde) Polymer from $OHCCH_2C\!=\!CCHO$, $CH_2C_6H_5$ $CH_2C_6H_5$ | A,3 | — | — | — | — | 160 |

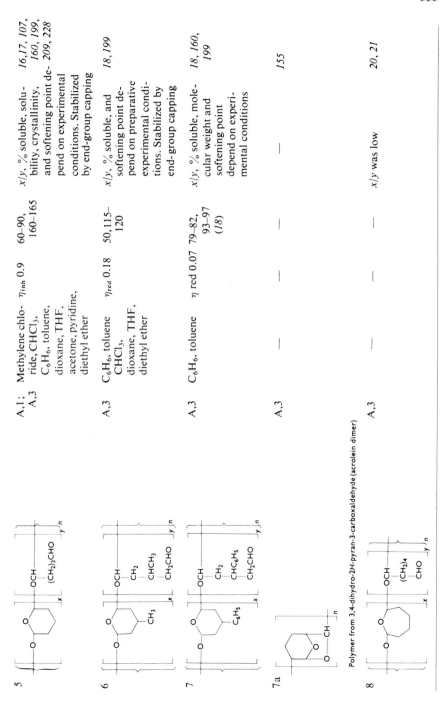

| No. | Structure | Method | Solvents | Viscosity | Softening point | Remarks | References |
|---|---|---|---|---|---|---|---|
| 5 | | A,1; A,3 | Methylene chloride, CHCl$_3$, C$_6$H$_6$, toluene, dioxane, THF, acetone, pyridine, diethyl ether | $\eta_{inh}$ 0.9 | 60–90, 160–165 | $x/y$, % soluble, solubility, crystallinity, and softening point depend on experimental conditions. Stabilized by end-group capping | 16,17, 107, 160, 199, 209, 228 |
| 6 | | A,3 | C$_6$H$_6$, toluene CHCl$_3$, dioxane, THF, diethyl ether | $\eta_{red}$ 0.18 | 50,115–120 | $x/y$, % soluble, and softening point depend on preparative experimental conditions. Stabilized by end-group capping | 18,199 |
| 7 | | A,3 | C$_6$H$_6$, toluene | $\eta$ red 0.07 | 79–82, 93–97 (18) | $x/y$, % soluble, molecular weight and softening point depend on experimental conditions | 18, 160, 199 |
| 7a | | A,3 | — | — | — | — | 155 |
| 8 | Polymer from 3,4-dihydro-2H-pyran-3-carboxaldehyde (acrolein dimer) | A,3 | — | — | — | $x/y$ was low | 20, 21 |

TABLE II.1—*continued*
Polydialdehydes

| No. | Structure | Method | Solubility | Molecular weight | $T_m$ (°C) | Remarks and property data | References |
|---|---|---|---|---|---|---|---|
| 9 | Copolymer from adipinaldehyde and chloral | A,3 | — | — | — | — | 20, 21 |
| 10 | Polysuberaldehyde | A,3 | — | — | >300 | Insoluble, and cross-linked. IR indicates some cyclization | 160, 161 |
| 11 | Poly(2-methyl-3-chlorosuberaldehyde) | A,3 | — | — | 105 | — | 160 |
| 12 | | A,3 | $C_6H_6$, $CHCl_3$, hexane | $[\eta]$ 0.1 | 143–148 | — | 226 |
| 12a | Poly(*trans*-1,2-cyclohexanedicarboxaldehyde) | A,3 | $CHCl_3$, $C_6H_5Cl$, trimethyl-benzene | $\eta_{red}$ 0.45 | 345.7 | Structure supported by IR and NMR data. Low polymerization temp. ($\sim$-40°C) yields incompletely cyclized polymer, m.p. 164°C | 156 |
| 13 | Polymer from 2,5-dimethyl-3,4-dihydro-2*H*-pyran-3-carboxaldehyde (methacrolein dimer) Poly(o-phthalaldehyde) | A,1; A,2; A,3; A,4 | Methylene chloride, $C_6H_6$, toluene, $CHCl_3$ | $[\eta]$ 0.31 by A,3 | 102–104, 139–154 | Softening points and geometry of the phthalane ring depend on experimental conditions. Structures studied by NMR | 28, 29 |

TABLE II.2
Polydiepoxides

| No. | Structure | Method | Solubility | Molecular weight | $T_m$ (°C) | Remarks and property data | References |
|---|---|---|---|---|---|---|---|
| 1 | Poly(isoprene oxide) | B,3 | — | Oligomer | — | — | 216 |
| 2 | Poly(1,2,5,6-diepoxyhexane) | B,3 | Phenol/s-tetrachloroethane (100/66, by wt.) | $\eta_{inh}$ 0.45 | 200d | Original polymer was 50% soluble in phenol/s-tetrachloroethane (100/66, by wt.). Data for soluble polymer. Crystalline (X-ray), can be enhanced via annealing at 125°C | 19, 25, 263 |
| 3 | Poly(1,2,5,6-diepoxyhexane) | B,4 | Toluene, "polar" solvents | $\eta_{inh}$ 0.27. Mol. wt. = 5000 (ebull.) | — | Structure supported by IR and NMR | 19, 25, 38 |
| 4 | Poly[o-di(epoxyethyl)benzene] | B,1 | THF | $\eta_{red}$ 0.1 | 120–125 | Original polymer was partly soluble in THF. All data are on the soluble fraction. Pendant epoxy groups were present by IR | 19, 25 |

TABLE II.2—*continued*
Polydiepoxides

| No. | Structure | Method | Solubility | Molecular weight | $T_m$ (°C) | Remarks and property data | References |
|-----|-----------|--------|-----------|------------------|-----------|---------------------------|-----------|
| 5 | Poly[o-di(epoxyethyl)benzene] | B,3 | THF | $\eta_{red}$ 0.11 | 135–145 | Original polymer was partly soluble in THF. All data are on the soluble fraction. Pendant epoxy groups were present by IR | 19, 25 |
| 6 | Poly[o-di(epoxyethyl)benzene] | B,4 | THF, CHCl$_3$ | $\eta_{red}$ 0.57 | 210 | IR indicates few epoxy groups in polymer. Cast from chloroform to brittle film | 19, 25 |
| 7 | Poly(N,N-diglycidylaniline) | B,4 | C$_6$H$_5$Cl | $\eta_{inh}$ 0.22. Mol. wt. (NMR) = 81,000 | — | NMR indicates no pendant oxirane rings in the polymer | 38 |

TABLE II.3
Poly(divinyl acetals) and Poly(divinyl ketals)

| No. | Structure | Method | Solubility | Molecular weight | $T_m$ (°C) | Remarks and property data | References |
|---|---|---|---|---|---|---|---|
| 1 | Poly(divinyl ether) | C,1 | $C_6H_6$, $CCl_4$ | $[\eta]$ 0.31 | 400 d | Structure by IR, NMR, and chemical data. Double-bond content (IR) = 19.4% | 30, 31 |
| | Copolymer [82] from divinyl ether with: | | | | | | |
| 2 | Acrylonitrile | C,1 | DMF | — | — | Composition of copolymer varied with monomer feed | 37, 71 |
| 3 | Fumaryl chloride | C,1 | "Soluble" | — | — | — | 71 |
| 4 | Maleic anhydride | C,1 | DMF, acetone, aqueous NaOH | $\eta_{inh}$ 0.50 | 350 d | Composition of copolymer was independent of monomer feed | 37, 61, 62, 64, 71 |
| 5 | Fumaronitrile | C,1 | DMF | $\eta_{red}$ 0.17 | — | Composition of copolymer was independent of monomer feed | 64, 71, 85 |
| 6 | Vinyl acetate | C,1 | "Soluble" | — | — | — | 71 |
| 7 | Tetracyanoethylene | C,1 | DMF | $\eta_{red}$ 0.07 | — | — | 85 |
| 8 | Dimethyl fumarate | C,1 | — | — | — | — | 64 |
| 9 | 4-Vinylpyridine | C,1 | DMF | $\eta_{red}$ 0.3 | — | Olefin more reactive than diene | 85 |
| 10 | Diethyl maleate | C,1 | "Soluble" | — | — | — | 64, 71 |

TABLE II.3—continued

Poly(divinyl acetals) and Poly(divinyl ketals)

| No. | Structure | Method | Solubility | Molecular weight | $T_m$ (°C) | Remarks and property data | References |
|---|---|---|---|---|---|---|---|
| 11 | N-Phenylmaleimide | C,1 | DMF | — | — | Composition of copolymer was independent of monomer feed | 37 |
| | *** | | | | | | |
| 12 | Polymer from $\left(CH_2{=}CCH_2\right)_2O$ with Cl | C,1 | — | — | — | — | 181 |

Poly(divinyl acetals) and poly(divinyl ketals)

| No. | Structure | Method | Solubility | Molecular weight | $T_m$ (°C) | Remarks and property data | References |
|---|---|---|---|---|---|---|---|
| 13 | $R_1 = R_2 =$ H | C,1 | HCOOH, CHCl$_3$, C$_6$H$_6$, chlorinated aliphatics and aromatics | $[\eta]$ 0.3 | 85–120 | $T_g$ = 74°C. Structure supported by IR, solubility, chemical reactions, and low degree of unsaturation (~6%). Dielectric properties and effective dipole moments | 11, 12, 34, 35, 157, 168, 174, 177, 190 |

Copolymer of 13 with:

| No. | Structure | Method | Solubility | Molecular weight | $T_m$ (°C) | Remarks and property data | References |
|---|---|---|---|---|---|---|---|
| 14 | Vinyl acetate | C,1 | C$_6$H$_6$, CCl$_4$ | — | — | Data on C$_6$H$_6$-soluble fraction. Cyclization supported by IR data. At 30% mole divinyl formal, $T_g$ = 22.5°C. At 50% mole divinyl formal, $T_g$ = 35°C | 191 |

| No. | Name | Type | Solvents | Viscosity | Temp. range | Description | References |
|---|---|---|---|---|---|---|---|
| 15 | Divinyl ethanal | C,1 | $C_6H_6$, acetone, $CHCl_3$ | — | — | Good adhesive properties | 168, 176 |
| 16 | Styrene | C,1 | $C_6H_6$, $CCl_4$ | $\eta_{sp}$ 0.034 | 78–94 | Several compositions prepared. Cyclization supported by IR. Data for copolymer containing 53% mole styrene | 179 |

\* \* \*

| No. | Name | Type | Solvents | Viscosity | Temp. range | Description | References |
|---|---|---|---|---|---|---|---|
| 17 | 2-Methyl-5-vinylpyridine | C,1 | Aromatics, chlorinated hydrocarbons | — | — | — | 192 |
| 18 | $R_1 = H$; $R_2 = CH_3$ | C,1 | Dioxane, $C_6H_6$, $CCl_4$, HCOOH, $CH_3CO_2H$ | $[\eta]$ 0.51. At $[\eta]$ 0.15, mol. wt. = 10,000 (ebull. in $C_6H_6$) | — | $T_g = 69°C$. Structure supported by IR, solubility, chemical reactions, and low degree of unsaturation (~6%). Dielectric properties and effective dipole moments | 11, 12, 34, 35, 157, 168, 177 |

Copolymer of 18 with:

| No. | Name | Type | Solvents | Viscosity | Temp. range | Description | References |
|---|---|---|---|---|---|---|---|
| 19 | Vinyl acetate | C,1 | $C_6H_6$, acetone, $CHCl_3$ | — | — | Several compositions prepared. Cyclization supported by IR | 178, 191, 192 |
| 20 | Styrene | C,1 | $C_6H_6$, $CCl_4$ | $\eta_{sp}$ 0.041 | 88–104 | Several compositions prepared. $T_m$ and viscosity for 57% m. of styrene copolymer. Cyclization supported by IR | 179 |

TABLE II.3—*continued*
Poly(divinyl acetals) and Poly(divinyl ketals)

| No. | Structure | Method | Solubility | Molecular weight | $T_m$ (°C) | Remarks and property data | References |
|---|---|---|---|---|---|---|---|
| 21 | 2-Methyl-5-vinylpyridine | C,1 | Aromatics, chlorinated hydrocarbons | — | — | — | 192 |
| 22 | Divinyl butyral *** | C,1 | $C_6H_6$, acetone, $CHCl_3$ | — | — | Good adhesive properties | 168, 176 |
| 23 | $R_1 = R_2 = CH_3$ | C,1 | $C_6H_6$, $CHCl_3$, dioxane | [$\eta$] 0.20 | — | $T_g = 60°C$. Structure supported by IR | 187 |
| 24 | $R_1 = H$; $R_2 = C_2H_5$ | C,1 | $C_6H_6$, $CHCl_3$ dioxane | — | 35–55 | Chemical data and infrared evidence support structure | 175 |
| 25 | $R_1 = CH_3$; $R_2 = C_2H_5$ | C,1 | $C_6H_6$, $CHCl_3$ dioxane | — | — | — | 187 |
| 26 | $R_1 = H$; $R_2 = n\text{-}C_3H_7$ | C,1 | Dioxane, $C_6H_6$, HCOOH, $CH_3CO_2H$ | [$\eta$] 0.42 | — | $T_g = 24°C$. Structure supported by IR, solubility, chemical reactions, and low degree of unsaturation (~8%). Dielectric properties and effective dipole moments | 12, 34, 35, 157, 168, 173, 177 |
| | Copolymer of 26 with: | | | | | | |
| 27 | Vinyl acetate | C,1 | $C_6H_6$, acetone, $CHCl_3$ | — | — | Several compositions were prepared | 168, 176, 178, 191, 192 |

| No. | Name | | Solvents | Viscosity | Range | Remarks | Ref. |
|---|---|---|---|---|---|---|---|
| 28 | Dimethyl crotonyl carbinol | C,1 | — | — | — | — | *192* |
| 29 | Styrene | C,1 | $C_6H_6$, $CCl_4$ | $\eta_{sp} \sim 0.05$ | — | Several compositions prepared. Viscosity for 75% mole styrene in copolymer | *168, 176, 179* |
| 30 | 2-Methyl-5-vinylpyridine | C,1 | Aromatics, chlorinated hydrocarbons | — | — | — | *192* |
| 31 | Dimethyl crotonyl carbinyl acetate | C,1 | — | — | — | — | *192* |
| | *** | | | | | | |
| 32 | $R_1 = H$; $R_2 = $ iso-$C_3H_7$ | C,1 | $C_6H_6$, $CHCl_3$, dioxane | — | 35–55 | Structure supported by chemical and infrared data | *175* |
| 33 | Copolymer of 32 with vinyl acetate | C,1 | $C_6H_6$, $CCl_4$ | — | — | Several compositions prepared. At 30% mole divinyl iso-butyral, $T_g = 26.5°C$ | *191* |
| 34 | $R_1 = H$; $R_2 = CH_3O(CH_2)_2$ | C,1 | $C_6H_6$, chlorinated solvents | $[\eta]$ 0.11 | — | — | *180* |
| 35 | $R_1 = H$; $R_2 = CH_2=CHO(CH_2)_2$ | C,1 | $C_6H_6$, chlorinated solvents | — | — | — | *180* |
| 36 | $R_1 = H$; $R_2 = Cl(CH_2)_2O(CH_2)_2$ | C,1 | $C_6H_6$, chlorinated solvents | — | — | — | *180* |

TABLE II.3—*continued*

Poly(divinyl acetals) and Poly(divinyl ketals)

| No. | Structure | Method | Solubility | Molecular weight | $T_m$ (°C) | Remarks and property data | References |
|---|---|---|---|---|---|---|---|
| 37 | $R_1 = H; R_2 = C_4H_9$ | C,1 | "Soluble" | — | — | — | 11 |
| 38 | $R_1 = H; R_2 = iso\text{-}C_4H_9$ | C,1 | $C_6H_6$, $CHCl_3$, dioxane | — | — | Rubbery, semisolid. Structure supported by chemical and IR evidence | 175 |
| 39 | Copolymer of 38 with vinyl acetate | C,1 | $C_6H_6$, $CCl_4$ | — | — | Several compositions prepared. At 70% mole vinyl acetate, copolymer has $T_g = 27°C$ | 191, 192 |
| 40 | $R_1 = H; R_2 = C_2H_5O(CH_2)_2$ | C,1 | $C_6H_6$, chlorinated solvents | — | — | — | 180 |
| 41 | $R_1 = R_2 = (CH_2)_5$ | C,1 | $C_6H_6$, $CHCl_3$, dioxane | [η] 0.13 | — | IR indicates absence of unsaturation | 187 |
| 42 | $R_1 = H; R_2 = CH_3(CH_2)_4$ | C,1 | $C_6H_6$, chlorinated solvents | [η] 0.12 | — | Solubility and chemical evidence support structure | 180 |
| 43 | $R_1 = H; R_2 = C_3H_7O(CH_2)_2$ | C,1 | $C_6H_6$, chlorinated solvents | [η] 0.11 | — | Solubility and chemical evidence support structure | 180 |

363

| No. | | | | | | | Ref. |
|---|---|---|---|---|---|---|---|
| 44 | $R_1 = H; R_2 = CH_3(CH_2)_5$ | C,1 | $C_6H_6$, chlorinated solvents | $[\eta]$ 0.17 | — | — | *180* |
| 45 | $R_1 = H; R_2 = (CH_3)_2CH(CH_2)_3CH(CH_3)CH_2-$ (R)(+) | C,1 | $CHCl_3$, $C_6H_6$ | "Low" | — | Viscous oil. IR in agreement with assigned structure | *1* |
| 46 | $R_1 = H; R_2 = (CH_3)_2CH(CH_2)_3CH(CH_3)CH_2-$ (R)(+) | C,3 | $CHCl_3$, $C_6H_6$ | $[\eta]$ 0.24 | 120–130 | Structure supported by solubility and IR. Amorphous (X-ray) | *1* |
| 47 | $R_1 = H; R_2 = o\text{-}ClC_6H_4$ | C,1 | — | — | — | — | *172* |
| 48 | $R_1 = H; R_2 = m\text{-}ClC_6H_4$ | C,1 | — | — | — | — | *172* |
| 49 | $R_1 = H; R_2 = p\text{-}ClC_6H_4$ | C,1 | — | — | — | — | *172* |
| 50 | $R_1 = H; R_2 = C_6H_5$ | C,1 | $C_6H_6$, $CHCl_3$, dioxane pyridine, DMF | Mol. wt. (ebull. in $C_6H_6$) = 19,260 | 80–100 | $T_g = 80°C$. Solubility, chemical evidence, and IR support cyclic structure | *7, 11, 170, 171* |
| 51 | $R_1 = H; R_2 = o\text{-}CH_3C_6H_4$ | C,1 | $C_6H_6$, $CHCl_3$ | Mol. wt. (ebull. in $C_6H_6$) = 10,000 | 85–95 | Structure supported by solubility, chemical and IR evidence | *170, 171* |
| 52 | $R_1 = H; R_2 = m\text{-}CH_3C_6H_4$ | C,1 | $C_6H_6$, chlorinated solvents | Mol. wt. (ebull. in $C_6H_6$) = 11,305 | 75–86 | Structure supported by solubility, chemical and IR evidence | *171* |

TABLE II.3—*continued*
Poly(divinyl acetals) and Poly(divinyl ketals)

| No. | Structure | Method | Solubility | Molecular weight | $T_m$ (°C) | Remarks and property data | References |
|---|---|---|---|---|---|---|---|
| 53 | $R_1 = H; R_2 = p\text{-}CH_3C_6H_4$ | C,1 | $C_6H_6$, $CHCl_3$ | Mol. wt. (ebull. in $C_6H_6$) = 16,800 | 95–115 | $T_g = 68°C$ | 170, 171 |
| 54 | $R_1 = H; R_2 = o\text{-}CH_3OC_6H_4$ | C,1 | $C_6H_6$, $CHCl_3$, dioxane | Mol. wt. (ebull. in $C_6H_6$) = 13,000 | 90–150 | Absence of olefinic unsaturation (IR) and solubility support cyclic structure | 184, 185 |
| 55 | $R_1 = H; R_2 = m\text{-}CH_3OC_6H_4$ | C,1 | $C_6H_6$, $CHCl_3$ | — | 90 | Cyclic, linear structure supported by solubility, chemical and IR evidence | 184 |
| 56 | $R_1 = H; R_2 = p\text{-}CH_3OC_6H_4$ | C,1 | $C_6H_6$, $CHCl_3$ | — | 150 | Cyclic linear structure supported by solubility, chemical and IR evidence | 184 |
| 57 | $R_1 = H; R_2 = 2,4\text{-}(CH_3)_2C_6H_3$ | C,1 | $C_6H_6$, other aromatics, chlorinated solvents | Mol. wt. (ebull. in $C_6H_6$) = 8,025 | 95–100 | $T_g = 67°C$. Solubility, absence of unsaturation, and IR support cyclic linear structure | 171 |

| | | | | | | |
|---|---|---|---|---|---|---|
| 58 | $R_1 = H; R_2 = 2,5\text{-}(CH_3)_2C_6H_3$ | C,1 | $C_6H_6$, other aromatics, chlorinated solvents | Mol. wt. (ebull. in $C_6H_6$) = 13,000 | 90–105 | — | 171 |
| 59 | $R_1 = H; R_2 = 3,4\text{-}(CH_3)_2C_6H_3$ | C,1 | $C_6H_6$, other aromatics, chlorinated solvents | Mol. wt. (ebull. in $C_6H_6$) = 8,387 | 90–100 | $T_g = 66°C$. Structure supported by solubility and IR data | 171 |
| 60 | $R_1 = H; R_2 = 2\text{-}(C_2H_5O)C_6H_4$ | C,1 | $C_6H_6$, $CHCl_3$, dioxane | Mol. wt. (ebull. in $C_6H_6$) = 6,500 | 85–135 | Absence of olefinic unsaturation (IR) and solubility support cyclic linear structure | 185 |
| 61 | $R_1 = H; R_2 = 2,4,6\text{-}(CH_3)_3C_6H_2$ | C,1 | $C_6H_6$, chlorinated solvents | Mol. wt. (ebull. in $C_6H_6$) = 17,300 | 105–115 | — | 171 |
| 62 | $R_1 = H; R_2 = o\text{-}(C_3H_7)C_6H_4$ | C,1 | $C_6H_6$, $CHCl_3$, dioxane | Mol. wt. (ebull. in $C_6H_6$) = 11,100 | 80–100 | Absence of olefinic unsaturation (IR) and solubility support cyclic linear structure | 185 |
| 63 | $R_1 = H; R_2 = o\text{-}(iso\text{-}C_3H_7O)C_6H_4$ | C,1 | $C_6H_6$, $CHCl_3$, dioxane | Mol. wt. (ebull. in $C_6H_6$) = 5,200 | 85–110 | Absence of olefinic unsaturation (IR) and solubility support cyclic linear structure | 185 |

TABLE II.3—*continued*
Poly(divinyl acetals) and Poly(divinyl ketals)

| No. | Structure | Method | Solubility | Molecular weight | $T_m$ (°C) | Remarks and property data | References |
|-----|-----------|--------|------------|------------------|------------|---------------------------|------------|
| 64 | $R_1 = H; R_2 = o-(C_4H_9O)C_6H_4$ | C,1 | $C_6H_6$, $CHCl_3$, dioxane | Mol. Wt. (ebull. in $C_6H_6$) = 9,300 | 70–85 | Absence of olefinic unsaturation (IR) and solubility support cyclic linear structure | 185 |
| 65 | $R_1 = H; R_2 = o-(iso-C_4H_9O)C_6H_4$ | C,1 | $C_6H_6$, $CHCl_3$, dioxane | — | 75–100 | Absence of olefinic unsaturation (IR) and solubility support cyclic linear structure | 185 |
| 66 | $R_1 = H; R_2 = o-(C_5H_{11}O)C_6H_4$ | C,1 | $C_6H_6$, $CHCl_3$, dioxane | — | 40–55 | Absence of olefinic unsaturation (IR) and solubility support cyclic linear structure | 185 |
| 67 | $R_1 = H; R_2 = o-(iso-C_5H_{11}O)C_6H_4$ | C,1 | $C_6H_6$, $CHCl_3$ dioxane | — | 60–75 | Absence of olefinic unsaturation (IR) and solubility support cyclic linear structure | 185 |
| 68 | $R_1 = R_2 = C_6H_5$ | C,1 | — | — | — | Absence of olefinic unsaturation (IR) and solubility support cyclic linear structure | 172 |

| No. | Structure | Type | Solubility | Mol. wt. | M.p. (°C) | Remarks | Ref. |
|---|---|---|---|---|---|---|---|
| 69 | $R_1$ = H; $R_2$ = α-Naphthyl | C,1 | | ~24,000 | — | Absence of olefinic unsaturation (IR) and solubility support cyclic linear structure | 172 |
| | *** | | | | | | |
| 70 | Polymer from divinyl acetal of furfural | C,1 | $CHCl_3$, $C_6H_6$ [η] 0.04 | | 125–145 | Structure when prepared with α,α'-azodiisobutyronitrile as initiator | 170, 231 |
| 71 | Polymer from divinyl acetal of furfural | C,1 | $CHCl_3$, $CCl_4$, dioxane [η] 0.09 | | Changes on standing | Structure when prepared with benzoyl peroxide or ultraviolet light and oxygen initiators | 231 |
| 72 | Poly(trivinyl orthoformate) | C,1 | $C_6H_6$, $CHCl_3$ | | 250–260 | Structure supported by solubility and low (3.6%) double-bond content, determined by IR and bromination | 183 |
| 73 | Poly(triallyl orthoformate). Contains an appreciable amount of a [4.4.2] bicyclic ring structure | C,1 | — | | — | | 272 |

TABLE II.4

Polymers from Unsaturated Glycidyl Esters

| No. | Structure | Method | Solubility | Molecular weight | $T_m$ (°C) | Remarks and property data | References |
|---|---|---|---|---|---|---|---|
| 1 | | D,3 | Dioxane, ethyl alcohol, acetone | Mol. wt. (cryos.) = 560 | — | Double-bond content = 5.42% | 5 |
| 2 | | D,3 | Dioxane, ethyl alcohol, acetone | [$\eta$] 0.057; mol. wt. (cryos.) = 800 | 120 | Contains 5–16% of C=C. No epoxide groups. Cross-links on heating at >70°C | 4, 5 |
| 3 | | D,4 | Acetone, dioxane | — | — | Contains 80% of epoxide groups. No olefinic unsaturation | 5 |
| 4 | | D,3 | $C_2H_5OH$, acetone $CH_3COOH$ | — | 130 | Contains olefinic unsaturation. Cross-links on heating | 240 |

## TABLE II.5
### Polymers from Diunsaturated Esters

| No. | Structure | Method | Solubility | Molecular weight | $T_m$ (°C) | Remarks and property data | References |
|---|---|---|---|---|---|---|---|
| 1 | (structure: $CH_2CH$–O–$COOCH=CH_2$)$_x$ ... $CH_2$ ... $CH_2$ ... $CH_2$)$_z$]$_n$ | D,1 | Acetone | $[\eta]$ 0.118 | 250–270 | $x$, $y$, $z$ depend on experimental conditions. IR supports structure and allows calculation of $x$, $y$, and $z$ | 210 |
| 2 | (structure: $CH_2CH$ $CH_2$ $OCOCH=CH_2$)$_{58\cdot6}$ ($CH_2CH$ $CO$ $OCH_2CH=CH_2$)$_{17\cdot2}$]$_{24\cdot2}$]$_n$ | D,1 | $C_6H_6$, pyridine | — | >320 (d) | Solubility and chemical evidence support structure | 181, 247, 276 |
| 3 | Poly(vinyl crotonate) | D,1 | — | $[\eta]$ 0.3 | — | Analysis indicates presence of 50% (–$CH_2$... $CH_3$... O... O units) | 11, 239 |

TABLE II.5—*continued*
Polymers from Diunsaturated Esters

| No. | Structure | Method | Solubility | Molecular weight | $T_m$ (°C) | Remarks and property data | References |
|---|---|---|---|---|---|---|---|
| 4 | Poly(vinyl methacrylate) | D,1 | $CHCl_3$, acetone | — | — | IR indicates cyclization of 50–60% | 152 |
| 5 | Poly(allyl methacrylate) | D,1 | — | — | — | $x$, $y$, $z$ vary with experimental conditions. At conversions <10%, polymer contains ~20% δ-lactone units | 275, 277, 278 |
| 6 | Poly(propargyl crotonate) | D,1 | Dioxane | — | — | Chemical and IR evidence support structure | 194 |

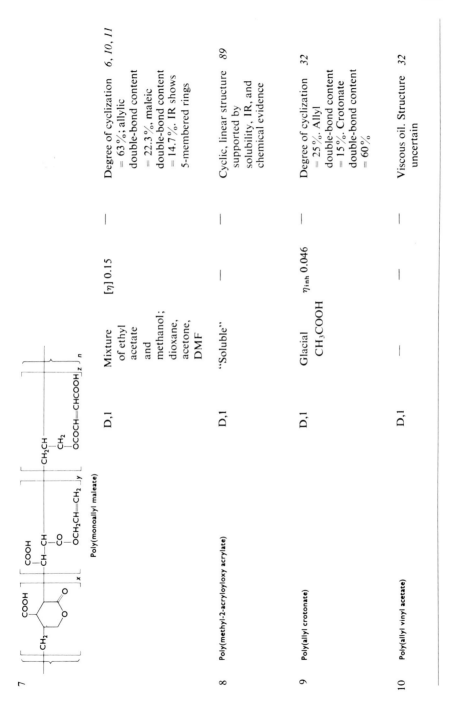

| 7 | Poly(monoallyl maleate) | D,1 | Mixture of ethyl acetate and methanol; dioxane, acetone, DMF | [η] 0.15 | — | Degree of cyclization = 63%; allylic double-bond content = 22.3%, maleic double-bond content = 14.7%. IR shows 5-membered rings | *6, 10, 11* |
| 8 | Poly(methyl-2-acryloyloxy acrylate) | D,1 | "Soluble" | | — | Cyclic, linear structure supported by solubility, IR, and chemical evidence | *89* |
| 9 | Poly(allyl crotonate) | D,1 | Glacial $CH_3COOH$ | $\eta_{inh}$ 0.046 | — | Degree of cyclization = 25%. Allyl double-bond content = 15%. Crotonate double-bond content = 60% | *32* |
| 10 | Poly(allyl vinyl acetate) | D,1 | — | | — | Viscous oil. Structure uncertain | *32* |

TABLE II.5—*continued*
Polymers from Diunsaturated Esters

| No. | Structure | Method | Solubility | Molecular weight | $T_m$ (°C) | Remarks and property data | References |
|---|---|---|---|---|---|---|---|
| 11 | 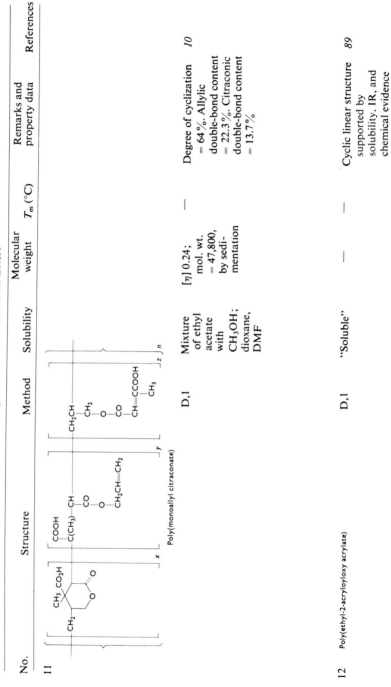 Poly(monoallyl citraconate) | D,1 | Mixture of ethyl acetate with $CH_3OH$; dioxane, DMF | $[\eta]$ 0.24; mol. wt. = 47,800, by sedimentation | — | Degree of cyclization = 64%. Allylic double-bond content = 22.3%. Citraconic double-bond content = 13.7% | 10 |
| 12 | Poly(ethyl-2-acryloyloxy acrylate) | D,1 | "Soluble" | — | — | Cyclic linear structure supported by solubility, IR, and chemical evidence | 89 |

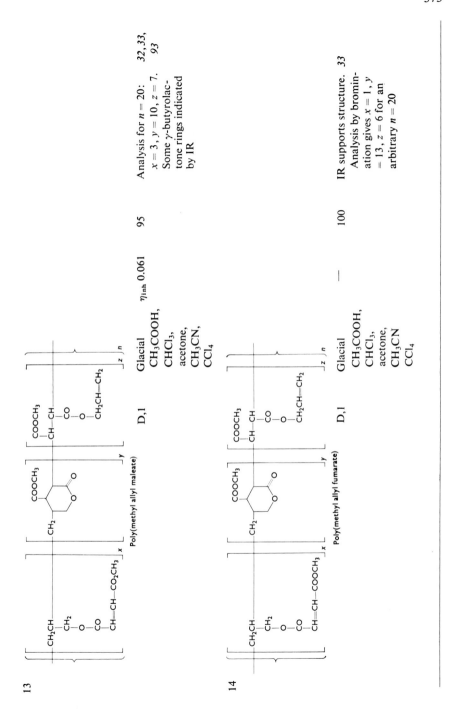

13    Poly(methyl allyl maleate)

D,1    Glacial CH₃COOH, CHCl₃, acetone, CH₃CN, CCl₄

η_inh 0.061    95

Analysis for $n = 20$: $x = 3$, $y = 10$, $z = 7$. Some γ-butyrolactone rings indicated by IR    *32, 33, 93*

14    Poly(methyl allyl fumarate)

D,1    Glacial CH₃COOH, CHCl₃, acetone, CH₃CN CCl₄

—    100

IR supports structure. Analysis by bromination gives $x = 1$, $y = 13$, $z = 6$ for an arbitrary $n = 20$    *33*

TABLE II.5—continued

Polymers from Diunsaturated Esters

| No. | Structure | Method | Solubility | Molecular weight | $T_m$ (°C) | Remarks and property data | References |
|---|---|---|---|---|---|---|---|
| 15 | Poly(methallyl crotonate) | D,1 | Glacial $CH_3COOH$ | $\eta_{inh}$ 0.039 | — | Degree of cyclization = 31%. Methallyl double-bond content = 11%. Crotonate double-bond content = 58% | 32 |
| 16 | Poly(methallyl vinyl acetate) | D,1 | — | — | — | Viscous oil. Structure uncertain | 32 |
| 17 | Poly(ethyl 2-methacryloyloxy acrylate) | D,1 | "Soluble" | — | — | Cyclic, linear structure supported by solubility, IR, and chemical evidence | 89 |
| 18 | Poly(methyl-2-butenyl maleate) | D,1 | Glacial $CH_3COOH$, $CHCl_3$, acetone, $CH_3CN$, $CCl_4$ | $\eta_{inh}$ 0.036 | 110 | Analysis for $n = 20$: $x = 4$, $y = 12$; $z = 4$. IR indicates $\gamma$-butyrolactone rings | 33 |

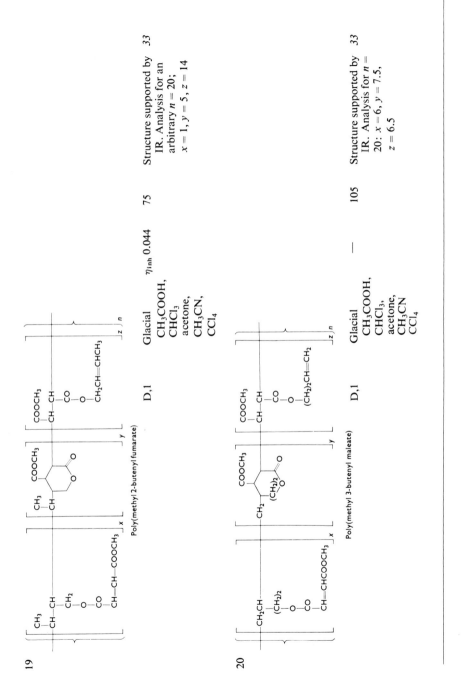

19

Poly(methyl 2-butenyl fumarate)

D,1    $\eta_{inh}$ 0.044    75

Glacial
$CH_3COOH$,
$CHCl_3$
acetone,
$CH_3CN$,
$CCl_4$

Structure supported by
IR. Analysis for an
arbitrary $n = 20$;
$x = 1$, $y = 5$, $z = 14$    33

20

Poly(methyl 3-butenyl maleate)

D,1    —    105

Glacial
$CH_3COOH$,
$CHCl_3$,
acetone,
$CH_3CN$
$CCl_4$

Structure supported by
IR. Analysis for $n =$
20: $x = 6$, $y = 7.5$,
$z = 6.5$    33

TABLE II.5—*continued*

Polymers from Diunsaturated Esters

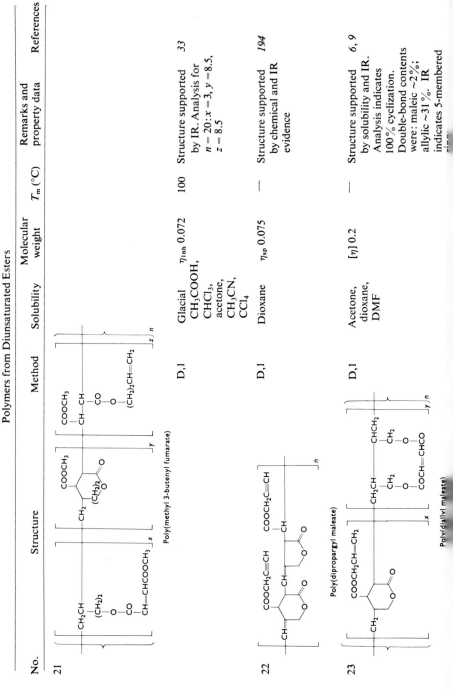

| No. | Structure | Method | Solubility | Molecular weight | $T_m$ (°C) | Remarks and property data | References |
|---|---|---|---|---|---|---|---|
| 21 | Poly(methyl 3-butenyl fumarate) | D,1 | Glacial $CH_3COOH$, $CHCl_3$, acetone, $CH_3CN$, $CCl_4$ | $\eta_{inh}$ 0.072 | 100 | Structure supported by IR. Analysis for $n = 20$: $x = 3$, $y = 8.5$, $z = 8.5$ | 33 |
| 22 | Poly(dipropargyl maleate) | D,1 | Dioxane | $\eta_{sp}$ 0.075 | — | Structure supported by chemical and IR evidence | 194 |
| 23 | Poly(diallyl maleate) | D,1 | Acetone, dioxane, DMF | [η] 0.2 | — | Structure supported by solubility and IR. Analysis indicates 100% cyclization. Double-bond contents were: maleic ~2%; allylic ~31%. IR indicates 5-membered | 6, 9 |

| 24 | Poly(*trans*-vinyl cinnamate) | D,1 | CHCl$_3$ | [η] 0.7 | — | Soluble polymer obtained at conversions <50%. IR indicates γ-lactone rings. Degree of cyclization = 80–85%. Noncyclized portion of polymer derived from vinyl polymerization | 11, 230, 238 |
| | Copolymer of 24 with: | | | | | | |
| 25 | Methacrylonitrile | D,1 | Acetone | — | — | Copolymers contained 54–95.5% mole of methacrylonitrile. Degree of cyclization of the *trans*-vinyl cinnamate was 18–60%. Copolymerization parameters. Copolymers did not contain appreciable vinyl unsaturation | 238 |
| 26 | Vinyl acetate | D,1 | Toluene | — | — | Copolymers contained 24–44% mole of vinyl acetate. Degree of cyclization of *trans*-vinyl cinnamate was 80–87.5%. Copolymerization parameters. Polymers did not contain appreciable vinyl unsaturation | 238 |

TABLE II.5—*continued*
Polymers from Diunsaturated Esters

| No. | Structure | Method | Solubility | Molecular weight | $T_m$ (°C) | Remarks and property data | References |
|---|---|---|---|---|---|---|---|
| 27 | Vinylpyrrolidone | D,1 | Acetone | — | — | Copolymers contained 6.5–48.5% mole of vinylpyrrolidone. Degree of cyclization of the *trans*-vinyl cinnamate was 81–89%. Copolymerization parameters. No significant vinyl unsaturation present | 238 |
| 28 | Styrene | D,1 | Butanone | — | — | Copolymers contained 34–88% mole of styrene; copolymerization parameters. IR indicated significant amounts of pendant vinyl groups | 238 |
| 29 | Poly(diallyl muconate) | D,1 | — | — | >230 | — | 8 |

TABLE II.6

Polymers from Diunsaturated Anhydrides

| No. | Structure | Method | Solubility | Molecular weight | $T_m$ (°C) | Remarks and property data | References |
|---|---|---|---|---|---|---|---|
| 1 | Poly(propargylic anhydride) | E,4 | DMF, $H_2SO_4$, aq. NaOH | 25,000 (end-group analysis) | — | Structure supported by IR and solubility. ESR signal ~5 × 10[17] spins/g; conductivity of ~10⁻⁹ ohm⁻¹ cm⁻¹ | 100, 284 |
| 2 | Poly(acrylic anhydride) | E,1 | DMF, DMSO, γ-butyro-lactone, N-methyl-pyrrolidone | 205,000 (viscosity measure-ments) | 220 | Low molecular weight polymer is soluble in acetone, dioxane, and ethyl acetate. Structure supported by solubility, IR, and chemical reactions | 94, 95, 139, 147, 148 |
|  | Copolymers of acrylic anhydride with: | | | | | | |
| 3 | Methacrylonitrile | E,1 | — | — | — | — | 253 |
| 4 | Styrene | E,1 | — | — | — | — | 253 |
| 5 | Poly(acrylic methacrylic anhydride) | E,1 | DMF, DMSO | $\bar{M}_w$ up to 850,000 obtained | — | Excellent adhesive for glass | 138, 139 |

**TABLE II.6—continued**

Polymers from Diunsaturated Anhydrides

| No. | Structure | Method | Solubility | Molecular weight | $T_m$ (°C) | Remarks and property data | References |
|-----|-----------|--------|------------|------------------|------------|---------------------------|------------|
| 6 | Poly(methacrylic anhydride) | E,1 | DMF, DMSO | $\overline{M}_v = 213{,}000$ (viscosity) | 172 to 200 | Structure supported by solubility, spectral, and chemical studies. Polymer prepared by bulk, "cast" polymerization had following properties: sp. gravity, 25°C/25°C = 1.256; refractive index, $n^{25} = 1.520$ Vicat softening temperature (10 mils) = 159°C; $T_g = 144$°C; Barcol hardness = 64 | 14, 57, 112, 113, 137, 139 |
| | Copolymers of methacrylic anhydride with: | | | | | | |
| 7 | Acrylonitrile | E,1 | — | — | — | Various compositions prepared. Depending on the copolymer, its composition, and the experimental conditions the materials had variable solubility in DMSO. Degree of cyclization varied with co-monomer. | 140 |
| 8 | Allyl chloride | E,1 | — | — | — | | 253 |
| 9 | Methacrylonitrile | E,1 | — | — | — | | 253 |
| 10 | Vinyl acetate | E,1 | — | — | — | | 140 |
| 11 | β-Chloroethyl vinyl ether | E,1 | — | — | — | | 140 |
| 12 | Allylurea | E,1 | — | — | — | Copolymerization parameters determined | 140 |

| | | | | | | | |
|---|---|---|---|---|---|---|---|
| 13 | Allyl chloroacetate | E,1 | — | E,1 | — | ⎱ Copolymerization parameters determined | 140 |
| 14 | Ethyl acrylate | E,1 | — | E,1 | — | Various compositions | 140 |
| 15 | Methyl methacrylate | E,1 | — | E,1 | — | prepared. Depending | 140 |
| 16 | Vinyl n-butyl sulfone | E,1 | — | E,1 | — | on the copolymer, its | 140 |
| 17 | Styrene | E,1 | — | E,1 | — | composition, and the experimental condi- | 140, 253 |
| 18 | Diisobutylene (mixture of isomeric trimethylpentenes) | E,1 | — | E,1 | — | tions, the materials had variable solubility in DMSO | 140 |
| 19 | Vinyl benzyl sulfide | E,1 | — | E,1 | — | ⎱ Copolymerization parameters determined | 140 |
| 20 | Allyl benzyl ether | E,1 | — | E,1 | — | | 140 |
| 21 | n-Hexyl methacrylate | E,1 | — | E,1 | — | | 140 |
| 22 | Lauryl methacrylate | E,1 | — | E,1 | — | — | 140 |
| 23 | Poly(methacrylic crotonic anhydride) | E,1 | — | — | — | Structure supported by IR | 153 |

* * *

Poly(methacrylic crotonic anhydride)

TABLE II.7

Polymers from Polyunsaturated Ammonium Salts and Amine Oxides

| No. | Structure | Method | Solubility | Molecular weight | $T_m$ (°C) | Remarks and property data | References |
|---|---|---|---|---|---|---|---|
| 1 | 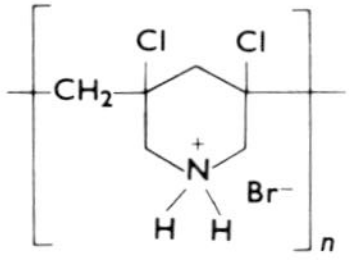 | G,1 | — | — | — | — | *73* |
| 2 | 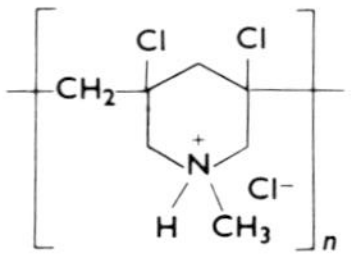 | G,1 | — | — | — | — | *59* |
| 3 | 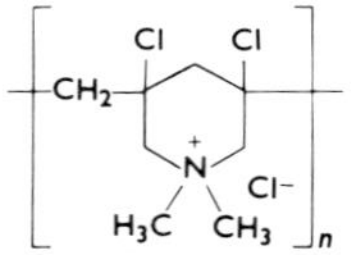 | G,1 | $H_2O$, aq. salts | — | — | — | *60, 69* |
| 4 | 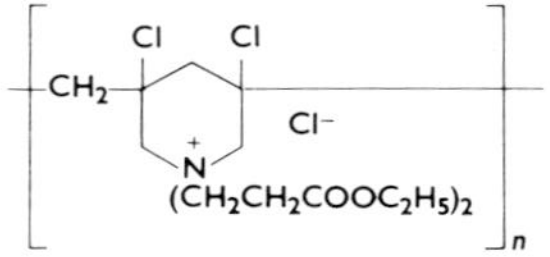 | G,1 | — | — | — | — | *60, 69* |
| 5 | 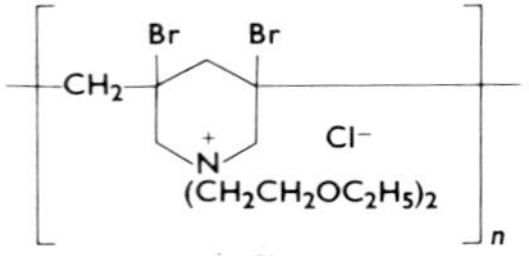 | G,1 | — | — | — | — | *60, 69* |

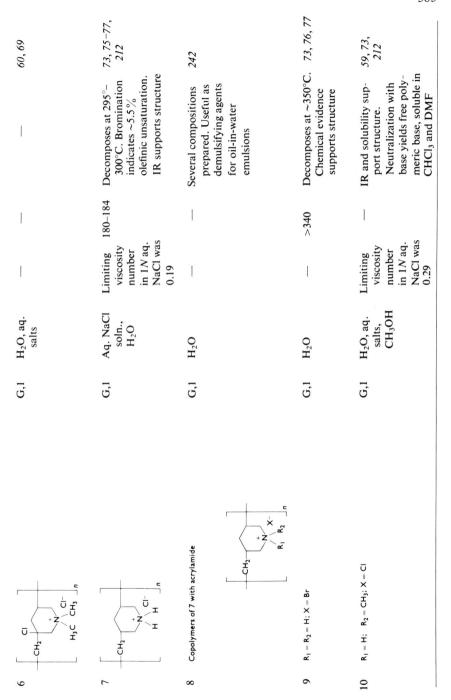

| No. | Structure | Method | Solvents | Viscosity/other | m.p./dec. | Remarks | References |
|---|---|---|---|---|---|---|---|
| 6 | | G,1 | H₂O, aq. salts | — | — | — | 60, 69 |
| 7 | | G,1 | Aq. NaCl soln., H₂O | Limiting viscosity number in 1N aq. NaCl was 0.19 | 180–184 | Decomposes at 295°–300°C. Bromination indicates ~5.5% olefinic unsaturation. IR supports structure | 73, 75–77, 212 |
| 8 | Copolymers of 7 with acrylamide | G,1 | H₂O | — | — | Several compositions prepared. Useful as demulsifying agents for oil-in-water emulsions | 242 |
| 9 | R₁ = R₂ = H; X = Br | G,1 | H₂O | — | >340 | Decomposes at ~350°C. Chemical evidence supports structure | 73, 76, 77 |
| 10 | R₁ = H; R₂ = CH₃; X = Cl | G,1 | H₂O, aq. salts, CH₃OH | Limiting viscosity number in 1N aq. NaCl was 0.29 | — | IR and solubility support structure. Neutralization with base yields free polymeric base, soluble in CHCl₃ and DMF | 59, 73, 212 |

TABLE II.7—*continued*

Polymers from Polyunsaturated Ammonium Salts and Amine Oxides

| No. | Structure | Method | Solubility | Molecular weight | $T_m$ (°C) | Remarks and property data | References |
|---|---|---|---|---|---|---|---|
| 11 | $R_1 = H; R_2 = C_2H_5; X = Cl$ | G,1 | $H_2O$, aq. salts, $CH_3OH$ | Limiting viscosity number in $1N$ aq. NaCl was 0.20 | — | IR and solubility support structure | 212 |
| 12 | $R_1 = H; R_2 = n\text{-}C_3H_7; X = Cl$ | G,1 | $H_2O$, aq. salts, $CH_3OH$ | Limiting viscosity number in $1N$ aq. NaCl was 0.15 | — | IR and solubility support structure | 212 |
| 13 | $R_1 = H; R_2 = n\text{-}C_4H_9; X = Cl$ | G,1 | $H_2O$, aq. salts, $CH_3OH$ | Limiting viscosity number in $1N$ aq. NaCl was 0.18 | — | IR and solubility support structure | 212 |
| 14 | $R_1 = R_2 = CH_3; X = Cl$ | G,1 | $H_2O$, aq. salts, $CH_3OH$ | $[\eta]$ 1.35 ($0.1N$ aq. KCl) | 348–352 | Structure supported by IR and chemical reactions. Contains residual unsaturation | *40, 60, 69, 72, 76, 77, 212, 213* |
| 15 | Copolymers of 14 with acrylonitrile | G,1 | $H_2O$, aq. salts, DMF | Up to 300,000 by viscosity measurements | — | Several compositions prepared. One (94% acrylonitrile) was spun to fibers | 244 |

| No. | Structure | | Solvent | Molecular weight | | Comments | References |
|---|---|---|---|---|---|---|---|
| 16 | Copolymers of 14 with acrylamide | G,1 | $H_2O$ | Up to 1,000,000 via viscosity measurements | — | Several compositions prepared. Strengthening agents for paper and for removal of suspended matter from water | 213, 244 |
| 17 | $R_1 = R_2 = CH_3$; X = Br | G,1 | $H_2O$, aq. salts | $[\eta]$ 0.17 (0.1$N$ aq. KCl) | — | IR supports structure | 60, 69, 76, 77, 212 |
| 18 | $R_1 = CH_3$; $R_2 = CH_2COO$: No X; copolymer with acrylonitrile and vinyl acetate | G,1 | $H_2O$, aq. salts, DMF | "High" (viscosity measurements) | — | Initial monomer charge was: 85% acrylonitrile, 5% vinyl acetate, 10% diallyl compound | 244 |
| 19 | $R_1 = CH_3$; $R_2 = CH_2CH_2CN$; X = $CH_3SO_3$ Copolymers with acrylamide | G,1 | — | — | — | — | 214 |
| 20 | $R_1 = CH_3$; $R_2 = CH_2CH_2CONH_2$; X = Cl | G,1 | $H_2O$ | — | — | Shows excellent flocculating activity on suspensions of colloidal clay | 214 |
| 21 | Copolymers of 20 with acrylamide | G,1 | — | — | — | — | 214 |
| 22 | $R_1 = CH_3$; $R_2 = CH_2CH_2COOCH_3$; X = $CH_3SO_3$ Copolymers with acrylamide | G,1 | — | — | — | — | 214 |
| 23 | $R_1 = CH_3$; $R_2 = \\-(CH_2)_3 \overset{+}{N}(CH_3)_3$; X = Cl | G,1 | $H_2O$, aq. salts | — | — | — | 60, 69 |

TABLE II.7—continued

Polymers from Polyunsaturated Ammonium Salts and Amine Oxides

| No. | Structure | Method | Solubility | Molecular weight | $T_m$ (°C) | Remarks and property data | References |
|---|---|---|---|---|---|---|---|
| 24 | $R_1 = CH_3$; $R_2 = +CH_2)_3 \overset{+}{N}(CH_3)_3$; $X = Cl$ | G,1 | $H_2O$, aq. salts | — | — | — | 60, 69 |
| 25 | $R_1 = CH_3$; $R_2 = C_6H_5$; $X = Cl$ | G,1 | $H_2O$, aq. salts | — | — | — | 60, 69 |
| 26 | $R_1 = R_2 = C_2H_5$; $X = Cl$ | G,1 | $H_2O$, aq. salts, ethanol, $CH_3OH$ | $[\eta]$ 1.32 (0.1N aq. KCl) | 346–354 d | Contains residual un-saturation. Quantitative hydrogenation indicates that one out of five units is not cyclized. IR supports structure | 60, 69, 72, 212 |
| 27 | $R_1 = R_2 = C_2H_5$; $X = Br$ | G,1 | $H_2O$, aq. salts, $CH_3OH$ | Limiting viscosity number in 1N aq. NaCl was 0.41 | — | Solubility and IR support structure | 212 |
| 28 | $R_1 = R_2 = CH_2CH_2OH$; $X = Cl$ | G,1 | $H_2O$, aq. salts | $[\eta]$ 1.28 | — | — | 60, 69 |
| 29 | $R_1 = R_2 = CH_2COCH_3$; $X = Cl$ | G,1 | — | — | — | — | 60, 69 |
| 30 | $R_1 = R_2 = CH_2CH(NO_2)CH_3$; $X = Cl$ | G,1 | — | — | — | — | 60, 69 |

| No. | Structure | | | | | | Ref. |
|---|---|---|---|---|---|---|---|
| 31 | $R_1 = R_2 = -(CH_2)_4F$; $X = Cl$ | G,1 | — | — | — | — | 60, 69 |
| 32 | $R_1 = R_2 = n\text{-}C_4H_9$; $X = Cl$ | G,1 | $H_2O$, aq. salts | — | — | — | 60, 69 |
| 33 | $R_1 = R_2 = CH_2SC_3H_7$; $X = Cl$ | G,1 | — | — | — | — | 60, 69 |
| 34 | $R_1 = R_2 = n\text{-}C_{12}H_{21}$; $X = Cl$ | G,1 | $H_2O$, aq. salts | — | — | — | 60, 69 |
| 35 | $R_1 = R_2 = CH_2$ [cyclohexyl, H]; $X = Cl$ | G,1 | — | — | — | — | 60, 69 |
| 36 | $R_1 = R_2 = $ Methoxyphenyl; $X = Cl$ | G,1 | $H_2O$, aq. salts | — | — | — | 60, 69 |
| 37 | $R_1 = R_2 = CH_2CH_2OC_6H_5$; $X = Cl$ | G,1 | — | — | — | — | 60, 69 |
| 38 | $R_1 = R_2 = $ Naphthoxyethyl; $X = Cl$ | G,1 | — | — | — | — | 60, 69 |
| 39 | * * * | G,1 | — | — | — | — | |
| 40 | [polymer structure] | G,1 | — | — | — | — | 60, 69 |

TABLE II.7—continued

Polymers from Polyunsaturated Ammonium Salts and Amine Oxides

| No. | Structure | Method | Solubility | Molecular weight | $T_m$ (°C) | Remarks and property data | References |
|---|---|---|---|---|---|---|---|
| 41 | | G,1 | $H_2O$, aq. salts | — | — | — | 60, 69 |
| 42 | | G,1 | $H_2O$, aq. salts | [η] 1.46 (0.1N aq. NaCl) | — | — | 60, 69, 212 |
| 43 | Bromide corresponding to polymer 42 | G,1 | $H_2O$, aq. salts | Limiting viscosity number in 1N aq. NaCl was 0.30 | — | IR and solubility support structure | 212 |
| 44 | | G,1 | — | — | — | — | 60, 69 |
| 45 | | G,1 | $H_2O$, aq. salts, $CH_3OH$ | [η] 1.26 (0.1N aq. KCl) | — | Structure supported by IR | 60, 69, 212 |

| | | | Limiting | Structure supported by IR | |
|---|---|---|---|---|---|
| 46 | Bromide corresponding to polymer 45 | G,1 | H₂O, CH₃OH, aq. salts | viscosity number in 1$N$ aq. NaCl was 0.14 | | 212 |

Poly(N-allyl-N-methyl-2-vinylpyrrolidinium chloride)

| | | | | | |
|---|---|---|---|---|---|
| 47 | | G,1 | H₂O, aq. salts | — | — | 60, 69 |

Poly (N-allyl-N-methyl-2-vinylmorpholinium chloride)

| | | | | | |
|---|---|---|---|---|---|
| 48 | | G,1 | H₂O, aq. salts | — | — | 60, 69 |

Poly(N-allyl-N-methyl-2-vinylpiperidinium chloride)

| | | | | | |
|---|---|---|---|---|---|
| 49 | | G,1 | H₂O, aq. salts | — | — | 60, 69 |

## TABLE II.7—*continued*

### Polymers from Polyunsaturated Ammonium Salts and Amine Oxides

| No. | Structure | Method | Solubility | Molecular weight | $T_m$ (°C) | Remarks and property data | References |
|---|---|---|---|---|---|---|---|
| 50 | | G,1 | H$_2$O, aq. salts | — | — | — | 60, 69 |
| | Poly(N,N-dimethyl-2,5-divinylpyrrolidinium chloride) | | | | | | |
| 51 | R = R′ = m-C$_2$H$_5$SC$_6$H$_4$ | G,1 | — | — | — | — | 60, 69 |
| 52 | R = C$_6$H$_5$SCH$_2$CH$_2$; R′ = HOCH$_2$CH$_2$ | G,1 | — | — | — | — | 60, 69 |
| 53 | R = R′ = CH$_3$; poly(N,N-dimethyl-2,5-divinylmor-pholinium chloride) | G,1 | H$_2$O, aq. salts | — | — | — | 60, 69 |

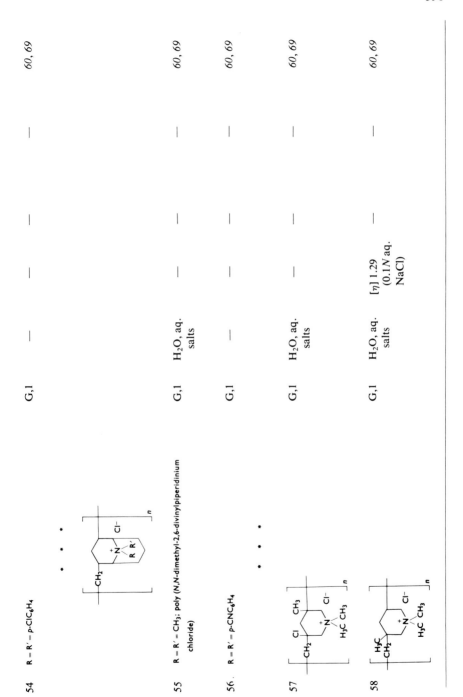

54  R = R' = p-ClC₆H₄

G,1 — — — — 60, 69

55  R = R = CH₃; poly (N,N-dimethyl-2,6-divinylpiperidinium chloride)

G,1 H₂O, aq. salts — — — 60, 69

56  R = R' = p-CNC₆H₄

G,1 — — — — 60, 69

57

G,1 H₂O, aq. salts — — — 60, 69

58

G,1 H₂O, aq. salts $[\eta]$ 1.29 (0.1$N$ aq. NaCl) — — 60, 69

## TABLE II.7—continued
### Polymers from Polyunsaturated Ammonium Salts and Amine Oxides

| No. | Structure | Method | Solubility | Molecular weight | $T_m$ (°C) | Remarks and property data | References |
|-----|-----------|--------|------------|------------------|------------|---------------------------|------------|
| 59 | | G,1 | — | — | — | — | 60, 69 |
| 60 | | G,1 | — | — | — | — | 73 |
| 61 | | G,1 | — | — | — | — | 59 |
| 62 |  $R_1 = R_2 = H; X = Cl$ | G,1 | — | — | 285–300 | — | 73, 186 |

| | | | | | | | |
|---|---|---|---|---|---|---|---|
| 63 | $R_1 = H; R_2 = CH_3; X = Cl$ | G,1 | — | — | — | — | *59* |
| 64 | $R_1 = R_2 = CH_3; X = Cl$ | G,1 | $H_2O$, aq. salts | — | — | — | *60, 69* |
| 65 | Copolymer of 64 with acrylonitrile and vinyl acetate | G,1 | Aq. NaSCN; DMF | — | — | Monomer charge used was 85% acrylonitrile, 5% vinyl acetate, 10% of the diallyl compound | *244* |
| 66 | $R_1 = R_2 = CNCH_2CH_2; X = Cl$ | G,1 | — | — | — | — | *60, 69* |
| 67 | $R_1 = R_2 = CH_2CH_2CONH_2; X = Cl$ | G,1 | — | — | — | — | *60, 69* |
| 68 | $R_1 = R_2 =$ Cyclopentylmethyl; $X = Cl$ | G,1 | — | — | — | — | *60, 69* |
| 69 | $R_1 = R_2 = CH_2CH_2SC_6H_5; X = Cl$ | G,1 | — | — | — | — | *60, 69* |
| | * * * | | | | | | |
| 70 | Poly(triallylamine hydrobromide) | G,1 | $H_2O$, DMF | $[\eta]$ 0.1 | 190–220 | Contains ~4.7% of residual unsaturation | *183* |
| 71 | Poly(trimethallylamine hydrochloride) | G,1 | "Soluble" | — | 120–130 | IR and solubility support cyclic structure | *186* |
| 72 | Poly(triallylethylammonium bromide) | G,1 | $H_2O$, $CH_3OH$ | $[\eta]$ 0.14 | — | Solubility and low residual unsaturation support cyclic structure | *272* |

TABLE II.7—*continued*

Polymers from Polyunsaturated Ammonium Salts and Amine Oxides

| No. | Structure | Method | Solubility | Molecular weight | $T_m$ (°C) | Remarks and property data | References |
|---|---|---|---|---|---|---|---|
| 73 | Poly(tetraallylammonium bromide) | G,l | $H_2O$, $CH_3OH$ | — | — | Solubility and low residual unsaturation support cyclic structure | 272 |
| 74 | Copolymer from diallylmethylamine N-oxide, acrylonitrile, and vinyl acetate | G,l | Aq. NaSCN, DMF | "High" (viscosity) | — | Cyclization of the diallyl moiety (supported by polymer solubility) is postulated. Monomer charge was 85% acrylonitrile, 5% vinyl acetate, and 10% diallyl compound | 244 |

**TABLE II.8**

Copolymers from Diunsaturated Ammonium Salts and Sulfur Dioxide

| No. | Diallylammonium salt | Method | Solubility | Molecular weight | $T_m$ (°C) | Remarks and property data | References |
|---|---|---|---|---|---|---|---|
| 1 | $R_1 = R_2 = H; X = Cl$ | G,1 | Aq. salts, DMSO, $H_2O$ | $\eta_{inh}$ 1.23 (0.1$N$ aq. NaCl) | — | Ratio $SO_2$/diallyl compound = 1. Cyclic structure supported by IR and solubility | 39, 127 |
| 2 | $R_1 = H; R_2 = CH_3; X = Cl$ | G,1 | 0.1$N$ aq. NaCl, $H_2O$ at pH ⩽ 7 | $\eta_{inh}$ 0.90 | >180 d | Ratio $SO_2$/diallyl compound = 1. For structural considerations, see text | 126 |
| 3 | $R_1 = H; R_2 = CH_3; X = CH_3COO$ | G,1 | — | — | — | Ratio $SO_2$/diallyl compound = 1. For structural considerations, see text | 39 |
| 4 | $R_1 = H; R_2 = C_2H_5; X = Cl$ | G,1 | 0.1$N$ aq. NaCl, $H_2O$ at pH ⩽ 7 | $\eta_{inh}$ 1.12 | >180 d | Ratio $SO_2$/diallyl compound = 1. For structural considerations, see text | 126 |
| 5 | $R_1 = H; R_2 = CH_2CH_2OH; X = Cl$ | G,1 | — | $\eta_{inh}$ 0.12 | — | Ratio $SO_2$/diallyl compound = 1 | 39 |

TABLE II.8—*continued*

Copolymers from Diunsaturated Ammonium Salts and Sulfur Dioxide

| No. | Diallylammonium salt | Method | Solubility | Molecular weight | $T_m$ (°C) | Remarks and property data | References |
|-----|----------------------|--------|------------|------------------|------------|---------------------------|------------|
| 6 | $R_1 = H$; $R_2 = n\text{-}C_3H_7$; $X = Cl$ | G,l | 0.1$N$ aq. NaCl, $H_2O$ at pH $\leqslant 7$ | $\eta_{inh}$ 1.90 | >180 d | Ratio $SO_2$/diallyl compound = 1. See text for structure of these copolymers. Stability at 190°C studied | *126* |
| 7 | $R_1 = H$; $R_2 = iso\text{-}C_3H_7$; $X = Cl$ | G,l | 0.1$N$ aq. NaCl, $H_2O$ at pH $\leqslant 7$. Slightly soluble in $CH_3OH$ | $\eta_{inh}$ 1.94 | >180 d | Ratio $SO_2$/diallyl compound = 1. See text for structure of these copolymers. Stability at 190°C studied | *126* |
| 8 | $R_1 = H$; $R_2 = n\text{-}C_4H_9$; $X = Cl$ | G,l | 0.1$N$ aq. NaCl, $H_2O$ at pH $\leqslant 7$, $CH_3OH$, DMSO | $\eta_{inh}$ 1.31 | >180 d | Ratio $SO_2$/diallyl compound = 1. See text for structure of these copolymers. Stability at 190°C studied | *126* |
| 9 | $R_1 = H$; $R_2 = tert\text{-}C_4H_9$; $X = Cl$ | G,l | 0.1$N$ aq. NaCl, $H_2O$ at pH $\leqslant 7$, $CH_3OH$ | $\eta_{inh}$ 1.45 | >180 d | Ratio $SO_2$/diallyl compound = 1. See text for structure of these copolymers. Stability at 190°C studied | *126* |

| No. | Structure | Type | Solubility | $\eta_{inh}$ | Stability | Remarks | Ref. |
|---|---|---|---|---|---|---|---|
| 10 | $R_1 = H$; $R_2 = CH_2CH_2CN$; $X = Cl$ | G,1 | 0.1$N$ aq. NaCl, $H_2O$ at pH ≤ 7, DMSO | $\eta_{inh}$ 0.1 | >180 d | Ratio $SO_2$/diallyl compound = 1. See text for structure of these copolymers. Stability at 190°C studied | 126 |
| 11 | $R_1 = H$; $R_2 = $ Cyclohexyl; $X = Cl$ | G,1 | 0.1$N$ LiCl soln. in DMSO, hot $H_2O$ at pH ≤ 7, $CH_3OH$ | $\eta_{inh}$ 0.78 | >180 d | Ratio $SO_2$/diallyl compound = 1. See text for structure of these copolymers. Stability at 190°C studied | 126 |
| 12 | $R_1 = H$; $R_2 = C_6H_5$; $X = CH_3COO$ | G,1 | — | $\eta_{inh}$ 0.62 | — | Ratio $SO_2$/diallyl compound = 1 | 39 |
| 13 | $R_1 = H$; $R_2 = C_6H_5CH_2$; $X = Cl$ | G,1 | 0.1$N$ LiCl soln. in DMSO, hot $H_2O$ at pH ≤ 7, DMSO, slightly soluble in DMF | $\eta_{inh}$ 0.45 | >180 d | Ratio $SO_2$/diallyl compound = 1. See text for structures of these copolymers. Stability at 190°C studied | 39, 126 |
| 14 | $R_1 = R_2 = CH_3$; $X = Cl$ | G,1 | 0.1$N$ aq. NaCl, $H_2O$ | $\eta_{inh}$ 1.22 | >180 d | Ratio $SO_2$/diallyl compound = 1. For structures of these copolymers, see text. Hydrolytic (aq. sodium hydroxide) and thermal stabilities studied | 39, 125 |

TABLE II.8—*continued*

Copolymers from Diunsaturated Ammonium Salts and Sulfur Dioxide

| No. | Diallylammonium salt | Method | Solubility | Molecular weight | $T_m$ (°C) | Remarks and property data | References |
|---|---|---|---|---|---|---|---|
| 15 | $R_1 = R_2 = C_2H_5$; $X = Cl$ | G,1 | 0.1$N$ aq. NaCl, $H_2O$, $CH_3OH$, $C_2H_5OH$, DMSO | $\eta_{inh}$ 1.61 | >180 d | Ratio $SO_2$/diallyl compound = 1. See text for structures of these copolymers. Stability at 190°C studied | 39, 126 |
| 16 | $R_1 = R_2 = C_2H_5$; $X = Br$ | G,1 | — | $\eta_{inh}$ 0.14 | — | Ratio $SO_2$/diallyl compound = 1 | 39 |
| 17 | | G,1 | 0.1$N$ aq. NaCl, $H_2O$ | $\eta_{inh}$ 0.83 | >180 d | Ratio $SO_2$/diallyl compound depends on composition of monomer feed. See text for structures of these copolymers. Thermal stability at 190°C studied | 126 |
| 18 | | G,1 | 0.1$N$ aq. NaCl, $H_2O$ | $\eta_{inh}$ 0.61 | >180 d | Ratio $SO_2$/diallyl compound depends on composition of monomer feed. See text for structures of these copolymers. Thermal stability at 190°C studied | 126 |

19

G,1 — $\eta_{inh}$ 0.45 — Ratio SO$_2$/diallyl compound = 1  39

TABLE II.9
Polymers from Diunsaturated Amines

| No. | Structure | Method | Solubility | Molecular weight | $T_m$ (°C) | Remarks and property data | References |
|---|---|---|---|---|---|---|---|
| 1 | Homopolymer from divinylaniline obtained in bulk | G,1 | $C_6H_6$ | $\eta_{inh}$ 0.09; mol. wt. ~5,000–7,000 (cryos.) | 160–170 | Structure supported by chemical evidence and solubility | 86 |
| 2 | Homopolymer from divinylaniline obtained in solution | G,1 | $C_6H_6$ | $\eta_{inh}$ 0.05 mol. wt ~4,300 (cryos.) | 150–160 | Contains negligible amount of residual unsaturation | 86 |
| | Copolymers from divinylaniline | | | | | | |
| 3 | With acrylonitrile: $R_1$ = H; $R_2$ = CN | G,1 | DMF | — | — | Polymer structure supported by IR, hydrogenation, and solubility. Copolymerization parameters studied | 86 |

| # | | G,1 | Solvent | Properties | Notes | Ref. |
|---|---|---|---|---|---|---|
| 4 | With vinyl acetate: $R_1 = H$; $R_2 = CH_3COO$ | G,1 | $C_6H_6$, DMF | — | — | Polymer structure supported by IR, hydrogenation, and solubility. Copolymerization parameters studied | *86* |
| 5 | With methyl methacrylate: $R_1 = CH_3$; $R_2 = COOCH_3$ | G,1 | $C_6H_6$, DMF | $\eta_{Inh}$ 0.153; mol. wt. ~10,000–22,000 (cryos.) | — | Polymer structure supported by IR, hydrogenation and solubility. Copolymerization parameters studied | *86* |
| 6 | With styrene: $R_1 = H$; $R_2 = C_6H_5$ | G,1 | $C_6H_6$, DMF | $\eta_{Inh}$ 0.36 | — | Polymer structure supported by IR, hydrogenation, and solubility. Copolymerization parameters studied | *86* |
| 7 | With p-methylstyrene: $R_1 = H$; $R_2 = p\text{-}CH_3C_6H_4$ | G,1 | $C_6H_6$, DMF | — | — | Polymer structure supported by IR, hydrogenation, and solubility. Copolymerization parameters studied | *86* |
| 8 | With diethyl fumarate | G,1 | — | — | — | — | *86* |

* * *

TABLE II.9—*continued*
Polymers from Diunsaturated Amines

| No. | Structure | Method | Solubility | Molecular weight | $T_m$ (°C) | Remarks and property data | References |
|---|---|---|---|---|---|---|---|
| 9 | Polydiallylamine | G,1 | DMSO, dilute HCl, toluene | $[\eta]$ 0.15 | — | — | 150 |
| | Copolymers from diallylamine with | | | | | | |
| 10 | Methyl methacrylate | G,1 | — | — | — | Ratio of co-monomers (molar), 1:1 | 150 |
| 11 | 2-Methyl-5-vinylpyridine | G,1 | — | — | — | Molar ratio of co-monomers, 1:1 | 150 |
| 12 | Acrylonitrile and vinyl acetate | G,1 | Aq. NaSCN, DMF | "High" (viscosity) | — | Cyclization of diallyl moiety postulated. Monomer charge was 85% acrylonitrile, 5% vinyl acetate, 10% diallylamine | 244 |
| 13 | Poly(N,N-diallylcyanamide) | G,1 | Phenol, m-cresol, hot pyridine | DP = 7 (cryos.) | 400–450 (PMT = 300) | Good stability up to 350°C. Bromination indicated 1 double bond per polymer chain | 183, 223, 282 |
| 14 | Copolymers from N,N-diallylurea with lauryl methacrylate | G,1 | Dioxane, $CHCl_3$, $C_6H_6$ | — | — | IR and reactivity ratio studies indicate cyclopolymerization of the N,N-diallylurea | 48 |

| No. | Polymers from | | Solvents | "High"/"Low" (viscosity) | | Comments | Ref. |
|---|---|---|---|---|---|---|---|
| 15 | Copolymer from diallyl methylamine, acrylonitrile, and vinyl acetate | G,1 | Aq. NaSCN, DMF | "High" (viscosity) | — | Cyclization of diallyl moiety postulated. Monomer charge was 85% acrylonitrile, 5% vinyl acetate, and 10% diallylamine | 244 |

Polymers from

| No. | R | | Solvents | | | Comments | Ref. |
|---|---|---|---|---|---|---|---|
| 16 | $R = SO_2CH_3$ | G,1 | "Variety of solvents" | $\bar{M}_n = 2,300$ | 116 | IR indicates absence of residual unsaturation | 96 |
| 17 | $R = CCl_3CO$ | G,1 | Acetone, CHCl$_3$, alcohols | "Low" | — | Cyclic structure supported by IR | 223 |
| 18 | $R = ClCH_2CO$ | G,1 | CHCl$_3$, C$_6$H$_6$ | — | 70–107 | Cyclic structure supported by IR | 182, 223 |
| 19 | $R = CH_2CN$ | G,1 | Acetone, CHCl$_3$, alcohols | "Low" | — | Cyclic structure supported by IR | 223 |
| 20 | $R = CH_3CO$ | G,1 | CHCl$_3$, C$_6$H$_6$ | Mol.wt. (ebull.) = 22,650 | 85–95 | Cyclic structure supported by IR | 182, 223 |

TABLE II.9—*continued*
Polymers from Diunsaturated Amines

| No. | Structure | Method | Solubility | Molecular weight | $T_m$ (°C) | Remarks and property data | References |
|-----|-----------|--------|-----------|------------------|------------|---------------------------|------------|
| 21 | R = CH₃OCO | G,1 | CHCl₃, C₆H₆ | Mol. wt. (ebull.) 13,680 | 76–98 | Cyclic structure supported by IR | *182* |
| 22 | R = SO₂C₂H₅ | G,1 | "Variety of solvents | $\bar{M}_n$ = 1,840 | 43 | Cyclic structure supported by IR | *96* |
| 23 | R = CH₂CH₂CN | G,1 | Acetone, chloroform, alcohols | "Low" | — | Cyclic structure supported by IR | *223* |
| 24 | R = CH₃CH₂CO | G,1 | CHCl₃, C₆H₆ | — | 66–78 | Cyclic structure supported by IR | *182* |
| 25 | R = SO₂C₆H₅ | G,1 | CHCl₃ | Mol. wt. = 13,750 (ebull.) | 103–125 | Cyclic structure supported by IR | *182* |
| 26 | R = SO₂C₆H₄CH₃-p | G,1 | CHCl₃ | Mol. wt. = 7,260 (ebull.) | 105–120 | Cyclic structure supported by IR | *182, 223* |

| No. | R | | Solvent | Mol. wt. / property | m.p. | Comments | Refs. |
|---|---|---|---|---|---|---|---|
| 27 | R = $C_6H_5CO$ | G,1 | $CHCl_3$, $C_6H_6$ | Mol. wt. = 24,000 (ebull.) | 98–116 | Cyclic structure supported by IR | *182, 223* |
| 28 | R = $p\text{-}NO_2C_6H_4CO$ | G,1 | Acetone, $CHCl_3$, alcohols | "Low" | — | Cyclic structure supported by IR | *223* |
| 29 | R = $C_6H_5\overset{\overset{S}{\|}}{N}HC—$ | G,1 | $CHCl_3$, $C_6H_6$ | — | 130–145 | Cyclic structure supported by IR | *186* |
| 30 | R = $C_6H_5CH_2$ | G,1 | Acetone, $CHCl_3$, alcohols | "Low" | — | Cyclic structure supported by IR | *223* |
| 31 | R = 3,5-Diaminotriazinyl (polydiallylmelamine) | G,1 | $CH_3COOH$, HCl, DMSO | $[\eta]$ 0.10–0.20 | >500 | IR and hydrogenation indicates less than 5% double bonds. Solubility in DMSO depends on polymerization conditions. Discolors at 250°C. Shows significant weight loss above that temperature | *114, 115* |
| 32 | Copolymers of 31 with acrylamide | G,1 | $H_2O$ | — | — | Various compositions prepared | *244* |

\*   \*   \*

TABLE II.9—*continued*
Polymers from Diunsaturated Amines

| No. | Structure | Method | Solubility | Molecular weight | $T_m$ (°C) | Remarks and property data | References |
|-----|-----------|--------|-----------|------------------|------------|---------------------------|------------|
| 33 | Cyclopolymers from<br><br>$R—N\left(CH_2—C=CH_2\atop \quad\quad\underset{CH_3}{\vert}\right)_2$<br><br>R = variety of groups | G,1 | "Soluble" | — | — | — | *186* |
| 34 | R = SO$_2$CH$_3$ | G,1 | "Soluble" | $\bar{M}_n = 1,920$ | 134 | IR indicates absence of residual unsaturation | *96* |
| 35 | R = SO$_2$C$_2$H$_5$ | G,1 | "Soluble" | $\bar{M}_n = 1,250$ | 138 | IR indicates absence of residual unsaturation | *96* |

TABLE II.10

Polymers from Diunsaturated Amides

| No. | Structure | Method | Solubility | Molecular weight | $T_m$ (°C) | Remarks and property data | References |
|---|---|---|---|---|---|---|---|
| 1 | Polydivinylurea | H,1 | Swells in p-chlorophenol | — | >300 | Contains some pendant vinyl groups. May be partly cross-linked | 141, 227 |
| 2 | Poly(N-allylacrylamide) | H,1 | — | — | — | Contradictory reports published. See text | 152, 181, 249, 273–275, 278 |
| 3 | Poly(N-allylmethacrylamide) | H,1 | — | — | — | Contradictory reports published. See text | 152, 181, 273–275, 278–280 |
| 4 | Poly(N-methyldiacrylamide) | H,1 | DMF | — | 110–120 | — | 103 |
| 5 | Poly(N-methyldiacrylamide) | H,4 | THF | — | 240–250 | — | 103 |
| 6 | Poly(dimethacrylamide) | H,1 | DMF | [η] 0.11 | 220 | Structure supported by infrared data | 122, 241, 256 |

TABLE II.10—*continued*
Polymers from Diunsaturated Amides

| No. | Structure | Method | Solubility | Molecular weight | $T_m$ (°C) | Remarks and property data | References |
|---|---|---|---|---|---|---|---|
| 7 | Poly(N-methyldimethacrylamide) | H,1 | $CHCl_3$, acetone, dioxane, acetic acid, THF, DMF, $C_6H_6$ | Mol. wt. $\cong 2,900$ (ebull.) | 155 | Contains residual unsaturation | 122, 241, 255, 256 |
| 8 | Poly(N-methyldimethacrylamide) | H,4 | — | — | — | IR supports structure. Isotactic configuration is favored | 122 |
| 9 | R = $C_2H_5$, Poly(N-ethyldimethacrylamide) | H,1 | Benzene | [η] 0.11 | 128 | IR supports structure | 241, 256 |
| 10 | R = n-$C_3H_7$, Poly(N-n-propyldimethacrylamide) | H,1 | Acetone | [η] 0.16; mol. wt. = 9,000 | — | IR supports structure | 241, 256 |

| | | | | | | | |
|---|---|---|---|---|---|---|---|
| 11 | R = C$_6$H$_5$. Poly(N-phenyldimethacrylamide) | H,1 | CHCl$_3$ | [η] 0.3 | 185 | IR supports structure | 241, 256 |
| 12 | Cyclopolymer from $CH_2=CHNHCOOCH_2CH=CH_2$ | H,1 | Pyridine, γ-butyrolactone, DMF, DMSO | — | >300 | See text | 247 |
| | Cyclopolymers from $CH_2=CHNHCOXR'XCONHCH=CH_2$ | | | | | | |
| 13 | XR'X = O(CH$_2$)$_2$O | H,1 | DMF, DMSO, pyridine, γ-butyrolactone | $\eta_{red} (\times 10^2)$ 0.93 | 300 | See text | 247, 248 |
| 14 | XR'X = O(CH$_2$)$_4$O | H,1 | DMF | $\eta_{red} (\times 10^2)$ 1.31 | 280 | See text | 248 |
| 15 | XR'X = O(CH$_2$)$_6$O | H,1 | DMF | $\eta_{red} (\times 10^2)$ 1.66 | 260 | See text | 248 |
| 16 | XR'X = O(CH$_2$)$_8$O | H,1 | DMF | $\eta_{red} (\times 10^2)$ 2.52 | 235 | See text | 248 |
| 17 | XR'X = O⟨cyclohexane, H, cis⟩O | H,1 | DMF | $\eta_{red} (\times 10^2)$ 1.14 | 300 d | See text | 248 |

TABLE II.10—*continued*
Polymers from Diunsaturated Amides

| No. | Structure | Method | Solubility | Molecular weight | $T_m$ (°C) | Remarks and property data | References |
|---|---|---|---|---|---|---|---|
| 18 | XR'X = *trans* | H,1 | Soluble in DMF, when prepared. Became insoluble on work-up | — | 300 d | See text | 248 |
| 19 | XR'X = o-OC₆H₄O | H,1 | DMF | $\eta_{red}$ (×10²) 0.84 | 260 d | See text | 248 |
| 20 | XR'X = m-OC₆H₄O | H,1 | DMF | $\eta_{red}$ (×10²) 2.60 | 260 d | See text | 248 |
| 21 | XR'X = p-OC₆H₄O | H,1 | DMF | $\eta_{red}$ (×10²) 3.29 | 260 d | See text | 248 |
| 22 | XR'X = | H,1 | Soluble in DMF when prepared. Became insoluble on work-up | — | 220 d | See text | 248 |

TABLE II.11

Cyclopolymers from Diisocyanates and Diisothiocyanates

| No. | Structure | Method | Solubility | Molecular weight | $T_m$ (°C) | Remarks and property data | References |
|---|---|---|---|---|---|---|---|
| | $\left[\text{CON} \underset{(CH_2)_m}{\overset{CO}{\diamond}} N \underset{(CH_2)_m}{\overset{CON}{\diamond}} \underset{(CH_2)_m}{\overset{N}{\diamond}} CO\right]_x \left[\begin{array}{c} C-O \\ \end{array}\right]_y \left[\begin{array}{c} CO-N \\ (CH_2)_m \\ NCO \end{array}\right]_z \Big]_n$ | | | | | | |
| 1 | $m = 1$: Poly(methylene diisocyanate) | I,4 | DMSO, $N$-methyl-pyrrolidone + LiCl, conc. $H_2SO_4$, hot pyridine | — | >350 d | Amorphous (X-ray). Structure by solubility, IR, and chemical reactions. Analysis: $x = 66\%$; $(y + z) = 34\%$. Polymer is stable to 300°C | 142, 145 |
| 2 | $m = 2$: Poly(1,2-ethylene diisocyanate) | I,4 | DMF, DMSO $H_2SO_4$ phenol/$S$-tetrachloroethane (6/4, by wt.) | $[\eta]$ 0.16 | ~250 d | Crystalline (X-ray). Structure by solubility, IR, and chemical reaction. Analysis: $x = 0\%$; $y = 100\%$; $z = 0\%$ | 142, 143, 154 |
| 3 | $m = 3$: Poly(1,3-propylene diisocyanate) | I,4 | $m$-Cresol $H_2SO_4$ (dec.) | $[\eta]$ 0.53 | >350 d | Crystalline (X-ray). Analaysis gave $x = 76\%$; $y = 24\%$; $z = 0\%$. Retains 93% of its weight at temperatures of up to 400°C. Treatment with alcohols or amines enhances the thermal stability | 49, 50, 142, 144, 203 |

TABLE II.11—*continued*

Cyclopolymers from Diisocyanates and Diisothiocyanates

| No. | Structure | Method | Solubility | Molecular weight | $T_m$ (°C) | Remarks and Property data | References |
|---|---|---|---|---|---|---|---|
| 4 | $m = 4$: Poly(1,4-butylenediisocyanate) | 1,4 | $H_2SO_4$ (dec.) ethylene carbonate, hot chloronaphthalene | $[\eta]$ 0.08 | >350 d | Crystalline (X-ray). Analysis gave $x = 40\%$; $y = 60\%$; $z = 0\%$. Treatment with alcohols or amines enhances polymer's stability | 142, 144 |
| 5 | Polymer from $OCNCH_2OCH_2NCO$ | 1,4 | Hot $H_2SO_4$ | — | 345 d | IR indicates low degree of cyclization | 154 |
| 6 | Polymer from $CH_3$—$CH$—$CH_2$ <br> NCO NCO | 1,4 | s-Tetrachloroethane/phenol (4/6, by wt.), HCOOH, $CH_3NO_2$, DMSO, DMF, $H_2SO_4$ | $\eta_{inh}$ 1.35. At $\eta_{inh}$ 0.56 osmometry gave mol. wt. = 45,000 | 287 (PMT) | Cyclized structure supported by IR and solubility | 154 |
| 7 | Polymer from $OCNCH_2CH(CH_2)_2NCO$ <br> Cl | 1,4 | m-Cresol | $[\eta]$ 0.68 | — | Stable up to 300°C. Treatment of polymer with amines enhances its thermal stability | 145 |

| No. | Compound | Linkage | Solvent | | | Remarks | Ref. |
|---|---|---|---|---|---|---|---|
| 8 | Polymer from OCNCH₂CHCH₂NCO with CH₃ | 1,4 | — | — | — | Stable to 270°C | 145 |
| 9 | Polymer from SCN(CH₂)₄NCS | 1,4 | — | — | — | Stable to 300°C. Treatment with amines increases thermal stability to 340°C | 144 |
| 10 | Polymer from (cis) NCO NCO Poly(1,2-cis-cyclopropane diisocyanate) | 1,4 | N-Methyl-pyrrolidone | $[\eta]$ 0.11 | 290 | Cyclic structure supported by IR. X-Ray indicates partial crystallinity | 222 |
| | Copolymer of 10 with: | | | | | | |
| 11 | C₂H₅NCO | 1,4 | N-Methyl-pyrrolidone | $[\eta]$ 0.07 | 265 | — | 222 |
| 12 | C₆H₅NCO | 1,4 | N-Methyl-pyrrolidone | $[\eta]$ 0.04 | 265 | — | 222 |
| 13 | OCN(CH₂)₂NCO | 1,4 | N-Methyl-pyrrolidone | $[\eta]$ 0.04 | 190 | — | 222 |
| 14 | Polymer from S ring with NCO NCO Poly(1,2-cyclohexylene diisocyanate) | 1,4 | s-Tetrachloro-ethane/phenol (4/6, by wt.), pyridine, CH₃NO₂, DMSO, H₂SO₄ | — | 303 (PMT) | Cyclic structure supported by IR and solubility | 154 |

TABLE II.11—*continued*

Cyclopolymers from Diisocyanates and Diisothiocyanates

| No. | Structure | Method | Solubility | Molecular weight | $T_m$ (°C) | Remarks and property data | References |
|---|---|---|---|---|---|---|---|
| 15 |  Poly(1,3-cyclohexylene diisocyanate) | I,4 | 98–100% HCOOH, *m*-cresol | $[\eta]$ 0.1 | 224–265 | Analysis gave $x = 65\%$; $y = 35\%$ | *92* |
| 16 |  Poly(glyceryl triisocyanate) | I,4 | $H_2SO_4$ | — | 365–400 | Cyclic structure supported by IR and solubility | *154* |

TABLE II.12

Cyclopolymers from Dinitriles

| No. | Structure | Method | Solubility | Molecular weight | Remarks and property data | References |
|---|---|---|---|---|---|---|
| 1 | Polyfumaronitrile | J,1 | Methane-sulfonic acid | $\eta_{inh}$ 0.35 | Structure supported by IR, UV, NMR, and chemical evidence. ESR and conductivity data | 163 |
| 2 | Polymaleonitrile | J,1 | Methane-sulfonic acid | $\eta_{inh}$ 0.31 | Structure supported by IR, UV, NMR, and chemical evidence. ESR and conductivity data | 163 |
| 3 | Polysuccinonitrile | J,1 | Methane-sulfonic acid | $\eta_{inh}$ 1.28 | Structure supported by IR, UV, NMR, and chemical evidence. ESR and conductivity data | 163 |
|  | Copolymer from: |  |  |  |  |  |
| 4 | Fumaronitrile and maleonitrile (1:1) | J,1 | Methane-sulfonic acid | $\eta_{inh}$ 0.18 | Conductivity data given | 163 |
| 5 | Maleonitrile and succinonitrile (1:1) | J,1 | Methane-sulfonic acid | $\eta_{inh}$ 0.67 | Conductivity data given | 163 |

TABLE II.12—*continued*
Cyclopolymers from Dinitriles

| No. | Structure | Method | Solubility | Molecular weight | Remarks and property data | References |
|---|---|---|---|---|---|---|
| 6 | Fumaronitrile and succinonitrile (1:1) | J,1 | Methane-sulfonic acid | $\eta_{inh}$ 0.41 | — | 163 |
| 7 | Fumaronitrile and succinonitrile (67:33) | J,1 | Methane-sulfonic acid | $\eta_{inh}$ 0.48 | — | 163 |
| 8 | Fumaronitrile and succinonitrile (33:67) | J,1 | Methane-sulfonic acid | $\eta_{inh}$ 0.45 | — | 163 |

TABLE II.13

Phosphorus-Containing Cyclopolymers

| No. | Structure | Method | Solubility | Molecular weight | $T_m$ (°C) | Remarks and property data | References |
|-----|-----------|--------|------------|------------------|------------|---------------------------|------------|
| 1 | Poly[bis(2-chloroallyl)methylphosphine] | K,1 | — | — | — | — | 65 |
| 2 | Copolymers of divinylphenylphosphine with acrylonitrile | K,1 | — | — | — | The phosphine moiety cyclopolymerized | 53 |
| 3 |  Poly(diallylphenylphosphine oxide) | K,1 | Alcohol, glacial acetic acid, 1,2-dichloro-ethane, diethylene glycol, dioxane | Mol. wt. (by light scattering) = 5,000 | — | Structure by solubility and infrared data. Useful as additive to engine oils. At [$\eta$] 0.26 has melting range of 85°–115°C | 44, 47, 259 |
| 4 | Copolymers of 3 with lauryl methacrylate | K,1 | $\beta,\beta'$-Dichloro-diethyl ether, benzene | — | — | Various compositions prepared. Copoly-merization parameters were determined. Utility similar to polymer 3 | 47, 259 |
| 5 |  R = CH₃  Poly(dimethallylmethylphosphine oxide) | K,1 | Alcohol, acetic acid | [$\eta$] 0.042 | 150–170 | Solubility and infrared support cyclic structure | 45 |

TABLE II.13—continued

Phosphorus-Containing Cyclopolymers

| No. | Structure | Method | Solubility | Molecular weight | $T_m$ (°C) | Remarks and property data | References |
|---|---|---|---|---|---|---|---|
| 6 | R = C₂H₅<br>Poly(dimethallylethylphosphine oxide) | K,1 | "Soluble" | — | 140–185 | Solubility and infrared support cyclic structure | 45 |
| 7 | R = C₆H₅<br>Poly(dimethallylphenylphosphine oxide) | K,1 | Alcohol, glacial acetic acid, DMF | [η] 0.04 | 130–165 | Solubility and infrared support cyclic structure. May contain five-membered rings | 44 |
| 8 | Copolymer of 7 with oleyl maleate<br>*  *  * | K,1 | — | — | — | — | 259 |
| 9 | Copolymers of diallyl phenyl phosphate with methyl methacrylate | K,1 | "Soluble" | — | — | Several compositions prepared. Copolymerization parameters were determined | 132 |
| 10 | Copolymers of diallyl phenyl phosphate with styrene | K,1 | "Soluble" | — | — | Several compositions prepared. Copolymerization parameters were determined | 132 |
| 11 | Poly(trivinyl phosphate) | K,1 | H₂O, DMF | [η] 0.16 | 155–190 | Contains ∼3–5% of residual unsaturation (IR) | 183 |

| | | | | | |
|---|---|---|---|---|---|
| 12 | Copolymers of diallyl n-butyl phosphonate with lauryl methacrylate | K,1 | Decalin, benzene | — | Based on IR and bromination, the phosphonate moiety is cyclized. Various compositions prepared. Copolymerization parameters were determined | 46 |
| 13 | Poly(allyl phenyl allyl phosphonate) | K,1 | Alcohol, $C_6H_6$ | Mol. wt. (by light-scattering) = 26,000 | Polymer could contain both units. Useful as additive to engine oils. At $[\eta]$ 0.09 has melting range of 92°–115°C | 44, 149, 258, 259 |

Copolymers of 13 with:

| | | | | | |
|---|---|---|---|---|---|
| 14 | Methyl methacrylate | K,1 | "Soluble" | — | Several compositions prepared. Copolymerization parameters determined. IR indicates cyclization of the diallyl moiety | 131 |
| 15 | Styrene | K,1 | "Soluble" | — | Several compositions prepared. Copolymerization parameters determined. IR indicates cyclization of the diallyl moiety. | 131 |

TABLE II.13—*continued*

Phosphorus-Containing Cyclopolymers

| No. | Structure | Method | Solubility | Molecular weight | $T_m$ (°C) | Remarks and property data | References |
|---|---|---|---|---|---|---|---|
| 16 | Lauryl methacrylate | K,1 | Decalin, benzene | — | — | Several compositions prepared. Copoly-merization parameters determined. IR in-dicates cyclization of the diallyl moiety. Useful as additive to engine oils | 46, 258, 259 |
| | * * * | | | | | | |
| 17 | Copolymer of allyl lauryl allyl phosphonate with vinyl stearate | K,1 | — | — | — | — | 259 |
| 18 | Poly(dimethallyl lauryl phosphonate) | K,1 | — | — | — | — | 259 |
| 19 | Poly(diunsaturated phosphonium salts) | K,1 | "Soluble" | — | — | See text | 53, 83 |

TABLE II.14

Sulfur-Containing Cyclopolymers

| No. | Structure | Method | Solubility | Molecular weight | $T_m$ (°C) | Remarks and property data | References |
|---|---|---|---|---|---|---|---|
| 1 | Cyclopolymer from $(CH_2=CClCH_2)_2S$ | L,1 | — | — | — | — | 181 |
| 2 | Cyclopolymer from $(CH_2=CClCH_2)_2SO_2$ | L,1 | — | — | — | — | 181 |
| 3 | Cyclopolymer from $(CH_2=CHCH_2)_2S$ | L,1; L,2; L,3 | — | "Low" | — | All methods yield high-boiling oils | 269 |
| 4 | Copolymer from $[(CH_2=CHCH_2)_2SCH_3]CH_3SO_4^-$ (10% wt.), acrylonitrile (85% wt.), and vinyl acetate (5% wt.) | L,1 | Aq. sodium thiocyanate | Viscosity of 1% soln. in 57.5% aq. NaSCN was 18.6 centipoises | — | Cyclization of diallyl-moiety was postulated | 244 |
| 5 | R = R' = H | L,1 | $CHCl_3$, dioxane, $C_6H_6$, THF | — | — | Chemical evidence and IR support structure. Oxidation yields the corresponding sulfone | 189, 237 |

TABLE II.14—continued
Sulfur-Containing Cyclopolymers

| No. | Structure | Method | Solubility | Molecular weight | $T_m$ (°C) | Remarks and property data | References |
|---|---|---|---|---|---|---|---|
| 6 | R = H; R' = CH$_3$ | L,1 | CHCl$_3$, dioxane, C$_6$H$_6$ | — | — | Chemical evidence and IR support structure. Oxidation yields the corresponding sulfone | 188, 189 |
| 7 | R = H; R' = n-C$_3$H$_7$ | L,1 | CHCl$_3$, dioxane C$_6$H$_6$ | — | — | Chemical evidence and IR support structure. Oxidation yields the corresponding sulfone | 188, 189 |
| 8 | R = R' = CH$_3$ | L,1 | CHCl$_3$, dioxane, C$_6$H$_6$ | — | — | Chemical evidence and IR support structure. Oxidation yields the corresponding sulfone | 188, 189 |
| 9 | Poly(allyl vinyl sulfonate) | L,1 | DMF, DMSO | $\eta_{red}$ 1.09 | 180–190 | Solubility and IR support structure. Contains residual unsaturation | 117 |
| 10 | Poly(N-allyl vinyl sulfonamide) | L,1 | DMF, DMSO, aq. NaOH | $\eta_{red}$ 0.15 | 235–240 | — | 118 |

| No. | Polymer | | Solvent | | mp | Remarks | Ref. |
|---|---|---|---|---|---|---|---|
| 11 | Poly(S,S'-divinyl dithiolocarbonate) | L,1 | "Soluble" | — | — | — | 225, 236 |
| 12 | Polymer from $CH_2=CHSCH_2CH_2SCOSCH=CH_2$ | L,1 | $C_6H_6$, $CHCl_3$, DMF | $\eta_{sp}$ 0.21 | 300–310 | IR shows low residual unsaturation | 225, 236 |
| 13 | Poly(S-vinyl-N-methyl vinyl thiocarbamate) | L,1 | $C_6H_6$, toluene, $CHCl_3$, $CCl_4$, DMF | $[\eta]$ 0.21 | — | Data on soluble portion of polymer obtained. $x$, $y$, and % soluble vary with experimental conditions. Polymer cited had 65% mole pendant double bonds (IR) | 229 |
| 14 | Poly(S-vinyl-N-methyl vinyl thiocarbamate) | L,3 | $C_6H_6$ | — | — | Structure supported by IR. Contains 42% pendant S-vinyl groups | 229 |
| 15 | Poly(S-vinyl-N-ethyl vinyl thiocarbamate) | L,1 | — | — | — | Soluble and insoluble polymer obtained. % soluble polymer ~90%; % double bonds in soluble material = 71.5% (IR) | 229 |
| 16 | Poly(S-vinyl-N-ethyl vinyl thiocarbamate) | L,3 | — | — | — | Tacky product | 229 |

TABLE II.14—*continued*
Sulfur-Containing Cyclopolymers

| No. | Structure | Method | Solubility | Molecular weight | $T_m$ (°C) | Remarks and property data | References |
|---|---|---|---|---|---|---|---|
| 17 | Poly(S-vinyl-N-n-butyl vinyl thiocarbamate) | L,1 | — | — | — | Soluble and insoluble polymer obtained. % soluble polymer ≅ 15% wt.; % double bonds in soluble material = 78% (IR) | 229 |
| 18 | Poly(S-vinyl-N-n-butyl vinyl thiocarbamate) | L,3 | — | — | — | Tacky product | 229 |
| 19 | Copolymer from 1,5-hexadiene and SO₂ | L,1 | Acetophenone, ketones, DMF DMSO | $\eta_{inh}$ 0.692; mol. wt. = 125,000 | >300 | Structure supported by IR, solubility, and elemental analysis | 64, 68, 265 |
| 20 | Copolymer from cis, cis-1,5-cyclooctadiene and SO₂ | L,1 | DMSO, tetramethylene sulfone, H₂SO₄, N-methyl-pyrrolidone, DMF containing 5% LiCl, N,N-dimethylacetamide containing 5% LiCl | $\eta_{inh}$ 2.1. At $\eta_{inh}$ 0.93 mol. wt. was 113,000, by light scattering | 260 d | Amorphous. Unstable in hot H₂SO₄ and to bases. Density = 1.5211 g/cc (30°C). Cast films are brittle, thermally, hydrolytically unstable, and flammable. Water absorption of film: 5.3%. Structure supported by IR and chemical evidence | 104–106 |

| No. | Structure | Type | Solvents | $\eta_{inh}$ | mp | Remarks | Ref. |
|---|---|---|---|---|---|---|---|
| 21 | 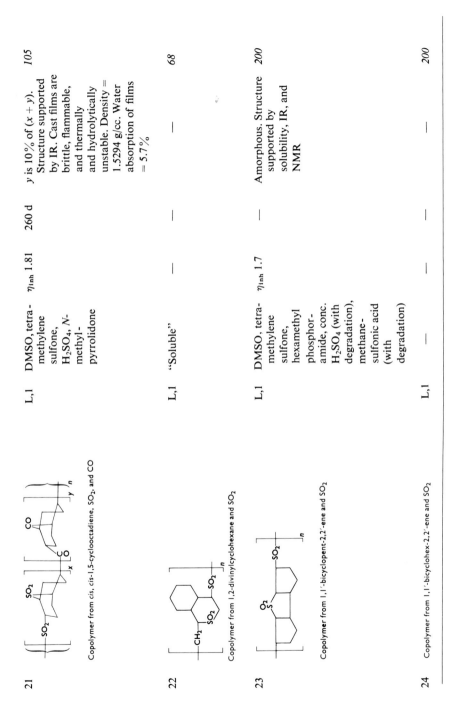Copolymer from cis, cis-1,5-cyclooctadiene, SO$_2$ and CO | L,1 | DMSO, tetra-methylene sulfone, H$_2$SO$_4$, N-methyl-pyrrolidone | $\eta_{inh}$ 1.81 | 260 d | $y$ is 10% of $(x+y)$. Structure supported by IR. Cast films are brittle, flammable, and thermally and hydrolytically unstable. Density = 1.5294 g/cc. Water absorption of films = 5.7% | 105 |
| 22 | Copolymer from 1,2-divinylcyclohexane and SO$_2$ | L,1 | "Soluble" | — | — | — | 68 |
| 23 | Copolymer from 1,1'-bicyclopent-2,2'-ene and SO$_2$ | L,1 | DMSO, tetra-methylene sulfone, hexamethyl phosphoramide, conc. H$_2$SO$_4$ (with degradation), methanesulfonic acid (with degradation) | $\eta_{inh}$ 1.7 | — | Amorphous. Structure supported by solubility, IR, and NMR | 200 |
| 24 | Copolymer from 1,1'-bicyclohex-2,2'-ene and SO$_2$ | L,1 | — | — | — | — | 200 |

TABLE II.14—*continued*
Sulfur-Containing Cyclopolymers

| No. | Structure | Method | Solubility | Molecular weight | $T_m$ (°C) | Remarks and property data | References |
|-----|-----------|--------|------------|------------------|------------|---------------------------|------------|
| 25 | Copolymer from divinyl sulfone and acrylonitrile | L,1 | DMF | $[\eta]$ up to 0.64 | — | Several compositions prepared. Reactivity ratios given. IR indicates low residual unsaturation | 81 |
| 26 | Copolymer from divinyl sulfone and maleic anhydride | L,1 | DMF, DMSO, 2N aq. NaOH | $[\eta]$ 0.064 | — | — | 62, 64, 71 |
| 27 | Copolymer from divinyl sulfone and dimethyl fumarate | L,1 | DMF, dichlorobenzene | — | 290 | — | 71 |

## Silicon-Containing Cyclopolymers

| No. | Structure | Method | Solubility | Molecular weight | $T_m$ (°C) | Remarks and property data | References |
|---|---|---|---|---|---|---|---|
| 1 | Copolymer from divinyldimethylsilane and acrylonitrile | M,1 | DMF | — | — | — | 71 |
| 2 | Copolymer from divinyldimethylsilane and fumaryl chloride | M,1 | Acetone, benzene, DMF, DMSO | [η] 0.142 | — | — | 64, 71 |
| 3 | Copolymer from divinyldimethylsilane and maleic anhydride | M,1 | DMF, DMSO, aq. NaOH | [η] 0.121 | >250 | — | 62, 64, 71 |
| 4 | Copolymer from divinyldimethylsilane and vinyl acetate | M,1 | DMF, other "common" solvents | [η] 0.063 | — | — | 64, 71 |
| 5 | Copolymer from divinylcyclopentamethylenesilane and maleic anhydride | M,1 | DMF, DMSO, dilute aq. NaOH | [η] 0.106 | 350 | — | 64, 71 |
| 6 | Polydivinyloxydimethylsilane | M,1 | — | — | — | Chemical and spectroscopic investigations support structure. Hydrolysis yields poly(vinyl alcohol) with "head-to-head" units | 109 |

TABLE II.15—*continued*
Silicon-Containing Cyclopolymers

| No. | Structure | Method | Solubility | Molecular weight | $T_m$ (°C) | Remarks and Property data | References |
|---|---|---|---|---|---|---|---|
| 7 | Poly[bis(α-chloroallyl)dimethylsilane] | M,1 | — | — | 80–100 | — | 87 |
| | Polymers of general structure Poly(diallyl dialkyl or diaryl silanes) | | | | | | |
| 8 | R = H; R′ = CH₃ | M,1 | Benzene, ether, CHCl₃, CCl₄ | [η] 0.2 | 70–95 | Structure supported by infrared and solubility. IR indicates maximum of 10% of residual unsaturation | 123 |
| 9 | R = H; R′ = C₂H₅ | M,1 | Benzene, ether, CHCl₃, CCl₄ | [η] 0.18 | 58–80 | Structure supported by infrared evidence and solubility | 123 |
| 10 | R = H; R′ = C₆H₅ | M,1 | Benzene, ether, CHCl₃, CCl₄ | [η] 0.18 | 100–110 | Structure supported by infrared evidence and solubility | 123 |
| 11 | R = R′ = CH₃ | M,1 | Benzene, ether, CHCl₃, CCl₄ | [η] 0.06 | 50–65 | Structure supported by infrared evidence and solubility | 65, 84, 123, 201 |

| | | | | | | |
|---|---|---|---|---|---|---|
| 12  R = R' = CH$_3$ | M,2 | C$_6$H$_6$ | [η] 0.13 | 80–110 | Structure supported by IR. No residual unsaturation in polymer (IR) | 52, 83, 84, 167, 271 |
| 13  Copolymer of 12 with propylene | M,2 | — | — | — | Several compositions were prepared | 211 |
| 14  R = R' = C$_2$H$_5$ | M,1; M,2 | C$_6$H$_6$, ether, CHCl$_3$, CCl$_4$ | [η] 0.08 | 50–65 | — | 123, 159, 271 |
| 15  R = CH$_3$; R' = C$_6$H$_5$ | M,1 | C$_6$H$_6$, ether, CHCl$_3$, CCl$_4$ | [η] 0.05 | 60–65 | — | 123 |
| 16  R = CH$_3$; R' = C$_6$H$_5$ | M,2 | C$_6$H$_6$ | [η] 0.08 | — | — | 52 |
| 17  Copolymer of 16 with propylene | M,2 | — | — | — | Several compositions were prepared | 211 |
| 18  R = R' = C$_6$H$_5$ | M,1; M,2 | C$_6$H$_6$ | [η] 0.065 | 125–155 | Structure supported by infrared studies which indicate absence of residual unsaturation | 84 |
| 19  Poly(N-methyl-sym-divinyltetramethyldisilazane) | M,4 | C$_6$H$_6$ | 2,600 | 70–75 | Structure supported by infrared studies | 266 |

TABLE II.15—*continued*

Silicon-Containing Cyclopolymers

| No. | Structure | Method | Solubility | Molecular weight | $T_m$ (°C) | Remarks and property data | References |
|---|---|---|---|---|---|---|---|
| 20 | <br>Poly(triallylmethylsilane) | M,1 | $C_6H_6$, $CCl_4$, methyl ethyl ketone | [η] 0.02 | — | Residual unsaturation (IR) = 18–20% | 124 |
| | | M,2 | — | — | — | Residual unsaturation = 38% | 272 |
| 21 | <br>Poly(triallylphenylsilane) | M,1 | $C_6H_6$, $CCl_4$, methyl ethyl ketone | [η] 0.016 | | Residual unsaturation (IR) = 13–15% | 124 |

TABLE II.16

Miscellaneous Cyclopolymers

| No. | Structure | Method | Solubility | Molecular weight | $T_m$ (°C) | Remarks and property data | References |
|-----|-----------|--------|-----------|------------------|-----------|---------------------------|-----------|
| 1 | 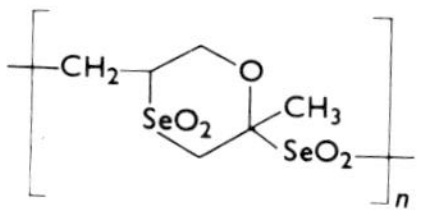 Copolymer from allyl isopropenyl ether and selenium dioxide | N,1 | "Soluble" | — | — | — | 68 |
| 2 | Poly(diallyldiethyllead) | N,1; N,2 | — | — | — | — | 65 |
|   | Poly(diallyldimethyltin) | N,1 | — | — | — | Contains "some" residual unsaturation | 63 |

REFERENCES

1. Abe, A., and Goodman, M., *J. Polym. Sci., Part A* **2**, 3491 (1964).
2. Aito, Y., Matsuo, T., and Aso, C., *Bull. Chem. Soc. Jap.* **40**, 130 (1967); *Chem. Abstr.* **66**, 8972, 95540a (1967).
3. Anand, C., Deshpande, A. B., and Kapur, S. L., *Chem. Ind.* (*London*) p. 1457 (1966); *Chem. Abstr.* **65**, 13833 (1966).
4. Arbuzova, I. A., and Efremova, V. N., *Vysokomol. Soedin.* **2**, 1586 (1960).
5. Arbuzova, I. A., Efremova, V. N., Eliseeva, A. G., and Zinder, M. F., *Vysokomol. Soedin.* **5**, 1819 (1963).
6. Arbuzova, I. A., Fedorova, E. F., Plotkina, S. A., and Minkova, R. M., *Vysokomol. Soedin., Ser. A* **9**, 189 (1967); *Chem. Abstr.* **66**, 8077, 86083b (1967).
7. Arbuzova, I. A., Kostikov, R. R., and Propp, L. N., *Vysokomol. Soedin.* **2**, 1402 (1960).
8. Arbuzova, I. A., Minkova, R. M., Plotkina, S. A., and Fedorova, E. F., *Vysokomol. Soedin., Ser. B* **9**, 638 (1967); *Chem. Abstr.* **67**, 10299, 109054e (1967).
9. Arbuzova, I. A., and Plotkina, S. A., *Vysokomol. Soedin.* **6**, 662 (1964); *Resins, Rubbers, Plastics* p. 507 (1965).
10. Arbuzova, I. A., Plotkina, S. A., and Sokolova, O. V., *Vysokomol. Soedin.* **4**, 843 (1962).
11. Arbuzova, I. A., and Rostovskii, E. N., *J. Polym. Sci.* **52**, 325 (1961).
12. Arbuzova, I. A., and Sultanov, K., *Vysokomol. Soedin.* **2**, 1077 (1960).
13. Aso, C., *J. Polym. Sci.* **39**, 475 (1959).
14. Aso, C., *J. Chem. Soc. Jap., Ind. Chem. Sect.* **63**, 363 (1960); *Chem. Abstr.* **56**, 3629 (1962).
15. Aso, C., *Kagaku Kyoto* **20**, 29 (1965); *Chem. Abstr.* **64**, 12812 (1966).
16. Aso, C., and Aito, Y., *Bull. Chem. Soc. Jap.* **35**, 1426 (1962); *Chem. Abstr.* **57**, 13967 (1962).
17. Aso, C., and Aito, Y., *Makromol. Chem.* **58**, 195 (1962); *Resins, Rubbers, Plastics* p. 1119 (1963).
18. Aso, C., and Aito, Y., *Bull. Chem. Soc. Jap.* **37**, 456 (1964); *Chem. Abstr.* **61**, 12097 (1964).
19. Aso, C., and Aito, Y., *Makromol. Chem.* **73**, 141 (1964).
20. Aso, C., and Aito, Y., *Pap., 13th Annu. Meet. High Polym. Div., Chem. Soc. Jap.* (1964).
21. Aso, C., and Aito, Y., *Kobunshi Kagaku* **23**, 564 (1966); *Chem. Abstr.* **67**, 6983, 73904r (1967).
22. Aso, C., Aito, Y., and Furuta, A., *Pap., 12th Annu. Meet. High Polym. Div., Chem. Soc. Jap.* (1963).
23. Aso, C., Aito, Y., and Furuta, A., *Pap., 14th Discuss. Meet. Soc. High Polym. Jap.* (1965).
24. Aso, C., Furuta, A., and Aito, Y., *Makromol. Chem.* **84**, 126 (1965); *Chem. Abstr.* **63**, 5750 (1965).
25. Aso, C., Kamao, H., and Aito, Y., *Kogyo Kagaku Zasshi* **67**, 974 (1964); *Chem. Abstr.* **62**, 10517 (1965).
26. Aso, C., and Miura, M., *J. Polym. Sci., Part B* **4**, 171 (1966).
27. Aso, C., and Miura, M., *Kobunshi Kagaku* **24**, 178 (1967); *Chem. Abstr.* **68**, 321, 3211j (1968).
28. Aso, C., and Tagami, S., *J. Poly. Sci., Part B* **5**, 217 (1967); *Chem. Abstr.* **66**, 10793, 115992e (1967).

29. Aso, C., and Tagami, S., *Polym. Prepr., Amer. Chem. Soc., Div. Polym. Chem.* **8**, 906 (1967).
30. Aso, C., and Ushio, S., *Kogyo Kagaku Zasshi* **65**, 2085 (1962).
31. Aso, C., Ushio, S., and Sogabe, M., *Makromol. Chem.* **100**, 100 (1967).
32. Barnett, M. D., and Butler, G. B., *J. Org. Chem.* **25**, 309 (1960).
33. Barnett, M. D., Crawshaw, A., and Butler, G. B., *J. Amer. Chem. Soc.* **81**, 5946 (1959); *Resins, Rubbers, Plastics* p. 375 (1960).
34. Barsamyan, S. T., *Vysokomol. Soedin., Ser. A* **9**, 749 (1967).
35. Barsamyan, S. T., Tolapchyan, L. S., and Pikalova, V. N., *Dokl. Akad. Nauk Arm. SSR*, **40**, 101 (1965); *Chem. Abstr.* **63**, 4409 (1965).
36. Barton, J. M., *J. Polym. Sci., Part B* **4**, 513 (1966).
37. Barton, J. M., Butler, G. B., and Chapin, E. C., *J. Polym. Sci., Part A* **3**, 501 (1965); *Chem. Abstr.* **62**, 14830 (1965).
38. Bauer, R. S., *J. Polym. Sci., Part A-1*, **5**, 2192 (1967).
39. Belgian Patent 664,427 (1965), issued to Nitto Boseki Co., Ltd.; *Chem. Abstr.* **65**, 2374 (1966).
40. Belgian Patent 672,370 (1966), issued to Penninsular Chemresearch Inc.
41. Belgian Patent 681,487 (1966), issued to Calgon Corporation.
42. Belgian Patent 682,255 (1966), issued to Calgon Corporation.
43. Belgian Patent 684,138 (1967), issued to Calgon Corporation.
44. Berlin, K. D., and Butler, G. B., *J. Amer. Chem. Soc.* **82**, 2712 (1960).
45. Berlin, K. D., and Butler, G. B., *J. Org. Chem.* **25**, 2006 (1960).
46. Beynon, K. I., *J. Polym. Sci., Part A* **1**, 3343 (1963).
47. Beynon, K. I., *J. Polym. Sci., Part A* **1**, 3357 (1963).
48. Beynon, K. I., and Hayward, E. J., *J. Polym. Sci., Part A* **3**, 1793 (1965).
49. Black, W. B., and Miller, W. L., Fr. Pat. 1,360,460 (1964); issued to Monsanto Chemical Co.; *Chem. Abstr.* **62**, 7777 (1965).
50. Black, W. B., and Miller, W. L., U.S. Pat. 3,163,624 (1964), issued to Monsanto Chemical Co.
51. Blout, E. R., and Ostberg, B. E., *J. Polym. Sci.* **1**, 230 (1946).
52. Bogomol'nyi, V. Ya., *Vysokomol. Soedin.* **1**, 1469 (1959); *Chem. Abstr.* **54**, 14753 (1960).
53. Bond, W. C., Jr., *Diss. Abstr.* **25**, 3260 (1964); *Chem. Abstr.* **62**, 13250 (1964).
54. Brady, W. T., and O'Neal, H. R., *J. Polym. Sci., Part B* **2**, 647 (1964).
55. Brady, W. T., and O'Neal, H. R., *J. Polym. Sci., Part A* **3**, 2337 (1965); *Chem. Abstr.* **63**, 5754 (1965).
56. Braun, D., *Angew. Chem.* **73**, 197 (1961); *Chem. Technol. Polym. (USSR)* No. 5, p. 3 (1962).
57. Bresler, S. E., Koton, M. M., Os'minskaya, A. T., Popov, A. G., and Savitskaya, M. N., *Vysokomol. Soedin.* **1**, 1070 (1959); *Polym. Sci. USSR* **1**, 393 (1960).
58. Bristow, G. M., *Trans. Faraday Soc.* **54**, 1064 (1958).
59. British Patent 905,831 (1962); issued to Penninsular Chemresearch, Inc.; *Chem. Abstr.* **58**, 4662 (1963).
60. British Patent 1,037,028 (1966), issued to Penninsular Chemresearch, Inc.; *Chem. Abstr.* **65**, 15611 (1966).
61. Butler, G. B., *Abstr. Pap., 133rd Meet., Amer. Chem. Soc.* p. 6R (1958).
62. Butler, G. B., *Abstr. Pap., 134th Meet., Amer. Chem. Soc.* p. 32T (1958).
63. Butler, G. B., quoted in Marvel (*165*).
64. Butler, G. B., *J. Polym. Sci.* **48**, 279 (1960).
65. Butler, G. B., U.S. Pat. 3,044,986 (1962), issued to Penninsular Chemresearch, Inc.
66. Butler, G. B., *Pure Appl. Chem.* **4**, 299 (1962).

67. Butler, G. B., *Encycl. Polym. Sci. Technol.* **4**, 568 (1966); *Chem. Abstr.* **65**, 17052 (1966).
68. Butler, G. B., U.S. Pat. 3,239,488 (1966), issued to Penninsular Chemresearch, Inc.
69. Butler, G. B., U.S. Pat. 3,288,770 (1966), issued to Penninsular Chemresearch, Inc.
70. Butler, G. B., *Polym. Prepr., Amer. Chem. Soc., Div. Polym. Chem.* **8**, 35 (1967).
71. Butler, G. B., U.S. Pat. 3,320,216 (1967), issued to Penninsular Chemresearch, Inc.; *Chem. Abstr.* **67**, 2143, 22329z (1967).
72. Butler, G. B., and Angelo, R. J., *J. Amer. Chem. Soc.* **79**, 3128 (1957).
73. Butler, G. B., Angelo, R. J., and Crawshaw, A., U.S. Pat. 2,926,161 (1960), issued to Penninsular Chemresearch, Inc.
74. Butler, G. B., and Brooks, T. W., *J. Org. Chem.* **28**, 2699 (1963).
75. Butler, G. B., and Brooks, T. W., *J. Macromol. Chem.* **1**, 231 (1966).
76. Butler, G. B., Crawshaw, A., and Miller, W. L., *Abstr. Pap., 132nd Meet. Amer. Chem. Soc.* p. 18T (1957).
77. Butler, G. B., Crawshaw, A., and Miller, W. L., *J. Amer. Chem. Soc.* **80**, 3615 (1958).
78. Butler, G. B., and Goette, R. L., *J. Amer. Chem. Soc.* **74**, 1939 (1952).
79. Butler, G. B., and Goette, R. L., *J. Amer. Chem. Soc.* **76**, 2418 (1954).
80. Butler, G. B., and Ingley, F. L., *J. Amer. Chem. Soc.* **73**, 895 (1951).
81. Butler, G. B., and Kasat, R. B., *J. Polym. Sci., Part A* **3**, 4205 (1965).
82. Butler, G. B., and Raymond, M. A., *J. Polym. Sci., Part A* **3**, 3413 (1965).
83. Butler, G. B., Skinner, D. L., and Stackman, R. W., *Conf. High Temp. Polym. Fluid Res.* (1959).
84. Butler, G. B., and Stackman, R. W., *J. Org. Chem.* **25**, 1643 (1960).
85. Butler, G. B., Vanhaeren, G., and Ramadier, M. F., *J. Polym. Sci., Part A-1*, **5**, 1265 (1967); *Chem. Abstr.* **67**, 3141, 32975w (1967).
86. Chang, E., and Price, C. C., *J. Amer. Chem. Soc.* **83**, 4650 (1961).
87. Chang, H.-C., Feng, H.-P., and Feng, H.-T., *Ko Fen Tzu T'ung Hsun* **6**, 487 (1964); *Chem. Abstr.* **63**, 16474 (1965).
88. Chang, H.-C., Ts'ao, W.-H., and Feng, H.-T., *Ko Fen Tzu T'ung Hsun* **6**, 61 (1964); *Chem. Abstr.* **63**, 11710 (1965).
89. Chiang, T.-C., Pao, H.-L., Sun, K.-H., and Feng, H.-T., *Ko Fen Tzu T'ung Hsun* **7**, 56 (1965); *Chem. Abstr.* **63**, 18270 (1965).
90. Clark, H. G., Miller, W. L., and Butler, G. B., *Meet.-in-Miniature, Amer. Chem. Soc., N. C. Sect.* (1961).
91. Cohen, S. G., Ostberg, B. E., Sparrow, D. B., and Blout, E. R., *J. Polym. Sci.* **3**, 264 (1948).
92. Corfield, G. C., and Crawshaw, A., *Chem. Commun.* p. 85 (1966); *Chem. Abstr.* **64**, 12563 (1966).
93. Crawshaw, A., Barnett, M. D., and Butler, G. B., *Abstr. Pap., 135th Meet., Amer. Chem. Soc.* p. 17S (1959).
94. Crawshaw, A., and Butler, G. B., *Abstr. Pap., 133rd Meet., Amer. Chem. Soc.* p. 8R (1958).
95. Crawshaw, A., and Butler, G. B., *J. Amer. Chem. Soc.* **80**, 5464 (1958).
96. Crawshaw, A., and Jones, A. G., *Chem. Ind.* (*London*) p. 2013 (1966).
97. Delmonte, D. W., and Hays, J. T., *Abstr. Pap., 136th Meet., Amer. Chem. Soc.* p. 1-T (1959).
98. DeWinter, W., *Rev. Macromol. Chem.* **1**, 329 (1966); *Chem. Abstr.* **66**, 5285, 55758w (1967).
99. Donati, M., and Farino, M., *Makromol. Chem.* **60**, 233 (1963).
100. Dvorko, G. F., and Yakhimovich, R. I., USSR Pat. 173,411 (1965), issued to Inst. Org. Chem., Acad. Sci. Ukr. SSR, and to Inst. Chem. Polym. and Monom., Acad. Sci. Ukr. SSR; *Chem. Abstr.* **64**, 836 (1966).

101. Erickson, J. G., U.S. Pat. 2,556,075 (1951), issued to American Cyanamid Co.; *Chem. Abstr.* **45**, 7377 (1951).
102. Feng, H.-T., Chang, H.-C., and Ts'ao, W. H., *Ko Fen Tzu T'ung Hsun* **7**, 120 (1965); *Chem. Abstr.* **64**, 3694 (1966).
103. Feng, H.-T., and Hung, H.-Y., *K'o Hsueh T'ung Pao* [*Foreign Lang. Ed.*] **17**, 163 (1966); *Chem. Abstr.* **67**, 3144, 34998f (1967).
104. Frazer, A. H., *J. Polym. Sci., Part A* **2**, 4031 (1964).
105. Frazer, A. H., *J. Polym. Sci., Part A* **3**, 3699 (1965).
106. Frazer, A. H., and O'Neill, W. P., *J. Amer. Chem. Soc.* **85**, 2613 (1963); *Resins, Rubbers, Plastics* p. 1145 (1964).
107. French Pat. 1,385,750 (1964), issued to Borg-Warner Corp.
108. French Pat. 1,494,438 (1967), issued to Calgon Corp.
109. Furue, M., Nozakura, S., and Murahashi, S., *Polym. Previews* **3**, 451 (1967).
110. Furue, M., Nozakura, S., and Murahashi, S., *Kobunshi Kagaku* **24**, 522 (1967); *Chem. Abstr.* **68**, 6715, 69390y (1968).
111. Gebelein, G. G., and Howard, E., Jr., *Abstr. Pap., Amer. Chem. Soc., 3rd Del. Valley Reg. Meet.* p. 79 (1960).
112. Gibbs, W. E., and Murray, J. T., *Abstr. Pap., IUPAC Meet. 1960* No. C31, ASD Techn. Rep. 61-409 (1960).
113. Gibbs, W. E., and Murray, J. T., *J. Polym. Sci.* **58**, 1211 (1962); *Resins, Rubbers, Plastics* p. 2063 (1962).
114. Gibbs, W. E., and Van Deusen, R. L., *Abstr. Pap., 140th Meet., Amer. Chem. Soc.* p. 23U (1961).
115. Gibbs, W. E., and Van Deusen, R. L., *J. Polym. Sci.* **54**, S1 (1961).
116. Gindin, L., Medvedev, S. S., and Fleshler, E., *Zh. Obshch. Khim.* **19**, 1694 (1946); *Chem. Abstr.* **44**, 1020 (1950). *J. Gen. Chem. USSR* **19**, a127 (1949); *Chem. Abstr.* **44**, 6387 (1950).
117. Goethals, E. J., *J. Polym. Sci., Part B* **4**, 691 (1966).
118. Goethals, E. J., Bombeke, J., and De Witte, E., *Makromol. Chem.* **108**, 312 (1967); *Chem. Abstr.* **67**, 11079, 117395g (1967).
119. Gordon, M., *J. Chem. Phys.* **22**, 610 (1954).
120. Gordon, M., and Roe, R.-J., *J. Polym. Sci.*, **21**, 27 (1956).
121. Gordon, M., and Roe, R.-J., *J. Polym. Sci.* **21**, 75 (1956).
122. Götzen, F., and Schröder, G., *Makromol. Chem.* **88**, 133 (1965); *Chem. Abstr.* **64**, 3694 (1966).
123. Gusel'nikova, L. E., Nametkin, N. S., Polak, L. S., and Chernysheva, T. I., *Vysokomol. Soedin.* **6**, 2002 (1964); *Chem. Abstr.* **62**, 6560 (1965).
124. Gusel'nikova, L. E., Nametkin, N. S., Polak, L. S., and Chernysheva, T. I., *Izv. Akad. Nauk SSSR, Ser. Khim.* p. 2072 (1964); *Chem. Abstr.* **62**, 9243 (1965). *Bull. Acad. Sci. USSR, Div. Chem. Sci.* p. 1967 (1964).
125. Harada, S., and Arai, K., *Makromol. Chem.* **107**, 64 (1967).
126. Harada, S., and Arai, K., *Makromol. Chem.* **107**, 78 (1967).
127. Harada, S., and Katayama, M., *Makromol. Chem.* **90**, 177 (1966).
128. Harada, S., and Katayama, M., U.S. Pat. 3,375,233 (1968), issued to Nitto Boseki Co., Ltd.
129. Harries, C., *Chem. Ber.* **35**, 1183 (1902).
130. Harries, C., and Tank, L., *Chem. Ber.* **41**, 1701 (1908).
131. Hashimoto, S., and Furukawa, I., *Makromol. Chem.* **82**, 298 (1965); *Chem. High Polym.* **21**, No. 234, 647 (1964).
132. Hashimoto, S., and Furukawa, I., *Makromol. Chem.* **89**, 288 (1965); *Chem. High Polym.* **22**, No. 240, 231 (1965).

133. Haward, R. N., *J. Polym. Sci.* **14**, 535 (1954).
134. Hay, J. M., and Kerr, C. M. L., *J. Polym. Sci. Part B* **3**, 19 (1965); *Chem. Abstr.* **62**, 6569 (1965).
135. Holt, T., and Simpson, W., *Proc. Roy. Soc., Ser. A* **238**, 154 (1956).
136. Hwa, J. C. H., *Abstr. Pap., 131st Meet., Amer. Chem. Soc.* p. 14S (1957).
137. Hwa, J. C. H., *J. Polym. Sci.* **60**, S12 (1962).
138. Hwa, J. C. H., U.S. Pat. 3,239,493 (1966), issued to Rohm and Haas Co.; *Chem. Abstr.* **64**, 17802 (1966).
139. Hwa, J. C. H., Fleming, W. A., and Miller, L., *J. Polym. Sci., Part A* **2**, 2385 (1964).
140. Hwa, J. C. H., and Miller, L., *J. Polym. Sci.* **55**, 197 (1961).
141. Ishida, S., High Temp. Polym., Tech. Doc. Rep. ML-TDR-64-80, Part II, p. 41 (1965).
142. Iwakura, Y., Uno, K., and Ichikawa, K., *J. Polym. Sci., Part A* **2**, 3387 (1964); *Resins, Rubbers, Plastics* p. 1013 (1965).
143. Iwakura, Y., Uno, K., and Ichikawa, K., Jap. Pat. 2157 (1967), issued to Asahi Chemical Ind. Co., Ltd.; *Chem. Abstr.* **67**, 2147, 22363f (1967).
144. Iwakura, Y., Uno, K., and Ichikawa, K., Jap. Pat. 7625 (1967), issued to Asahi Chemical Ind. Co., Ltd.; *Chem. Abstr.* **68**, 1329, 13693g (1968).
145. Iwakura, Y., Uno, K., and Ichikawa, K., Jap. Pat. 7627 (1967), issued to Asahi Chemical Ind. Co., Ltd.; *Chem. Abstr.* **68**, 1329, 13694h (1968).
146. Japanese Patent 14,587 (1966), issued to Nitto Boseki Co., Ltd.
147. Jones, J. F., Ital. Pat. 563,941 (1957), issued to B. F. Goodrich Co.
148. Jones, J. F., *J. Polym. Sci.* **33**, 15 (1958).
149. Kamai, G., and Kukhtin, V. A., *Zh. Obshch. Khim.* **25**, 1875 (1955); *Chem. Abstr.* **50**, 8502 (1956).
150. Katayama, M., Harada, S., Seno, T., and Miyamichi, K., Jap. Pat. 13,674 (1965), issued to Nitto Boseki Co.; *Chem. Abstr.* **63**, 18300 (1965).
151. Kawai, W., *Nippon Kagaku Zasshi* **87**, 75 (1966); *Chem. Abstr.* **65**, 15514 (1966).
152. Kawai, W., *J. Polym. Sci., Part A-1* **4**, 1191 (1966).
153. Kawai, W., *Makromol. Chem.* **93**, 255 (1966).
154. King, C., *J. Amer. Chem. Soc.* **86**, 437 (1964).
155. Kitahama, Y., Ohama, H., and Kobayashi, H., *Int. Symp. Macromol. Chem. (Tokyo-Kyoto), Prepr. Pap., 1966* p. 1 (1966).
156. Kitahama, Y., Ohama, H., and Kobayshi, H., *J. Polym. Sci., Part B* **5**, 1019 (1967); *Chem. Abstr.* **67**, 11077, 117379e (1967).
157. Kocharyan, N. M., Barsamyan, S. T., Matsoyan, S. G., Pikalova, V. N., Tolapchyan, L. S., and Voskanyan, M. G., *Izv. Akad. Nauk. Arm. SSR, Khim. Nauki* **18**, 441 (1965); *Chem. Abstr.* **64**, 12799 (1966).
158. Kolesnikov, G. S., and Davydova, S. L., *Russ. Chem. Rev.* **29**, 679 (1960).
159. Kolesnikov, G. S., Davydova, S. L., and Ermolaeva, T. I., *Vysokomol. Soedin.* **1**, 1493 (1959); *Chem. Abstr.* **54**, 17940 (1960).
160. Koral, J. N., and Lucas, H. R., U.S. Pat. 3,215,675 (1965), issued to American Cyanamid Co.; *Chem. Abstr.* **64**, 3723 (1966).
161. Koral, J. N., and Smolin, E. M., *J. Polym. Sci., Part A* **1**, 2831 (1963); *Resins, Rubbers, Plastics*, p. 1137 (1964).
162. Koton, M. M., Mulyar, P. A., and Kamenetskaya, N. M., *Zh. Prikl. Khim.* **29**, 311 (1956); *Chem. Abstr.* **50**, 13815e (1956); *J. Appl. Chem. USSR* **29**, 347 (1956); *Chem. Abstr.* **51**, 2651 (1957).
163. Liepins, R., Campbell, D., and Walker, C., *Polym. Prepr., Amer. Chem. Soc., Div. Polym. Chem.* **9**, 765 (1968).
164. Luessi, H., *Kolloid-Z. Z. Polym.* **212**, 24 (1966); *Chem. Abstr.* **66**, 1837, 18884p (1967).

165. Marvel, C. S., *J. Polym. Sci.* **48**, 101 (1960).
166. Marvel, C. S., and Garrison, W. E., *J. Amer. Chem. Soc.* **81**, 4737 (1959).
167. Marvel, C. S., and Woolford, R. G., *J. Org. Chem.* **25**, 1641 (1960).
168. Matsoyan, S. G., *J. Polym. Sci.* **52**, 189 (1961).
169. Matsoyan, S. G., *Usp. Khim.* **35**, 70 (1966); *Chem. Abstr.* **64**, 12799 (1966).
170. Matsoyan, S. G., and Akopyan, L. M., *Vysokomol. Soedin.* **3**, 1311 (1961).
171. Matsoyan, S. G., and Akopyan, L. M., *Vysokomol. Soedin* **5**, 1329 (1963); *Resins, Rubbers, Plastics* p. 1193 (1964).
172. Matsoyan, S. G., and Akopyan, L. M., *Izv. Akad. Nauk. Arm. SSR, Khim. Nauki* **16**, 51 (1963); *Chem. Abstr.* **59**, 11367 (1963).
173. Matsoyan, S. G., and Avetyan, M. G., USSR Patent 126,264 (1960); *Chem. Abstr.* **54**, 16024 (1960).
174. Matsoyan, S. G., and Avetyan, M. G., *Zh. Obshch. Khim.* **30**, 697 (1960); *J. Gen. Chem. USSR* **30**, 719 (1960).
175. Matsoyan, S. G., Avetyan, M. G., Akopyan, L. M., Voskanyan, M. G., Morlyan, N. M., and Eliazyan, M. A., *Vysokomol. Soedin.* **3**, 1010 (1961).
176. Matsoyan, S. G., Avetyan, M. G., and Voskanyan, M. G., *Vysokomol. Soedin.* **2**, 314 (1960).
177. Matsoyan, S. G., Avetyan, M. G., and Voskanyan, M. G., *Vysokomol. Soedin.* **3**, 562 (1961); *Resins, Rubbers, Plastics* p. 2659 (1962).
178. Matsoyan, S. G., Avetyan, M. G., and Voskanyan, M. G., *Vysokomol. Soedin.* **3**, 1140 (1961).
179. Matsoyan, S. G., Avetyan, M. G., and Voskanyan, M. G., *Vysokomol. Soedin.* **4**, 882 (1962); *Resins, Rubbers, Plastics* p. 2303 (1962).
180. Matsoyan, S. G., Eliazyan, M. A., and Gevorkyan, E. Z., *Vysokomol. Soedin.* **4**, 1515 (1962); *Resins, Rubbers, Plastics* p. 2541 (1962).
181. Matsoyan, S. G., Pogosyan, G. M., and Cholakyan, A. A., *Izv. Akad. Nauk Arm. SSR, Khim. Nauki* **18**, 178 (1965); *Chem. Abstr.* **63**, 14693 (1965).
182. Matsoyan, S. G., Pogosyan, G. M., Dzhagalyan, A. O., and Mushegyan, A. V., *Vysokomol. Soedin.* **5**, 854 (1963); *Resins, Rubbers, Plastics* p. 381 (1964).
183. Matsoyan, S. G., Pogosyan, G. M., and Eliazyan, M. A., *Vysokomol. Soedin.* **5**, 777 (1963); *Resins, Rubbers, Plastics* p. 2721 (1963).
184. Matsoyan, S. G., Pogosyan, G. M., and Saakyan, A. A., *Karbotsepnye Vysokomol. Soedin.*, *(Suppl. Issue)* **3–6** (1963); *Resins, Rubbers, Plastics* p. 2257 (1964).
185. Matsoyan, S. G., Pogosyan, G. M., and Saakyan, A. A., *Vysokomol. Soedin.* **5**, 1334 (1963); *Resins, Rubbers, Plastics* p. 1197 (1964).
186. Matsoyan, S. G., Pogosyan, G. M., and Zhamkochyan, G. A., *Izv. Akad. Nauk Arm. SSR, Khim. Nauki* **17**, 62 (1964); *Chem. Abstr.* **61**, 4489 (1964).
187. Matsoyan, S. G., and Saakyan, A. A., *Vysokomol. Soedin.* **3**, 1317 (1961).
188. Matsoyan, S. G., and Saakyan, A. A., *Vysokomol. Soedin.* **3**, 1755 (1961).
189. Matsoyan, S. G., and Saakyan, A. A., *Izv. Akad. Nauk Arm. SSR, Khim. Nauki* **15**, 463 (1962); *Chem. Abstr.* **59**, 7655 (1963).
190. Matsoyan, S. G., and Voskanyan, M. G., *Izv. Akad. Nauk Arm. SSR, Khim. Nauki* **16**, 151 (1963).
191. Matsoyan, S. G., Voskanyan, M. G., and Cholakyan, A. A., *Vysokomol. Soedin.* **5**, 1035 (1963); *Resins, Rubbers, Plastics* p. 627 (1964).
192. Matsoyan, S. G., Voskanyan, M. G., Gevorkyan, E. Z., and Cholakyan, A. A., *Izv. Akad. Nauk Arm. SSR, Khim. Nauki* **17**, 420 (1964); *Chem. Abstr.* **62**, 1748 (1965).
193. Matsoyan, S. G., Voskanyan, M. G., and Saakyan, A. A., *Izv. Akad. Nauk Arm. SSR, Khim. Nauki* **16**, 455 (1963); *Chem. Abstr.* **60**, 10796 (1964).

194. Medvedeva, L. I., Fedorova, E. F., and Arbuzova, I. A., *Vysokomol. Soedin., Ser. A* **9**, 2042 (1967); *Chem. Abstr.* **67**, 11078, 117383b (1967).
195. Mercier, J., *Ind. Chim. Belge* **25**, 359 (1960); *Resins, Rubbers, Plastics* p. 1437 (1961).
196. Mercier, J., and Smets, G., *J. Polym. Sci.* **57**, 763 (1962).
197. Mercier, J., and Smets, G., *J. Polym. Sci., Part A* **1**, 1491 (1963); *Resins, Rubbers, Plastics* p. 2291 (1963).
198. Meyersen, K., Dissertation, University of Mainz, Germany (1961).
199. Meyersen, K., Schulz, R. C., and Kern, W., *Makromol. Chem.* **58**, 204 (1962); *Resins, Rubbers, Plastics* p. 1357 (1963).
200. Meyersen, K., and Wang, J. Y. C., *J. Polym. Sci., Part A-1* **5**, 1827 (1967).
201. Mikulasova, D., and Hrivik, A., *Chem. Zvesti* **11**, 641 (1957); *Chem. Abstr.* **52**, 9028 (1958).
202. Miller, W. L., U.S. Pat. 3,218,283 (1965), issued to Monsanto Chemical Co.
203. Miller, W. L., and Black, W. B., *Polym. Prepr. Amer. Chem. Soc., Div. Polym. Chem.* **3**, 345 (1962).
204. Miller, W. L., Brey, W. S., Jr., and Butler, G. B., *J. Polym. Sci.* **54**, 329 (1961).
205. Minoura, Y., and Mitoh, M., *J. Polym. Sci., Part A* **3**, 2149 (1965); *Chem. Abstr.* **63**, 5749 (1965).
206. Minsk, L., and Chechak, J. J., U.S. Pat. 2,632,004 (1953), issued to Eastman Kodak Co.; *Chem. Abstr.* **47**, 12071 (1953).
207. Mitake, T., *Kogyo Kagaku Zasshi* **64**, 1272 (1961); *Resins, Rubbers, Plastics* p. 1379 (1962).
208. Morimoto, G., Kawazura, H., and Yoshie, Y., *Nippon Kagaku Zasshi* **82**, 1464 (1961); *Chem. Abstr.* **57**, 4220 (1962).
209. Moyer, W. W., Jr., and Grev, D. A., *J. Polym. Sci., Part B* **1**, 29 (1963).
210. Murahashi, S., Nozakura, S., Fuji, S., and Kikukawa, K., *Bull. Chem. Soc. Jap.* **38**, 1905 (1965); *Chem. Abstr.* **64**, 6765 (1966).
211. Nametkin, N. S., Topchiev, A. V., Dugar'yan, S. G., and Tolchinskii, I. M., *Vysokomol. Soedin.* **1**, 1739 (1959); *J. Polym. Sci.* **44**, 287 (1960); *Chem. Abstr.* **54**, 14767 (1960).
212. Negi, Y., Harada, S., and Ishizuka, D., *J. Polym. Sci., Part A-1*, **5**, 1951 (1967).
213. Netherlands Patent Application 6,514,783 (1966), issued to Penninsular Chemresearch, Inc.; *Chem. Abstr.* **68**, 333, 3330x (1968).
214. Netherlands Patent Application 6,609,899 (1967), issued to Calgon Corp.; *Chem. Abstr.* **67**, 1155, 11974e (1967).
215. Netherlands Patent Application 6,611,834 (1967), issued to Calgon Corp.
216. Nikolaev, A. F., Balaev, G. A., Meiya, N. V., and Dreiman, N. A., *Vysokomol. Soedin., Ser. B* **9**, 651 (1967); *Chem. Abstr.* **67**, 11074, 117355u (1967).
217. Noma, K., Yosomiya, R., and Sakurada, I., *Kobunshi Kagaku* **22**, 166 (1965); *Chem. Abstr.* **65**, 805 (1966).
218. Oiwa, M., and Ogata, Y., *Nippon Kagaku Zasshi* **79**, 1506 (1958); *Chem. Abstr.* **54**, 4488 (1960).
219. Okada, M., Hayashi, K., Hayashi, K., and Okamura, S., *Nippon Hoshasen Kobunshi Kenkyu Kyokai Nempo* **5**, 95 (1963); *Chem. Abstr.* **63**, 7112 (1965).
220. Okada, M., Hayashi, K., Hayashi, K., and Okamura, S., *Kobunshi Kagaku* **22**, 441 (1965); *Chem. Abstr.* **63**, 16477 (1965).
221. Okamura, S., Hayashi, K., and Mori, K., *10th Annu. Meet. High Polym. Div.*, *Chem. Soc. Jap., 1961*; see also Furukawa, I., and Saegusa, T., "Polymerization of Aldehydes and Oxides." Wiley, New York, 1963.
222. Oku, A., Shono, T., and Oda, R., *Makromol. Chem.* **100**, 224 (1967); *Chem. Abstr.* **66**, 9860, 105260a (1967).

223. Ostroverkhov, V. G., Brunovskaya, L. A., and Kornienko, A. A., *Vysokomol. Soedin.* **6**, 925 (1964).

224. Otsu, T., Goto, K., and Imoto, M., *Kobunshi Kagaku* **21**, 703 (1964); *Chem. Abstr.* **62**, 14825 (1965); *Resins, Rubbers, Plastics* p. 1549 (1965).

225. Overberger, C. G., and Daly, W. H., *J. Org. Chem.* **29**, 757 (1964).

226. Overberger, C. G., and Ishida, S., *Polym. Prepr., Amer. Chem. Soc., Div. Polym. Chem.* **5**, 210 (1964); *Chem. Abstr.* **64**, 2170 (1966).

227. Overberger, C. G., and Ishida, S., *J. Polym. Sci., Part B* **3**, 789 (1965).

228. Overberger, C. G., Ishida, S., and Ringsdorf, H., *J. Polym. Sci.* **62**, S1 (1962).

229. Overberger, C. G., Ringsdorf, H., and Avchen, B., *J. Org. Chem.* **30**, 3088 (1965); *Chem. Abstr.* **63**, 11467 (1965).

230. Paesschen, G. van, Janssen, R., and Hart, R., *Makromol. Chem.* **37**, 46 (1960).

231. Panayotov, I. M., Schopov, I., and Obreshkov, A., *Makromol. Chem.* **100**, 41 (1967).

232. Panayotov, I. M., Schopov, I., and Obreshkov, A., *Izv. Inst. Org. Khim., Bulg. Akad. Nauk.* **3**, 17 (1967); *Chem. Abstr.* **68**, 5803, 59929s (1968).

233. Price, J. A., U.S. Pat. 2,871,229 (1959); issued to American Cyanamid Co., *Chem. Abstr.* **53**, 8651 (1959).

234. Ramp, F. L., *Polym. Prepr., Amer. Chem. Soc., Div. Polym. Chem.* **7**, 582 (1966).

235. Reinmöller, M., and Fox, T. G., *Polym. Prepr., Amer. Chem. Soc., Div. Polym. Chem.* **7**, 1005 (1966).

236. Ringsdorf, H., and Overberger, C. G., *Makromol. Chem.* **44/46**, 418 (1961); *Resins, Rubbers, Plastics* p. 193 (1962).

237. Ringsdorf, H., and Overberger, C. G., *J. Polym. Sci.* **61**, S11 (1962)l

238. Roovers, J., and Smets, G., *Makromol. Chem.* **60**, 89 (1963); *Resins, Rubbers, Plastics* p. 2385 (1963).

239. Rostovskii, E. N., and Barinova, A. N., *Vysokomol. Soedin.* **1**, 1707 (1959); *J. Polym. Sci.* **44**, 286 (1960); *Chem. Abstr.* **54**, 17952 (1960).

240. Rostovskii, E. N., Lis, A. L., and Arbuzova, I. A., *Vysokomol. Soedin.* **7**, 1792 (1965); *Chem. Abstr.* **64**, 5215 (1966); *Polym. Sci. USSR* **7**, 1973 (1965).

241. Rudkovskaya, G. D., Sokolova, T. A., and Koton, M. M., *Dokl. Akad. Nauk SSSR* **164**, 1069 (1965); *Chem. Abstr.* **64**, 3693 (1966); *Dokl. Chem.* **164**, 984 (1965).

242. Sackis, J. J., U.S. Pat. 3,316,181 (1967), issued to Nalco Chem. Co.

243. Sakurada, I., Iwagaki, T., and Sakaguchi, Y., *Kobunshi Kagaku* **21**, 270 (1964); *Makromol. Chem.* **78**, 241 (1964); *Chem. Abstr.* **62**, 6583 (1965).

244. Schuller, W. H., Price, J. A., Moore, S. T., and Thomas, W. M., *J. Chem. Eng. Data* **4**, 273 (1959).

245. Shultz, A. R., *Abstr. Pap., 131st Meet., Amer. Chem. Soc.* p. 14S (1957).

246. Schulz, R. C., *Kolloid. Z. Z. Polym.* **216/217**, 309 (1967).

247. Schulz, R. C., Marx, M., and Hartmann, H., *Makromol. Chem.* **44/46**, 281 (1961); *Resins, Rubbers, Plastics* p. 191 (1962).

248. Schulz, R. C., and Stenner, R., *Makromol. Chem.* **91**, 10 (1966).

249. Shcherbina, F. F., and Fedorova, I. P., *Ukr. Khim. Zh.* **33**, 394 (1967); *Chem. Abstr.* **67**, 5160, 54694f (1967).

250. Shostakovskii, M. F., Khomutov, A. M., Chekulaeva, I. A., and Khomutova, N. M., *Bull. Acad. Sci. USSR, Div. Chem. Sci.* p. 1933 (1961).

251. Simpson, W., and Holt, T., *J. Polym. Sci.* **18**, 335 (1955).

252. Simpson, W., Holt, T., and Zetie, R. J., *J. Polym. Sci.* **10**, 489 (1953).

253. Smets, G., Deval, N., and Hous, P., *J. Polym. Sci., Part A* **2**, 4835 (1964); *Chem. Abstr.* **62**, 6557 (1965).

254. Smets, G., Hous, P., and Deval, N., *J. Polym. Sci., Part A* **2** 4825 (1964); *Chem. Abstr.* **62**, 6557 (1965).

255. Sokolova, T. A., and Rudkovskaya, G. D., *Vysokomol. Soedin.* **3**, 706 (1961).
256. Sokolova, T. A., and Rudkovskaya, G. D., *J. Polym. Sci., Part C* **16**, 1157 (1967).
257. South African Patent 65/6178 (1966), issued to Penninsular Chemresearch, Inc.
258. Spooncer, W. W., Belg. Pat. 608,759 (1962), issued to Shell Research. Ltd.; *Chem. Abstr.* **60**, 15666 (1964).
259. Spooncer, W. W., U.S. Pat. 3,239,492 (1966), issued to Shell Oil Co.; *Chem. Abstr.* **65**, 7399 (1966).
260. Staudinger, H., and Heuer, W., *Chem. Ber.* **67**, 1159 (1934).
261. Stille, J. K., *Abstr. Pap., 139th Meet., Amer. Chem. Soc.* p. 12Q (1961).
262. Stille, J. K., "Introduction to Polymer Chemistry," p. 229. Wiley, New York, 1962.
263. Stille, J. K., and Culbertson, B. M., *J. Polym. Sci., Part A* **2**, 405 (1964).
264. Stille, J. K., and Hillman, J. J., *J. Polym. Sci., Part A-1* **5**, 2067 (1967).
265. Stille, J. K., and Thomson, D. W., *J. Polym. Sci.* **62**, S118 (1962); *Resins, Rubbers, Plastics* p. 187 (1963).
266. Stober, M. R., Michael, K. W., and Speier, J. L., *J. Org. Chem.* **32**, 2740 (1967).
267. Sultanov, K., and Arbuzova, I. A., *Uzb. Khim. Zh.* **9**, 38 (1965); *Chem. Abstr.* **64**, 12799 (1966).
268. Sumi, M., Nozakura, S., and Murahashi, S., *Kobunshi Kagaku* **24**, 512 (1967); *Chem. Abstr.* **68**, 6715, 69389e (1968).
269. Tabushi, I., Sato, S., and Oda, R., *Kogyo Kagaku Zasshi* **67**, 478 (1964); *Chem. Abstr.* **61**, 16002 (1964).
270. Tiers, G. V. D., and Bovey, F. A., *J. Polym. Sci.* **47**, 479 (1960).
271. Topchiev, A. V., Nametkin, N. S., Durgar'yan, S. G., and Dyankov, S. S., *Khim. Prakt. Primen. Kremneorg. Soedin., Tru. Konf., 2nd, 1958* No. 2, p. 118 (1958); *Chem. Abstr.* **53**, 8686 (1959).
272. Trifan, D. S., and Hoglen, J. J., *J. Amer. Chem. Soc.* **83**, 2021 (1961).
273. Trosarelli, L., Guaita, M., and Priola, A., *Ric. Sci., Parte 2, Sez. A* **8** [3], 429 (1965); *Chem. Abstr.* **64**, 5272 (1966).
274. Trosarelli, L., Guaita, M., and Priola, A., *Ric. Sci.* **35**, II-A 429 (1965).
275. Trosarelli, L., Guaita, M., and Priola, A., *IUPAC Int. Symp. Macromol. Chem., 1965* Preprints, p. 442.
276. Trosarelli, L., Guaita, M., and Priola, A., *Ric. Sci.* **36**, 993 (1966); *Chem. Abstr.* **66**, 7179, 76328w (1967).
277. Trosarelli, L., Guaita, M., and Priola, A., *Ann. Chim. (Rome)* **56**, 1065 (1966); *Chem. Abstr.* **66**, 2808, 29189s (1967).
278. Trosarelli, L., Guaita, M., and Priola, A., *J. Polym. Sci., Part B* **5**, 129 (1967).
279. Trosarelli, L., Guaita, M., and Priola, A., *Makromol. Chem.* **100**, 147 (1967).
280. Trosarelli, L., Guaita, M., Priola, A., and Saini, G., *Chim. Ind. (Milan)* **46**, 1173 (1964); *Chem. Abstr.* **62**, 6562 (1965).
281. Tschamler, H., and Leutner, R., *Monatsh. Chem.* **83**, 1502 (1952).
282. Uno, K., Tsuruoka, K., and Iwakura, Y., *J. Polym. Sci., Part A-1* **6**, 85 (1968); *Chem. Abstr.* **68**, 6741, 69656q (1968).
283. Walling, C., *J. Amer. Chem. Soc.* **67**, 441 (1945).
284. Yakhimovich, R. I., Shilov, E. A., and Dvorko, G. F., *Dokl. Akad. Nauk SSSR* **166**, 388 (1966); *Chem. Abstr.* **64**, 12800 (1966); *Dokl. Chem.* **166**, 98 (1966).
285. Yokota, K., Ito, Y., and Ishii, Y., *Kogyo Kagaku Zasshi* **66**, 1112 (1963); *Chem. Abstr.* **60**, 9362 (1964).
286. Yokota, K., Suzuki, Y., and Ishii, Y., *Kogyo Kagaku Zasshi* **68**, 2459 (1965); *Chem. Abstr.* **65**, 13835 (1966).

# $\alpha, \beta$-Unsaturated Aldehyde Polymerizations

Unsaturated aldehydes can yield ring-containing structures on polymerization. Consideration of the structure of acrolein [1], the parent member of the class, illustrates the many polymer possibilities (*109, 110, 112*).

$$\overset{4}{C}H_2=\overset{3}{C}H-\overset{2}{C}H=\overset{1}{O} \qquad \text{(III-1)}$$

**[1]**

Polymers and copolymers with repeat units [2], [3], and [4] are conceivable.

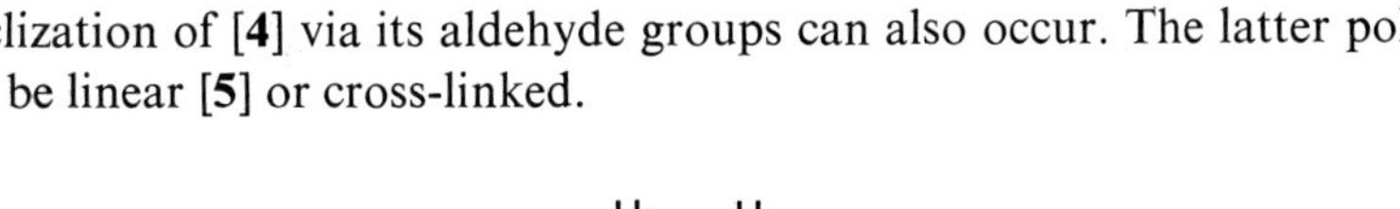

$$\left[\begin{array}{c} CH-O \\ | \\ CH=CH_2 \end{array}\right]_n \qquad \left[ CH_2-CH=CH-O \right]_n \qquad \left[\begin{array}{c} CH_2-CH \\ | \\ CHO \end{array}\right]_n \qquad \text{(III-2)}$$

**[2]**                  **[3]**                **[4]**

1,2-Polymerization      1,4-Polymerization      3,4-Polymerization

Cyclization of [4] via its aldehyde groups can also occur. The latter polymer can be linear [5] or cross-linked.

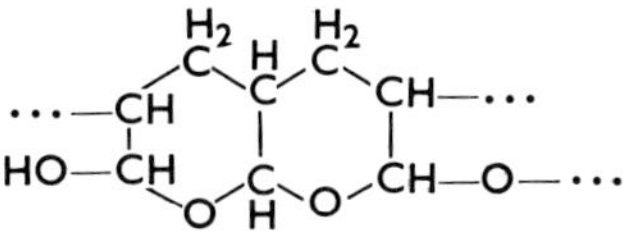

$$\text{(III-3)}$$

**[5]**

Self-polymerization by a Diels-Alder reaction is another example of a ring-forming polymerization that this monomer can undergo. Finally, polycondensation via Michael-type reactions could also occur with this type of monomer.

441

OHC—//  +  ⟨ring⟩$_O$  ⟶  OHC—⟨ring⟩—O  $\xrightarrow{n\text{CH}_2=\text{CHCHO}}$  Ring-containing polymer

(III-4)

[1]    [1]    [6]    [7]

The greatest part of the published work on unsaturated aldehyde polymerization was performed on acrolein. Thus, its behavior will be considered first.


## A. Acrolein Polymers

### 1. POLYMERIZATION

The first report about acrolein polymers dates back to 1843. The observation that freshly prepared acrolein spontaneously changes on standing to an insoluble solid was made by Redtenbacher (92), who gave the polymer the name "disacryle." Since then, polyacroleins have been prepared by radical, anionic, Ziegler, and cationic initiation. Numerous initiators are described in the literature. Representative examples are summarized in Table III.1.

Several reviews on acrolein polymerization have been published (32, 58, 90, 104, 106, 107, 112).


### 2. PROPERTIES

#### a. Molecular Weight

Table III.1 shows that a variety of initiators can be used to polymerize acrolein. Unfortunately, the utility of many of these catalysts is difficult to assess since molecular weight data are not always available. The molecular weights of the polyacroleins vary as a function of the particular method by which they are prepared. The high molecular weight nature of disacryle, the polymer first observed by Redtenbacher (92), was demonstrated by Schulz and co-workers (106). High molecular weight polymers ($\bar{M} \sim 600,000$) are also obtained by redox initiation (49). Values of 3000–7000 are quoted for polyacrolein prepared by the action of butyllithium or tritylsodium in tetrahydrofuran (133). Polymerization with alkali metal cyanides in dimethylformamide, tetrahydrofuran, or toluene gives polymers with reduced viscosities of 0.5–10 (137). In another instance, the polyacrolein prepared with gaseous boron trifluoride in bulk was partly soluble in benzene. The intrinsic viscosity ([η]) of the soluble fraction was 0.17 in benzene (5). Polymerization with manganic pyrophosphate (150) was claimed to yield polymers with intrinsic viscosities up to 3.0 dl/g (aqueous sulfurous acid solution). Intrinsic viscosities of 6.0 dl/g (in aqueous sulfurous acid) corresponding to molecular

## TABLE III.1
### Initiator Systems for Acrolein Polymerization

| Initiator | References |
|---|---|
| METHOD 1. FREE-RADICAL INITIATION | |
| Redox systems in aqueous solutions, preferably at 0°–40°C | *16, 48, 49, 58, 91,* |
| Those used included: $AgNO_3/K_2S_2O_8$, $Fe^{2+}/K_2S_2O_8$, | *104, 106, 113, 141* |
| $Na_2S_2O_5/K_2S_2O_8$, $TiCl_3/K_2S_2O_8$, $NaNO_2/H_2O_2$, $Fe^{2+}/H_2O_2$, | |
| $Fe^{2+}/tert$-$C_4H_9OOH$, $TiCl_3/tert$-$C_4H_9OOH$, $Na_2SO_3/tert$- | |
| $C_4H_9OOH$, Ag/peroxides | |
| Redox systems in aqueous emulsions, preferably at 0°–40°C | |
| $AgNO_3/K_2S_2O_8$/emulsifier | *29, 30* |
| Best performance obtained with sodium poly(vinyl sulfonate) | |
| and the complex polyacrolein·$SO_2$ as the emulsifiers | |
| Polyacrolein·$SO_2/tert$-$C_4H_9OOH$ | *100* |
| Polyacrolein·$SO_2/K_2S_2O_8$ | *8* |
| Polyacrolein·$SO_2/Fe(NO_3)_3$ | *8* |
| Polyacrolein·$SO_2$/peroxides | *84* |
| Redox systems in suspension. Preferred temperatures: | |
| 25°–30°C | |
| $AgNO_3/K_2S_2O_8/H_2O/H_3PO_4/n$-heptane | *154, 155* |
| $Na_2S_2O_5/tert$-$C_4H_9OOH/H_2O/CH_3COOH$/isopentane | *10, 83* |
| Redox systems in $H_2O$/acetone. Preferred temperatures: | |
| 20°–25°C | |
| $SO_2$/peroxides | *11, 99* |
| Polyacrolein·$SO_2$/peroxide | *11, 99* |
| Combinations of redox and basic catalysts. Preferred | |
| temperature: 50°C | |
| $Fe^{2+}/H_2O_2/NH_2NH_2/H_2O$ | *9, 82* |
| Peroxides, $\alpha,\alpha'$-azodiisobutyronitrile, in bulk, at $\leqslant$35°C | *12, 55, 85* |
| Peroxides, $\alpha,\alpha'$-azodiisobutyronitrile in solvents (benzene, | |
| xylene, dioxane) at ~60°C, in the presence of a primary | |
| alcohol | *75* |
| $\alpha,\alpha'$-Azodiisobutyronitrile in benzene, at 50°C | *58, 104* |
| $\alpha,\alpha'$-Azodiisobutyronitrile in dimethylformamide, at | |
| 40–50°C | *134* |
| $\gamma$-Rays with solid acrolein, at $-196$°C | *36* |
| $\gamma$-Rays, in bulk at 20°C | *50* |
| $\gamma$-Rays in methanol, ethanol, or acetone, at $-78$°C | *36* |
| $\gamma$-Rays in methanol or water, at room temperature | *50* |
| UV radiation | *1, 22, 54* |
| High-energy radiation in aqueous medium at pH ~1–5 | *17* |
| High-energy radiation in vapor suspension at ~0–40°C | *151* |
| METHOD 2. THERMAL INITIATION | |
| Thermal | *101* |

TABLE III.1—*continued*

| Initiator | References |
|---|---|

METHOD 3. CATIONIC INITIATION

Gaseous $BF_3$ in bulk or in methylene chloride solution
  at $-40°$ to $-20°C$ — 5
$BF_3 \cdot$ether in bulk, $20°C$ — *104, 133*
Manganic pyrophosphate in aqueous solution, pH $\sim$1–5,
  at $0$–$35°C$ — *150*

METHOD 4. ANIONIC INITIATION

Aqueous sodium hydroxide and other bases — *44, 76, 78, 135, 143*
Amines — *77*
Amines and amides in conjunction with sulfur in dioxane
  solvent — *94*
Organic bases in ester-type solvents — *95*
Phosphines — *33, 51, 56, 138, 140*
NaOH/Polyacrolein$\cdot$$SO_2$/$H_2O$ — *23*
$Na_2SO_3$ or other salts of $H_2SO_3$ at pH $>4.0$ — *38*
Adduct of polyacrolein with $NaHSO_3$ in aqueous solution
  or in bulk — *97, 98*
Na, K, or Li metal in THF, dioxane, $CH_2Cl_2$ or *n*-heptane,
  at $-20°$ to $+60°C$ — *26, 152*
$K_2CO_3$ in bulk — *58, 104*
Sodium, potassium, and lithium cyanides in THF, toluene,
  and dimethylformamide at $\geqslant-40°C$ — *13, 137, 138*
Metal oxides ($Ag_2O$, PbO, $SnO_2$, BeO) in bulk, at $\sim$40°C — *139*
Metal alkoxides — *158*
Sodium and lithium naphthalene in $C_6H_6$ or THF at $-40°$
  to $+20°C$ — *71, 138*
Butyllithium, naphthyllithium, tritylsodium, $NaNH_2$ in bulk
  or in petroleum ether, methylal, THF, acetonitrile, and
  1,2-dimethoxyethane at $-60°$ to $0°C$ — *133, 138*
Grignard reagents in ether, hexane, toluene, or petroleum
  ether at $-20°$ to $+25°C$, optionally in the presence of
  Nylon-6 or other N-containing polymers — *41, 142*
Combination of an alkali metal or hydride with an
  organometallic derivative at $-70°$ to $-20°C$; e.g.
  Na + $Al(C_2H_5)_3$ or NaH + $Zn(C_2H_5)_2$ — *24*

weights of 4,000,000 are claimed in another patent (*151*) in which high-energy radiation was the initiator.

### b. Solubility and Softening Points

Examination of solubility and softening point data reveals important variations from one type of polyacrolein to another. Undoubtedly, these variations reflect differences in the chemical structure of the polymer. This was demonstrated by Schulz and co-workers (*106*), who measured the carbonyl contents of the various materials. The information is summarized in the accompanying table.

Properties and Carbonyl Content of Polyacroleins[a]

| Initiator | Softening point (°C) | Solubility | Carbonyl content (%)[b] |
|---|---|---|---|
| BF$_3$·ether, bulk | 220–240 | Soluble[c] | 14–16 |
| H$_2$O$_2$/Fe$^{2+}$, H$_2$O Soln. | Infusible; becomes yellow-brown at 180–220 | Slightly Sol.[d] | 64 |
| K$_2$CO$_3$, bulk | 100 | Soluble[e] | 20 |
| α, α'-Azodiisobutyronitrile, C$_6$H$_6$ solution | Infusible; yellows at 200–220 | Insoluble[f] | — |
| None (disacryle) | Infusible; yellows at 210–230 | Insoluble[g] | 64–66 |

[a] References (*58, 104, 106*).
[b] Measured by oximation of the polymers.
[c] Soluble in nitrobenzene, DMF, formic acid, pyridine, dioxane, ethyl acetate, glycol monoethyl ether, and acetone.
[d] Soluble in DMF, formic acid, and pyridine.
[e] Soluble in benzene, nitrobenzene, formic acid, DMF, pyridine, dioxane, ethyl acetate, ethyl and *n*-butyl alcohols, glycol monoethyl ether, and acetone.
[f] Swells in formic acid, DMF, and 2$N$ NaOH.
[g] Swells in pyridine and 2$N$ NaOH.

An interesting feature of the data in the table is the similarity of disacryle and the redox polymer (*107*). Both possess a high carbonyl content and are readily soluble in aqueous sulfurous acid and sodium bisulfite. Both have limited solubility in organic solvents and are infusible. On the other hand, the polyacrolein prepared with potassium carbonate as the initiator was soluble in several solvents, softened at 100°C, and displayed a relatively low carbonyl content. Similar properties, but a higher softening point, were shown by the polymer prepared by cationic initiation.

That the physical properties of acrolein polymers depend on preparative conditions is further illustrated by Table III.2. The data show convincingly that slight changes in the polymerization conditions may lead to polymers with completely different properties and, *ipso facto*, different chemical struc-

TABLE III.2

Properties of Various Polyacroleins

| No. | Initiator | Softening point (°C) | Solubility | References |
|-----|-----------|---------------------|------------|------------|
| 1 | Gaseous BF$_3$; bulk, $-40°$ to $-20°$C | 270–320 | Partly in benzene, at conversions <25–30%. Soluble in aq. H$_2$SO$_3$ | (5) |
| 2 | $\gamma$-Rays, $-78°$C, in solvents (alcohol, acetone) | — | DMF, pyridine | (36) |
| 3 | $\gamma$-Rays, $-196°$C, bulk | — | Acetone, carbon tetrachloride, petr. ether, acetonitrile, toluene, glacial acetic acid, $n$-butanol | (36) |
| 4 | $\gamma$-Rays, 20°C, bulk | — | Insoluble in all solvents tested including aq. H$_2$SO$_3$ | (50) |
| 5 | $\gamma$-Rays, 20°C, in solvents (alcohols) | — | Soluble in aq. H$_2$SO$_3$ | (50) |
| 6 | Peroxides, $\alpha,\alpha'$-azodiisobutyronitrile in solvents (benzene, xylene, dioxane), in the presence of a primary alcohol, $\sim 65°$C | — | "Soluble" | (75) |
| 7 | BuLi, tritylsodium, NaNH$_2$ in THF, acetonitrile, and other solvents at $-60°$ to 0°C | 90–110 or 130–150 (depends on particular initiator and solvent) | Benzene, acetone, CCl$_4$, DMF, dioxane THF | (133) |
| 8 | Alkali metal cyanides in THF, DMF, toluene, at $-40°$ to $+40°$C | 80 to 220 depending on experimental conditions. Softening points increase with the increase in polymerization temperature. | DMF, THF, dioxane when polymerization temp. is low. Increase in polymerization temp. leads to insoluble polymer | (137) |

tures. The most dramatic illustration of this phenomenon is provided by polymers 4 and 5 (Table III.2). Polymer 4 is insoluble in all solvents, including aqueous sulfurous acid. This could be due to the fact that it is cross-linked or possesses a very low carbonyl group content. Polymer 5 is soluble in aqueous sulfurous acid and therefore must possess an appreciable amount of available aldehyde groups. Again (cf. polymers 7 and 8) the data show that anionic initiation leads to soluble polymers. However, an important effect of polymerization temperature on the solubility and softening point of polymer 8 was observed, pointing to a change in reaction mechanism.

The solubility characteristics of redox polyacrolein were studied in great detail (21, 25, 40, 49, 122). The solubility changed with the particular redox system employed (107) and with the molecular weight of the polymer (122).

## c. Crystallinity

No crystallinity was detectable in redox polyacrolein or in disacryle (102, 107). X-Ray investigation showed some crystallinity in the polymer prepared via initiation with Grignard reagents in ether or hydrocarbon solvent (142).

## d. Particle Size and Density

The following values were reported (107) for redox polyacrolein and disacryle:

| Type of polymer | Particle size ($\mu$) | Density |
|---|---|---|
| Redox | 0.3–0.5 | Greater than that of disacryle. Depends on initiator |
| Disacryle | 2.5–3 | 1.322 g/cc |

The higher chemical reactivity of the redox polymer was explained by the differences in the particle size (107).

## e. Mechanical Properties and Uses

High pressures (800–18,000 kg/cm$^2$) and high temperatures (~150°C) were used to press redox polyacrolein into transparent sheets on which mechanical properties can be measured (37, 102). The properties are listed in the accompanying table. A variety of uses are claimed for polyacrolein. These include coatings (91), castings (91), improving the wet strength of paper (10, 57), impregnating agents for paper, cloth, and leather (91), adhesives for wood (11, 99), sedimentation aid for purification of water (42), and insolubilizing agents for starch and other materials (11, 99). The treatment of polyacrolein in emulsion with sodium hydroxide and formaldehyde yields a gel dispersion.

Mechanical Properties of Polyacrolein[a]

| | |
|---|---|
| Tensile strength (psi) | 6,000–10,000 |
| Flexural strength (psi) | 7,900 |
| Heat deflection temp. (°C) | 91–132 |
| Izod impact (ft lb/in.$^3$) | 0.35–0.51 |
| Water absorption (%) | 0.45 |
| Transparency (molded pieces) | Fair |
| Acid resistance | Good |
| Alkali resistance | Poor |
| Weatherometer aging | Degradation starts after 100 hr. Polymer is badly degraded after 500 hr of exposure |

[a] Reference (37).

The latter was described as a useful sedimentation agent for sewage purification (96). Stabilization of the polymer with epoxides (91) and aromatic amines (156) is also described.

### f. Infrared Spectra

The infrared spectra of polyacroleins have been studied (36, 50, 102, 104, 106, 107, 137). Important differences were found for polymers of different origin. Redox polyacrolein and disacryle (107) display the following characteristic absorption bands:

| Absorption (μ) | Source |
|---|---|
| 2.9 | Hydroxyl groups, probably hydrated aldehyde, hemiacetal or alcohol—the band remains even after the most careful drying of the polymer |
| 3.6 | Acetal groups |
| 5.8 | Carbonyl absorption |
| 6.1 | Olefinic unsaturation |
| 8.5–11 | Bands in this region characteristic of acetal groups |

The infrared spectrum of polyacrolein obtained by polymerization with sodium cyanide (−48°C, tetrahydrofuran) showed a complete absence of carbonyl absorption. Strong bands were present at 6.1, 10.2, and 10.75 μ, indicative of appreciable olefinic unsaturation (137).

## g. *Chemical Reactions*

Redox polyacrolein and disacryle are soluble in aqueous sulfurous acid and sodium bisulfite, but not in water (*59, 105*). This implies reaction of free or masked aldehyde groups that results in a water-soluble derivative. Several other reactions that are characteristic of aldehydes were performed with redox polyacrolein and disacryle (*58, 106, 107, 112, 121, 129, 130*). The reactions are listed in the accompanying table. An interesting common feature of many of these transformations is the already-mentioned fact that soluble polymeric derivatives are often formed. This indicates that some cross-linking sites that were present originally in the polyacrolein are broken and that linear derivatives are formed. Some interesting polymer–polymer reactions are also reported

Reactions of Redox Polyacrolein and Disacryle

| Reaction | References |
|---|---|
| Formation of oximes | *63, 117* |
| Formation of hydrazones | *119* |
| Formation of acetals | *62, 118* |
| Formation of mercaptals | *132* |
| Oxidation | *14, 15, 108, 126* |
| Reduction | *108, 116, 123* |
| Cannizzaro reaction | *108, 111, 131* |
| Condensation with formaldehyde | *27, 88, 124* |
| Condensation with active methylene compounds | *128* |
| Condensation with Grignard reagents | *141* |

(*61, 108, 111, 112, 125*). For example, reaction of redox polyacrolein with poly(vinyl alcohol) yields a cross-linked, insoluble material. Similar transformations were effected with cellulose, gelatin, collagen, and other polymers.

An interesting feature of redox polyacrolein and disacryle is their stability toward autooxidation (*112*). Solutions of redox polyacrolein in aqueous sulfurous acid absorb oxygen in the presence of catalytic amounts of heavy metal salts (*127*). However, analysis of the reaction mixture indicates that the only species oxidized under these conditions is the sulfurous acid. A clear water solution of a linear polyacrolein is obtained along with sulfuric acid. The polymer gives all the reactions discussed previously (see preceding table), but is much more reactive than the usual insoluble form. On evaporation of the water, the starting insoluble material reforms. The free, water-soluble linear modification can also be obtained by dialyzing solutions of polyacrolein in aqueous sulfurous acid (*111, 112*). These facts show that breaking and forming the cross-links is reversible.

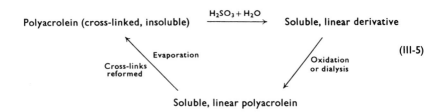

The chemical properties of "anionic" polyacrolein are considerably different (*133*). The polymer is insoluble, even under severe conditions, in aqueous sulfurous acid or sodium bisulfite. Furthermore, anionically prepared poly-acrolein is very prone toward autooxidation (133); polymers of this type can also be cross-linked with ultraviolet light (*46*).

### 3. STRUCTURE

Various structures have been proposed for the polyacroleins, but until the fundamental studies by Schulz and co-workers (*58, 104, 106, 107, 112*), little experimental work was available to support any one structure. Some of the early theories are discussed below.

#### a. Early Theories

The reaction of aqueous acrolein with sodium hydroxide was studied by Gilbert and Donleavy (*44*). Products soluble in ketones and dioxane were obtained. The materials softened with decomposition from 80° to 125°C. A Michael-type polymerization mechanism was postulated (III.6). Structure

$$CH_2{=}CHCHO + H_2O \longrightarrow HOCH_2CH_2CHO \xrightarrow{CH_2{=}CHCHO}$$
$$[8] \qquad\qquad\qquad [9]$$

$$\underset{[10]}{\overset{\displaystyle HOCH_2CHCHO}{\underset{\displaystyle CH_2CH_2CHO}{|}}} \xrightarrow{Monomer} \underset{[11]}{\overset{\displaystyle HOCH_2CHCHO}{\underset{\displaystyle CH_2CH_2CHO}{\overset{|}{(CH_2CHCHO)_n}}}} \qquad (III\text{-}6)$$

**[11]** was supported by oxidation studies and formation of hydrazones and oximes. By cryoscopy, the polymer was shown to be of low molecular weight, probably a pentamer.

According the Scherlin and co-workers (*101*) the thermal polymerization of acrolein is a Diels-Alder polycondensation (III.7) This theory, although

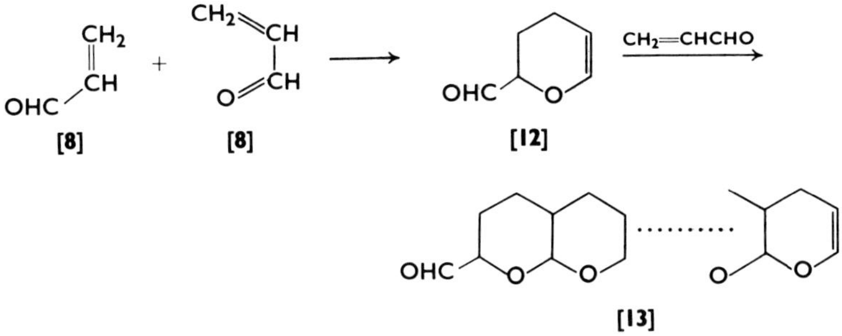

(III-7)

quite attractive, has a serious drawback since it postulates that the number of available aldehyde groups decreases as the polymer molecular weight increases. Experimental evidence does not verify this assumption. It was shown that the carbonyl contents of disacryle and redox polyacrolein are independent of their molecular weights (*106*). It should be noted that the dimerization of

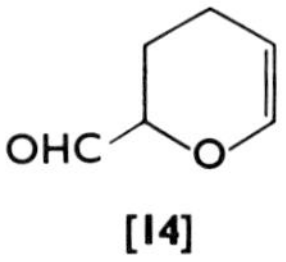

(III-8)

acrolein to [14] is known (*2*). Possibly, the high-boiling residues of this reaction may contain products with Scherlin's structure. Further investigation of these products would be desirable.

The polymerization of acrolein in methanol solution, in the presence of alkali metal hydroxides or carbonates, yielded an insoluble, cross-linked polymer (*135*). The following reactions were postulated [III.9 and III.10]:

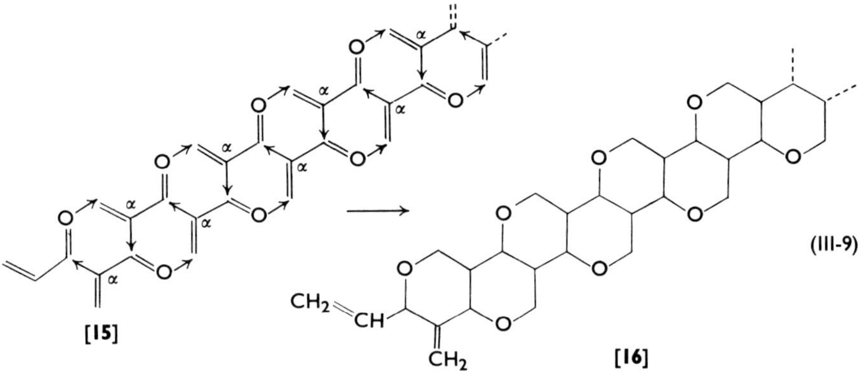

(III-9)

or

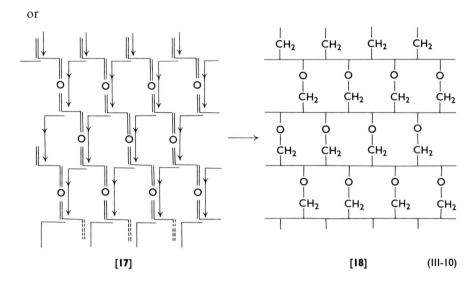

[17]          [18]      (III-10)

No proof for structures [16] and [18] was offered.

### b. Current Status

The studies of Schulz on disacryle and redox polyacrolein support structure [19] for these polymers (*58, 106, 107, 112, 121*). The polymer is formed by a

$$\text{(III-11)}$$

[19]

vinyl polymerization followed by cyclization. Cross-linking of the intermediate polymer [20] occurs by formation of acetal linkages. The structure is in agreement with the chemical and spectroscopic data discussed earlier. The

$$\left[\begin{array}{c} CH_2-CH \\ | \\ CHO \end{array}\right]_n$$

$$\text{(III-12)}$$

[20]

majority of the aldehyde groups are masked but are available under appropriate reaction conditions (53). Due to the hemiacetal structure, the carbonyl contents as measured by oximation are high and independent of molecular weight. The free carboxyl and alcohol groups shown in the formula are probably due to disproportionation by a Cannizzaro-type reaction. Oxidation of free aldehyde groups by redox initiators may also cause carboxyl formation. The nearly complete absence of free aldehyde groups explains the resistance of these polymers towards autooxidation. Infrared examination of these polyacroleins shows that they contain olefinic unsaturation (not shown in formula [19]). The origin of this unsaturation is not clear. It may be due to polymerization via the carbonyl groups, to a 1,4-polymerization, or to the termination step of the growing macroradicals. Formula [19] shows the existence of an acetal-type cross-link (c). Cleavage of that site during characteristic aldehyde reactions will occur. As a result, soluble linear polyacrolein derivatives are formed.

Formula [19] does not specifically indicate the amount of the various structural units in the overall polymer chain. Obviously, one is dealing with a complex copolymer. Changes in the polymerization conditions will lead to a different copolymer with different physical properties. A very elegant attempt to determine quantitatively the amounts of the various structural units was made (52). Low-temperature pyrolysis was used and an empirical polyacrolein formula proposed.

The discussion of polymer properties in the preceding section points out the great differences between free-radical and anionic polyacroleins. Schulz and his students demonstrated (112) that polymerization across the carbonyl group to polymers or copolymers of the type [21] and [22] takes place under

$$\left[\begin{array}{c} CH-O \\ | \\ CH=CH_2 \end{array}\right] \qquad \left\{\left[\begin{array}{c} CH_2-CH \\ | \\ CHO \end{array}\right]_x \left[\begin{array}{c} CH-O \\ | \\ CH=CH_2 \end{array}\right]_y\right\}_n \qquad \text{(III-13)}$$

[21]                               [22]

anionic conditions (13, 133, 136, 137). Carbonyl versus vinyl polymerization depends on the reaction conditions (152). An excellent illustration of changes in the polymerization mechanism has been reported by Schulz et al. (137, 138). The polymerization of acrolein initiated by sodium cyanide in tetrahydrofuran proceeds only through the carbon–oxygen double bond at temperatures below −10°C. At higher temperatures copolymers [22] are obtained. This was substantiated by infrared and chemical evidence. Furthermore, as the polymerization temperatures were increased, the softening points of the polymers increased. They changed from the original value of 80°C (polymer [21]—all carbonyl polymerization) to 220°C (copolymer [22] in which the value of $x$ is substantial). Other investigations (26, 142) provide further evidence that acrolein polymerizes substantially through the carbon–oxygen double bond under anionic initiation conditions. Therefore, it is understandable that anionic polyacrolein can be cross-linked with ultraviolet light (46). Since copolymers of type [22] are probably random, little cyclization to tetrahydro-furan nuclei occurs. Free aldehyde groups are pendant to the backbone. This explains the autooxidation tendency of these materials (133).

Comparatively little work has been reported on the structure of cationically prepared polyacrolein. The polymerization with gaseous boron trifluoride (bulk or methylene chloride solution, at −20° to −40°C) was claimed (5) to yield polymer [23]. There was no discussion of the reasons for this structural

(III-14)

[23]

assignment. It is not certain whether one deals with Scherlin's or Schulz's structure. The fact that the polymer was soluble in aqueous sulfurous acid would indicate that the proposed structure was that of redox polyacrolein (5). According to another source (133), the polyacrolein prepared at room tem-perature, in bulk, in the presence of boron trifluoride etherate possessed approximately 30% olefinic unsaturation. This result indicates a structure similar to that of the anionically prepared polymer.

Depending on experimental conditions, products with widely different properties result when acrolein is polymerized by γ-rays (Table III.2, polymers 2–5). The structure of redox polyacrolein was assigned to polymers 2 and 5 (Table III.2) prepared in solvents (36, 50). Polymerization in bulk at −196°C (polymer 3) yielded a soluble material. The infrared spectrum of the latter showed that it contained three types of units, [24], [25], and [26] (36).

[24]                    [25]                    [26]        (III-15)

An insoluble product was obtained at 20°C in bulk (polymer 4) (50). Its structure was not established, but might be of type [16] or [18].

A recent patent (24) claims that polymerization of acrolein initiated by a mixture of an alkali metal or its hydride, in combination with an organo-metallic compound, proceeds predominantly through the carbonyl groups.

Catalysts such as sodium and triethylaluminum or sodium hydride and diethyl-zinc were used at $-20°$ to $-70°C$. Vinyl-type polymerization accounted for less than $10\%$ under these conditions.

## 4. COPOLYMERS

Copolymers of acrolein with vinyl monomers are prepared under free-radical conditions (112, 157). Comonomers that were used include allyl alcohol (93), acrylonitrile (43, 45, 47, 60, 75, 81, 99, 103, 114, 120), acrylic acid (31, 81), acrylamide (114, 120), methyl acrylate (114), ethyl acrylate (81, 99), methacrylonitrile (120), methyl methacrylate (6, 75, 99), methacrolein (112), alkyl vinyl ketones (28, 75), vinyl acetate (75, 114), styrene (75), sodium p-styrene sulfonate (31), and 2-vinylpyridine (120). Terpolymers from acrolein with ethyl acrylate and acrylonitrile useful for coloring leather were also prepared (86). Initiators used were typical free-radical catalysts in bulk (93) or solution (75, 81, 120). Redox systems were also employed (28, 114). Copolymerization parameters were determined for several acrolein/comonomer pairs (114, 120). Infrared studies were carried out on some of the copolymers that were synthesized (115). It is generally believed (112) that carbon–carbon double bond polymerization of acrolein takes place under these conditions. Copolymers containing more than $20\%$ of acrolein units were generally insoluble in organic solvents. A notable exception to this rule was the behavior of the copolymers based on acrolein/methacrylonitrile and acrolein/methacrolein. The latter were soluble at acrolein contents as high as 70 mole $\%$ (112). It was also reported (75) that soluble derivatives are obtained by free-radical copolymerization in solvents in the presence of a primary aliphatic alcohol. The role of the alcohol in this polymerization (formation of a soluble acetal?) is not fully understood.

The copolymerization of acrolein with aryl vinyl ethers, catalyzed by stannous chloride, was also studied (144–149). Relatively low molecular weight materials are obtained (500–2000). They are light yellow powders, soluble in acetone, diethyl ether, and petroleum ether, but somewhat less soluble in aromatic hydrocarbons. Infrared analysis showed the absence of free aldehyde groups in these products. The following polymerization path was postulated (III.16).

(III-16)

[1]          [27]               [28]                    [29]

A recent patent (74) disclosed copolymers of acrolein with formaldehyde and other aldehydes. Lithium hydride in n-heptane at −40°C was the catalyst. The resulting polymers possessed molecular weights of 17,000–58,000. Polymerization took place via the carbonyl functions. The polymers possess vinyl groups and could be cured to useful elastomeric materials.

Grafting of acrolein onto poly(methyl methacrylate) is achieved in the presence of γ-rays (50, 112). By proper choice of conditions, a thin poly-acrolein layer can be deposited on a film of the poly(methyl methacrylate). An insoluble coating can thus be formed on the surface of the soluble substrate. The polyacrolein layer contained available aldehyde groups and several typical reactions could be performed on it.

## B. Polymers from α-Substituted Acroleins

### 1. POLYMERIZATIONS OF α-ALKYL ACROLEINS

α-Alkyl acroleins are polymerized by a wide variety of initiators. The properties and structures of the polymers are very sensitive to the reaction conditions. In general, the polymerization behavior of these acroleins parallels that of acrolein. However, some exceptions were noted. Pertinent data are summarized in Table III.3. Examination of the data of Table III.3 shows that polymers resulting from vinyl and/or carbonyl polymerizations can be obtained under suitable experimental conditions.

The use of a redox system on methacrolein in aqueous solution led to polymer 1 (Table III.3) that possessed up to 92% of available carbonyl groups (4, 72). A structure [30] similar to that of redox polyacrolein was postulated for this material.

(III-17)

[30]

Chemical and spectroscopic evidence support [30]. On the other hand, the use of sodium cyanide in tetrahydrofuran at −40°C (13, 138) gave the poly-methacrolein (17, Table III.3) that possessed both polyacetal links and pendant aldehyde groups. On the basis of chemical and infrared analysis, structure [31] was assigned (138) to this polymer.

Copolymers resulting from both vinyl and carbonyl polymerizations were obtained (4, 73) from methacrolein by the action of sodium naphthalene in tetrahydrofuran solution. As judged from the double-bond content of the

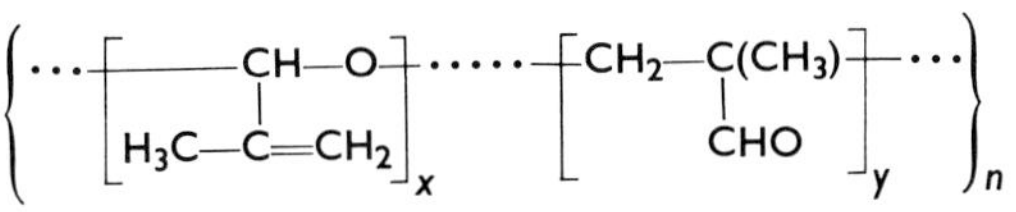

$$\left\{ \cdots \left[ \begin{array}{c} -CH-O- \\ | \\ H_3C-C=CH_2 \end{array} \right]_x \cdots \left[ \begin{array}{c} CH_2-C(CH_3)- \\ | \\ CHO \end{array} \right]_y \cdots \right\}_n \quad \text{where} \quad x > y \qquad (III\text{-}18)$$

[31]

copolymer, carbonyl polymerization is favored at low temperatures. This behavior was explained by the following mechanism:

*Initiation:*

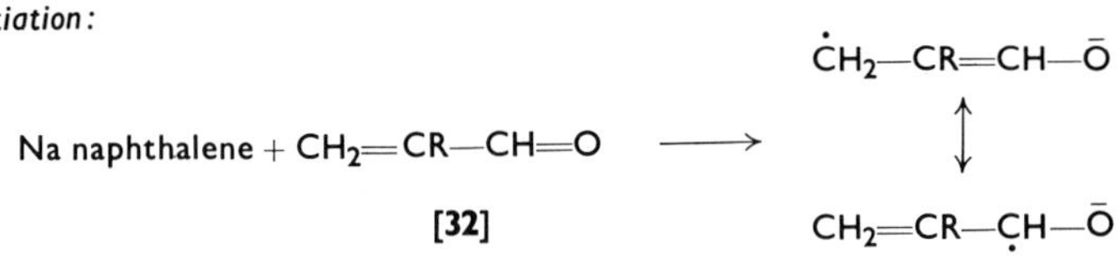

$$\text{Na naphthalene} + CH_2=CR-CH=O \longrightarrow \begin{array}{c} \overset{\cdot}{C}H_2-CR=CH-\bar{O} \\ \updownarrow \\ CH_2=CR-\underset{\cdot}{CH}-\bar{O} \end{array} \qquad (III\text{-}19)$$

[32]       [33]

*Propagation:*

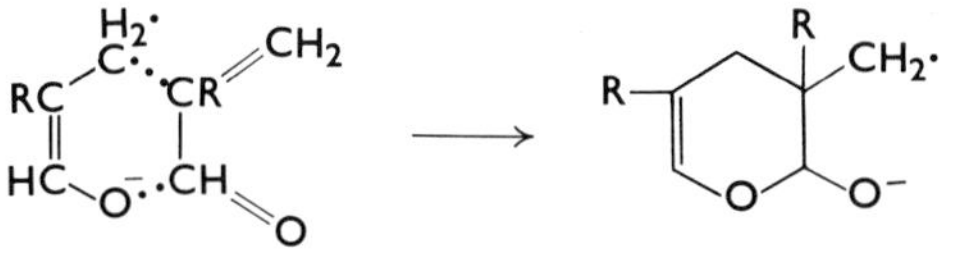

(III-20)

[34]      [35]

*Termination:*

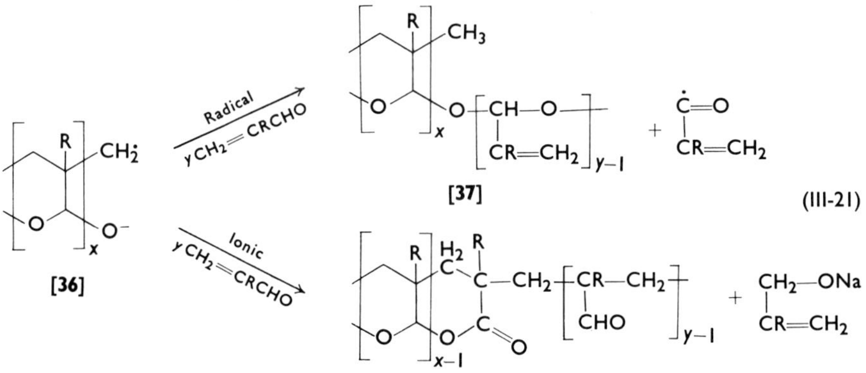

(III-21)

[36]   [37]   [38]

Ionic termination was believed predominant at 0°C and higher.

Structure [39] was postulated (5) for polymethacrolein (polymer 5, Table III.3) obtained by the action of gaseous boron trifluoride at −20° to −40°C, in bulk or methylene chloride solution.

$$
\begin{bmatrix} \text{CH}_3 & \text{CH}_3 \\ & \\ \text{O} & \text{O} \end{bmatrix}_n
$$

(III-22)

[39]

The foregoing discussion shows that there is a clear parallel in the polymerization behavior of acrolein and its homologs. However, exceptions have been encountered. For example, the polymer obtained from acrolein and boron trifluoride etherate at room temperature in bulk possessed 29.9 % of free double bonds; a similar polymerization of methacrolein gave polymer 6 with no olefinic unsaturation (*133*).

## 2. POLY(α-CHLOROACROLEINS)

The polymerization of α-chloroacrolein is effected under a variety of conditions. Ultraviolet light, peroxides, redox systems, Lewis acids, and alkali metals are useful polymerization initiators (*19, 20*). Copolymers with vinyl ethers were prepared (*18*). Polymer properties varied considerably with the experimental conditions. For instance, polymerization of α-chloroacrolein in the presence of hydrogen peroxide yields an amber-colored powder. Its molecular weight is about 3000, it is soluble in acetone, and softens at 115°C. The polymer possesses a high carbonyl content and a high double-bond content (*20*). These facts, coupled with infrared analysis, are evidence for

$$
\left[ -(CH_2CCl{=}CH{-}O{-})_3 \; (CH_2{-}CCl)_2 \atop \qquad\qquad\qquad\qquad CHO \right]_n
$$

(III-23)

[40]

structure [40] (*20*). Redox polymerization [silver nitrate/potassium persulfate] gave very high molecular weight (76,000) poly(α-chloroacrolein). The latter

$$
\begin{bmatrix} -CH_2{-}CCl- & & Cl \\ \quad OHC & O & \\ & Cl & CHO \end{bmatrix}_n
$$

(III-24)

[41]

softened at 190°C and was slightly soluble in the usual organic solvents, but was very soluble in dimethylformamide and pyridine. Results of chemical, infrared, and X-ray analyses (*7*, *20*) were rationalized (*20*) in terms of structure [41]. A different polymer that softened at 150°C was obtained by polymerization with stannous chloride (*20*). Structure [42] was assigned to it (*20*).

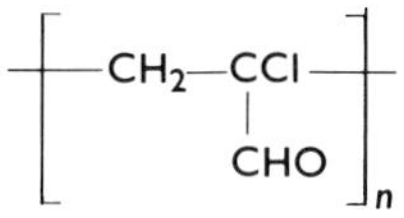

$$\left[\begin{array}{c} -CH_2-CCI- \\ | \\ CHO \end{array}\right]_n \qquad (III\text{-}25)$$

[42]

Similar variations in properties are encountered also in the copolymer area (*18*).

## C. Polymerization of β-Substituted Acroleins

### 1. CROTONALDEHYDE

Crotonaldehyde [43] polymerizes in the presence of amines (*80*), phosphines (*65–67, 69, 80*), aluminum alkoxides (*89*), sodium salts of organic and inorganic acids (*13, 80*), and sodium naphthalene (*3*). Both vinyl and carbonyl

$$CH_3CH{=}CH{-}CHO \qquad (III\text{-}26)$$

[43]

polymerizations are observed. The reaction with phosphines in bulk or solution (methylene chloride, toluene, acetonitrile) was studied in detail (*65–67, 69*). Relatively low molecular weight polymers ($\bar{M}_n \sim 3500$) were obtained. Chemical and spectroscopic evidence indicate that they possess structure [44].

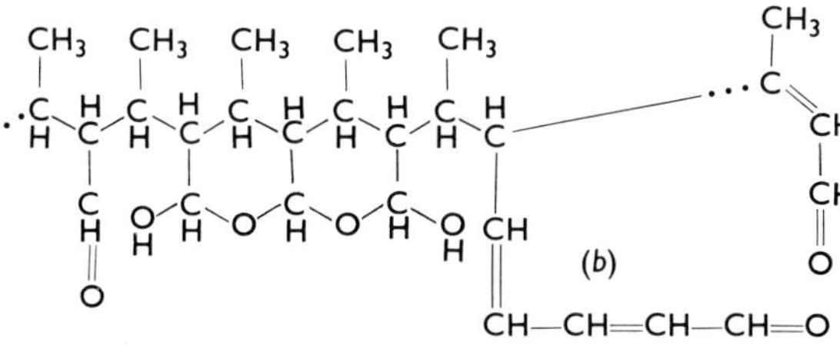

[44]

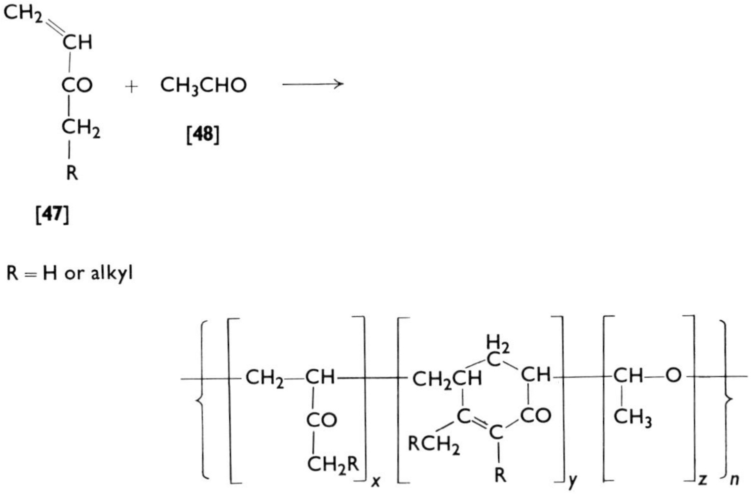

$$CH_2\!=\!\!\overset{\displaystyle\ \ }{\underset{\displaystyle\ }{CH}}$$

CH₂
╲
CH
|
CO    +    CH₃CHO    ⟶
|
CH₂         [48]
|
R

[47]

(III-30)

R = H or alkyl

$$\left\{\left[CH_2\!-\!CH\!-\!\underset{\displaystyle CO}{\underset{\displaystyle |}{\phantom{CO}}}\right]_x\!\left[\cdots\right]_y\!\left[CH\!-\!O\right]_z\right\}_n$$

[49]

TABLE III.3

Polymerization of α-Alkyl Acroleins

| No. | α-Substituted Acrolein | Polymerization conditions | Mol.wt. | Solubility | Softening pt. (°C) | Remarks and property data | References |
|---|---|---|---|---|---|---|---|
| 1 | $CH_2=C-CHO$ $\mid$ $CH_3$ | $AgNO_3/K_2S_2O_8/H_2O$ | $[\eta]$ 0.5 | DMF, pyridine | — | IR and NMR studied. Carbonyl content determined by oximation varied from 75 to 92% depending on polymer molecular weight. 12 to 30% of the carbonyl groups were present in "free" (CHO) state. The remainder were cyclized. A Tishchenko-type reaction was observed in pyridine solution. It was catalyzed by $AgNO_3$ | 39, 72 |
| 2 | $CH_2=C-CHO$ $\mid$ $CH_3$ | "Spontaneous" in the presence of air (bulk) | — | — | — | Product is of "low molecular weight." Chemical evidence indicates that peroxide structures of the type $$\cdots -CH_2-\overset{\overset{\displaystyle CH_3}{\mid}}{\underset{\underset{\displaystyle CHO}{\mid}}{C}}-O-OCH_2-\overset{\overset{\displaystyle CH_3}{\mid}}{\underset{\underset{\displaystyle CHO}{\mid}}{C}}-\cdots$$ are partially present in the polymer | 58 |
| 3 | $CH_2=C-CHO$ $\mid$ $CH_3$ | $K_2S_2O_8/H_2O$, 50°C, 6 hr | $Z_\eta$ 0.04 | Pyridine, DMF, aniline, γ-butyrolactone, m-cresol, quinoline, picoline, nitrobenzene | | Can be oxidized ($KMnO_4$/pyridine or $H_2O_2$/pyridine) to poly(methacrylic acid). Forms polymeric oximes and diacetates. Does not form polyacetals nor addition compounds with sodium bisulfite. Gives a partially lactonized product when heated with aqueous sodium hydroxide at high temperature and pressure | 34, 35, 58, 87 |

| No. | Monomer | Conditions | [η] / State | Solvent | Temp (°C) | Remarks | Ref. |
|---|---|---|---|---|---|---|---|
| 4 | $CH_2{=}\underset{CH_3}{C}{-}CHO$ | Peroxides, $\alpha,\alpha'$-azodi-isobutyronitrile in DMF solvent, at 40°–50°C | — | — | — | Kinetics of polymerization studied | 35, 134 |
| 5 | $CH_2{=}\underset{CH_3}{C}{-}CHO$ | Gaseous $BF_3$ at −20° to −40°C in bulk or in $CH_2Cl_2$ | Insoluble in organic solvents. Swells in $H_2SO_3$ | — | — | Carbonyl content: 9% | 5 |
| 6 | $CH_2{=}\underset{CH_3}{C}{-}CHO$ | $BF_3 \cdot (C_2H_5)_2O$, bulk, room temp. | — | — | — | No olefinic unsaturation found in the polymer | 133 |
| 7 | $CH_2{=}\underset{CH_3}{C}{-}CHO$ | $BF_3 \cdot (C_2H_5)_2O$, $AlCl_3$, $SnCl_4$ in "various solvents" | $Z\,\eta\,0.02$ | Pyridine | — | Polymers contain reactive aldehyde groups | 58 |
| 8 | $CH_2{=}\underset{CH_3}{C}{-}CHO$ | $BF_3 \cdot (C_2H_5)_2O$, at 30°C in chloroform or 1,2-dichloroethane | — | — | — | A fraction $[\eta]$ 0.06–0.09 ($CH_3OH$) was isolated from the initial reaction product. It softened at ~220°C. The polymerization was considered to proceed by a combination of 1,2, 3,4, and 1,4 addition processes | 79 |
| 9 | $CH_2{=}\underset{CH_3}{C}{-}CHO$ | Manganic pyrophosphate in aqueous solution, pH 1–5, at 0–35°C | — | — | — | — | 150 |
| 10 | $CH_2{=}\underset{CH_3}{C}{-}CHO$ | Aq. sodium hydroxide | Pentamer | — | — | Polymerization proceeds via a Michael-type polycondensation | 58 |
| 11 | $CH_2{=}\underset{CH_3}{C}{-}CHO$ | Na metal, at 0° to +20°C, in $C_6H_6$ | $[\eta]\,0.08$ | $C_6H_6$ | 120–170 | Polymerization believed to proceed across vinyl and carbonyl groups. Polymer is partly cyclized | 73 |

TABLE III.3—*continued*
Polymerization of α-Alkyl Acroleins

| No. | α-Substituted Acrolein | Polymerization Conditions | Mol.Wt. | Solubility | Softening pt. (°C) | Remarks and Property Data | References |
|---|---|---|---|---|---|---|---|
| 12 | $CH_2{=}C{-}CHO$<br>$\quad\ \ CH_3$ | Na metal, at room temperature, in bulk | — | — | — | No olefinic unsaturation found in the polymer | *133* |
| 13 | $CH_2{=}C{-}CHO$<br>$\quad\ \ CH_3$ | Na metal, $C_6H_6$ solution at reflux, 4 hr | $[\eta]\ 0.08$ | $C_6H_6$ | 162 | Polymer contained 36.8% of double bonds | *70* |
| 14 | $CH_2{=}C{-}CHO$<br>$\quad\ \ CH_3$ | Na metal, $C_6H_6$ solution at reflux, 8-24 hr | — | $C_6H_6$ and "other" organic solvents | 95 | — | *153* |
| 15 | $CH_2{=}C{-}CHO$<br>$\quad\ \ CH_3$ | Na naphthalene, THF solution room temperature, 30 min | — | — | — | No polymer formed | *133* |
| 16 | $CH_2{=}C{-}CHO$<br>$\quad\ \ CH_3$ | Na naphthalene or tritylsodium, THF solution, −10°C, 15-16 hr | $[\eta]\ 0.14$ | $C_6H_6$ | 285-300 | Structural analysis gave following results:<br><br>% $-\!\left[CH_2\!-\!\underset{\underset{CH_3}{\overset{\mid}{|}}}{\overset{\overset{CHO}{\mid}}{C}}\right]\!-$ units = 27-28<br><br>% $-\!\left[CH\!-\!O\!-\!\underset{\underset{CH_2}{\parallel}}{\overset{\overset{CH_3}{\mid}}{C}}\right]\!-$ units = 11-12<br><br>% (ring structure) units = 57-60 | *73* |

| No. | Monomer | Catalyst/conditions | $\eta_{red}$ | Solvent | $T_m$ (°C) | Remarks | Ref. |
|---|---|---|---|---|---|---|---|
| 17 | $CH_2{=}C(CH_3){-}CHO$ | NaCN, THF solution at $-40°C$ | $\eta_{red}$ 4.7 | Dioxane, DMF, THF | 165–175 | Polymer was shown (IR and chemical evidence) to result from both carbonyl and vinyl polymerizations | 13, 138 |
| 18 | $CH_2{=}C(CH_3){-}CHO$ | $(n\text{-}C_4H_9)_3P$, bulk, or THF or DMF solution at $-50°$ to $0°C$ | $\eta_{red}$ 11.5 | Dioxane, DMF, THF (for polymer prepared in bulk; use of solvents led to insoluble polymers) | 165–175 | The "bulk" polymer was shown (IR and chemical evidence) to result from both carbonyl and vinyl polymerizations | 138 |
| 19 | Copolymer of $CH_2{=}C(CH_3){-}CHO$ with: Acrolein | $K_2S_2O_8/AgNO_3/H_2O$ | — | — | — | Soluble in "organic solvents" when acrolein content not higher than 70% mole | 112 |
| 20 | Acrylonitrile | $K_2S_2O_8/AgNO_3/H_2O$ or $\alpha,\alpha'$-azodiiso-butyronitrile in $C_6H_6$ or DMF | — | — | — | Polymers contain 50–100% carbonyl groups. They can be cross-linked with hexamethylene-diamine, $Fe^{3+}$, $Cu^{2+}$, or $Ni^{2+}/NH_2OH$ to thermally stable systems | 64 |

TABLE III.3—*continued*
Polymerization of α-Alkyl Acroleins

| No. | α-Substituted Acrolein | Polymerization Conditions | Mol. Wt. | Solubility | Soft. Pt. (°C) | Remarks and Property Data | References |
|---|---|---|---|---|---|---|---|
| 21 | Methacrylonitrile | $\alpha,\alpha'$-Azodiisobutyronitrile in DMF or dioxane | — | — | — | Copolymerization parameters determined | 120 |
| 22 | Ethyl acrylate | Redox or free-radical catalyst | — | — | — | — | 35 |
| 23 | $CH_2{=}C{-}CHO$ with $C_2H_5$ | Gaseous $BF_3$, $-20°$ to $-40°C$, bulk or $CH_2Cl_2$ solution | $[\eta]\ 0.065$ | $C_6H_6$ | — | — | 5 |
| 24 | $CH_2{=}C{-}CHO$ $C_2H_5$ | Na metal, $C_6H_6$ solution, reflux, 4 hr | $[\eta]\ 0.08$ | $C_6H_6$ | 100 | Polymer contained 25.2% double bonds | 70 |
| 25 | $CH_2{=}C{-}CHO$ $C_2H_5$ | Na metal, $C_6H_6$ solution, reflux, ~8–24 hr | — | — | 40 | Pale yellow resinous product | 153 |
| 26 | $CH_2{=}C{-}CHO$ $C_2H_5$ | Na naphthalene, $+20°C$, THF solution, 21 hr | $[\eta]\ 0.06$ | $C_6H_6$ | — | Contains 20–28% of units | 73 |

467

| | | | | | | | |
|---|---|---|---|---|---|---|---|
| 27 | CH$_2$=C—CHO<br>　　n-C$_3$H$_7$ | Na metal, C$_6$H$_6$ solution, reflux 16 hr | [η] 0.011 | C$_6$H$_6$ | 115 | Contains 19.6% double bonds | 70 |
| 28 | CH$_2$=C—CHO<br>　　iso-C$_3$H$_7$ | Na metal, C$_6$H$_6$ solution, reflux | [η] 0.013 | C$_6$H$_6$ | 91 | Contains 20.3% double bonds | 70 |
| 29 | CH$_2$=C—CHO<br>　　n-C$_4$H$_9$ | Na metal, C$_6$H$_6$ solution, reflux, 20 hr | [η] 0.017 | C$_6$H$_6$ | 82 | Contains 11.2% double bonds | 70 |

## REFERENCES

1. Adelson, D. E., and Dannenberg, H., U.S. Pat. 2,326,736 (Shell Development Co.) (1943); *Chem. Abstr.* **38**, 1051 (1944).
2. Alder, K., and Rüden, E., *Chem. Ber.* **74**, 920 (1941).
3. Amerik, V. V., Krentsel, B. A., and Shishkina, M. V., *Vysokomol. Soedin.* **7**, 1713 (1965); *Polym. Sci. USSR* **7**, 1887 (1965); *Chem. Abstr.* **64**, 3698 (1966).
4. Andreeva, I. V., Koton, M. M., Getmanchuk, Yu. P., Madorskaya, L. Ya., Pokrovskii, E. I., and Kolstov, A. I., *J. Polym. Sci., Part C* No. 16, 1409 (1967); *Chem. Abstr.* **67**, 4164, 44142k (1967).
5. Andreeva, I. V., Koton, M. M., and Kovaleva, K. A., *Vysokomol. Soedin.* **4**, 528 (1962); *Resins, Rubbers, Plastics* p. 1689 (1962).
6. Annenkova, V. M., Annenkova, V. Z., Gaitseva, E. A., and Borodin, L. I., *Izv. Sib. Otd. Akad. Nauk. SSSR, Ser. Khim. Nauk* No. 1, p. 113 (1967); *Chem. Abstr.* **67**, 5138, 54449e (1967).
7. Annenkova, V. Z., Ivanova, L. T., and Vasil'ev, E. K., *Izv. Sib. Otd. Akad. Nauk SSSR, Ser. Khim. Nauk* No. 3, p. 65 (1966); *Chem. Abstr.* **67**, 2135, 22241q (1967).
8. Bäder, E., Rink, K. H., and Trautwein, H., *Makromol. Chem.* **92**, 198 (1966).
9. Belgian Patent 656,896 (1964), issued to Deutsche Gold und Silber.
10. Belgian Patent 657,656 (1964), issued to Shell Int. Res. Mij., N.V.
11. Belgian Patent 657,657 (1964), issued to Shell Int. Res. Mij., N.V.
12. Belgian Patent 663,266 (1965), issued to Dynamit Nobel A.G.
13. Belgian Patent 675,412 (1966), issued to Deutsche Gold und Silber.
14. Belgian Patent 691,351 (1966), issued to Dynamit Nobel A.G.
15. Belgian Patent 691,564 (1966), issued to Dynamit Nobel A.G.
16. Belgian Patent 691,724 (1966), issued to Dynamit Nobel A.G.
17. Bell, E. R., Campanile, V. A., and Bergman, E., U.S. Pat. 3,105,801 (Shell Oil Co.) (1963); *Chem. Abstr.* **59**, 14133 (1963).
18. Belyaev, V. I., Annenkova, V. Z., Ivanova, L. T., and Ugryumova, G. S., *Izv. Sib. Otd. Akad. Nauk SSSR, Ser. Khim. Nauk* No. 2, p. 114 (1966); *Chem. Abstr.* **66**, 2809, 29194q (1967).
19. Belyaev, V. I., Annenkova, V. Z., Ivanova, L. T., Ugryumova G. S., and Kuryaev, B. S., *Izv. Sib. Otd. Akad Nauk SSSR, Ser. Khim. Nauk* No. 1, p. 144 (1965); *Chem. Abstr.* **63**, 13423 (1965).
20. Belyaev, V. I., Annenkova, V. Z., Ugryumova, G. S., and Ivanova, L. T., *Bull. Acad. Sci. USSR, Div. Chem. Sci.* p. 1202 (1966).
21. Bergman, E., Tsatsos, W. T., and Fischer, R. F., *J. Polym. Sci. Part A* **3**, 3485 (1965).
22. Blacet, F. E., Fielding, G. H., and Roof, J. G., *J. Amer. Chem. Soc.* **59**, 2375 (1937).
23. British Patent 990,263 (1965), issued to Deutsche Gold und Silber; *Chem. Abstr.* **63**, 4413 (1965).
24. British Patent 1,029,689 (1965), issued to Mitsubishi Petrochem. Co.
25. British Patent 1,040,870 (1965), issued to Dynamit Nobel A.G.
26. British Patent 1,051,858 (1966), issued to Mitsubishi Petrochem. Co.; *Chem. Abstr.* **66**, 4447, 46778a (1967).
27. British Patent 1,064,860 (1967), issued to Dynamit Nobel; *Chem. Abstr.* **67**, 1157, 11989p (1967).
28. Campanile, V. A., and Tsatsos, W. T., U.S. Pat. 3,277,057 (Shell Oil Co.) (1966).
29. Cherdron, H., *Kunststoffe* **50**, 568 (1960).
30. Cherdron, H., Schulz, R. C., and Kern, W., *Makromol. Chem.* **32**, 197 (1959).

31. D'Alelio, G. F., and Huemmer, T. F., *J. Polym. Sci.*, *Part A-1* **5**, 77 (1967); *Chem. Abstr.* **66**, 6222, 65861p (1967).

32. DeWinter, W., *Rev. Macromol. Chem.* **1**, 329 (1966); *Chem. Abstr.* **66**, 5285, 55758w (1967).

33. Eifert, R. L., and Marks, B. M., Brit. Pat. 872,331 (1958); issued to E. I. duPont de Nemours and Co.

34. Eifert, R. L., and Marks, B. M., U.S. Pat. 3,000,862 (E. I. duPont de Nemours and Co.) (1961); *Chem. Abstr.* **56**, 1610 (1962).

35. Eifert, R. L., and Marks, B. M., U.S. Pat. 3,118,860 (E. I. duPont de Nemours and Co.) (1964); *Chem. Abstr.* **61**, 744 (1964).

36. Finkelshtein, E. I., and Abkin, A. D., *Proc. Acad. Sci. USSR* p. 356 (1965).

37. Fischer, R. F., and Stewart, A. T., Jr., *J. Polym. Sci. Part A* **3**, 3495 (1965); *Chem. Abstr.* **64**, 870 (1966).

38. French Patent 1,375,851 (1964), issued to Deutsche Gold und Silber; *Chem. Abstr.* **62**, 11935 (1965).

39. French Patent 1,455,044 (1964), issued to Chemical Investors S.A.

40. French Patent 1,455,795 (1965), issued to Dynamit Nobel A.G.

41. Fukui, K., Kagiya, T., Shimizu, T., and Sumitomo, Y., Jap. Pat. 951 (1967), issued to Kokoku Rayon and Pulp Co., Ltd.; *Chem. Abstr.* **67**, 1157, 11990g (1967).

42. German Patent 1,229,045 (1963), issued to Deutsche Gold und Silber.

43. German Patent 1,231,435 (1966), issued to Deutsche Gold und Silber.

44. Gilbert, E. E., and Donleavy, J. J., *J. Amer. Chem. Soc.* **60**, 1911 (1938).

45. Goeltner, W., and Schlack, P., Ger. Pat. 1,231,435 (Degussa) (1966); *Chem. Abstr.* **66**, 3694, 38435w (1967).

46. Gole, J., and Calvayroc, H., *C. R. Acad. Sci.* **260**, 163 (1965); *Chem. Abstr.* **62**, 13266 (1965).

47. Göltner, W., and Schlack, P., Belg. Pat. 593,116 (1960), issued to Deutsche Gold und Silber.

48. Gorokhovskaya, A. S., *Sb. Nauch.-Issled. Rabot Khim. Khim. Tekhnol., Vysokomol. Soedin., Tashkentsk. Tekstil'n. Inst.* No. 1, p. 84 (1964); *Chem. Abstr.* **64**, 2165 (1966).

49. Hank, R., *Makromol. Chem.* **52**, 108 (1962).

50. Henglein, A., Schnabel, W., and Schulz, R. C., *Makromol. Chem.* **31**, 181 (1959).

51. Horner, L., Jurgeleit, W., and Klüpfel, K., *Justus Liebigs Ann. Chem.* **591**, 116 (1955).

52. Hunter, L., and Forbes, J. W., *J. Polym. Sci., Part A* **3**, 3471 (1965).

53. Hunter, L., and Nixon, A. C., U.S. Pat. 3,322,725 (Shell Oil Co.) (1967); *Chem. Abstr.* **67**, 5165, 54738y (1967).

54. Joliot, F., Fr. Pat. 966,760 (1940).

55. Joshi, R. M., *Makromol. Chem.* **55**, 48 (1962).

56. Jurgeleit, W., and Heisenberg, E., Ger. Pat. 961,131 (Vereinigte Glanzstoff-Fabriken) (1957); *Chem. Abstr.* **53**, 14585 (1959).

57. Kekish, G. T., U.S. Pat. 3,317,370 (Nalco Chem. Co.) (1967).

58. Kern, W., and Schulz, R. C., *Angew. Chem.* **69**, 153 (1957); *Resins, Rubbers, Plastics* p. 805 (1957).

59. Kern, W., and Schulz, R. C., Ger. Pat. 1,016,020 (Degussa) (1957); *Chem. Abstr.* **53**, 16589 (1959).

60. Kern, W., Schulz, R. C., and Cherdron, H., U.S. Pat. 3,036,978 (Deutsche Gold und Silber) (1962); *Chem. Abstr.* **57**, 10050 (1962).

61. Kern, W., Schulz, R. C., and Löflund, I., U.S. Pat. 3,271,334 (Deutsche Gold und Silber) (1966).

62. Kern, W., Schweitzer, O., and Schulz, R. C., U.S. Pat. 3,206,433 (Deutsche Gold und Silber) (1965); *Chem. Abstr.* **64**, 3722 (1966).
63. Kern, W., Schweitzer, O., and Schulz, R. C., U.S. Pat. 3,234,164 (Deutsche Gold und Silber) (1966); *Chem. Abstr.* **64**, 12843 (1966).
64. Kol'k, A. R., Konkin, A. A., and Rogovin, Z. A., *Khim. Volokna* p. 15 (1966); *Chem. Abstr.* **66**, 3722, 38747z (1967).
65. Koral, J. N., *J. Polym. Sci.* **61**, S37 (1962).
66. Koral, J. N., *Polym. Prepr., Amer. Chem. Soc., Div. Polym. Chem.* **3**, No. 2, 293 (1962).
67. Koral, J. N., *Makromol. Chem.* **62**, 148 (1963); *Chem. Abstr.* **59**, 768 (1963).
68. Koral, J. N., U.S. Pat. 3,293,215 (American Cyanamid Co.) (1966); *Chem. Abstr.* **66**, 3693, 38434v (1967).
69. Koral, J. N., U.S. Pat. 3,293,216 (American Cyanamid Co.) (1966); *Chem. Abstr.* **66**, 3686, 38357x (1967).
70. Koton, M. M., Andreeva, I. V., and Getmanchuk, Yu. P., *Vysokomol. Soedin.* **4**, 1537 (1962); *Resins, Rubbers, Plastics* p. 2635 (1962).
71. Koton, M. M., Andreeva, I. V., and Getmanchuk, Yu. P., *Dokl. Akad. Nauk SSSR* **155**, 836 (1964); *Dokl. Chem. (English transl.)* **155**, 308 (1964).
72. Koton, M. M., Andreeva, I. V., Getmanchuk, Yu. P., Madorskaya, L. Ya., and Pokrovskii, E. I., *Vysokomol. Soedin.* **8**, 1389 (1966).
73. Koton, M. M., Andreeva, I. V., Getmanchuk, Yu. P., Madorskaya, L. Ya., Pokrovskii, E. I., Kol'stov, A. I., and Filatova, V. A., *Vysokomol. Soedin.* **7**, 2039 (1965); *Polym. Sci. USSR* **7**, 2232 (1965).
74. Mark, H. F., and Atlas, S. M., Fr. Pat. 1,438,201 (1966); *Chem. Abstr.* **66**, 2877, 29874m (1967).
75. Miller, H. C., and Rothrock, H. S., U.S. Pat. 2,657,192 (E. I. duPont de Nemours and Co.) (1953); *Chem. Abstr.* **48**, 3723 (1954).
76. Moureau, C., and Dufraisse, C., *C. R. Acad. Sci.* **169**, 621 (1919).
77. Moureau, C., and Dufraisse, C., Brit. Pat. 141,058 (1920).
78. Moureau, C., and Dufraisse, C., *C. R. Acad. Sci.* **175**, 127 (1922).
79. Nagai, Y., and Nakajima, T., *Kogyo Kagaku Zasshi* **66**, 1905 (1963); *Chem. Abstr.* **61**, 8181 (1964).
80. Nagai, Y., and Nakajima, T., *Kogyo Kagaku Zasshi* **69**, 495 (1966); *Chem. Abstr.* **65**, 12297 (1966).
81. Neher, H. T., and Woodward, C. F., U.S. Pat. 2,416,536 (Rohm and Haas Co.) (1947); *Chem. Abstr.* **41**, 4006 (1947).
82. Netherlands Patent Application 6,411,783 (1965), issued to Deutsche Gold und Silber; *Chem. Abstr.* **63**, 18299 (1965).
83. Netherlands Patent Application 6,415,218 (1965), issued to Shell International Maatsch. Research, N.V.; *Chem. Abstr.* **63**, 18294 (1965).
84. Netherlands Patent Application 6,500,192 (1965), issued to Dynamit Nobel A.G.; *Chem. Abstr.* **63**, 18299 (1965).
85. Netherlands Patent Application 6,505,495 (1965), issued to Dynamit Nobel A.G.; *Chem. Abstr.* **64**, 14312 (1966).
86. Netherlands Patent Application 6,603,821 (1966), issued to Ciba, Ltd.; *Chem. Abstr.* **67**, 1156, 11975f (1967).
87. Netherlands Patent Application 6,608,759 (1966), issued to Gulf Research and Development Co.; *Chem. Abstr.* **66**, 9881, 105461s (1967).
88. Netherlands Patent Application 6,614,286 (1967), issued to Dynamit Nobel A.G.
89. Ota, T., *Nippon Kagaku Zasshi* **86**, 850 (1965); *Chem. Abstr.* **65**, 15523 (1966).
90. Overberger, C. G., *Abstr. 148th Meet. Amer. Chem. Soc., Div. Chem. Educ.*, 9F–20 (1964).

91. Pannell, C. E., U.S. Pat. 3,244,651 (Shell Oil Co.) (1966).
92. Redtenbacher, J., *Ann. Chem. Pharm.* **47**, 113 (1843).
93. Reinhard, K. H., U.S. Pat. 2,891,037 (Monsanto Chemical Co.) (1959); *Chem. Abstr.* **53**, 17581 (1959).
94. Rink, K. H., Ger. Pat. 1,059,661 (Degussa) (1959); *Chem. Abstr.* **55**, 7920 (1961).
95. Rink, K. H., Ger. Pat. 1,059,662 (Degussa) (1959); *Chem. Abstr.* **55**, 7921 (1961).
96. Rink, K. H., and Baeder, E., Ger. Pat. 1,229,045 (Degussa) (1966); *Chem. Abstr.* **66**, 3715, 38666x (1967).
97. Rink, K. H., and Schweitzer, O., Ger. Pat. 1,138,546 (Deutsche Gold und Silber) (1962); *Chem. Abstr.* **58**, 4661 (1963).
98. Rink, K. H., and Schweitzer, O., U.S. Pat. 3,313,750 (Deutsche Gold und Silber) (1967).
99. Ryder, E. E., Jr., and June, R. K., U.S. Pat. 3,310,531 (Shell Oil Company) (1967).
100. Ryder, E. E., Jr., and Pezzaglia, P., *J. Polym. Sci., Part A* **3**, 3459 (1965); *Chem. Abstr.* **64**, 2166 (1966).
101. Scherlin, S. M., *Zh. Obshch. Khim.* **8**, 22 (1938); *Chem. Zentralbl.* I, 1971 (1939).
102. Schilling, H., *Kolloid-Z.* **175**, 110 (1961).
103. Schlack, P., and Goeltner, W., Ger. Pat. 1,232,750 (Deutsche Gold und Silber) (1967); *Chem. Abstr.* **66**, 7194, 76480q (1967).
104. Schulz, R. C., *Makromol. Chem.* **17**, 62 (1955).
105. Schulz, R. C., *Chem.-Ing.-Tech.* **28**, 296 (1956).
106. Schulz, R. C., *Kunststoffe* **47**, 303 (1957); *Resins, Rubbers, Plastics* p. 1041 (1957).
107. Schulz, R. C., *Kunststoffe* **48**, 257 (1958).
108. Schulz, R. C., *Kunstst.-Plast.* **6**, 32 (1959); *Chem. Abstr.* **53**, 17563 (1959).
109. Schulz, R. C., *Kunststoffe* **51**, 778 (1961).
110. Schulz, R. C., *Angew. Chem.* **73**, 777 (1961).
111. Schulz, R. C., *Kolloid-Z.* **182**, 99 (1962).
112. Schulz, R. C., *Angew. Chem., Int. Ed. Engl.* **3**, 416 (1964).
113. Schulz, R. C., Cherdron, H., and Kern, W., *Makromol. Chem.* **24**, 141 (1957); *Resins, Rubbers, Plastics* p. 273 (1958).
114. Schulz, R. C., Cherdron, H., and Kern, W., *Makromol. Chem.* **28**, 197 (1958).
115. Schulz, R. C., Cherdron, H., and Kern, W., *Makromol. Chem.* **29**, 190 (1959).
116. Schulz, R. C., and Elzer, P., *Makromol. Chem.* **42**, 205 (1961).
117. Schulz, R. C., Fauth, H., and Kern, W., *Makromol. Chem.* **20**, 161 (1956).
118. Schulz, R. C., Fauth, H., and Kern, W., *Makromol. Chem.* **21**, 227 (1956).
119. Schulz, R. C., Holländer, R., and Kern, W., *Makromol. Chem.* **40**, 16 (1960).
120. Schulz, R. C., Kaiser, E., and Kern, W., *Makromol. Chem.* **58**, 160 (1962); *Resins, Rubbers, Plastics* p. 1209 (1963).
121. Schulz, R. C., and Kern, W., *Makromol. Chem.* **18/19**, 4 (1955).
122. Schulz, R. C., Kovacs, J., and Kern, W., *Makromol. Chem.* **52**, 236 (1962).
123. Schulz, R. C., Kovacs, J., and Kern, W., *Makromol. Chem.* **54**, 146 (1962).
124. Schulz, R. C., Kovacs, J., and Kern, W., *Makromol. Chem.* **67**, 187 (1963).
125. Schulz, R. C., and Löflund, I., *Angew. Chem.* **72**, 771 (1960).
126. Schulz, R. C., Löflund, I., and Kern, W., *Makromol. Chem.* **28**, 58 (1958).
127. Schulz, R. C., Löflund, I., and Kern, W., *Makromol. Chem.* **32**, 209 (1959).
128. Schulz, R. C., Meyersen, K., and Kern, W., *Makromol. Chem.* **53**, 58 (1962).
129. Schulz, R. C., Meyersen, K., and Kern, W., *Makromol. Chem.* **54**, 156 (1962).
130. Schulz, R. C., Meyersen, K., and Kern, W., *Makromol. Chem.* **59**, 123 (1963).
131. Schulz, R. C., Müller, E., and Kern, W., *Naturwissenschaften* **45**, 440 (1958).
132. Schulz, R. C., Müller, E., and Kern, W., *Makromol. Chem.* **30**, 39 (1959).

133. Schulz, R. C., and Passmann, W., *Makromol. Chem.* **60**, 139 (1963); *Resins, Rubbers, Plastics* p. 2069 (1963).
134. Schulz, R. C., Suzuki, S., Cherdron, H., and Kern, W., *Makromol. Chem.* **53**, 145 (1962).
135. Schulz, H., and Wagner, H., *Angew. Chem.* **62**, 105 (1950).
136. Schulz, R. C., and Wegner, G., *Makromol. Chem.* **104**, 185 (1967).
137. Schulz, R. C., Wegner, G., and Kern, W., *Makromol. Chem.* **100**, 208 (1967).
138. Schulz, R. C., Wegner, G., and Kern, W., *J. Polym. Sci., Part C* **16**, 989 (1967).
139. Shokal, E. C., U.S. Pat. 2,819,252 (Shell Development Co.) (1958); *Chem. Abstr.* **52**, 6843 (1958).
140. Shokal, E. C., Fr. Pat. 1,138,854 (Shell Development Co.) (1959).
141. Shostakovskii, M. F., Belyaev, V. I., and Ivanova, L. T., *Izv. Sib. Otd. Akad. Nauk SSSR, Ser. Khim. Nauk* No. 3, p. 110 (1964); *Chem. Abstr.* **63**, 4411 (1965).
142. Shostakovskii, M. F., Belyaev, V. I., Okladnikova, Z. A., Vasil'eva, L. V., and Serebrennikova, E. V., *Izv. Sib. Otd. Akad. Nauk SSSR, Ser. Khim. Nauk* No. 1, p. 88 (1965); *Chem. Abstr.* **63**, 13423 (1965).
143. Shostakovskii, M. F., Belyaev, V. I., Okladnikova, Z. A., Vasil'eva, L. V., and Zavalei, V. M., *Izv. Sib. Otd. Akad. Nauk SSSR, Ser. Khim. Nauk* No. 3, p. 115 (1964); *Chem. Abstr.* **63**, 4412 (1965).
144. Shostakovskii, M. F., Skvortsova, G. G., Samoilova, M. Ya., and Zapunnaya, K. V., *Izv. Sib. Otd. Akad. Nauk SSSR, Ser. Khim. Nauk* p. 37 (1961).
145. Shostakovskii, M. F., Skvortsova, G. G., Samoilova, M. Ya., and Zapunnaya, K. V., *Izv. Sib. Otd. Akad. Nauk SSSR, Ser. Khim. Nauk* p. 50 (1961).
146. Shostakovskii, M. F., Skvortsova, G. G., and Zapunnaya, K. V., *Vysokomol. Soedin.* **5**, 767 (1963).
147. Shostakovskii, M. F., Skvortsova, G. G., and Zapunnaya, K. V., *Izv. Akad. Nauk SSSR, Ser. Khim.* p. 2032 (1965); *Bull. Acad. Sci. USSR, Div. Chem. Sci.* p. 1996 (1965).
148. Shostakovskii, M. F., Skvortsova, G. G., Zapunnaya, K. V., and Kozyrev, V. G., *Dokl. Akad. Nauk SSSR*, **162**, 124 (1965); *Dokl. Chem. (English transl.)* **162**, 451 (1965); *Chem. Abstr.* **63**, 7113 (1965).
149. Shostakovskii, M. F., Skvortsova, G. G., Zapunnaya, K. V., and Kozyrev, V. G., *Vysokomol. Soedin., Ser. A* **9**, 704 (1967); *Chem. Abstr.* **67**, 2129, 22181v (1967).
150. Sobolev, I., U. S. Pat. 3,215,674 (Shell Oil Co.) (1965).
151. Stewart, A. T., Jr., and Nixon, A. C., U.S. Pat. 3,215,612 (Shell Oil Co.) (1965).
152. Tanaka, S., Mabuchi, K., Shimazaki, N., and Shimizu, M., Jap. Pat. 19,912 (Mitsubishi Petrochem. Co.) (1966); *Chem. Abstr.* **66**, 7200, 76539r (1967).
153. Weiler, W., Ger. Pat. 889,227 (BASF) (1953).
154. Welch, F. J., U.S. Pat. 3,069,389 (Union Carbide Corp.) (1962); *Chem. Abstr.* **58**, 4661 (1963).
155. Welch, F. J., Fr. Pat. 1,325,213 (Union Carbide Corp.) (1963).
156. Welch, F. J., U.S. Pat. 3,225,000 (Union Carbide Corp.) (1965).
157. Wolz, H., and Bock, W., Ger. Pat. 855,162 (1952).
158. Zilkha, A., Feit, B., and Frankel, M., *Proc. Chem. Soc., London* p. 255 (1958).

## CHAPTER IV

# Miscellaneous Ring-Forming Polymerizations

## A. Polydiazadiphosphetidines

Polydiazadiphosphetidines are prepared by the method shown in Eq. (IV-1) (*36*). The reactions are performed in bulk at 225°–285°C. Several

$$
H_2NArNH_2 \xrightarrow[\text{[2]}]{2[(CH_3)_2N]_3PO/\text{Heat}} \left[ \begin{array}{c} -P(O)-NAr- \\ | \quad\quad | \\ -N-\!-\!-P(O)- \end{array} \right]_n \quad\quad \text{(IV-1)}
$$

[1]                                                                                  [3]

aromatic diamines were used. Fractionation of the products yielded samples in which the values of *n* were as high as 18–20. The latter polymers were insoluble in DMF. On the other hand, solubility in that solvent was achieved at lower molecular weights (*n* = 11–13). The materials reportedly displayed stability up to 500°C and were claimed to be useful in the preparation of composites for "environmental extremes" (*36*).

## B. Polytetrazadiborines

The reaction of *p*-phenylenediboronic acid [4] with hydrazine yields polyphenylenetetrazadiborine (*35*). The oxygen analog of [6] was obtained by a similar condensation starting from pyroboretetraacetate [8] (*35*). Both polymerizations were conducted in bulk. Polymers [6] and [9] were insoluble in most organic solvents. They were hydrolytically unstable; their structures

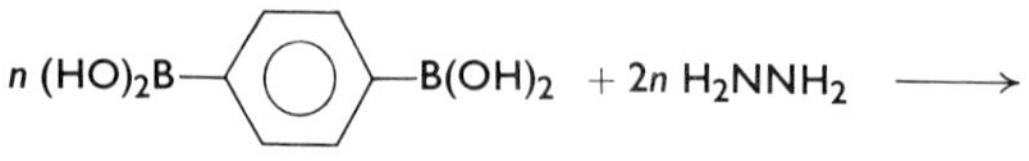

$$n \; (HO)_2B-\!\!\left\langle\bigcirc\right\rangle\!\!-B(OH)_2 \;+\; 2n \; H_2NNH_2 \;\longrightarrow$$

[4]                 [5]

(IV-2)

$$\left[ -\!\!\left\langle\bigcirc\right\rangle\!\!-B \!\! \begin{array}{c} NH\!\!-\!\!NH \\ \diagup \quad\quad \diagdown \\ NH\!\!-\!\!NH \end{array} \!\! B- \right]_n \;+\; 4n \; H_2O$$

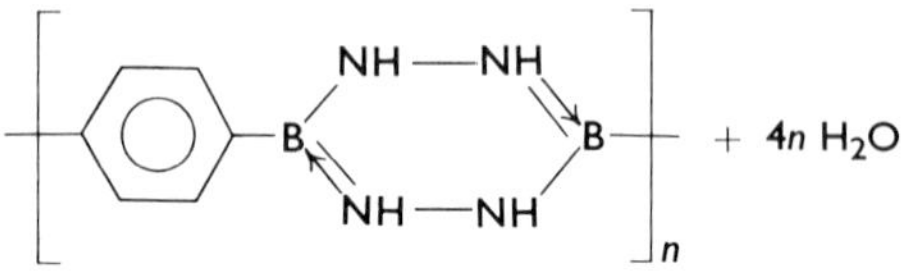

[6]                 [7]

were supported by hydrolysis and infrared data. No molecular weights were reported (*35*).

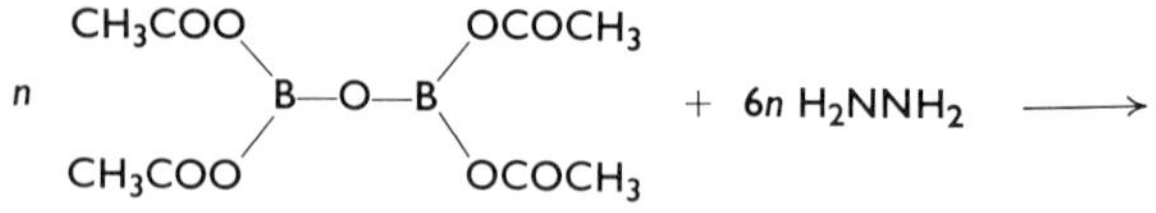

$$n \; \begin{array}{c} CH_3COO \\ \diagdown \\ \quad\quad B\!\!-\!\!O\!\!-\!\!B \\ \diagup \\ CH_3COO \end{array} \!\! \begin{array}{c} OCOCH_3 \\ \diagup \\ \\ \diagdown \\ OCOCH_3 \end{array} \;+\; 6n \; H_2NNH_2 \;\longrightarrow$$

[8]                 [5]

(IV-3)

$$\left[ -O\!\!-\!\!B \!\! \begin{array}{c} NH\!\!-\!\!NH \\ \diagup \quad\quad \diagdown \\ NH\!\!-\!\!NH \end{array} \!\! B- \right]_n \;+\; 4n \; CH_3COOH\!\cdot\!H_2NNH_2$$

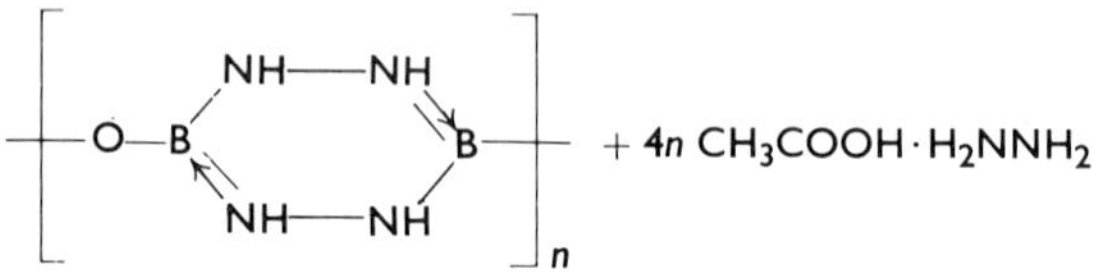

[9]                 [10]

    Somewhat similar materials were obtained by the interaction of dihydrazides and dihydrazines with bis(dimethylamino) phenyl boron [12] (*2, 21*). The reactions are shown in Eqs. (IV-4) and (IV-5). Polymerizations of the

$$n \; H_2NNHCORCONHNH_2 \;+\; 2n \; C_6H_5B[N(CH_3)_2]_2 \;\longrightarrow$$

[11]                       [12]

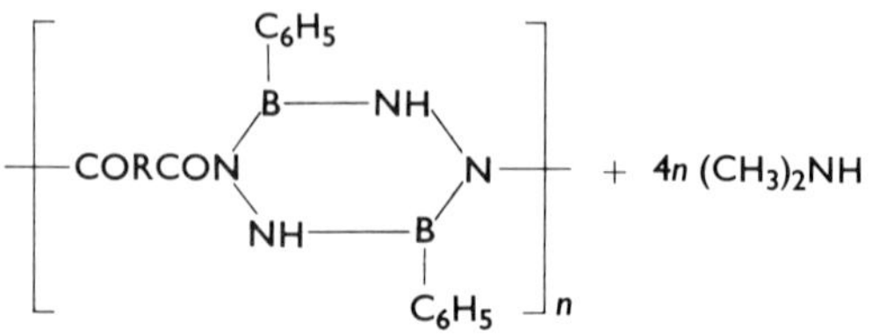

$$\left[ -CORCON \!\! \begin{array}{c} \overset{\displaystyle C_6H_5}{\underset{|}{B}}\!\!-\!\!NH \\ \diagup \quad\quad\quad \diagdown \\ NH\!\!-\!\!\underset{|}{\underset{\displaystyle C_6H_5}{B}} \end{array} \!\! N- \right]_n \;+\; 4n \; (CH_3)_2NH \quad (IV-4)$$

[13]                 [14]

$n$ H$_2$NNHRNHNH$_2$ + 2$n$ C$_6$H$_5$B[N(CH$_3$)$_2$]$_2$ $\longrightarrow$

[15] [12]

$$\left[ -R-N \underset{NH-B}{\overset{B-NH}{\underset{\underset{C_6H_5}{|}}{\overset{\overset{C_6H_5}{|}}{\phantom{x}}}}} N- \right]_n + 4n\ (CH_3)_2NH \quad (IV\text{-}5)$$

[16] [14]

dihydrazides [Eq. (IV-4)] were carried out in DMF at room temperature followed by short heating to about 50°C. Molecular weights ranging from a few hundred to several thousand were achieved. Both aromatic and aliphatic dihydrazides could be used (*21*). The polymers were hydrolytically unstable.

The condensation involving the dihydrazides [Eq. (IV-5)] was done in bulk at 120°–250°C. In some instances, better results were obtained by first pre-reacting the ingredients in benzene solution, removing the solvent, and then polymerizing in the melt at 200°–300°C. Polymers [16] in which the R group was

$-(CH_2)_6-$ and (aromatic biphenyl structure)

[17] [18]

were prepared. They were dark powders, readily soluble in DMF, DMSO, and cresols. Their structures were supported by elemental analyses and infrared evidence (*2*).

## C. Polytetrazoles

Polytetrazoles were prepared by the condensation of diazides with dinitriles. A typical reaction is illustrated in Eq. (IV-6) (*4*). The polymerizations are carried out in bulk at 150°C. Several diazides and dinitriles were used. The

$n$ N$_3$(CH$_2$)$_6$N$_3$ + $n$ NC(CF$_2$)$_3$CN $\longrightarrow$

[19] [20]

$$\left[ \underset{C}{\overset{N\diagdown N}{\underset{|}{\overset{N\diagup\diagdown N}{}}}} -N-(CH_2)_6-N- \underset{C}{\overset{N\diagdown N}{\overset{N\diagup\diagdown N}{}}}-(CF_2)_3 \right]_n \quad (IV\text{-}6)$$

[21]

materials obtained had molecular weights of about 3500 and were claimed to be useful as thermally stable propellant binders and plastics (4).

A method for the synthesis of the closely related polyaminotetrazoles [22] has been described (5, 11). It proceeds via addition of hydrazoic acid to poly-(carbodiimides).

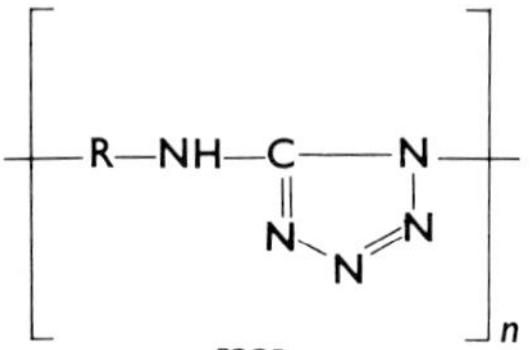

[22]

## D. Polybenzboroxazinones

The reaction of pyroboretetraacetate [8] with 3,3′-dicarboxybenzidine [23] yielded the polybenzboroxazinone [24] (6). The molecular weight of [24] was not reported. It hydrolyzed readily (6).

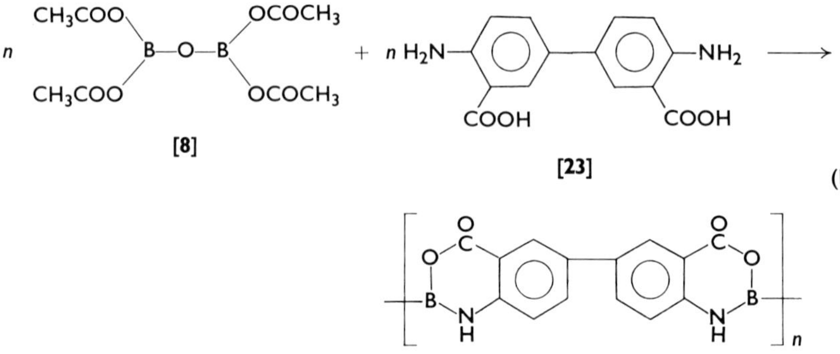

[8]

[23]

(IV-7)

[24]

## E. Polysulfimides and Related Polymers

The polymerizations shown in Eqs. (IV-8), (IV-9), and (IV-10) were recently reported (8).

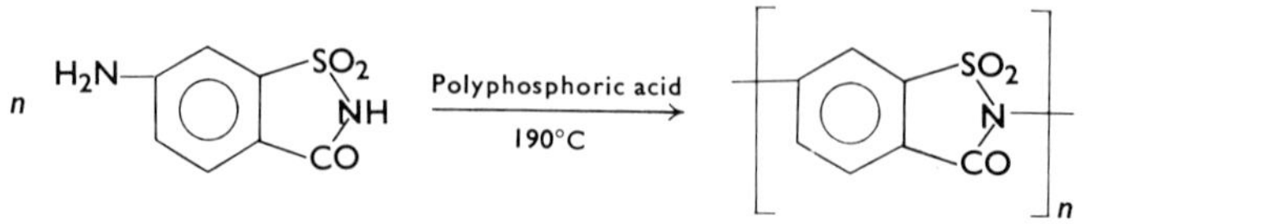

[25]

Polyphosphoric acid
190°C

[26]

(IV-8)

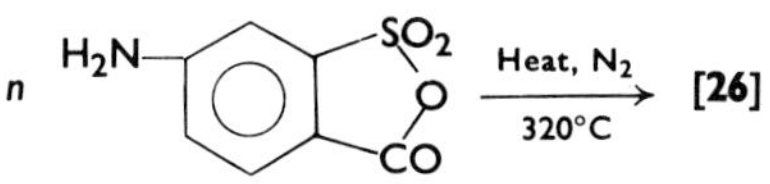

(IV-9)

[27]

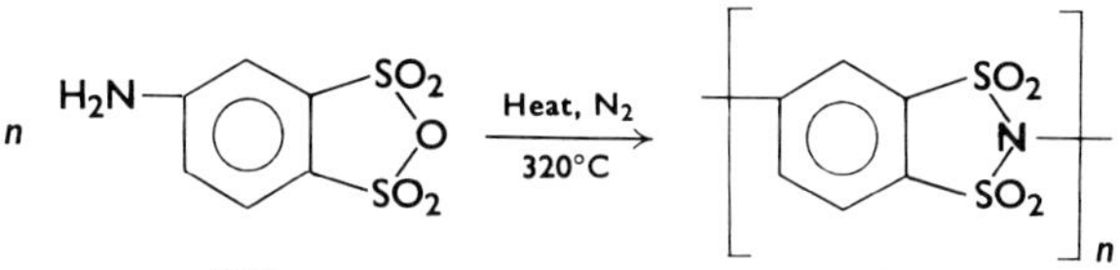

(IV-10)

[28]          [29]

The polymers were dark powders of low molecular weight. Polymer [26] was slightly soluble in concentrated sulfuric acid. Good thermal stability was claimed for these products (8).

Poly(isoindolonebenzothiadiazine dioxides) [33] are a class of related, sulfur-containing polymers. They are formed by the condensation of aromatic dianhydrides with diaminodisulfonamides (14). An example is shown in Eq. (IV-11). In the first step the polyamic acid [32] is prepared. The latter

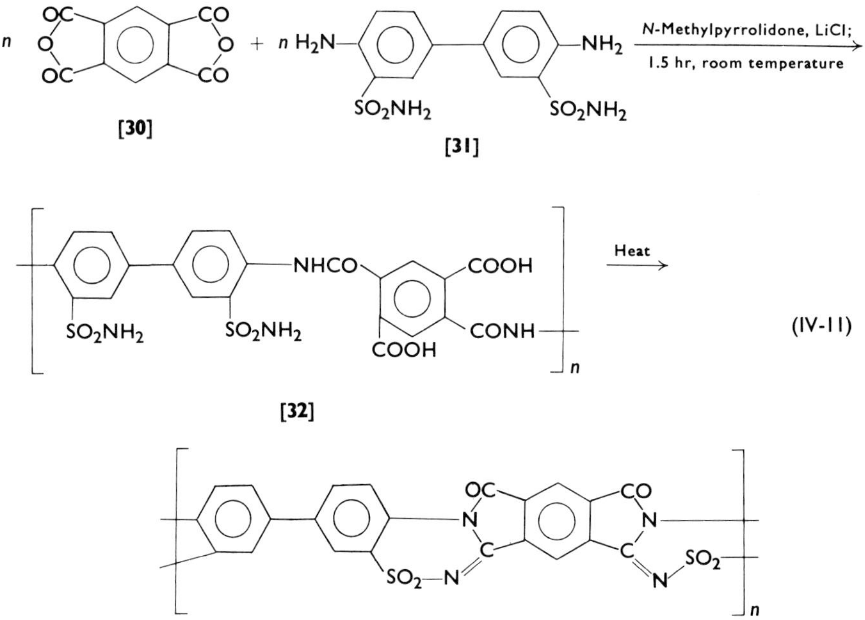

[30]          [31]

(IV-11)

[32]

[33]

is dehydrated on heating to give polymer [33]. The reaction is quite general and is applicable to other dianhydrides and diaminodisulfonamides. The resins are reported to possess good heat resistance and to be useful in films, paints, and adhesives (*14*).

## F. Poly(4-phosphoniapyran salts)

The very interesting, ring-forming polymerizations shown in Eq. (IV-12) and (IV-13) are carried out in a mixture of hexamethylphosphoramide and benzene (*34*). Infrared spectra of materials [36] and [38] were reported and discussed (*34*).

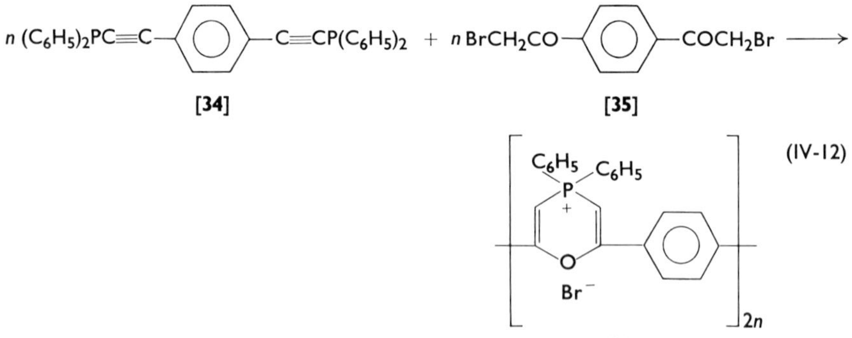

$$n\,(C_6H_5)_2PC \equiv C - \langle\bigcirc\rangle - C \equiv CP(C_6H_5)_2 \;+\; n\,BrCH_2CO - \langle\bigcirc\rangle - COCH_2Br \longrightarrow$$

[34]                                      [35]

(IV-12)

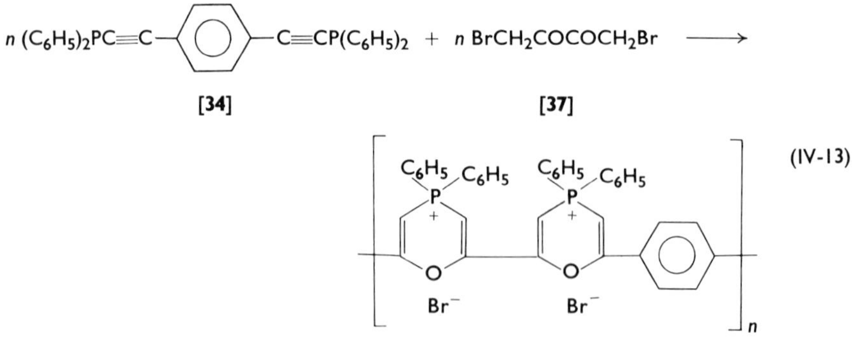

[36]

$$n\,(C_6H_5)_2PC \equiv C - \langle\bigcirc\rangle - C \equiv CP(C_6H_5)_2 \;+\; n\,BrCH_2COCOCH_2Br \longrightarrow$$

[34]                                      [37]

(IV-13)

[38]

## G. Poly(amide-indoles)

The condensation of 2,4-diaminotoluene [39] with isophthaloyl chloride [40] yielded the polyamide [41]. The latter could be cyclized to the poly(amide-indole) [42] (23). Best results were obtained by heating a cast film of [41] with

(IV-14)

sodium ethoxide or isobutoxide at 340°C under vacuum. Acetic anhydride or its mixtures with pyridine or sodium isobutoxide were also useful; however, lower degrees of cyclization were observed under these conditions. The structure of the poly(amide-indole) was supported by infrared evidence. It was insoluble and displayed good thermal stability. Thermogravimetric analysis (air, 2–3°C/minute) indicated an initial decomposition temperature of 400°C; at 460°C, the weight loss was 15% (23).

## H. Polyxanthones

The Friedel-Crafts reaction of 4,4'-dihydroxybiphenyl [43] with 2,5-dichloroterephthalic acid [44] gave the polyketone [45]. Dehydrohalogenation of the latter led to the polyxanthone [46] (9, 10). The first step was performed in the presence of hydrofluoric acid and boron trifluoride. Mixing of the acid, the biphenol, and the catalyst was done at −60°C; the reaction was completed by heating to 75°C. The dehydrohalogenation of the polyketone was accomplished by treatment with KOH in DMSO/benzene at reflux. Polymer [46] was soluble in concentrated sulfuric acid, its inherent viscosity was 0.6 dl/g. Clear cast films were obtained (9).

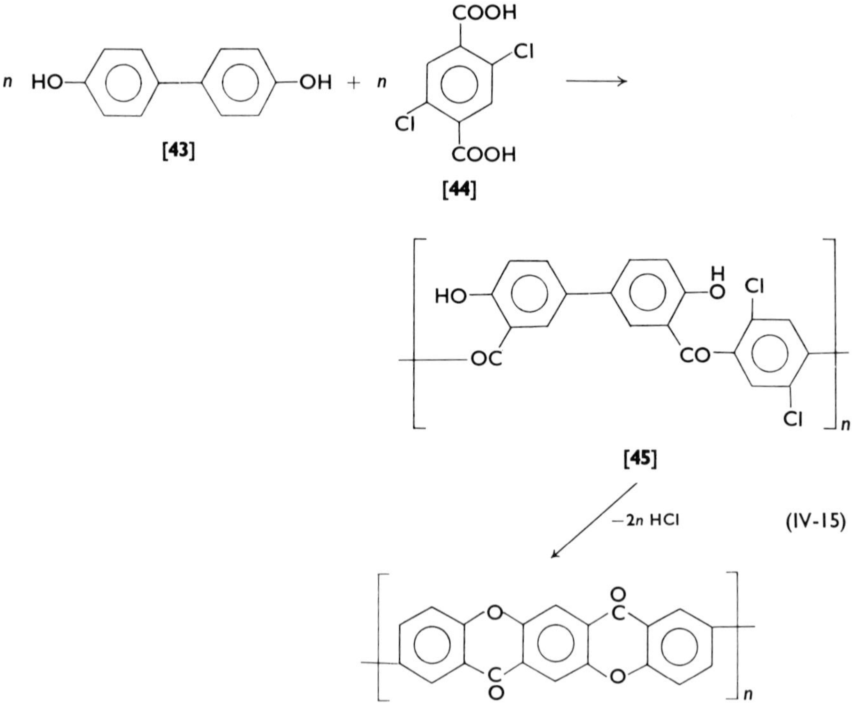

[43] + [44]

[45]

$-2n$ HCl        (IV-15)

[46]

## I. Copolymerization of 1,4-Butynediol with Acrolein

The reaction of 1,4-butynediol with acrolein was studied in the presence of cationic catalysts such as ferric and stannic chloride (*31*). The reaction was carried out in bulk at 20°C for 2–45 days, and the amount of catalyst was 0.05% based on the sum of monomers. The copolymers that were obtained were of low molecular weights (700–2550); they were fractionated by precipitation with water from THF solution. All fractions except one (the last) were soluble in ethanol, dioxane, and chloroform. Analysis and infrared investigation showed that the soluble polymers corresponded to formula [47]. No aldehyde

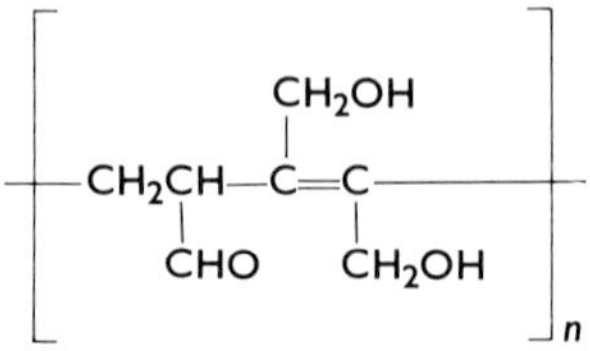

[47]

or hydroxyl groups were detected in the insoluble product; it was probably a cyclized material possessing the ladder structure [48] (*31*).

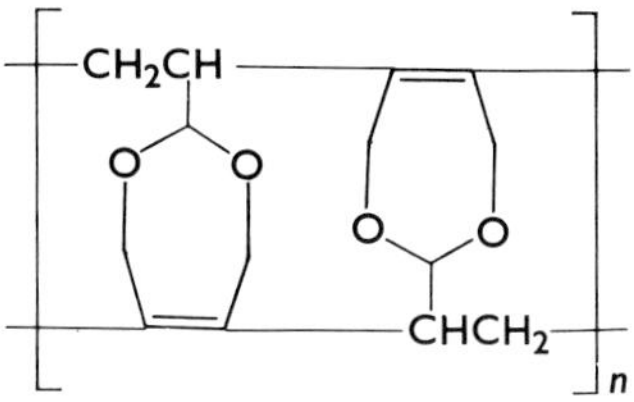

[48]

## J. Polymerizations to Fused Pyridine Ring-Containing Polymers

### 1. POLYQUINOLINES

Polyquinolines were prepared by the reaction sequence shown in Eq. (IV-16) (*22*). The polymers were insoluble, and were claimed to be "mechanically

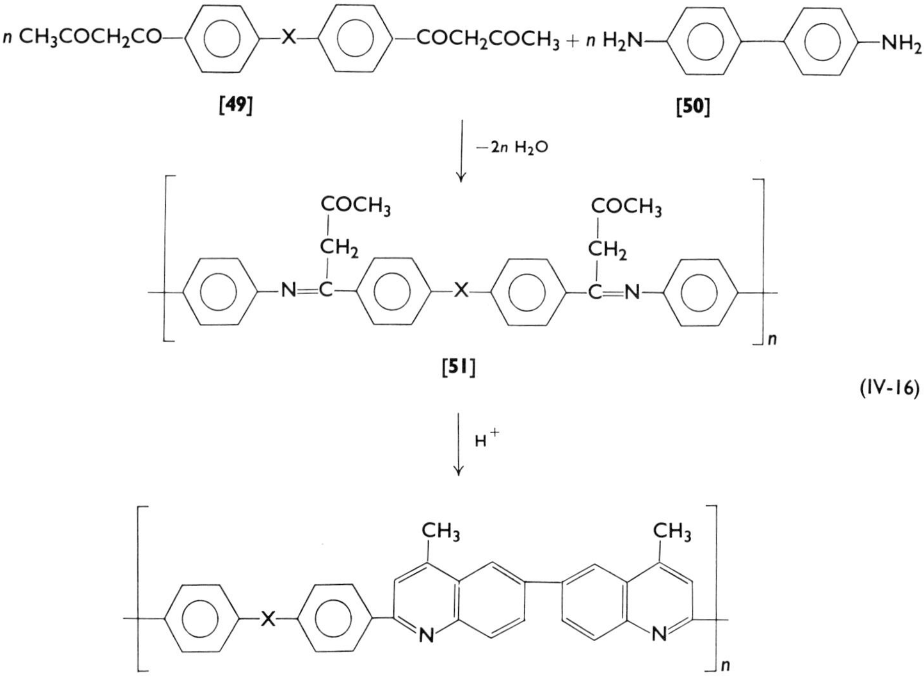

[49]     [50]

$-2n\ H_2O$

[51]

(IV-16)

$H^+$

[52]

X = O or —CH$_2$CH$_2$—

(IV-17)

strong and stable up to 400°C." A recent report (*30*) describes another synthesis of polyquinolines based on the Pfitzinger reaction. Polymer [**55**] was a dark brown powder, soluble in DMF, pyridine, DMSO, and aqueous sodium hydroxide. On heating to 300°C it lost carbon dioxide and gave the poly-quinoline [**56**]. The structure of the latter was supported by infrared evidence. No molecular weights were reported.

## 2. POLYQUINOLONES

Polyquinolones or polyhydroxyquinolines were synthesized from diamino-diesters and diketone bis(ketals) (*12*, *32*, *33*). An example is shown in Eq. (IV-18). The scope of the reaction appears to be broad. In addition to di-

phenyl ether, *m*-cresol was also a good polymerization solvent. For best results, the polymers obtained after the refluxing step were first dried and then heated progressively under vacuum (0.1 mm) at 300°, 350°, and 400°C (2 hours at each temperature). The thermal treatment completed the reaction as indicated by elemental analyses and infrared spectral data. The polyhydroxy-quinolines are soluble in formic acid and hexamethyl phosphoramide. They are dark-colored powders. Apparently, quite high molecular weights were attained since the inherent viscosity of polymer [**59**] was 0.41 (HCOOH). Good thermal stability was indicated by thermogravimetric analysis (1°C/minute), with a weight loss of less than 5% at 500°C (inert atmosphere); in air, decomposition started at 300°C and the weight loss was 25% at 400°C.

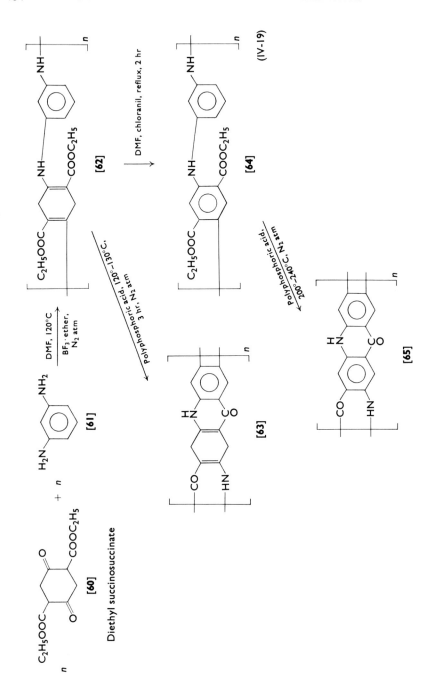

A different approach to polyquinolones was described in reference (*17*). It is illustrated in Eq. (IV-19). The condensation of diethyl succinosuccinate [60] with *m*-phenylenediamine [61] yielded polymer [62]. The latter was oxidized with chloranil to the completely aromatic form [64]. Both [62] and [64] could be cyclized to the quinolones [63] and [65] in polyphosphoric acid. The polymers were dark-colored powders. Polyquinolones [63] and [65] were insoluble and infusible. Infrared spectra were in agreement with the proposed structures. The synthetic scheme of Eq. (IV-19) was applied to other diamines as well. Generally, the polyquinolones possessed good thermal stability. Typical weight losses incurred after 1 hour of heating at 500°C are given in the accompanying table.

| Polymer | Weight loss (%) | |
| --- | --- | --- |
| | Air | Nitrogen |
| [63] | 23.7 | 10.8 |
| [65] | 17.7 | 6.3 |

## 3. POLYACRIDINES

Benzaldehyde was condensed in a two-step reaction with *m*-phenylenediamine dihydrochloride to give the polyacridine [68] (*15, 16*). Other aldehydes could also be used. In addition, tetraamines could be employed. An example

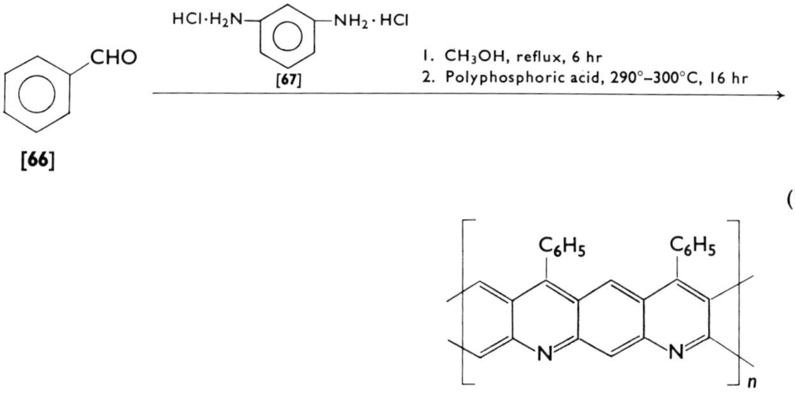

(IV-20)

[68]

is shown in Eq. (IV-21). The polymers were claimed to be useful as semi-conductors and catalysts (*16*).

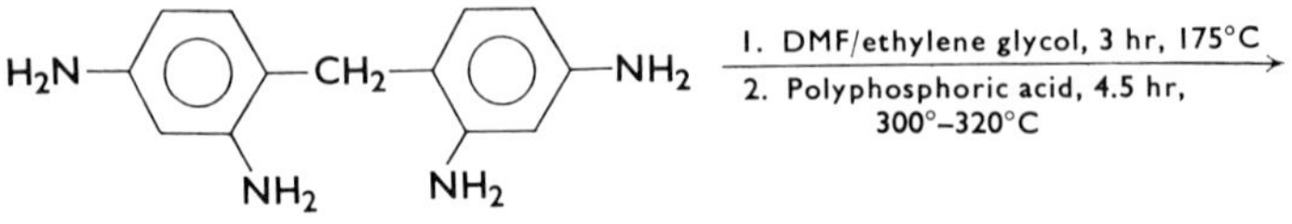

**[69]**                            (IV-21)

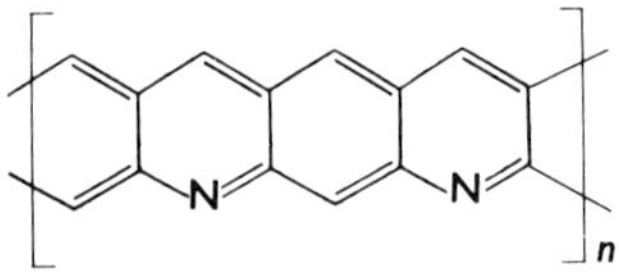

**[70]**

## 4. POLYANTHRAZOLINES

The Friedlander condensation of a diketone with a diamino dialdehyde is reported to give polyanthrazoline **[73]** (*3*). Relatively low molecular weight polymers (highest inherent viscosity in sulfuric acid = 0.24) were obtained due to premature precipitation. The materials were soluble in $H_2SO_4$. They displayed good thermal stability; TGA (10°C/minute) indicated a decomposition temperature of 450°C.

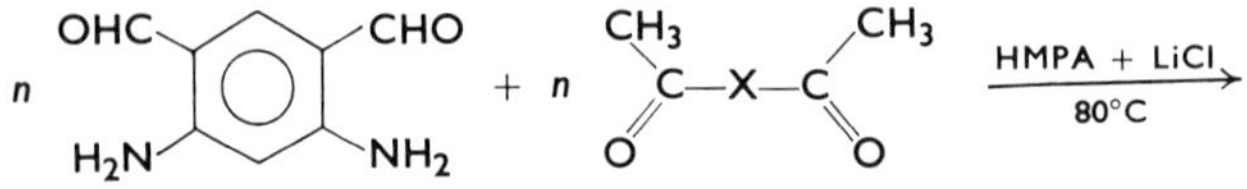

**[71]**              **[72]**                       (IV-22)

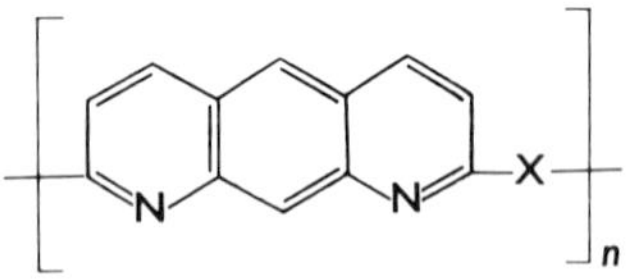

**[73]**

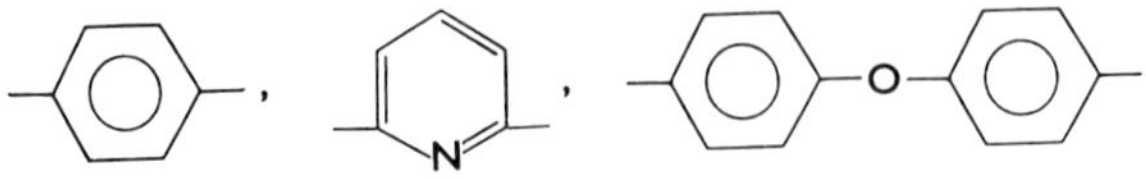

## 5. POLYQUINACRIDONES

The following interesting reaction sequence has been described (*29, 37*).

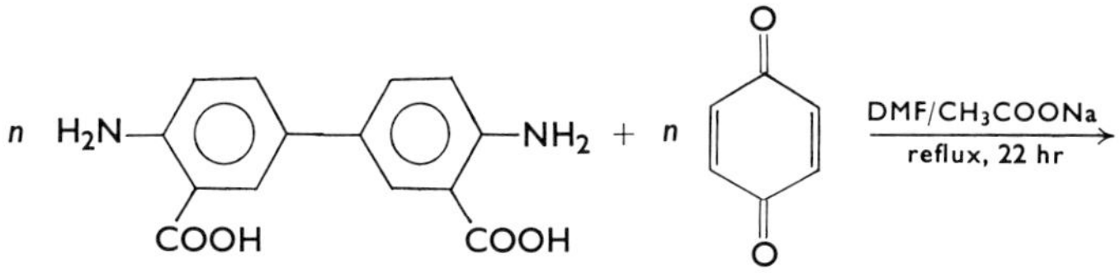

[23]          [74]

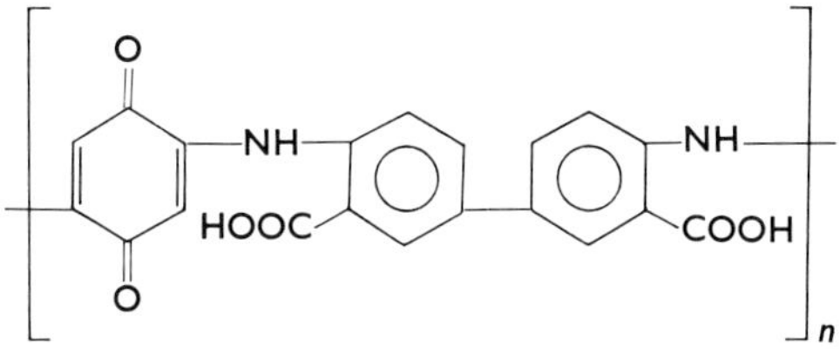

[75]

(IV-23)

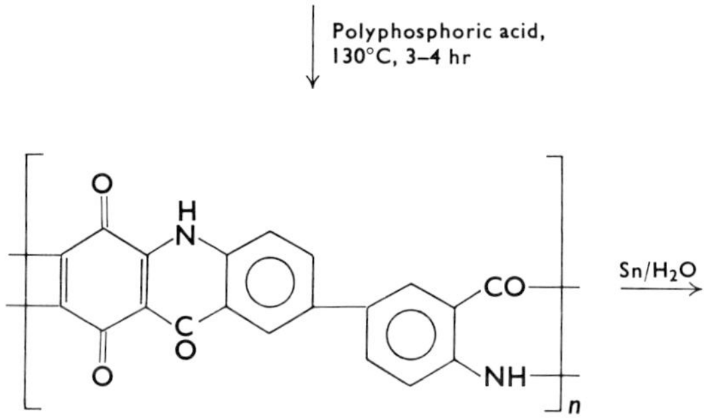

Polyphosphoric acid,
130°C, 3–4 hr

[76]

Sn/H₂O

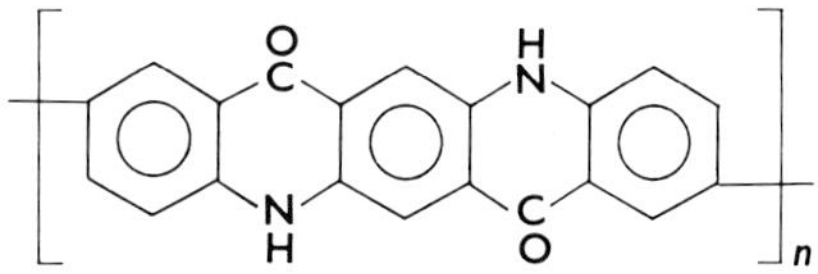

[77]

The condensation of 3,3'-dicarboxybenzidine [23] with $p$-benzoquinone gave polymer [75]. The latter was cyclized in polyphosphoric acid to [76], which in turn was reduced by tin/$H_2O$ to the polyquinacridone [77]. The reduced material was soluble in $H_2SO_4$ with an inherent viscosity of 1.86. Good thermal stability was claimed for this polymer.

## 6. POLYPHENANTHRIDINES AND POLY(4,9-DIAZAPYRENES)

Polyphenanthridines [81] were prepared by the cyclodehydration of poly-amides based on 2,4-diaminobiphenyl [78] (*18–20*). The synthetic sequence is shown in Eq. (IV-24). In a similar manner, the cyclodehydration of poly-

amides derived from 2,2'-diaminobiphenyl [82] gave poly(4,9-diazapyrenes) [84] (*18–20*). Phosphorus oxychloride in nitrobenzene was found to be the best dehydrating agent. Thermal, noncatalyzed cyclization of the polyamides gave unsatisfactory results (20). Polymers [81] and [84], in which the Ar group was [18] and [85]–[87] were prepared. Their structures were supported by infrared

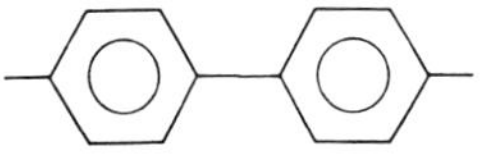

[18]         [85] 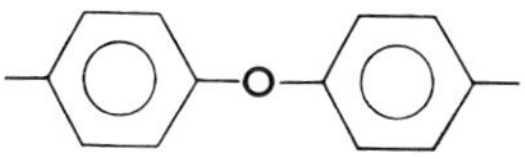        [86]

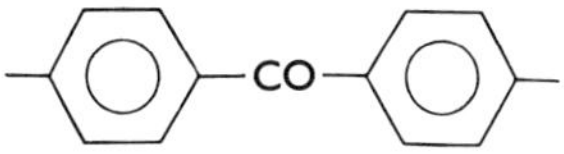

[87]

data which also were a useful tool in determining the degree of cyclization. Apparently, the materials were of high molecular weight, since under well-controlled conditions little degradation accompanied the dehydration step.

The polyphenanthridines that were synthesized from terephthaloyl chloride were insoluble in all solvents. Those that were made from the other diacid chlorides were soluble in DMF, DMA, N-methylpyrrolidone, and sulfuric acid. On the other hand, all of the poly(4,9-diazapyrenes) were soluble in these latter solvents (20). Polyphenanthridines and poly(4,9-diazapyrenes) are high softening polymers. Thermogravimetric analysis in air indicated good stability to about 350°C for both classes of materials (20).

Interestingly, the above polymers can also be obtained by a one-step reaction as exemplified in Eq. (IV-26) (26). However, this approach was not studied in great detail.

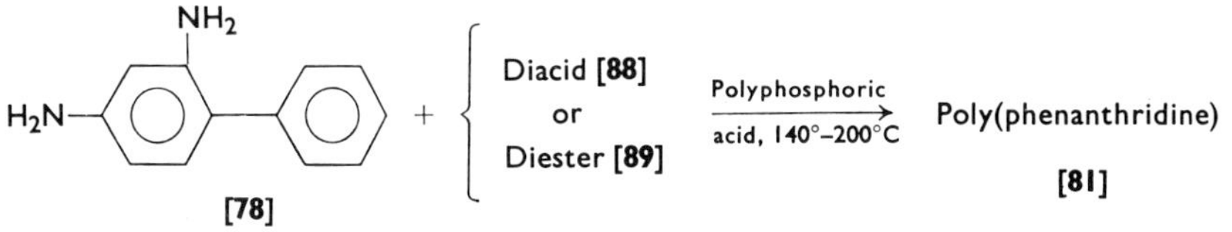

(IV-26)

## K. Polydiazepines

The condensation of 2,2',6,6'-tetraamino-4,4'-bis(methoxycarbonyl)biphenyl [90] with the diethyl ester of oxaldiimidic acid [91] has been reported (28). A yellow polymer [92] that was soluble in DMF and dioxane was obtained. Cryoscopic measurements performed in dioxane indicated a molecular weight of about 1000.

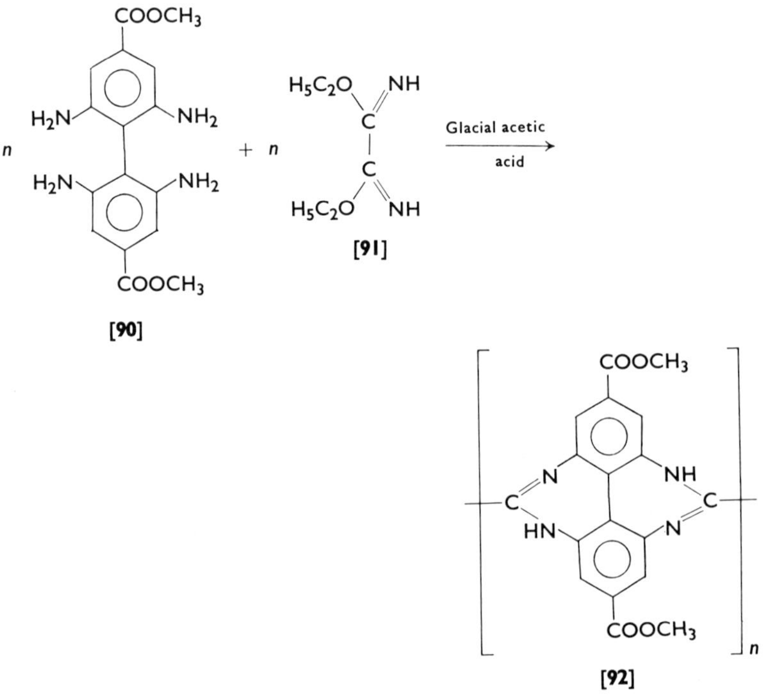

(IV-27)

## L.  Polydisalicylides

The preparation of these polymers is illustrated in Eq. (IV-28) (*25*). Formula

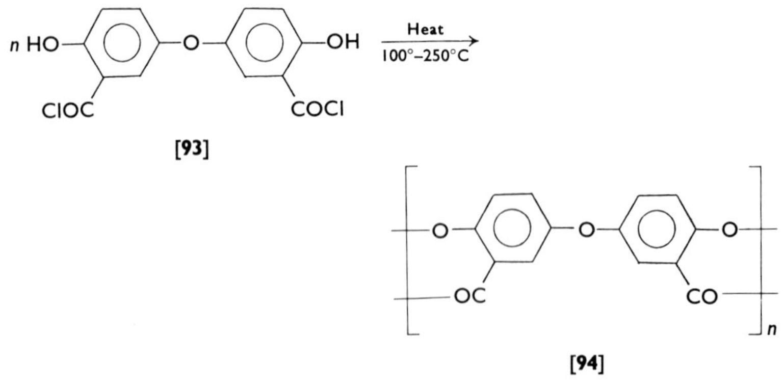

(IV-28)

[94] is an idealized representation of the product. In the first step the reaction yields a linear polyester with pendant hydroxyl and acid chloride groups. Continued heating promotes cyclization to the ladder structure [94]. Simultaneously, intermolecular elimination of hydrogen chloride takes place and

leads to a cross-linked material. Low molecular weight, linear, soluble polymers ($\eta_{inh}$ 0.14) were obtained, when the reactions were conducted in solution. Various solvents were used; sulfolane, nitrobenzene, benzophenone, and triethylene glycol dimethyl ether were the best. Analysis of the materials obtained under these conditions indicated that they were only partially cyclized.

It is noteworthy that bulk reactions yielded cross-linked products. In addition to the free dihydroxy derivatives, their diacetylated analogs were also useful monomers. The polydisalicylides were reportedly thermally stable and yielded excellent quality glass laminates (25).

## M. Polyferrocenes and Polytitanocenes

Soluble ladder polymers were obtained from organosilicon ferrocene derivatives possessing two butadienyl or isoprenyl groups (13). The polymerizations were initiated anionically and stopped at low (~15%) conversions. At higher conversions cross-linked products were obtained. The reaction is illustrated in Eq. (IV-29) using 1,1′-bis(dimethyl-1-butadienylsilyl) ferrocene [95] as the monomer. Along related lines, the following preparation of a

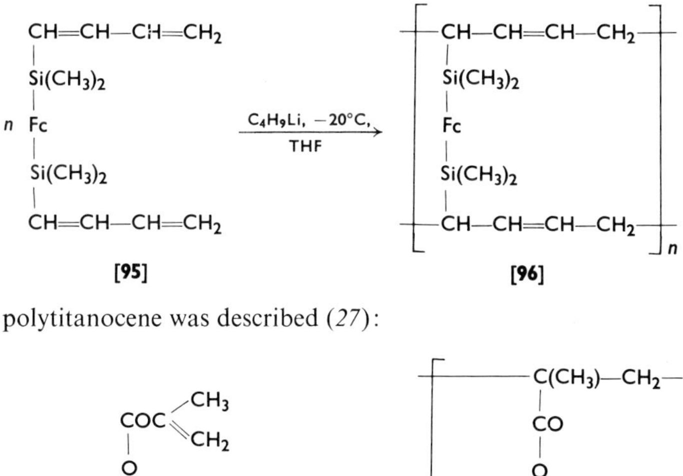

(IV-29)

polytitanocene was described (27):

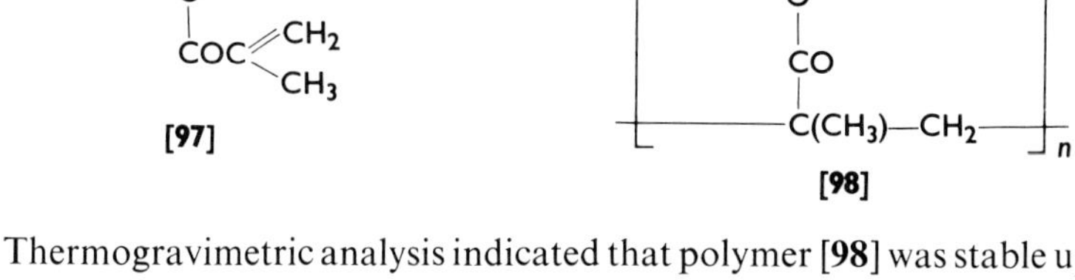

(IV-30)

Thermogravimetric analysis indicated that polymer [98] was stable up to 220°C.

### N. Miscellaneous Polymerizations

The polycondensation of bis(o-aminocarboxylic acids) to heat-stable, ring-containing polymers was described in a patent (24). The reactions were carried out at 130°–160°C in polyphosphoric acid in the presence of an aromatic amine. A nitrogen atmosphere was used. A brown polymer, melting at 350°C, soluble in concentrated sulfuric acid was obtained from 3,3'-dicarboxybenzidine [23]. Its inherent viscosity was 1.13 ($H_2SO_4$). Infrared spectroscopy indicated the presence of an eight-membered ring containing two lactam links in the repeat units of the material.

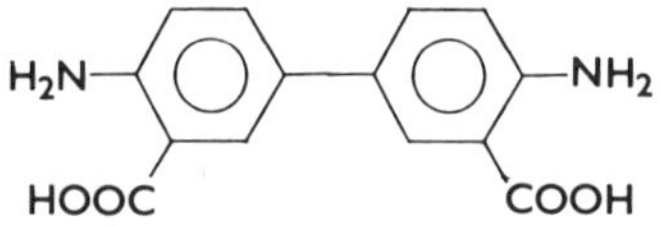

[23]

Along similar lines, heat-resistant polymers were also obtained by the polycondensation of bis(amino acids) such as represented by structure [23] with bis(acylamino) compounds (1). The polymerizations were performed in polyphosphoric acid and gave high molecular weight polymers that had melting points about 500°C.

A recent paper (7) described the condensation of phthalic [99] and sulfobenzoic [100] anhydrides with several dihydroxyphenols and their dimethyl

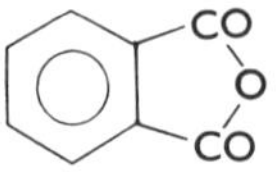

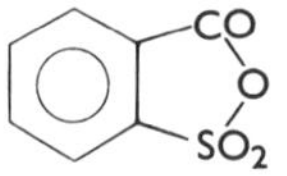

[99]              [100]

ethers. Structures of the type [101] were obtained; their semiconducting and paramagnetic properties were reported (7).

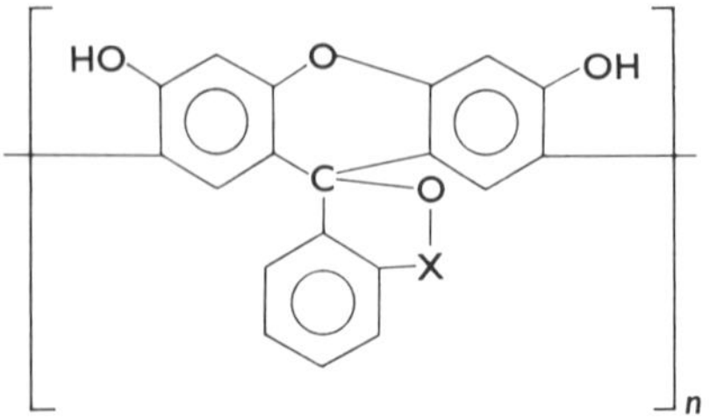

[101]

X = CO or $SO_2$

## REFERENCES

1. Baba, Y., and Yoda, N., Jap. Pat. 68/02,476 (Toyo Rayon Co.) (1968); *Chem. Abstr.* **69**, 28122q (1968).
2. Bekasova, N. I., Korshak, V. V., and Prigozhina, M. P., *Vysokomol. Soedin., Ser. B* **11**, 366 (1969); *Chem. Abstr.* **71**, 39505j (1969).
3. Brake, W., *Macromolecules* **2**, 286 (1969).
4. Carpenter, W. R., U.S. Pat. 3,386,968 (United States Dept. of the Navy) (1968); *Chem. Abstr.* **69**, 28106n (1968).
5. Christie, P. A., *Diss. Abstr. B* **29**, 933 (1968); *Chem. Abstr.* **70**, 12030t (1969).
6. Cihodaru, S., and Ungurenasu, C., *Tetrahedron Lett.* p. 3361 (1968); *Chem. Abstr.* **69**, 28050q (1968).
7. Dabrowski, R., Witkiewicz, Z., Leibler, K., Baginski, J., and Konopczynski, A., *Rocz. Chem.* **44**, 621 (1970); *Chem. Abstr.* **73**, 99281j (1970).
8. D'Alelio, G. F., U.S. Pat. 3,518,233 (United States Dept. of the Air Force) (1970); *Chem. Abstr.* **73**, 56760x (1970).
9. Darms, R., Ger. Pat. 1,806,419 (E. I. duPont de Nemours and Co.) (1969); *Chem. Abstr.* **71**, 50696t (1969).
10. Darms, R., Belg. Pat. 723,232 (E. I. duPont de Nemours and Co.) (1969).
11. Dyer, E., and Christie, P. A., *J. Polym. Sci., Part A-1* **6**, 729 (1968).
12. French Patent 1,468,677 (Institut Français du Petrole, des Carburants et Lubrifiants) (1967).
13. Greber, G., and Hallensleben, M. L., *Makromol. Chem.* **104**, 90 (1967); *Chem. Abstr.* **67**, 100476m (1967).
14. Hagiwara, Y., Kurihara, M., and Yoda, N., Jap. Pat. 69/28,715 (Toyo Rayon Co., Ltd.) (1969); *Chem. Abstr.* **72**, 67488v (1970).
15. Idelson, A. L., and Litt, M. H., U.S. Pat. 3,418,261 (Allied Chemical Corp.) (1968); *Chem. Abstr.* **70**, 38303j (1969).
16. Idelson, A. L., and Litt, M. H., U.S. Pat. 3,535,284 (Allied Chemical Corp.) (1970); *Chem. Abstr.* **74**, 32177g (1971).
17. Kimura, S., *Makromol. Chem.* **117**, 203 (1968); *Chem. Abstr.* **69**, 107168n (1968).
18. Kolesnikov, G. S., Fedotova, O. Ya., and Matvelashvili, G. S., *Vysokomol. Soedin.* **8**, 1135 (1966).
19. Kolesnikov, G. S., Fedotova, O. Ya., and Matvelashvili, G. S., *Vysokomol. Soedin. Ser. B.* **9**, 819 (1967); *Chem. Abstr.* **68**, 30220t (1968).
20. Kolesnikov, G. S., Fedotova, O. Ya., and Matvelashvili, G. S., *Vysokomol. Soedin., Ser. A* **12**, 536 (1970); *Chem. Abstr.* **73**, 4231f (1970).
21. Korshak, V. V., Bekasova, N. I., and Prigozhina, M. P., *Vysokomol. Soedin., Ser. B* **9**, 903 (1967); *Chem. Abstr.* **68**, 59957z (1968).
22. Korshak, V. V,, Kronganz, E. S., Berlin, A. M., and Travnikova, A. P., *Vysokomol. Soedin., Ser. B* **9**, 171 (1967); *Chem. Abstr.* **67**, 11787w (1967).
23. Kudryavtsev, G. I., Odnoralova, V. N., Nazimova, N., and Shablygin, M. V., *Vysokomol. Soedin., Ser. B* **12**, 371 (1970); *Chem. Abstr.* **73**, 36018h (1970).
24. Kurihara, M., and Yoda, N., Jap. Pat. 68/634 (Toyo Rayon Co., Ltd.) (1968); *Chem. Abstr.* **68**, 115221u (1968).
25. Loughran, G. A., and Burkett, J. L., U.S. Pat. 3,515,695 (United States Dept. of the Air Force) (1970); *Chem. Abstr.* **73**, 26238r (1970).
26. Matvelashvili, G. S., Fedotova, O. Ya., and Kolesnikov, G. S., *Tr. Mosk. Khim.-Tekhnol. Inst.* No. 57, 177 (1968); *Chem. Abstr.* **70**, 97286d (1969).

27. Ralea, R., Ungureanu, C., and Maxim, I., *Rev. Roum. Chim.* **12**, 523 (1967); *Chem. Abstr.* **68**, 78656s (1968).
28. Ried, W., and Braeutigam, J., *Chem. Ber.* **99**, 3304 (1966); *Chem. Abstr.* **66**, 10914j (1967).
29. Shimojo, S., Tanimot, S., Okano, M., and Oda, R., *Kobunshi Kagaku* **25**, 16 (1968); *Chem. Abstr.* **69**, 19630b (1968).
30. Shopov, I., *Vysokomol. Soedin., Ser. B* **11**, 248 (1969); *Chem. Abstr.* **71**, 22392x (1969).
31. Shostakovskii, M. F., Kuznetsova, T. S., and Annenkova, V. Z., *Vysokomol. Soedin., Ser. B* **12**, 848 (1970); *Chem. Abstr.* **74**, 32046p (1971).
32. Sillion, B., and de Gaudemaris, G., *J. Polym. Sci., Part C*, No. 16, Part 8, 4653 (1969); *Chem. Abstr.* **71**, 22418k (1969).
33. Sillion, B., and de Gaudemaris, G., Fr. Pat. 1,468,677 (Institut Français du Petrole, des Carburants et Lubrifiants) (1967); *Chem. Abstr.* **67**, 74020m (1967).
34. Strzelecki, L., Maric, B., and Simalty, M., *C. R. Acad. Sci. (Paris) Ser. C.* **269**, 382 (1969); *Chem. Abstr.* **71**, 102305m (1969).
35. Ungurenasu, C., Cihodaru, S., and Popescu, I., *Tetrahedron Lett.* p. 1435 (1969).
36. Washburn, R. M., Karle, D. W., Hunter, R. W., and Marantz, L. B., *Mater. Process., Nat. SAMPE (Soc. Aerosp. Mater. Process Eng.), Symp. Exhib. 15th* p. 319 (1969): *Chem. Abstr.* **73**, 15666c (1970).
37. Yoda, N., and Kurihara, M., Jap. Pat. 70/11,518 (Toyo Rayon Co., Ltd.) (1970); *Chem. Abstr.* **73**, 67228f (1970).

**Ring-Forming Polymerizations Volume B, Part 2 Supplementary Reference List**

**Chapter I**

SECTION A (p. 1)

1. *Polyimides, Preparation* (6–9, 19, 27, 30–33, 37, 38, 44, 45, 48, 50, 56, 63, 64, 68, 70, 74, 76, 82, 83, 85, 91, 92, 100, 105, 107, 109, 110, 112–116, 118, 120, 124, 129, 131, 133, 134, 142, 143, 147, 148, 150, 151, 153–155, 158, 163, 164, 167, 169–171, 174, 176, 177, 180, 181, 183, 185, 186, 188, 189, 196, 199, 200, 212, 219, 220, 227–229, 239, 242–244, 247, 248, 251, 255–257, 263, 268, 273, 275, 301, 302, 307–311, 314, 315, 317–320, 324–328, 330, 334, 337, 339–347, 349, 352, 356, 357, 366, 374, 378, 379, 382–385, 389, 391, 403–405, 408, 414, 416, 419–421, 428, 436, 438, 440, 442, 444–449, 452, 464, 474–476, 478, 479, 483, 486, 489, 492, 496, 500, 512, 515, 516, 520, 521, 523–526, 531, 533, 536, 537, 542–544, 549–552, 555, 557, 561, 572, 592–595).

2. *Polyimides, Properties* (6–9, 12, 13, 19, 25, 27, 29, 37, 48, 56, 57, 60, 62, 67, 74, 83, 89, 91–94, 100, 103, 105, 107, 109, 110, 113, 115, 117, 118, 128, 135, 151, 156, 158, 163, 188, 189, 203, 205, 206, 212, 220, 228, 229, 239, 243, 244, 248, 252, 258, 296, 297, 300, 302, 304, 310, 311, 315, 333, 337, 340, 342, 343, 345, 349–351, 355, 356, 366, 378, 383–385, 390, 391, 405, 416, 417, 419–421, 441, 442, 444, 468–470, 472–476, 481, 482, 484, 489, 492, 515, 516, 520, 521, 523–527, 531, 536, 537, 542, 543, 550, 556, 557, 560, 562, 597).

3. *Polyimides, Applications* (4, 5, 10, 18, 28, 34, 43, 65, 66, 70–72, 84, 86, 87, 90, 96, 98, 99, 102, 104, 109–112, 119, 126, 127, 132, 137–143, 152, 157, 159, 166, 173, 181, 189, 198, 202, 209–211, 214, 232, 237, 240, 252, 259–261, 269, 273, 288, 297, 306, 312, 313, 344, 348, 364–366, 368–373, 375, 376, 383–388, 392, 393, 398, 400–402, 406, 419, 421, 422, 430–434, 437, 443, 445, 451, 453–455, 461, 473, 478, 495, 497–499, 502, 513, 514, 516–518, 532, 548, 553, 558, 563, 564, 596).

4. *Poly(Amide-Imides), Preparation* (1, 2, 39, 41, 42, 49, 50, 52, 75, 77, 79, 80, 88, 97, 101, 106, 144, 149, 161, 162, 165, 167, 172, 175, 184, 190, 192, 195, 207, 230, 233, 234, 245, 246, 249, 250, 265, 267, 275, 284, 293, 294, 316, 321–323, 329, 353, 377, 394, 396, 397, 410, 418, 424, 426, 427, 449, 463, 474, 493, 503–505, 508, 509, 534, 539, 547, 568, 569, 571, 577, 581, 591).

5. *Poly(Amide-Imides), Properties* (1, 3, 41, 42, 49, 75, 79, 88, 97, 106, 165, 190, 207, 230, 233, 234, 316, 418, 463, 471, 482, 493, 503, 505, 508, 509, 534, 568, 569).

6. *Poly(Amide-Imides), Applications* (21, 41, 43, 47, 52, 73, 77, 80, 97, 136, 190, 192, 240, 265, 270, 289, 295, 407, 424, 425, 460, 477, 480, 501, 503, 519, 534, 545, 546, 568, 569).

7. *Poly(Ester-Imides), Preparation* (20, 40, 59, 81, 145, 187, 191, 201, 215, 230, 231, 264, 272, 277, 283, 354, 367, 490, 491, 494, 507, 510, 511).

8. *Poly(Ester-Imides), Properties* (20, 40, 81, 215, 230, 272, 354, 490, 507, 510, 511).

9. *Poly(Ester-Imides), Applications* (40, 59, 81, 187, 191, 197, 201, 264, 272, 277, 282, 285, 367, 450, 490, 497, 554, 559).

10. *Copolymers Containing Imide Groups* (11, 14, 35, 51, 53–55, 108, 121–123, 130, 168, 213, 221–226, 235, 236, 241, 253, 254, 276, 278–281, 286, 287, 292, 303, 305, 331, 332, 335, 338, 359, 363, 399, 409, 411–413, 415, 421, 456–459, 488, 529, 530, 538, 540, 541, 578–580, 582–586, 588, 589).

SECTION B (p. 69).

*Polymers Containing Pyrrole and Related Rings*

Polymers from Nitriles and Amines (204).
Poly(Pyrrolines) (535).

SECTION C (p. 73).

Poly(1,3-Oxazinones) (178, 208, 291, 360, 435, 573, 575, 576, 587).
Poly(1,3-Oxazindiones) (36, 61, 69, 193, 361, 590).

SECTION D (p. 78).

Poly(Oxazines) (274).
Poly(Thiazines) (17, 439).

SECTION E (p. 81).

Poly(Pyrazines) (218).
Poly(Quinoxalines) (15, 16, 26, 58, 160, 216, 217, 238, 262, 290, 395, 465, 506, 522, 565–567, 570).

SECTION G (p. 87).

Poly(Pyrimidines) (336).

SECTION H (p. 89).

Poly(Quinazolones) (46, 78, 95, 146, 179, 182, 194, 362, 380, 381, 462, 466, 467, 485, 487, 528).
Poly(Quinazolinediones) (266, 271, 358, 423, 574).

SECTION I (p. 96).
Poly(Ketals) (22–24).

SECTION J (p. 98).

Poly(Quinones) (125).

## REFERENCES

1. Adduci, J. M., Ramirez, R. S., and Sikka, S. K., *Polym. Prepr., Amer. Chem. Soc., Div. Polym. Chem.* **12** (1), 611 (1971).

2. Adrova, N. A., Koton, M. M., Maricheva, T. A., and Rudakov, A. P., *Vysokomol. Soedin., Ser. B* **11**, 507 (1969); *Chem. Abstr.* **71**, 102314p (1969).

3. Allard, P., Ger. Pat. 1,965,378 (Société Rhodiaceta) (1970); *Chem. Abstr.* **73**, 78462d (1970).

4. Almouli, J. A., and Bruins, P. F., *Plast. Elec. Insul.* p. 155 (1968); *Chem. Abstr.* **71**, 125556h (1969).

5. Amstutz, H. C., *J. Biomed. Mater. Res.* **3**, 547 (1969); *Chem. Abstr.* **72**, 22357k (1970).

6. Angelo, R. J., U.S. Pat. 3,437,636 (E. I. duPont de Nemours & Co.) (1969); *Chem. Abstr.* **70**, 115841n (1969).

7. Angelo, R. J., U.S. Pat. 3,533,997 (E. I. duPont de Nemours & Co.) (1970); *Chem. Abstr.* **73**, 131697n (1970).

8. Angelo, R. J., U.S. Pat. 3,546,175 (E. I. duPont de Nemours & Co.) (1970);

9. Arai, S., Kotani, N., Nagoya, A., Kurihara, M., and Yoda, N., *Ger. Pat.* 1,934,077 (Toyo Rayon Co., Ltd.) (1970); *Chem. Abstr.* **72**, 80015h (1970).

10. Ardolino, E. J., Bova, J. D., and Hoover, D. P., U.S. Pat. 3,519,510 (1970); *Chem. Abstr.* **73**, 56792j (1970).

11. Arnold, F. E., *J. Polym. Sci. Part A-1* **8**, 2079 (1970); *Chem. Abstr.* **73**, 77640y (1970).

12. Asahara, T., Seno, M., and Fukui, M., *Kogyo Kagaku Zasshi* **72**, 1389 (1969); *Chem. Abstr.* **71**, 91959z (1969).

13. Asahara, T., Seno, M., and Fukui, M., *Kogyo Kagaku Zasshi* **72**, 1923 (1969); *Chem. Abstr.* **72**, 22110z (1970).

14. Ashida, K., Okamoto, M., and Hamo, K., Ger. Pat. 1,933,621 (Toray Ind., Inc.) (1970); *Chem. Abstr.* **72**, 101832q (1970).

15. Augl, J. M., *J. Polym. Sci., Part A-1* **8**, 3145 (1970); *Chem. Abstr.* **74**, 42702p (1971).

16. Augl, J. M., and Wrasidlo, W. J., AD 1969, AD 699888 (1969); *U.S. Govt. Res. & Develop. Rep.* **70**, 64 (1970); *Chem. Abstr.* **73**, 35805a (1970).

17. Augl, J. M., and Wrasidlo, W. J., *J. Polym. Sci., Part A-1* **8**, 63 (1970); *Chem. Abstr.* **72**, 67333r (1970).

18. Australian Patent 68/32357 (Monsanto Co.) (1969).

19. Azarov, V. I., Korshak, V. V., and Tseitlin, G. M., Russ, Pat. 250,453 (Mendeleev Institute for Chemical Technology (Moscow)) (1970).

20. Azarova, M. T., Dyurnbaum, V. S., and Bogdanov, M. N., *Vysokomol. Soedin., Ser. B* **11**, 136 (1969); *Chem. Abstr.* **70**, 97278c (1969).

21. Bach, H. J., U.S. Pat. 3,475,212 (Mobil Oil Corp.) (1969).

22. Bailey, W. J., Beam, C. F., Jr., and Haddad, I., *Polym. Prepr., Amer. Chem. Soc., Div. Polym. Chem.* **12** (1), 169 (1971).

23. Bailey, W. J., and Hinrichs, R. L., *Polym. Prepr., Amer. Chem. Soc., Div. Polym. Chem.* **11** (2), 598 (1970).

24. Bailey, W. J., and Volpe, A. A., *J. Polym. Sci., Part A-1* **8**, 2109 (1970); *Chem. Abstr.* **73**, 77750j (1970).

25. Ball, G., and Boettner, E. A., *Polym. Prepr., Amer. Chem. Soc., Div. Polym. Chem.* **12** (1), 766 (1971).

26. Banishemi, A., and Marvel, C. S., *J. Polym. Sci., Part A-1* **8**, 3211 (1970); *Chem. Abstr.* **74**, 42686m (1971).

27. Bargain, M., Fr. Pat. 1,537,135 (Société des Usines Chimiques Rhone-Poulenc) (1968); *Chem. Abstr.* **71**, 13725f (1969).
28. Barry, A. J., and Plueddemann, E. P., U.S. Pat. 3,481,815 (Dow Corning Corp.) (1969); *Chem. Abstr.* **72**, 44499k (1970).
29. Barton, J. M., and Critchley, J. P., *Polymer* **11**, 212 (1970); *Chem. Abstr.* **73**, 25987x (1970).
30. Behr, E., Fr. Pat. 2,012,288 (Dynamit Nobel A.-G.) (1970); *Chem. Abstr.* **73**, 131561p (1970).
31. Belgian Patent 718,563 (Farbenfabriken Bayer, A. G.) (1968).
32. Belgian Patent 719,082 (Farbenfabriken Bayer, A. G.) (1969).
33. Belgian Patent 719,083 (Farbenfabriken Bayer, A. G.) (1969).
34. Belgian Patent 721,350 (E. I. duPont de Nemours & Co.) (1969).
35. Belgian Patent 723,772 (Farbenfabriken Bayer, A. G.) (1969).
36. Belgian Patent 726,620 (Farbenfabriken Bayer, A. G.) (1969).
37. Belgian Patent 727,682 (Kureha Kagaku Kogyo K.K.) (1969).
38. Belgian Patent 727,741 (E. I. duPont de Nemours & Co.) (1969).
39. Belgian Patent 731,419 (Farbenfabriken Bayer, A. G.) (1969).
40. Belgian Patent 733,357 (Westinghouse Electric Corp.) (1969).
41. Belgian Patent 733,651 (Société des Usines Chimiques Rhone-Poulenc, S.A.) (1969).
42. Belgian Patent 739,399 (Société Rhodiaceta) (1970).
43. Belgian Patent 741,192 (Westinghouse Electric Corp.) (1970).
44. Belgian Patent 741,815 (Société Rhodiaceta) (1970).
45. Belgian Patent 742,309 (Farbenfabriken Bayer, A. G.) (1970).
46. Belgian Patent 742,311 (Farbenfabriken Bayer, A. G.) (1970),
47. Belgian Patent 743,366 (Société des Usines Chimiques Rhone-Poulenc, S.A.) (1970).
48. Belgian Patent 745,023 (Institut Français du Petrole, des Carburants, et Lubrifiants) (1970).
49. Belgian Patent 745,715 (ICI, Ltd.) (1970).
50. Belgian Patent 746,668 (Veba-Chemie, A. G.) (1970).
51. Belgian Patent 749,151 (Institut Français du Petrole, des Carburants, et Lubrifiants) (1970).
52. Belgian Patent 749,449 (Phelps Dodge Magnet Wire Corp.) (1970).
53. Bell, V. L., Jr., *NASA Tech Note*, **NASA TN D-559**. (1969).
54. Bell, V. L., Jr., and Pezdirtz, G. F., U.S. Pat. 3,532,673 (U.S. National Aeronautics and Space Administration) (1970); *Chem. Abstr.* **74**, 43070t (1971).
55. Bentz, F., and Bodesheim, F., Ger. Pat. 1,811,167 (Farbenfabriken Bayer, A. G.) (1970); *Chem. Abstr.* **73**, 46064j (1970).
56. Berg, C. J., U.S. Pat. 3,536,666 (Minnesota Mining and Manufacturing Co.) (1970).
57. Berlin, A. A., Liogon'kii, B. I., and Shamraev, G. M., Russ. Pat. 275,376 (Institute of Chemical Physics, Academy of Sciences, USSR) (1970); *Chem. Abstr.* **74**, 4250g (1971).
58. Berlin, A. M., *Usp. Khim.* **39**, 158 (1970); *Chem. Abstr.* **72**, 111826e (1970).
59. Bescos, J., Fontan, Y. J., and Babe, S. G., *Rev. Plast. Mod.* **20**, 987–992, 994–996, and 1001–1006 (1969); *Chem. Abstr.* **73**, 15864r (1970).
60. Bikson, B. R., and Freimanis, J., *Vysokomol. Soedin.*, *Ser. A* **12**, 69 (1970); *Chem. Abstr.* **72**, 101173a (1970).
61. Binsack, R., Bottenbruch, L., and Schnell, H., U.S. Pat. 3,541,048 (Farbenfabriken Bayer, A. G.) (1970).
62. Bishop, D. P., and Smith, D. A., *J. Appl. Polym. Sci.* **14**, 345 (1970); *Chem. Abstr.* **72**, 79741k (1970).

63. Boldebuck, E. M., and Gaertner, R. F., Fr. Pat. 1,533,258 (General Electric Co.) (1968); *Chem. Abstr.* **71**, 13727h (1969).
64. Boldebuck, E. M., and Gaertner, R. F., U.S. Pat. 3,440,202 (General Electric Co.) 1969).
65. Boldebuck, E. M., and Sonnenberg, J., U.S. Pat. 3,440,203 (General Electric Co.) (1969).
66. Bond, H. M., U.S. Pat. 3,486,934 (Minnesota Mining and Manufacturing Co.) (1969).
67. Bondarenko, E. M., Rode, V. V., and Korshak, V. V., *Usp. Khim. Fiz. Polim.*, p. 206 (1970); *Chem. Abstr.* **74**, 31976e (1971).
68. Boram, W. R., and Acle, L., Jr., U.S. Pat. 3,506,583 (International Harvester Co.) (1970); *Chem. Abstr.* **72**, 133587e (1970).
69. Bottenbruch, L., *Angew. Makromol. Chem.* **13**, 109 (1970); *Chem. Abstr.* **73**, 99270e (1970).
70. Bower, G. M., U.S. Pat. 3,526,610 (Westinghouse Electric Corp.) (1970).
71. Bragaw, C. G., Jr., Fr. Pat. 1,542,129 (E. I. duPont de Nemours & Co.) (1968); *Chem. Abstr.* **71**, 39975n (1969).
72. Bragaw, C. G., Jr., and Herman, A. L., Fr. Pat. 1,543,128 (E. I. duPont de Nemours & Co.) (1968); *Chem. Abstr.* **71**, 22649m (1969).
73. Bratton, F. H., Fick, H. J., and Klimish, K. W., U.S. Pat. 3,449,193 (Schjeldahl, G.T., Co.) (1969); *Chem. Abstr.* **71**, 71599d (1969).
74. British Patent 1,147,856 (E. I. duPont de Nemours & Co.) (1969); *Chem. Abstr.* **70**, 115746k (1969).
75. British Patent 1,155,230 (Dr. Beck and Co.) (1969).
76. British Patent 1,159,760 (American Cyanamid Co.) (1969).
77. British Patent 1,160,097 (Mobil Oil Corp.) (1969).
78. British Patent 1,162,041 (Institut Français du Petrole, des Carburants et Lubrifiants) (1969).
79. British Patent 1,175,555 (Sumitomo Electric Industries, Ltd.) (1969); *Chem. Abstr.* **72** 112226w (1970).
80. British Patent 1,181,446 (Hitachi Chemical Co., Ltd.) (1970).
81. British Patent 1,184,139 (Dr. Beck and Co.) (1970).
82. British Patent 1,189,496 (Monsanto Co.) (1970).
83. British Patent 1,192,147 (Toray Industries Inc.) (1970).
84. British Patent 1,198,856 (Minnesota Mining and Manufacturing Co.) (1970).
85. British Patent 1,214,860 (E. I. duPont de Nemours & Co.) (1970); *Chem. Abstr.* **74**, 43494j (1971).
86. Browning, C. E., and Marshall, J. A., *Proc. Anniv. Tech. Conf. SPI (Soc. Plast. Ind.) Reinf. Plast. Div.*, 25th, 1970 p. 19-C (1970); *Chem. Abstr.* **73**, 15644u (1970).
87. Browning, C. E., and Marshall, J. A., *J. Compos. Mater.* No. 4, 390 (1970); *Chem. Abstr.* **74**, 23309z (1971).
88. Buckley, A., and Seddon, J. D., Brit. Pat. 1,188,936 (ICI, Ltd.) (1970); *Chem. Abstr.* **73**, 4818c (1970).
89. Buketova, N. I., Rafikov, S. R., and Arkhipova, I. A., *Izv. Akad. Nauk Kaz. SSR, Ser. Khim.* **19**, 77 (1969); *Chem. Abstr.* **72**, 67633p (1970).
90. Burns, G. C., German Patent 1,925,092 (E. I. duPont de Nemours & Co.) (1969); *Chem. Abstr.* **72**, 32853h (1970).
91. Burns, R. L., *J. Oil Colour Chem. Ass.* **53**, 52 (1970); *Chem. Abstr.* **72**, 44363m (1970).
92. Burnside, J. Y., *Advan. Struct. Compos. Soc. Aerosp. Mater. Process Eng. Nat. Symp. Exhib.*, 12th, 1967 P-4: 21 pp.; *Chem. Abstr.* **70**, 78682p (1969).

93. Butta, E., De Petris, S., and Pasquini, M., *J. Appl. Polym. Sci.* **13**, 1073 (1969); *Chem. Abstr.* **71**, 92019e (1969).
94. Canadian Patent 801,802 (Minnesota Mining and Manufacturing Co.) (1968).
95. Canadian Patent 819,179 (Celanese Corp.) (1969).
96. Cannizzaro, J. R., *Advan. Tech. Mater. Invest. Fabr. Soc. Aerosp. Mater. Process Eng. Nat. Symp. Exhib.*, *14th*, *1968* 11-3B-1.; *Chem. Abstr.* **71**, 92307x (1969).
97. Chudina, L. I., Litovchenko, S. I., Spirina, T. N., and Chukurov, A. M., *Plast. Massy*, No. 8, p. 12 (1970); *Chem. Abstr.* **73**, 99219v (1970).
98. Clancy, H. M., Kaumeyer, R. A., and Yoshino, S. Y., *Advan. Tech. Mater. Invest. Fabr., Soc. Aerosp. Mater. Process. Eng. Nat. Symp. Exhib.*, *14th*, *1968* II-4B-4; *Chem. Abstr.* **71**, 92319c (1969).
99. Cohen, C., Giuliani, P., Rabilloud, G., Sillion, B., and deGaudemaris, G., Fr. Addn. 94,205 [to French Patent 1,530,445; *Chem. Abstr.* **71**, 22607w (1969)] (Institut Français du Petrole, des Carburants et Lubrifiants) (1969); *Chem. Abstr.* **72**, 56246p (1970).
100. Cohen, C., Giuliani, P., and Sillion, B., Ger. Pat. 2,005,587 (Institut Français du Petrole, des Carburants et Lubrifiants) (1970); *Chem. Abstr.* **74**, 13590r (1971).
101. Collardeau, G., and Leblanc, J. C., Fr. Pat. 1,575,839 (Société des Usines Chimiques Rhone-Poulenc) (1969); *Chem. Abstr.* **72**, 67523c (1970).
102. Copeland, R. L., Beeler, D. R., and Chase, V. A., *Proc.*, *Anniv. Tech. Conf.*, *SPI* (*Soc. Plast. Ind.*) *Reinf. Plast. Div.*, *23rd 1968* pp. 3C-I-3C-9 (1968); *Chem. Abstr.* **70**, 88524c (1969).
103. Coulehan, R. E., and Pickering, T. L., *Polym. Prepr.*, *Amer. Chem. Soc.*, *Div. Polym. Chem.* **12** (1), 305 (1971).
104. Cray, M. C., *Proc.*, *Anniv. Tech. Conf.*, *SPI* (*Soc. Plast. Ind.*) *Reinf. Plast. Div.*, *25th. 1970* p. 19-B (1970); *Chem. Abstr.* **73**, 15642s (1970).
105. Critchley, J. P., McLoughlin, V. C. R., Thrower, J., and White, I. M., *Chem. Ind.* (*London*) p. 934 (1969); *Chem. Abstr.* **71**, 61852s (1969).
106. Culbertson, B. M., U.S. Pat. 3,453,236 (Ashland Oil and Refining Co.) (1969); *Chem. Abstr.* **71**, 50688s (1969).
107. D'Alelio, G. F., *NASA Contract. Rep.* **NASA CR-1057528** (1969); *Chem. Abstr.* **72**, 112181c (1970).
108. D'Alelio, G. F., and Kieffer, H. E., *J. Macromol. Sci., Chem.* **2**, 1275 (1968); *Chem. Abstr.* **70**, 20426s (1969).
109. De Brunner, R. E., U.S. Pat. 3,502,712 (Monsanto Res. Corp.) (1970).
110. De Brunner, R. E., U.S. Pat. 3,530,074 (Monsanto Res. Corp.) (1970); *Chem. Abstr.* **73**, 99586n (1970).
111. De Brunner, R. E., U.S, Pat. 3,542,703 (Monsanto Res. Corp.) (1970); *Chem. Abstr.* **74**, 23342e (1971).
112. De Brunner, R. E., and Fincke, J. K., U.S. Pat. 3,511,790 (Monsanto Res. Corp.) (1970); *Chem. Abstr.* **73**, 15758j (1970).
113. Delvigs, P., *NASA Tech. Note* **NASA TN D-5832** (1970); *Chem. Abstr.* **73**, 77697x (1970).
114. Delvigs, P., Hsu, L-C., and Serafini, T. T., *NASA Tech. Note* **NASA TN D-5184** (1969); *Chem. Abstr.* **71**, 13430f (1969).
115. DiLeone, R. R., U.S. Pat. 3,501,443 (American Cyanamid Co.) (1970); *Chem. Abstr.* **72**, 133538q (1970).
116. Dine-Hart, R. A., and Seddon, J. D., S. Afr. Pat. 68/02,346 (National Research Development Corp.) (1969). *Chem. Abstr.* **71**, 39660f (1969).

117. Dine-Hart, R. A., and Wright, W. W., U.S. Clearinghouse, Fed. Sci. Tech. Inform., AD 1969, AD 700878 (1970); from *U.S. Govt. Res. & Develop. Rep.* **70**, 109 (1970); *Chem. Abstr.* **73**, 56669z (1970).

118. Ditrych, Z., and Huml, J., *Plast. Hmoty Kauc.*, **10**, 302 (1969); *Chem. Abstr.* **72**, 13244u (1970).

119. Dixon, D. R., Rose, J. B., and Turton, C. N., Ger. Pat. 2,002,549 (ICI, Ltd.) (1970); *Chem. Abstr.* **73**, 78065b (1970).

120. Dixon, D. R., Rose, J. B., and Turton, C. N., S. Afr. Pat. 68/05,198 (ICI, Ltd.) (1970); *Chem. Abstr.* **73**, 56758c (1970).

121. Dogoshi, N., Kurihara, M., and Yoda, N., Jap. Pat. 69/19,876 (Toyo Rayon Co., Ltd.) (1969); *Chem. Abstr.* **71**, 113445h (1969).

122. Dogoshi, N., Kurihara, M., Toyama, S., Ikeda, K., and Yoda, N., Jap. Pat. 69/19,879 (Toyo Rayon Co., Ltd.) (1969); *Chem. Abstr.* **71**, 113442e (1969).

123. Dogoshi, N., Toyama, S., Fujita, S., Kurihara, M., and Yoda, N., *J. Polym. Sci., Part A-1* **8**, 2197 (1970); *Chem. Abstr.* **73**, 77694u (1970).

124. Ducloux, M., and Gruffaz, M., Ger. Pat. 1,933,880 (Rhone-Poulenc S.A.) (1970); *Chem. Abstr.* **72**, 79812j (1970).

125. Dulov, A. A., Gurov, A. A., Liogon'kii, B. I., and Berlin, A. A., *Vysokomol. Soedin., Ser. A* **12**, 74 (1970); *Chem. Abstr.* **72**, 90938a (1970).

126. Dunphy, J. F., and Parish, D. J., U.S. Pat. 3,505,168 (E. I. duPont de Nemours & Co.) (1970).

127. Durst, G., and Schneider, K. E., U.S. Pat. 3,454,445 (Texas Instruments Inc.) (1969); *Chem. Abstr.* **71**, 71631h (1969).

128. Ehlers, G. F. L., Fisch, K. R., and Powell, W. R., *J. Polym. Sci., Part A-1* **8**, 3511 (1970); *Chem. Abstr.* **74**, 42826g (1971).

129. Ellinger, L. P., Ger. Pat. 2,001,692 (BP Chemicals, U.K., Ltd.) (1970); *Chem. Abstr.* **74**, 32184g (1971).

130. Evers, R. C., *Polym Prepr., Amer. Chem. Soc., Div. Polym. Chem.* **12** (1), 240 (1971).

131. Farrissey, W. J. Jr., McLaughlin, A., and Rose, J. S., U.S. Pat. 3,562,189 (Upjohn Co.) (1971).

132. Farrissey, W. J. Jr., Rose, J. S., and Carleton, P. S., *J. Appl. Polym. Sci.*, **14**, 1093 (1970); *Chem. Abstr.* **72**, 122301f (1970).

133. Fedorova, E. F., Pokrovskii, E. I., Florinskii, F. S., Koton, M. M., Bessonov, M. I., and Rudakov, A. P., *Spectrosk. Polim.* p. 114 (1968); *Chem. Abstr.* **71**, 3772e (1969).

134. Fedotova, O. Ya., Kolesnikov, G. S., and Matvelashvili, G. S., Russ. Pat. 250,452 (1970).

135. Fick, H. J., Ger. Pat. 1,293,456 (Schjeldahl, G. T., Co.) (1969); *Chem. Abstr.* **71**, 4243v (1969).

136. Fillius, R. H., U.S. Pat. 3,513,134 (Anaconda Wire and Cable Co.) (1970).

137. Fincke, J. K., DeBrunner, R. E., and Wilson, G. R., Brit. Pat. 1,211,633 (Monsanto Co.) (1970); *Chem. Abstr.* **74**, 13814w (1971).

138. Fincke, J. K., DeBrunner, R. E., and Wilson, G. R., U.S. Pat. 3,558,350 (Monsanto Res. Corp.) (1971).

139. Fincke, J. K., and Wilson, G. R., *Mod. Plast.* **46** (4), 108–109, 112, and 117 (1969); *Chem. Abstr.* **71**, 4067r (1969).

140. Fincke, J. K., and Wilson, G. R., Ger. Pat. 1,808,128 (Monsanto Co.) (1969); *Chem. Abstr.* **71**, 71438a (1969).

141. Florkoski, S. P., and Halperin, B. I., *Advan. Tech. Mater. Invest. Fabr., Soc. Aerosp. Mater. Process Eng. Nat. Symp. Exhib., 14th, 1968* 11-48-1; *Chem. Abstr.* **71**, 92318b (1969).

142. Flowers, R. G., and Sherer, T. L., Fr. Pat. 1,511,318 (General Electric Co.) (1968); *Chem. Abstr.* **70**, 68909d (1969).
143. Flowers, R. G., and Sherer, T. L., U.S. Pat. 3,468,852 (General Electric Co.) (1969).
144. Fontan, Y., DeAbajo, J., and Babe, S. G., *Rev. Plast. Mod.* **21**, 177 (1970); *Chem. Abstr.* **73**, 88235w (1970).
145. French Patent Addn. 91,327 [to Fr. Pat. 1,437,746 (Dr. Beck and Co.) (1968)].
146. French Patent Addn. 91,854/1,423,631 (Institut Français du Petrole, des Carburants et Lubrifiants) (1968).
147. French Patent Addn. 93,795 [to Fr. Pat. 1,529,727) (Société des Usines Chimiques Rhone-Poulenc) (1969)]; *Chem. Abstr.* **71**, 125232z (1969).
148. French Patent Addn. 94,379 [to Fr. Pat. 1,560,555) (Monsanto Chemicals Ltd.) (1969)]; *Chem. Abstr.* **72**, 101394y (1970).
149. French Patent Addn. 94,881 [to Fr. Pat. 1,485,931 (Société Rhodiaceta) (1970)]; *Chem. Abstr.* **73**, 99542v (1970).
150. French Patent 200,824 (Toyo Rayon Co., Ltd.) (1969); *Chem. Abstr.* **72**, 44336e (1970).
151. French Patent 1,527,947 (Toyo Rayon Co., Ltd.) (1968); *Chem. Abstr.* **71**, 31059w (1969).
152. French Patent 1,528,074 (E. I. duPont de Nemours & Co.) (1968); *Chem. Abstr.* **71**, 4245x (1969).
153. French Patent 1,528,318 (General Electric Co.) (1968).
154. French Patent 1,537,135 (Société des Usines Chimiques Rhone-Poulenc) (1968).
155. French Patent 1,539,074 (Minnesota Mining and Manufacturing Co.) (1968).
156. French Patent 1,543,501 (Schjeldahl, G. T., Co.) (1968); *Chem. Abstr.* **71**, 13691s (1969).
157. French Patent 1,549,101 (Politechnika Slaska, Poland) (1968).
158. French Patent 1,550,044 (Monsanto Co.) (1968).
159. French Patent 1,556,405 (Fileca, S.A.) (1969).
160. French Patent 1,556,975 (Institut Français du Petrole, des Carburants et Lubrifiants) (1969).
161. French Patent 1,557,641 (Institut Français du Petrole, des Carburants et Lubrifiants) (1969).
162. French Patent 1,559,357 (Dr. Beck and Co.) (1969).
163. French Patent 1,560,555 (Monsanto Chemicals Ltd.) (1969); *Chem. Abstr.* **71**, 113485w (1969).
164. French Patent 1,561,519 (Farbenfabriken Bayer, A. G.) (1969); *Chem. Abstr.* **71**, 113488z (1969).
165. French Patent 1,565,876 (General Electric Co.) (1969).
166. French Patent 1,566,821 (Minnesota Mining and Manufacturing Co.) (1969).
167. French Patent 1,567,863 (Institut Français du Petrole, des Carburants et Lubrifiants) (1969).
168. French Patent 1,569,875 (Toyo Rayon Co., Ltd.) (1969); *Chem. Abstr.* **72**, 32747b (1970).
169. French Patent 1,575,176 (Farbenfabriken Bayer, A. G.) (1969); *Chem. Abstr.* **72**, 79671n (1970).
170. French Patent 1,576,319 (ICI, Ltd.) (1969); *Chem. Abstr.* **72**, 80228e (1970).
171. French Patent 1,576,844 (Farbenfabriken Bayer, A. G.) (1969); *Chem. Abstr.* **72**, 101382t (1970).
172. French Patent 1,577,526 (Sumitomo Electric Industries, Ltd.) (1969).
173. French Patent 1,579,035 (Fileca, S.A.) (1969).
174. French Patent 1,582,899 (Société des Usines Chimiques Rhone-Poulenc, S.A.) (1969).

175. French Patent 1,582,973 (Société des Usines Chimiques Rhone-Poulenc, S.A.) (1969).
176. French Patent 2,000,601 (Toyo Rayon Co., Ltd.) (1969); *Chem. Abstr.* **72**, 44621u (1970).
177. French Patent 2,000,824 (Toyo Rayon Co., Ltd.) (1969); *Chem. Abstr.* **72**, 91006q (1970).
178. French Patent 2,010,603 (Farbenfabriken Bayer, A. G.) (1970); *Chem. Abstr.* **73**, 46056h (1970).
179. French Patent 2,014,273 (Farbenfabriken Bayer, A. G.) (1970); *Chem. Abstr.* **74**, 32173c (1971).
180. French Patent 2,014,752 (Glanzstoff, A. G.) (1970); *Chem. Abstr.* **74**, 4018n (1971).
181. Fujimoto, Y., and Koiwa, Y., Ger. Pat. 1,931,153 (Kyowa Hakko Kogyo Co., Ltd.) (1970).
182. Fukami, A., *Kogyo Kagaku Zasshi* **73**, 1239 (1970); *Chem. Abstr.* **73**, 110186e (1970).
183. Fukui, K., Kagiya, T., Izu, M., Hatta, M., and Matsuda, T., Jap. Pat. 69/28,314 (Seitetsu Chemical Industry Co., Ltd.) (1969); *Chem. Abstr.* **72**, 56100m (1970).
184. Fukukawa, S., Nishioka, T., Ishibashi, I., and Suzuki, Y., Jap. Pat. 68/13,229 (Nitto Electric Industrial Co., Ltd.) (1968); *Chem. Abstr.* **70**, 38314p (1969).
185. Funner, R. E., Fr. Pat. 2,001,022 (E. I. duPont de Nemours & Co.) (1969). *Chem. Abstr.* **72**, 44623w (1970).
186. Funner, R. E., U.S. Pat. 3,516,967 (E. I. duPont de Nemours & Co.) (1970).
187. Gagliani, J., U.S. Pat. 3,505,272 (Schjeldahl, G. T., Co.) (1970).
188. Gay, F. P., and Agolini, F., U.S. Pat. 3,455,879 (E. I. duPont de Nemours & Co.) (1969).
189. Genz, R. H., *Rass. Chim.* **21**, 249 (1969); *Chem. Abstr.* **72**, 79724g (1970).
190. George, N. J., U.S. Pat. 3,554,984 (P. D. George Co.) (1970).
191. German Patent 1,495,100 (Dr. Beck and Co.) (1969).
192. German Patent 1,495,216 (Dr. Beck and Co.) (1969).
193. German Patent 1,495,839 (Farbenfabriken Bayer, A. G.) (1970).
194. German Patent 1,805,955 (Farbenfabriken Bayer, A. G.) (1970).
195. German Patent 1,812,357 (Institut Français du Petrole, des Carburants et Lubrifiants) (1969).
196. German Patent 1,902,875 (Toray Industries, Inc.) (1970).
197. German Patent 1,956,409 (General Electric Co.) (1970).
198. German Patent 1,957,598 (Minnesota Mining and Manufacturing Co.) (1970).
199. German Patent 2,019,436 (General Electric Co.) (1970).
200. Gilch, H., and Schnell, H., U.S. Pat. 3,503,928 (Farbenfabriken Bayer, A. G.) (1970).
201. Gilch, H., Schnell, H., and Bottenbruch, L., Ger. Pat. 1,901,028 (Farbenfabriken Bayer, A. G.) (1970); *Chem. Abstr.* **73**, 99415f (1970).
202. Gordon, N. R., and Langley, N. R., *U.S. At. Energy Comm.* **BNWL-866** (1968); avail. CFSTI. from *Nucl. Sci. Abstr.* **22**, 49865 (1968); *Chem. Abstr.* **70**, 68951m (1969).
203. Gordon, S. E., *Therm. Anal., Proc. Int. Conf., 2nd, 1968* Vol. I, p. 667 (1969); *Chem. Abstr.* **71**, 125382y (1969).
204. Graham, J., and Packham, D. I., *Polymer* **10**, 645 (1969); *Chem. Abstr.* **72**, 44267h (1970).
205. Gribkova, P. N., Rode, V. V., and Korshak, V. V., *Izv. Akad. Nauk SSSR, Ser. Khim.* p. 568 (1970); *Chem. Abstr.* **73**, 26169u (1970).
206. Gribkova, P. N., Rode, V. V., Vygodskii, Ya. S., Vinogradova, S. V., and Korshak, V. V., *Vysokomol. Soedin., Ser. A* **12**, 220 (1970); *Chem. Abstr.* **72**, 79596s (1970).
207. Grundschober, F., and Cervini, C., Ger. Pat. 1,949,281 (Société Rhodiaceta) (1970). *Chem. Abstr.* **73**, 15534h (1970).

208. Hagiwara, Y., Kurihara, M., and Yoda, N., Jap. Pat. 69/31,351 (Toyo Rayon Co., Ltd.) (1969); *Chem. Abstr.* **72**, 101243y (1970).
209. Haller, J. R., Fr. Pat. 1,507,751 (Minnesota Mining and Manufacturing Co.) (1967); *Chem. Abstr.* **70**, 29836h (1969).
210. Haller, J. R., Fr. Pat. 1,559,097 (Minnesota Mining and Manufacturing Co.) (1969); *Chem. Abstr.* **71**, 102827h (1969).
211. Haller, J. R., U.S. Pat. 3,494,823 (Minnesota Mining and Manufacturing Co.) (1970); *Chem. Abstr.* **72**, 80227d (1970).
212. Hansen, R. L., U.S. Pat. 3,476,705 (Minnesota Mining and Manufacturing Co.) (1969).
213. Harris, M., U.S. Pat. 3,502,625 (Eastman Kodak Co.) (1970); *Chem. Abstr.* **72**, 112276n (1970).
214. Harruff, P. W., and Moorefield, S. A., *Advan. Struct. Compos., Soc. Aerosp. Mater. Process Eng. Nat. Symp. Exhib., 12th, 1967* P-2: 10 pp.; *Chem. Abstr.* **70**, 78698y (1969).
215. Hathaway, C. E., Jr., DeBrunner, R. E., and Butler, J. M., U.S. Pat. 3,435,004 (Monsanto Res. Corp.) (1969).
216. Hergenrother, P. M., and Kiyohara, D. E., *Macromolecules* **3**, 387 (1970); *Chem. Abstr.* **73**, 77653e (1970).
217. Hergenrother, P. M., and Levine, H. H., *J. Appl. Polym. Sci.* **14**, 1037 (1970); *Chem. Abstr.* **72**, 122231h (1970).
218. Higgins, J., Jones, J. F., and Thornburgh, A., *Macromolecules* **2**, 558 (1969); *Chem. Abstr*, **72**, 3840w (1970).
219. Hiratsuka, K., and Kamiyama, S., Jap. Pat. 69/23,510 (Mitsubishi Edogawa Chemical Co., Ltd.) (1969); *Chem. Abstr.* **72**, 44354j (1970).
220. Hirsch, A., Can. Pat. 808,252 (Canadian Technical Tape Ltd.) (1969).
221. Hirsch, S. S., Fr. Pat. 1,550,044 (Monsanto Co.) (1968); *Chem. Abstr.* **71**, 40130q (1969).
222. Hirsch, S. S., Fr. Pat. 1,550,601 (Monsanto Co.) (1968); *Chem. Abstr.* **71**, 51178n (1969).
223. Hirsch, S. S., *J. Polym. Sci., Part A-1* **7**, 15 (1969).
224. Hirsch, S. S., U.S. Pat. 3,448,080 (Monsanto Co.) (1969).
225. Hirsch, S. S., and Holsten, J. R., *Appl. Polymer Symp.* **9**, 187 (1969); *Chem. Abstr.* **72**, 33137q (1970).
226. Hoback, J. T., and Holub, F. F., Fr. Pat. 2,008,042 (General Electric Co.) (1970).
227. Hoback, J. T., and Holub, F. F., Ger. Pat. 1,937,388 (General Electric Co.) (1970).
228. Hoegger, E. F., U.S. Pat. 3,436,372 (E. I. duPont de Nemours & Co.) (1969); *Chem. Abstr.* **70**, 107113m (1969).
229. Hoegger, E. F., U.S. Pat. 3,442,861 (E. I. duPont de Nemours & Co.) (1969); *Chem. Abstr.* **71**, 31005a (1969).
230. Holub, F. F., U.S. Pat. 3,435,002 (General Electric Co.) (1969); *Chem. Abstr.* **70**, 107100e (1969).
231. Holub, F. F., U.S. Pat. 3,440,215 (General Electric Co.) (1969).
232. Holub, F. F., U.S. Pat. 3,507,765 (General Electric Co.) (1970).
233. Holub, F. F., and Gaertner, R. F., U.S. Pat. 3,448,068 (General Electric Co.) (1969).
234. Holub, F. F., and Hoback, J. T., Ger. Pat. 1,922,339 (General Electric Co.) (1970); *Chem. Abstr.* **72**, 79925y (1970).
235. Holub, F. F., and Hoback, J. T., Fr. Demande 2,008,042 (General Electric Co.) (1970); *Chem. Abstr.* **73**, 46209k (1970).
236. Holub, F. F., and Hoback, J. T., U.S. Pat. 3,534,003 (General Electric Co.) (1970).
237. Horning, A. E., U.S. Pat. 3,523,056 (General Electric Co.) (1970); *Chem. Abstr.* **73**, 78243h (1970).

238. Hoyt, J. M., and Koch, K., U.S. Pat. 3,509,079 (National Distillers and Chemical Corp.) (1970).

239. Hsu, L.-C., and Serafini, T. T., *NASA Tech. Note*, **NASA TN D-5184** (1969).

240. Hubbuch, L. P., U.S. Pat. 3,444,183 (E. I. duPont de Nemours & Co.) (1969).

241 Ida, N., Kurihara, M., Toyama, S., Dokoshi, N., and Nakanishi, Y., Jap. Pat. 70/24,593 (Toray Co., Ltd.) (1970); *Chem. Abstr.* **74**, 13595a (1971).

242. Imai, T., *J. Polym. Sci., Part B* **8**, 555 (1970); *Chem. Abstr.* **74**, 42701n (1971).

243. Imai, T., *Makromol. Chem.* **138**, 293 (1970); *Chem. Abstr.* **74**, 3908r (1971).

244. Imai, T., Hara, S., and Uchida, M., Jap. Pat. 70/27,995 (Teijin, Ltd.) (1970); *Chem. Abstr.* **74**, 4185q (1971).

245. Imai, T., and Uchiyama, H., *J. Polym. Sci., Part B* **8**, 559 (1970); *Chem. Abstr.* **74**, 42685k (1971).

246. Imai, T., and Yamamoto, M., *Kobunshi Kagaku* **27**, 384 (1970); *Chem. Abstr.* **73**, 110193e (1970).

247. Imai, T., Yamamoto, A., and Uchida, M., Jap. Pat. 70/07,756 (Teijin Co., Ltd.) (1970); *Chem. Abstr.* **72**, 133652x (1970).

248. Imai, T., Yamamoto, A., and Uchida, M., Jap. Pat. 70/07,757 (Teijin Co., Ltd.) (1970); *Chem. Abstr.* **72**, 133642u (1970).

249. Imai, T., Yamamoto, A., and Uchida, M., Jap. Pat. 70/14,315 (Teijin Co., Ltd.) (1970); *Chem. Abstr.* **73**, 35967e (1970).

250. Imai, T., Yamamoto, A., and Uchida, M., Jap. Pat. 70/22,352 (Teijin Co., Ltd.) (1970); *Chem. Abstr.* **73**, 88651d (1970).

251. Imai, T., Yoshida, T., and Uchida, M., Jap. Pat. 70/27,994 (Teijin Co., Ltd.) (1970); *Chem. Abstr.* **74**, 4184p (1971).

252. Irwin, R. S., Def. Publ. U.S. Pat. Off. 647,320 (E. I. duPont de Nemours & Co.) (1969); *Chem. Abstr.* **71**, 14104q (1969).

253. Itoga, M., and Fujita, S., Jap. Pat. 69/08,235 (Toyo Rayon Co., Ltd.) (1969); *Chem. Abstr.* **71**, 81924v (1969).

254. Itoga, M., and Fujita, S., Jap. Pat. 69/16,671 (Toyo Rayon Co., Ltd.) (1969); *Chem. Abstr.* **71**, 125425q (1969).

255. Ivanov, S. S., Koton, M. M., Zakharchuk, G. A., and Volkov, A. P., Russ. Pat. 238,153 (Institute of High-Molecular-Weight Compounds, Academy of Sciences, USSR) (1969); *Chem. Abstr.* **71**, 50695s (1969).

256. Iwakura, Y., and Hayano, F., *J. Polym. Sci., Part A-1* **7**, 597 (1969); *Chem. Abstr.* **71**, 13432h (1969).

257. Iwakura, Y., Kurita, K., and Hayano, F., *J. Polym. Sci., Part A-1* **7**, 609 (1969); *Chem. Abstr.* **71**, 13435m (1969).

258. Izard, E. F., and Lindsey, W. B., Fr. Pat. 1,509,269 (E. I. duPont de Nemours & Co.) (1968); *Chem. Abstr.* **71**, 13726g (1969).

259. Izumi, M., Shima, H., Matsumura, S., and Asano, N., Jap. Pat. 68/17,200 (Sumitomo Electric Industries, Ltd.) (1968); *Chem. Abstr.* **70**, 20519z (1969).

260. Izumi, M., Shima, H., Matsumura, S., and Asano, N., Jap. Pat. 68/22,998 (Sumitomo Electric Industries, Ltd.) (1968); *Chem. Abstr.* **70**, 68912z (1969).

261. Izumi, M., Shima, H., Matsumura, S., and Asano, N., Jap. Pat. 69/29,072 (Sumitomo Electric Industries, Ltd.) (1969); *Chem. Abstr.* **72**, 91013g (1970).

262. Jackson, W. G., and Schroeder, W., U.S. Pat. 3,484,387 (Burdick and Jackson Laboratories, Inc.) (1969); *Chem. Abstr.* **73**, 99417h (1970).

263. Japanese Patent 68/28,836 (Toyo Rayon Co., Ltd.) (1968).

264. Japanese Patent 69/12,621 (Showa Electr. Wire and Cable Co., Ltd.) (1969);

265. Japanese Patent 69/28,065 (Tokyo Shibaura Electric Co.) (1969).

266. Japanese Patent 70/01,631 (Toyo Rayon Co., Ltd.) (1970).
267. Japanese Patent 70/06,012 (Hitachi Kasei Kogyo) (1970).
268. Japanese Patent 70/06,555 (Toray Industries, Inc.) (1970).
269. Japanese Patent 70/07,894 (Hitachi Wire and Cable Co., Ltd.) (1970).
270. Japanese Patent 70/08,148 (Sumitomo Elec. Ind., Ltd.) (1970).
271. Japanese Patent 70/08,993 (Toray Industries, Inc.) (1970).
272. Japanese Patent 70/09,192 (Densen Denran KKS) (1970).
273. Japanese Patent 70/09,547 (Tokyo Shibaura Electric Co., Ltd.) (1970).
274. Japanese Patent 70/11,518 (Toray Industries, Inc.) (1970).
275. Japanese Patent 70/13,115 (Toray Industries, Inc.) (1970).
276. Japanese Patent 70/13,117 (Toray Industries, Inc.) (1970).
277. Japanese Patent 70/13,597 (Hitachi Chem. Ind. Co., Ltd.) (1970).
278. Japanese Patent 70/15,990 (Toray Industries, Inc.) (1970).
279. Japanese Patent 70/15,991 (Toray Industries, Inc.) (1970).
280. Japanese Patent 70/15,992 (Toray Industries, Inc.) (1970).
281. Japanese Patent 70/15,993 (Toray Industries, Inc.) (1970).
282. Japanese Patent 70/18,312 (Tokyo Shibaura Electric Co., Ltd.) (1970).
283. Japanese Patent 70/18,316 (Hitachi Kasei Industries, Ltd.) (1970).
284. Japanese Patent 70/18,678 (Hitachi Chem. Industry Co., Ltd.) (1970).
285. Japanese Patent 70/21,846 (Showa Electr. Wire and Cable Co., Ltd.) (1970).
286. Japanese Patent 70/24,594 (Tore Co., Ltd.) (1970).
287. Japanese Patent 70/24,791 (Toray Industries, Inc.) (1970).
288. Japanese Patent 70/25,792 (Adachi Veneer KK) (1970).
289. Japanese Patent 70/29,690 (Hitachi Chem. Ind. Co., Ltd.) (1970).
290. Japanese Patent 70/30,132 (Matsushita Elec. Ind. Co., Ltd.) (1970).
291. Japanese Patent 70/32,629 (Toray Industries, Inc.) (1970).
292. Japanese Patent 70/33,670 (Toray Industries, Inc.) (1970).
293. Japanese Patent 70/35,073 (Nitto Electr. Co.) (1970).
294. Japanese Patent 70/35,557 (Mitsubishi Electr. Co.) (1970).
295. Japanese Patent 70/37,902 (Hitachi Chemical Industries, Ltd.) (1970).
296. Johnston, T. H., and Gaulin, C. A., *J. Macromol. Sci., Chem.* 3, 1161 (1969); *Chem. Abstr.* 71, 92154v (1969).
297. Jones, J. F., Vaughan, R. W., and Burns, E. A., *Nat. SAMPE (Soc. Aerosp. Mater. Process Eng.) Symp. Exhib., 15th, 1969* pp. 163–169 (1969); *Chem. Abstr.* 73, 15648y (1970).
298. Jones, J. I., Fr. Pat. 1,544,354 (National Res. Development Corp.) (1968); *Chem. Abstr.* 71, 39743k (1969).
299. Jones, J. I., *J. Polym. Sci., Part C* 22, 773 (1969).
300. Josten, F., Scholz, T., and Hessling, H., Ger. Pat. 1,301,114 (Rheinpreussen A. G. für Bergbau und Chemie) (1969); *Chem. Abstr.* 71, 102954e (1969).
301. Juveland, O. O., U.S. Pat. 3,472,815 (Standard Oil Co., Indiana) (1969); *Chem. Abstr.* 72, 3936g (1970).
302. Katsarava, R. D., Korshak, V. V., and Rusanov, A. L., Russ. Pat. 245,359 (Institute of Heterocyclic Organic Compounds, Academy of Sciences, USSR) (1969).
303. Kawashima, H., Kikkawa, M., and Masuko, T., Ger. Pat. 1,804,461 (Hitachi, Ltd.) (1970).
304. Kazaryan, L. G., Lur'e, E. G., and Igonin, L. A., *Vysokomol. Soedin., Ser. B* 11, 779 (1969); *Chem. Abstr.* 72, 44245z (1970).
305. Kersten, H., and Meyer, G., *Makromol. Chem.* 138, 265 (1970); *Chem. Abstr.* 74, 3913p (1971).

306. Kerwin, R. E., and Goldrick, M. R., *Tech. Pap.*, *Reg. Tech. Conf.*, *Soc. Plast. Eng.*, *Mid-Hudson Sect.* pp. 44–50 (1970); *Chem. Abstr.* **74**, 4332k (1971).

307. Khar'kov, S. N., Lavrova, Z. N., and Chegolya, A. S., Russ. Pat. 238,155 (1969); *Chem. Abstr.* **71**, 61896j (1969).

308. Khar'kov, S. N., Lavrova, Z. N., Chegolya, A. S., and Krasnov, E. P., *Vysokomol. Soedin.*, *Ser. B* **12**, 349 (1970); *Chem. Abstr.* **73**, 46148q (1970).

309. Khofbauer, E. I., and Nesterova, E. I., Russ. Pat. 231,798 (1968); *Chem. Abstr.* **70**, 68911y (1969).

310. Kitamura, K., and Minami, S., Jap. Pat. 70/06,555 (Toyo Rayon Co., Ltd.) (1970); *Chem. Abstr.* **72**, 122372e (1970).

311. Klebanskii, A. L., Tsukerman, N. Ya., Borovikova, N. A., Dolgopolskii, I. M., Shvedova, V. N., Selivanovskaya, G. A., and Alekseichuk, G. A., Russ. Pat. 248,208 (S. V. Lebedev All-Union Scientific-Res. Institute of Synthetic Rubber) (1969); *Chem. Abstr.* **72**, 44360h (1970).

312. Knapp, E. C., Markhart, A. H., and Serlin, I., Ger. Pat. 1,912,551 (Monsanto Co.) (1969); *Chem. Abstr.* **72**, 22312s (1970).

313. Knapp, E. C., Markhart, A. H., and Serlin, I., U.S. Pat. 3,554,935 (Monsanto Co.) (1970).

314. Kobayashi, F., Sakata, A., and Mizoguchi, T., Jap. Pat. 69/28,112 (Japan Rayon Co., Ltd.) (1969); *Chem. Abstr.* **72**, 56269y (1970).

315. Kobayashi, F., Sakata, A., Mizoguchi, T., and Suyama, N., Jap. Pat. 69/27,674 (Japan Rayon Co., Ltd.) (1969); *Chem. Abstr.* **72**, 56598e (1970).

316. Kobayashi, F., Sakata, A., Mizoguchi, T., and Yamaguchi, R., Jap. Pat. 69/19,272 (Japan Rayon Co., Ltd.) (1969); *Chem. Abstr.* **71**, 113660z (1969).

317. Kobayashi, F., Sakata, A., Mizoguchi, T., and Yamaguchi, R., Jap. Pat. 69/27,675 (Japan Rayon Co., Ltd.) (1969); *Chem. Abstr.* **72**, 56084j (1970).

318. Kolesnikov, G. S., Fedotova, O. Ya., and Al-Sufi, H. H. M. A., *Vysokomol. Soedin.*, *Ser. B* **10**, 742 (1968); *Chem. Abstr.* **70**, 20575q (1969).

319. Kolesnikov, G. S., Fedotova, O. Ya., Al-Sufi, H. H. M. A., and Bolevskii, S. F., *Vysokomol. Soedin,. Ser. A* **12**, 323 (1970); *Chem. Abstr.* **73**, 4210y (1970).

320. Kolesnikov, G. S., Fedotova, O. Ya., and Fedyna, V. N., *Vysokomol. Soedin.*, *Ser. A* **11**, 691 (1969); *Chem. Abstr.* **71**, 13437p (1969).

321. Kolesnikov, G. S., Fedotova, O. Ya., Fedyna, V. N., and Firsova, E. V., Russ. Pat. 226,846 (1968); *Chem. Abstr.* **70**, 48214v (1969).

322. Kolesnikov, G. S., Fedotova, O. Ya., and Fedyna, V. N., Russ. Pat. 245,360 (1969).

323. Kolesnikov, G. S., Fedotova, O. Ya., Fedyna, V. N., and Lenskaya, A. V., Russ. Pat. 245,360 (1969); *Chem. Abstr.* **72**, 4029a (1970).

324. Kolesnikov, G. S., Fedotova, O. Ya., and Hofbauer, E. I., *Vysokomol. Soedin.*, *Ser. B* **11**, 617 (1969); *Chem. Abstr.* **71**, 124962a (1969).

325. Kolesnikov, G. S., Fedotova, O. Ya., and Matvelashvili, G. S., Russ. Pat. 250,452 (1969); *Chem. Abstr.* **72**, 79657n (1970).

326. Kolesnikov, G. S., Fedotova, O. Ya., and Paresishvili, O. I., *Vysokomol. Soedin.*, *Ser. B* **10**, 781 (1968); *Chem. Abstr.* **70**, 20391b (1969).

327. Kolesnikov, G. S., Fedotova, O. Ya., and Paresishvili, O. I., Russ. Pat. 228,271 (1968); *Chem. Abstr.* **70**, 58629y (1969).

328. Kolesnikov, G. S., Fedotova, O. Ya., Paresishvili, O. I., and Belevskii, S. F., *Vysokomol. Soedin.*, *Ser. A* **12**, 317 (1970); *Chem. Abstr.* **73**, 4241j (1970).

329. Korshak, V. V., USSR Patent 214,803 (1969); Buryatskii Res. Inst.

330. Korshak, V. V., Doroshenko, Yu. E., Teplyakov, M. M., Fedorova, R. D., and Volkov, B. V., *Vysokomol. Soedin.*, *Ser. A* **12**, 677 (1970); *Chem. Abstr.* **72**, 133255v (1970).

331. Korshak, V. V., Izyneev, A. A., and Fokin, E. P., Russ. Pat. 214,803 (Buryat Complex Scientific Res. Institute) (1968); *Chem. Abstr.* **70**, 20522v (1969).
332. Korshak, V. V., Mamaev, V. P., Vinogradova, S. V., Vygodskii, Ya. S., and Fokin, E. P., (Institute of Heteroorganic Compounds, Academy of Sciences, USSR and Institute of Organic Chemistry, Academy of Sciences, USSR Russ. Pat. 267,893) (1970); *Chem. Abstr.* **73**, 67232c (1970).
333. Korshak, V. V., Pavlova, S. A., Boiko, L. V., Babchinitser, T. M., Vinogradova, S. V., Vygodskii, Ya. S., and Golubeva, N. A., *Vysokomol. Soedin, Ser. A* **12**, 56 (1970); *Chem. Abstr.* **72**, 90918u (1970).
334. Korshak, V. V., and Rusanov, A. L., *Izv. Akad. Nauk. SSSR, Ser. Khim.* p. 2418 (1968); *Chem. Abstr.* **70**, 20584s (1969).
335. Korshak, V. V., and Rusanov, A. L., *Izv. Akad. Nauk. SSSR, Ser. Khim.* p. 2661 (1968); *Chem. Abstr.* **70**, 78423e (1969).
336. Korshak, V. V., and Rusanov, A. L., *Izv. Akad. Nauk SSSR, Ser. Khim.* p. 1917 (1970); *Chem. Abstr.* **73**, 88229x (1970).
337. Korshak, V. V., Rusanov, A. L., Katsarava, R. D., and Tugushi, D. S., Russ. Pat. 245,359 (Institute of Heteroorganic Compounds, Academy of Sciences, USSR) (1969); *Chem. Abstr.* **72**, 13229t (1970).
338. Korshak, V. V., Tseitlin, G. M., and Al-Haider, Z. T., *Izv. Akad. Nauk. SSSR, Ser. Khim.* p. 2160 (1970); *Chem. Abstr.* **73**, 131383g (1970).
339. Korshak, V. V., Tseitlin, G. M., and Azarov, V. I., *Vysokomol. Soedin., Ser. B* **11**, 225 (1969); *Chem. Abstr.* **70**, 115622s (1969).
340. Korshak, V. V., Tseitlin, G. M., Azarov, V. I., Akutin, M. S., Elin, I. O., and Lyubyashkina, E. G., Russ. Pat. 250,453 (D. I. Mendeleev Chemical-Technological Institute, Moscow) (1969); *Chem. Abstr.* **72**, 101431h (1970).
341. Korshak, V. V., Vinogradova, S. V., and Vygodskii, Ya. S., Russ. Pat. 250,454 (Institute of Heteroorganic Compounds, Academy of Sciences, USSR) (1969); *Chem. Abstr.* **72**, 79655k (1970).
342. Koton, M. M., *Khim. Volokna* p. 15 (1969); *Chem. Abstr.* **71**, 92094a (1969).
343. Koton, M. M., *Zh. Prikl. Khim. (Leningrad)* **42**, 1841 (1969); *Chem. Abstr.* **72**, 3944h (1970).
344. Koton, M. M., *et al.*, Brit. Pat, 1,183,306 (Institute of High-Molecular-Weight Compounds, Academy of Sciences, USSR) (1970); *Chem. Abstr.* **72**, 112677u (1970).
345. Koton, M. M., Adrova, N. A., Florinskii, F. S., Bessonov, M. I., Rudakov, A. P., and Moskvina, E. M., Russ. Pat. 249,625 (Institute of High-Molecular-Weight Compounds, Academy of Sciences, USSR) (1969); *Chem. Abstr.* **72**, 44494e (1970).
346. Koton, M. M., Florinskii, F. S., Adrova, N. A., Dubnova, A. M., Bessonov, M. I., and Rudakov, A. P., Russ. Pat. 224,056 (Institute of High-Molecular-Weight Compounds, Academy of Sciences, USSR) (1968); *Chem. Abstr.* **70**, 48055u (1969).
347. Koton, M. M., Florinskii, F. S., Bessonov, M. I., and Rudakov, A. P., Russ. Pat. 257,010 (Institute of Heteroorganic Compounds, Academy of Sciences, USSR) (1969); *Chem. Abstr.* **73**, 4399s (1970).
348. Koton, M. M., Frenkel, S. Ya., and Adrova, N. A., Brit. Pat. 1,183,306 (Institute of High-Molecular-Weight Compounds, Academy of Sciences, USSR) (1970).
349. Kousaka, T., Shono, T., and Oda, R., *Kogyo Kagaku Zasshi* **71**, 1738 (1968); *Chem. Abstr.* **70**, 47950v (1969).
350. Kovarskaya, B. M., Annenkova, N. G., Al-Sufi, H. H. M. A., Gur'yanova, V. V., Fedotova, O. Ya., Aimbetov, Zh. V., and Kolesnikov, G. S., *Plast. Massy* No. 8, p. 41 (1969); *Chem. Abstr.* **71**, 113414x (1969).

351. Kransov, E. P., Aksenova, V. P., Khar'kov, S. N., and Baranova, S. A., *Vysokomol. Soedin.*, *Ser. A* **12**, 873 (1970); *Chem. Abstr.* **73**, 15469r (1970).
352. Kreuz, J. A., Fr. Pat. 1,516,515 (E. I. duPont de Nemours & Co.) (1968); *Chem. Abstr.* **70**, 78734g (1969).
353. Kruger, A., Ger. Pat. 1,905,590 (1970).
354. Kubota, T., Tsuda, T., Minami, S., and Watanabe, M., Jap. Pat. 70/01,831 (Toyo Rayon Co., Ltd.) (1970); *Chem. Abstr.* **72**, 101379x (1970).
355. Kudryavtsev, G. I., Balakleitseva, L. F., Shchetinin, A. M., and Chikurina, L. V., *Vysokomol. Soedin.*, *Ser. A* **12**, 2205 (1970); *Chem. Abstr.* **74**, 13540d (1971).
356. Kudryavtsev, V. V., Koton, M. M., and Svetlichnyi, V. M., *Vysokomol. Soedin.*, *Ser. B* **11**, 280 (1969).
357. Kudryavtsev, V. V., Koton, M. M., and Svetlichnyi, V. M., *Vysokomol. Soedin.*, *Ser. B* **11**, 280 (1969); *Chem. Abstr.* **71**, 30709w (1969).
358. Kurihara, M., *Macromolecules* **3**, 722 (1970); *Chem. Abstr.* **74**, 42681f (1971).
359. Kurihara, M., Dokoshi, N., Toyama, S., Ikeda, K., Yoda, N., and Nakanishi, R., Jap. Pat. 69/23,109 (Toyo Rayon Co., Ltd.) (1969); *Chem. Abstr.* **72**, 22152q (1970).
360. Kurihara, M., and Hagiwara, Y., *Polym. J.* **1**, 425 (1970); *Chem. Abstr.* **73**, 110188g (1970).
361. Kurihara, M., Hagiwara, Y., and Yoda, N., Jap. Pat. 69/28,517 (Toyo Rayon Co., Ltd.) (1969); *Chem. Abstr.* **72**, 101233v (1970).
362. Kurihara, M., Hagiwara, Y., and Yoda, N., Jap. Pat. 69/31,350 (Toyo Rayon Co., Ltd.) (1969); *Chem. Abstr.* **72**, 112073u (1970).
363. Kurihara, M., Toyama, S., Dokoshi, N., Hagiwara, Y., and Yoda, N., Jap. Pat. 70/15,274 (Toray Industries, Inc.) (1970); *Chem. Abstr.* **73**, 67102k (1970).
364. Lanza, V. L., and Stivers, E. C., Belg. Pat. 739,060 (Raychem Corp.) (1970).
365. Lanza, V. L., and Stivers, E. C., Ger. Pat. 1,947,030 (Raychem Corp.) (1970); *Chem. Abstr.* **73**, 67406n (1970).
366. Lapierre, A., *Rev. Gen. Caout. Plast.* **47**, 1185 (1970); *Chem. Abstr.* **74**, 31966b (1971).
367. Lavin, E., Markhart, A. H., and Kass, R. E., U.S. Pat. 3,518,219 (Monsanto Co.) (1970); *Chem. Abstr.* **73**, 46329z (1970).
368. Lavin, E., and Serlin, I., Fr. Pat. 1,540,432 (Monsanto Co.) (1968); *Chem. Abstr.* **71**, 22594q (1969).
369. Lavin, E., and Serlin, I., Fr. Pat. 1,540,433 (Monsanto Co.) (1968); *Chem. Abstr.* **71**, 22590k (1969).
370. Lavin, E., and Serlin, I., U.S. Pat. 3,483,144 (Monsanto Co.) (1969).
371. Lavin, E., and Serlin, I., U.S. Pat. 3,554,939 (Monsanto Co.) (1970).
372. Lee, J., *NASA Contract Rep.* **NASA CR-72563** (1969); *Sci. Tech. Aerosp. Rep.* **8**, 118 (1970); *Chem. Abstr.* **73**, 15886z (1970).
373. Lewis, R. B., *Lubric. Eng.* **25**, 356 (1969); *Chem. Abstr.* **71**, 125071w (1969).
374. Libackyj, A., U.S. Pat. 3,541,036 (E. I. duPont de Nemours and Co.) (1970).
375. Lidorenko, N. S., Gindin, L. G., Egorov, B. N., Kondratenkov, V. I., Ravich, I. Ya., and Toroptseva, T. N., *Dokl. Akad. Nauk. SSSR* **187**, 581 (1969); *Chem. Abstr.* **71**, 113542n (1969).
376. Lidorenko, N. S., Ravich, I. Ya., Gindin, L. G., Kagan, A. S., Koval'skii, A. E., Toroptseva, T. N., and Zhigareva, N. K., *Dokl. Akad. Nauk SSSR* **182**, 1087 (1968); *Chem. Abstr.* **70**, 29637u (1969).
377. Loncrini, D. F., and Kruh, D., Ger. Pat. 1,922,316 (General Electric Co.) (1969); *Chem. Abstr.* **72**, 44353h (1970).
378. Loncrini, D. F., and Witzel, J. M., *J. Polym. Sci.*, *Part A-1* **7**, 2185 (1969); *Chem. Abstr.* **71**, 113332u (1969).

379. Long, F., and Ensor, G. R., U.S. Pat. 3,554,969 (Monsanto Chemicals Ltd.) (1970).
380. Lorenz, G., Gallus, M., Giessler, W., Bodesheim, F., Wieden, H., and Nischk, G. E., *Makromol. Chem.* **130**, 65 (1969); *Chem. Abstr.* **72**, 44284m (1970).
381. Loudas, B. L., U.S. Pat. 3,503,929 (3M Co.) (1970).
382. Lovejoy, E. R., Ger. Pat. 1,929,439 (E. I. duPont de Nemours & Co.) (1969); *Chem. Abstr.* **72**, 79683t (1970).
383. Lubowitz, H. R., *Polym. Prepr., Amer. Chem. Soc., Div. Polym. Chem.* **12** (1), 329 (1971).
384. Lubowitz, H. R., Burns, E. A., and Jones, J. F., Fr. Pat. 2,001,579 (TRW, Inc.) (1969); *Chem. Abstr.* **72**, 44496g (1970).
385. Lubowitz, H. R., Kendrick, W. P., Jones, J. F., and Burns, E. A., Fr. Pat. 1,581,983 (TRW, Inc.) (1969); *Chem. Abstr.* **73**, 15730v (1970).
386. Lubowitz, H. R., Kendrick, W. P., Jones, J. F., Thorpe, R. S., and Burns, E. A., Fr. Pat. 1,580,456 (TRW, Inc.) (1969); *Chem. Abstr.* **73**, 46229s (1970).
387. Lubowitz, H, R., and Vaughan, R. W., Ger. Pat. 1,913,515 (TRW, Inc.) (1970); *Chem. Abstr.* **73**, 56789p (1970).
388. Lubowitz, H. R., Wilson, E. R., Kendrick, W. P., and Burns, E. A., Fr. Pat. 1,572,798 (TRW, Inc.) (1969); *Chem. Abstr.* **72**, 56085k (1970).
389. Lucas, H. R., U.S. Pat. 3,489,725 (American Cyanamid Co.) (1969).
390. Lur'e, E. G., Kazaryan, L. G., Kovriga, V. V., Lebedinskaya, M. L., Uchastkina, E. L., Dobrokhotova, M. L., and Emel'yanova, L. N., *Plast. Massy* p. 59 (1970); *Chem. Abstr.* **73**, 88473x (1970).
391. Lynch, E. R., Brit. Pat. 1,168,494 (Monsanto Chemicals Ltd.) (1969); *Chem. Abstr.* **72**, 22150n (1970).
392. Mahon, J., Grumman Aircraft Corp. N69-17604 (1968).
393. Mahon, J., *U.S. At. Energy Comm.* **FSR-AD8-05-68.1** (1968); *Sci. Tech. Aerosp. Rep.* **7**, 1158–1159 (1969); *Chem. Abstr.* **71**, 71306f (1969).
394. Marin, G., and Rio, A., Fr. Pat, 1,525,338 (Société des Usines Chimiques Rhone-Poulenc) (1968); *Chem. Abstr.* **70**, 115742f (1969).
395. Marvel, C. S., U.S. Pat. 3,563,917 (Research Corporation) (1971).
396. Mashkevich, S. A., Zhubanov, B. A., and Rafikov, S. R., *Tr. Inst. Khim. Nauk Akad. Nauk Kaz. SSR* **28**, 78 (1970); *Chem. Abstr.* **73**, 35818g (1970).
397. Mashkevich, S. A., Zhubanov, B. A., Rafikov, S. R., and Kurmangazieva, Zh. M., *Izv. Akad. Nauk Kaz. SSR, Ser. Khim.* **20**, 56 (1970); *Chem. Abstr.* **74**, 42717x (1971).
398. Matsubara, K., Watanabe, M., and Karasawa, M., *Junkatsu* **14**, 43 (1969); *Chem. Abstr.* **70**, 115780s (1969).
399. Matvelashvili, G. S., Fedotova, O. Ya., and Kolesnikov, G. S., *Vysokomol. Soedin., Ser. B* **11**, 53 (1969); *Chem. Abstr.* **70**, 88341r (1969).
400. May, L. C., and Hertz, J., *Proc. Anniv. Tech. Conf., SPI (Soc. Plast. Ind.) Reinf. Plast. Div., 25th, 1970* p. 12-B (1970); *Chem. Abstr.* **73**, 15881u (1970).
401. May, L. C., and Hertz, J., *SAMPE Quart.* **1**, 15 (1970); *Chem. Abstr.* **74**, 32405e (1971).
402. McKeown, J. J., and Toner, M. E., U.S. Pat. 3,520,845 (Minnesota Mining and Manufacturing Co.) (1970); *Chem. Abstr.* **73**, 67225c (1970).
403. Meyers, R. A., *Polym. Prepr., Amer. Chem. Soc., Div. Polym. Chem.* **10** (1), 186 (1969); *Chem. Abstr.* **73**, 131390g (1970).
404. Meyers, R. A., *J. Polym. Sci., Part A-1* **7**, 2757 (1969); *Chem. Abstr.* **72**, 13117e (1970).
405. Meyers, R. A., Ger. Pat, 1,810,467 (TRW, Inc.) (1970); *Chem. Abstr.* **73**, 78075e (1970).
406. Miller, G. W., U.S. Pat. 3,489,696 (Mobay Chemical Co.) (1970).
407. Miller, W. A., and Fowler, F. D., *Plast. Des. Process.* **10**, 17 (1970); *Chem. Abstr.* **73**, 46072k (1970).

408. Minami, M., Jap. Pat. 69/19,878 (Toyo Rayon Co., Ltd.) (1969); *Chem. Abstr.* **72**, 44495f (1970).
409. Minami, M., Jap. Pat. 69/28,712 (Toyo Rayon Co., Ltd.) (1969); *Chem. Abstr.* **72**, 56244m (1970).
410. Minami, M., Ger. Pat. 1,946,475 (Toyo Rayon Co.) (1970).
411. Minami, M., Jap. Pat. 70/20,508 (Toray Ind., Inc.) (1970); *Chem. Abstr.* **73**, 88598s (1970).
412. Minami, M., Jap. Pat. 70/32,715 (Toray Ind., Inc.) (1970); *Chem. Abstr.* **74**, 43057u (1971).
413. Minami, M., and Kitamura, K., Jap. Pat. 70/03,396 (Toyo Rayon Co., Ltd.) (1970); *Chem. Abstr.* **72**, 101411b (1970).
414. Minami, M., and Kitamura, K., Jap. Pat. 70/26,752 (Toray Ind., Inc.) (1970); *Chem. Abstr.* **74**, 13601z (1971).
415. Minami, M., and Kitamura, K., Jap. Pat. 70/32,716 (Toray Ind., Inc.) (1970); *Chem. Abstr.* **74**, 43051n (1971).
416. Minami, M., Taniguchi, M., Tsutsumi, T., Murakami, M., and Yoshii, T., Ger. Pat. 1,944,213 (Toyo Rayon Co., Ltd.) (1970); *Chem. Abstr.* **72**, 133695p (1970).
417. Mod'yugin, B. G., and Fefelov, P. A., *Mekh. Polim.* **5**, 1111 (1969); *Chem. Abstr.* **72**, 122152h (1970).
418. Morello, E. F., Ger. Pat. 1,931,602 (Standard Oil Co., Indiana) (1970); *Chem. Abstr.* **72**, 67535h (1970).
419. Mortillaro, L., *Mater. Plast. Elastomeri* **34**, 305 (1968); *Chem. Abstr.* **70**, 20435u (1969).
420. Mozgova, K. K., Korshak, V. V., and Levitskaya, S. G., *Plast. Massy* No. 10, p. 14 (1968); *Chem. Abstr.* **70**, 20397h (1969).
421. Muramatsu, A., and Tamura, S., Jap. Pat. 69/14,954 (Tokyo Shibaura Electric Co., Ltd.) (1969); *Chem. Abstr.* **71**, 102583a (1969).
422. Muramoto, A., and Tamura, S., Jap. Pat. 69/31,834 (Tokyo Shibaura Electric Co., Ltd.) (1969); *Chem. Abstr.* **72**, 122524f (1970).
423. Nakanishi, R., Yoda, N., Kurihara, M., and Toyama, S., Jap. Pat. 69/27,756 (Toyo Rayon Co., Ltd.) (1969); *Chem. Abstr.* **72**, 56300b (1970).
424. Nakano, M., and Koyama, T., U.S. Pat. 3,541,038 (Hitachi Chemical Co., Ltd.) (1970).
425. Nakano, M., Oyama, T., and Endo, T., Jap. Pat. 70/12,985 (Hitachi Chemical Co., Ltd.) (1970); *Chem. Abstr.* **73**, 88677s (1970).
426. Naselow, A. B., U.S. Pat. 3,485,796 (E. I. duPont de Nemours & Co.) (1969).
427. Naselow, A. B., and Strugar, D., Fr. Pat. 1,515,066 (E. I. duPont de Nemours & Co.) (1968); *Chem. Abstr.* **70**, 78560x (1969).
428. Nehring, R., and Seeliger, W., *Angew. Chem., Int. Ed. Engl.* **9**, 461 (1970); *Chem. Abstr.* **73**, 56483j (1970).
429. Netherlands Patent Application 6,500,698 (E. I. duPont de Nemours & Co.) (1965); *Chem. Abstr.* **64**, 839g (1966).
430. Netherlands Patent Application 6,805,968 (American Cyanamid Co.) (1968).
431. Netherlands Patent Application 6,814,003 (E. I. duPont de Nemours & Co.) (1969).
432. Netherlands Patent Application 6,817,104 (TRW, Inc.) (1969).
433. Netherlands Patent Application 6,817,284 (E. I. duPont de Nemours & Co.) (1969).
434. Netherlands Patent Application 6,901,939 (TRW, Inc.) (1969).
435. Netherlands Patent Application 6,903,051 (Institut Français du Petrole, des Carburants et Lubrifiants) (1969).
436. Netherlands Patent Application 6,904,339 (Institut Français du Petrole, des Carburants et Lubrifiants) (1969).
437. Netherlands Patent Application 6,909,368 (TRW, Inc.) (1969).

438. Netherlands Patent Application 6,917,578 (E. I. duPont de Nemours & Co. ) (1970).
439. Netherlands Patent Application 6,917,804 (Farbenfabriken Bayer, A. G.) (1969).
440. Netherlands Patent Application 7,001,648 (American Cyanamid Co.) (1970).
441. Nishioka, T., Ishibashi, I., Suzuki, Y., and Hamada, Y., Jap. Pat. 70/06,554 (Nitto Electric Industrial Co., Ltd.) (1970); Chem. Abstr. 72, 133604h (1970).
442. Oda, R., and Tabushi, Y., Jap. Pat. 70/21,957 (Idemitsu Kosan Co., Ltd.) (1970); Chem. Abstr. 73, 121213n (1970).
443. Olson, E. H., Arndt, R. P., and Filius, R. H., U.S. Pat. 3,493,413 (Anaconda Wire and Cable Co.) (1970).
444. Oprits, Z. G., Kudryavtsev, G. I., Korzhavin, L. N., Ginzburg, B. M., and Frenkel, S. Ya., Khim. Volokna No. 3, 61 (1970); Chem. Abstr. 73, 46506e (1970).
445. Oprits, Z. G., Lazutkina, T. P., and Kudryavtsev, G. I., Khim. Volokna No. 4, p. 10 (1970); Chem. Abstr. 73, 99866d (1970).
446. Packham, D. I., Brit. Polym. J., 2, 280 (1970); Chem. Abstr. 74, 23146u (1971).
447. Packham, D. I., Davies, J. D., and Rackley, F. A., Polymer 11, 533 (1970); Chem. Abstr. 73, 131375f (1970).
448. Parish, D. J., U.S. Pat. 3,492,270 (E. I. duPont de Nemours & Co.) (1970).
449. Parsons, J. F., U.S. Pat. 3,507,683 (E. I. duPont de Nemours & Co.) (1970); Chem. Abstr. 72, 133570v (1970).
450. Payette, L. J., U.S. Pat. 3,567,673 (General Electric Co.) (1970).
451. Pecka, J. T., U.S. Pat. 3,485,734 (E. I. duPont de Nemours & Co.) (1969); Chem. Abstr. 72, 44579m (1970).
452. Peterson, M. A., Ger. Pat. 2,008,747 (General Electric Co.) (1970); Chem. Abstr. 73, 131647w (1970).
453. Petker, I., Sakakura, R. T., and Segimoto, M., Proc. Anniv. Tech. Conf., SPI (Soc. Plast. Ind.) Reinf. Plast. Div., 23rd, 1968 pp. 17B-1-17B6 (1968); Chem. Abstr. 70, 78603p (1969).
454. Phillips, L. N., and Rogers, K. F., Composites 1, 286 (1970); Chem. Abstr. 73, 121150q (1970).
455. Pike, R. A., and Maynard, C. P., Aircr. Struct. Mater. Appl. Nat. SAMPE (Soc. Aerosp. Mater. Process Eng.) Tech. Conf., 1969 p. 331; Chem. Abstr. 73, 26149n (1970).
456. Preston, J., and Black, W. B., Appl. Polym. Symp. 9, 107 (1969); Chem. Abstr. 72, 13669m (1970).
457. Preston, J., Black, W. B., and DeWinter, W., Appl. Polym. Symp. 9, 145 (1969); Chem. Abstr. 72, 22507j (1970).
458. Preston, J., DeWinter, W., and Black, W. B., J. Polym. Sci., Part A-1 7, 283 (1969); Chem. Abstr. 71, 3754a (1969).
459. Preston, J., DeWinter, W., Black, W. B., and Hoffenbert, W. L., Jr., J. Polym. Sci., Part A-1 7, 3027 (1969); Chem. Abstr. 72, 21972v (1970).
460. Preston, J. A., and Otis, H. R., U.S. Pat. 3,553,154 (Essex International, Inc.) (1970).
461. Prokhorov, O., Korzhavin, L., Florinskii, F., and Frenkel, S., Khim. Volokna No. 5, 73 (1970); Chem. Abstr. 73, 131860k (1970).
462. Rabilloud, G., Sillion, B., deGaudemaris, G., Belg. Pat. 738,776 (Institut Français du Petrole, des Carburants et Lubrifiants) (1969).
463. Rabilloud, G., Sillion, B., and deGaudemaris, G., Ger. Pat. 1,812,357 (Institut Français du Petrole, des Carburants et Lubrifiants) (1969); Chem. Abstr. 71, 50687r (1969).
464. Rabilloud, G., Sillion, B., and deGaudemaris, G., Ger. Pat. 1,913,280 (Institut Français du Petrole, des Carburants et Lubrifiants) (1969); Chem. Abstr. 71, 125621a (1969).
465. Rabilloud, G., Sillion, B., and deGaudemaris, G., Ger. Pat. 1,946,943 (Institut Français du Petrole, des Carburants et Lubrifiants) (1970); Chem. Abstr. 72, 133426b (1970).

466. Rabilloud, G., Sillion, B., and deGaudemaris, G., U.S. Pat. 3,461,096 (Institut Français du Petrole, des Carburants et Lubrifiants) (1969).
467. Radlmann, E., Lorenz, G., Wolf, G. D., and Nischk, G., Ger. Pat. 1,805,955 (Farbenfabriken Bayer, A. G.) (1970); *Chem. Abstr.* **73**, 88407d (1970).
468. Rafikov, S. R., Arkhipova, I. A., and Buketova, N. I., *Vysokomol. Soedin., Ser. B* **12**, 234 (1970); *Chem. Abstr.* **73**, 15330p (1970).
469. Rafikov, S. R., Arkhipova, I. A., Buketova, N. I., and L'dokova, G. M., *Tr. Inst. Khim. Nauk, Akad. Nauk Kaz. SSR* **28**, 107 (1970); *Chem. Abstr.* **73**, 15662y (1970).
470. Rafikov, S. R., Zhubanov, B. A., Almabekov, O. A., and Derevyanchenko, V. P., *Izv. Akad. Nauk Kaz. SSR, Ser. Khim.* **18**, 51 (1968); *Chem. Abstr.* **70**, 68822v (1969).
471. Rafikov, S. R., Zhubanov, B. A., Mashkevitch, S. A., Prokofiev, K. V., and Atanazevitch, E. I., *Vysokomol. Soedin., Ser. B* **11**, 165 (1969); *Chem. Abstr.* **70**, 115623t (1969).
472. Ravich, I. Ya., Neklyudova, O. V., Toroptseva, T. N., and Gindin, L. G., *Zh. Prikl. Spektrosk.* **9**, 674 (1968); *Chem. Abstr.* **70**, 20585t (1969).
473. Reibach, H. I., Fr. Pat, 1,573,966 (E. I. duPont de Nemours & Co.) (1969); *Chem. Abstr.* **72**, 122373f (1970).
474. Reimschuessel, H. K., *Advan. Chem. Ser.* No. 91, pp. 717–733 (1969); *Chem. Abstr.* **72**, 21958v (1970).
475. Reimschuessel, H. K., Klein, K. P., and Schmitt, G. J., *Macromolecules* **2**, 567 (1969); *Chem. Abstr.* **72**, 32300a (1970).
476. Reimschuessel, H. K., and Pascale, J. V., U.S. Pat. 3,542,744 (Allied Chemical Corp.) (1970); *Chem. Abstr.* **74**, 42889e (1971).
477. Reinhard, D. L., U.S. Pat. 3,440,204 (General Electric Co.) (1969).
478. Reynolds, R. J. W., and Seddon, J. D., Brit. Pat. 1,207,577 (ICI, Ltd.) (1970); *Chem. Abstr.* **74**, 13834c (1971).
479. Rio, A., Fr. Pat. 1,529,727 (Société des Usines Chimiques Rhone-Poulenc) (1968); *Chem. Abstr.* **70**, 115747m (1969).
480. Rochina, V., and Allard, P., Ger. Pat. 1,928,435 (Société Rhodiaceta) (1969); *Chem. Abstr.* **72**, 44976p (1970).
481. Rode, V. V., Gribkova, P. N., Vinogradov, A. V., Korshak, V. V., Tseitlin, G. M., and Azarov, V. I., *Vysokomol. Soedin., Ser., A* **11**, 1617 (1969); *Chem. Abstr.* **71**, 102407w (1969).
482. Rode, V. V., Gribkova, P. N., Vygodskii, Ya. S., Vinogradova, S. V., and Korshak, V.V., *Vysokomol. Soedin., Ser. A* **12**, 1566 (1970); *Chem. Abstr.* **73**, 77731d (1970).
483. Rogers, F. F., Jr., U.S. Pat. 3,450,678 (E. I. duPont de Nemours & Co.) (1969).
484. Rudakov, A. P., Bessonov, M. I., Tuichiev, Sh., Koton, M. M., Florinskii, F. S., Ginzburg, B. M., and Frenkel, S. Ya., *Vysokomol. Soedin., Ser. A* **12**, 641 (1970); *Chem. Abstr.* **72**, 133203b (1970).
485. Russo, M., *Mater. Plast. Elastomeri* **36**, 886 (1970); *Chem. Abstr.* **74**, 42633s (1971).
486. Russo, M., and Mortillaro, L., *J. Polym. Sci. Part A-1* **7**, 3337 (1969); *Chem. Abstr.* **72**, 44201g (1970).
487. Saga, M., Hachihama, Y., and Shono, T., *J. Polym. Sci., Part A-1* **8**, 2265 (1970); *Chem. Abstr.* **73**, 77686t (1970).
488. Salle, R., Sillion, B., and deGaudemaris, G., Fr. Pat. 1,567,863 (Institut Français du Petrole, des Carburants et Lubrifiants) (1969); *Chem. Abstr.* **71**, 125457b (1969).
489. Sandler, S. R., Berg, F. R., and Kitazawa, G., U.S. Pat. 3,531,436 (Borden, Inc.) (1970); *Chem. Abstr.* **74**, 13782j (1971).
490. Sattler, F. A., Fr. Pat. 2,009,052 (Westinghouse Electric Corp.) (1970); *Chem. Abstr.* **73**, 36238c (1970).
491. Schmidt, K., and Hansch, F., U.S. Pat. 3,459, 829 (Dr. Beck and Co.) (1969).

492. Schuller, W. H., and Lawrence, R. V., U.S. Pat. 3,503,998 (U.S. Dept. of Agriculture) (1970); *Chem. Abstr.* **72**, 122388q (1970).
493. Schuller, W. H., and Lawrence, R. V., U.S. Pat. 3,522,211 (U.S. Dept of Agriculture) (1970); *Chem. Abstr.* **73**, 67198w (1970).
494. Schweitzer, F. E., U.S. Pat. 3,459,706 (E. I. duPont de Nemours & Co.) (1969).
495. Sears, L. A., U.S. Pat. 3,523,773 (E. I. duPont de Nemours & Co.) (1970); *Chem. Abstr.* **73**, 78231c (1970).
496. Seddon, J. D., Brit. Pat. 1,192,001 (ICI, Ltd.) (1970); *Chem. Abstr.* **73**, 26076m (1970).
497. Serlin, I., Markhart, A. H., and Lavin, E., *Proc., Anniv. Tech. Conf., SPI (Soc. Plast. Ind.) Reinf. Plast. Div., 23rd, 1968* pp. 10F-1-10F-4 (1968); *Chem. Abstr.* **70**, 88625m (1969).
498. Serlin, I., Markhart, A. H., and Lavin, E., *Mod. Plast.* **47**, 120 (1970); *Chem. Abstr.* **74**, 13703j (1971).
499. Serlin, I., Markhart, A. H., and Lavin, E., Proc., Anniv. Tech. Conf., *SPI (Soc. Plast. Ind.) Reinf. Plast. Div., 25th, 1970*, p. 19-A (1970); *Chem. Abstr.* **73**, 15656z (1970).
500. Shamraev, G. M., Dulov, A. A., Liogon'kii, B. I., and Berlin, A. A., *Vysokomol. Soedin., Ser. A* **12**, 401 (1970); *Chem. Abstr.* **72**, 133365f (1970).
501. Sheffer, H. E., and Zielinski, G. C., U.S. Pat. 3,518,230 (Schenectady Chemicals, Inc.) (1970).
502. Shelton, C. F., Jr., U.S. Pat. 3,425,865 (Cerro Corp.) (1969); *Chem. Abstr.* **70**, 78903m (1969).
503. Sherer, T. L., and Flowers, R. G., Fr. Pat. 1,511,317 (General Electric Co.) (1968); *Chem. Abstr.* **70**, 58529r (1969).
504. Sherer, T. L., and Flowers, R. G., U.S. Pat. 3,471,444 (General Electric Co.) (1969).
505. Shono, T., Mitani, M., and Oda, R., *Kogyo Kagaku Zasshi* **72**, 1392 (1969); *Chem. Abstr.* **71**, 113331t (1969).
506. Shopov, I., *Izv. Otd. Khim. Nauki, Bulg. Akad. Nauk* **3**, 47 (1970); *Chem. Abstr.* **74**, 32028j (1971).
507. Sillion, B., and deGaudemaris, G., Fr. Pat. 1,540,930 (Institut Français du Petrole, des Carburants et Lubrifiants) (1968); *Chem. Abstr.* **71**, 31015d (1969).
508. Silverstone, G. A., Brit. Pat. 1,184,013 (V. Wolf, Ltd.) (1970); *Chem. Abstr.* **72**, 112330a (1970).
509. Silverstone, G. A., Brit. Pat. 1,184,014 (V. Wolf, Ltd.) (1970); *Chem. Abstr.* **72**, 112332c (1970).
510. Sloan, M. F., U.S. Pat. 3,554,981 (Hercules, Inc.) (1970).
511. Sloan, M. F., U.S. Pat. 3,554,982 (Hercules, Inc.) (1971).
512. Slonimskii, G. L., Korshak, V. V., Mzhel'skii, A. I., Askadskii, A. A., Vygodskii, Ya. S., and Vinogradova, S. V., *Dokl. Akad. Nauk SSSR* **182**, 851 (1968); *Chem. Abstr.* **70**, 20583r (1969).
513. Smith, M. B., U.S. Clearinghouse, Fed. Sci. Tech. Inform. AD 1968, AD-834698 (1968); *Chem. Abstr.* **72**, 22339f (1970).
514. Sokolov, N. A., *et al.*, *Tr. Vses Elektrotekhn. Inst.* No. 77, p. 45 (1968); *Chem. Abstr.* **73**, 66928x (1970).
515. Sroog, C. E., *Macromol. Syn.* **3**, 83–86 (1969); *Chem. Abstr.* **73**, 36026j (1970).
516. Sroog, C. E., *Encycl. Polym. Sci. Technol.* **11**, 247 (1969); *Chem. Abstr.* **72**, 55907t (1970).
517. Standage, A. E., and Turner, W. N., Fr. Pat. 1,532,018 (Rolls-Royce, Ltd.) (1968); *Chem. Abstr.* **71**, 4249b (1969).
518. Starr, F. C., Jr., Ger. Pat. 1,911,614 (E. I. duPont de Nemours & Co.) (1969); *Chem. Abstr.* **72**, 122530e (1970).

519. Stephens, J. R., U.S. Pat. 3,451,848 (Standard Oil Co. Indiana) (1968).
520. Stickrodt, J., and Koenig, U., Ger. Pat. 1,904,396 (Veba-Chemie Nord, G.m.b.H.) (1970); *Chem. Abstr.* **73**, 77875d (1970).
521. Stickrodt, J., and Koenig, U., Ger. Pat. 1,904,988 (Veba-Chemie Nord, G.m.b.H.) (1970); *Chem. Abstr.* **73**, 77877f (1970).
522. Stille, J. K., AD 867,031, Iowa University, Iowa City (1970).
523. Stivers, E. C., Belg. Pat. 739,061 (Raychem Corp.) (1970).
524. Stivers, E. C., Belg. Pat. 739,062 (Raychem Corp.) (1970).
525. Stivers, E. C., Ger. Pat. 1,946,925 (Raychem Corp.) (1970); *Chem. Abstr.* **73**, 110489f (1970).
526. Stivers, E. C., Ger. Pat. 1,947,029 (Raychem Corp.) (1970); *Chem. Abstr.* **73**, 110459w (1970).
527. Suzuki, S., Kaneda, I., Takahashi, M., and Nagai, H., Ger. Pat. 1,904,857 (Kureha Chemical Industry Co., Ltd.) (1969); *Chem. Abstr.* **71**, 125220u (1969).
528. Takekoshi, T., U.S. Pat. 3,541,054 (General Electric Co.) (1970); *Chem. Abstr.* **74**, 32180c (1971).
529. Takenaka, T., Nishi, E., and Ohta, K., Jap. Pat. 70/01,832 (Toaka Dyestuffs Manufg. Co., Ltd.) (1970); *Chem. Abstr.* **72**, 101412c (1970).
530. Takenaka, T., Nishi, E., and Ohta, W., Jap. Pat. 70/08,435 (Toaka Dyestuffs Manufg. Co., Ltd.) (1970); *Chem. Abstr.* **73**, 67099q (1970).
531. Takenaka, T., and Dota, W., Jap. Pat. 70/09,393 (Taoka Dyestuffs Manufg. Co., Ltd.) (1970); *Chem. Abstr.* **73**, 15799y (1970).
532. Talykov, V. A., Golubkov, G. E., Ikonnikova, A. I., and Belyaeva, A. P., *Vysokomol. Soedin.*, *Ser. A* **11**, 1303 (1969).
533. Tatsuno, M., Kubota, T., and Tsuda, Y., Jap. Pat. 68/28,836 (Toyo Rayon Co., Ltd.) (1968); *Chem. Abstr.* **70**, 78562z (1969).
534. Terney, S., Keating, J., Zielinski, J., Hakala, J., and Sheffer, H., *J. Polym. Sci., Part A-1* **8**, 683 (1970); *Chem. Abstr.* **72**, 111889c (1970).
535. Thomas, C. L., U.S. Pat. 3,505,353 (Sun Oil Co.) (1970).
536. Tocker, S., U.S. Patent 3,437,635 (E. I. duPont de Nemours & Co.) (1969); *Chem. Abstr.* **71**, 3934j (1969).
537. Tokarev, A. V., Bogdanov, M. N., Mandrosova, F. M., Smirnova, A. I., Chikurina, L. V., and Kudryavtsev, G. I., Russ. Pat. 256,238 (1969); *Chem. Abstr.* **72**, 101772v (1970).
538. Tokarev, A. V., Chikurina, L. V., Kudryavtsev, G. I., Smirnova, A. I., and Litovchenko, G. D., *Khim. Volokna* No. 5, 20 (1970); *Chem. Abstr.* **74**, 32572g (1971).
539. Tokuno, I., and Nakamura, N., Jap. Pat, 69/32,065 (Asahi Electro-Chemical Co., Ltd.) (1969); *Chem. Abstr.* **72**, 79685v (1970).
540. Toyama, S., Dogoshi, N., Ikeda, K., Kurihara, M., and Yoda, N., Jap. Pat. 69/20,110 (Toyo Rayon Co., Ltd.) (1969); *Chem. Abstr.* **72**, 4045c (1970).
541. Toyama, S., Dogoshi, N., Ikeda, K., Kurihara, M., Yoda, N., Nakanishi, R., and Watanabe, M., Jap. Pat. 69/23,107 (Toyo Rayon Co., Ltd.) (1969); *Chem. Abstr.* **72**, 22151p (1970).
542. Trostyanskaya, E. B., Venkova, E. S., and Mikhailin, Yu. A., *Vysokomol. Soedin., Ser. B* **10**, 841 (1968); *Chem. Abstr.* **70**, 68961q (1969).
543. Trostyanskaya, E. B., Venkova, E. S., and Mikhailin, Yu. A., Russ. Pat. 256,237 (1969); *Chem. Abstr.* **72**, 101470v (1970).
544. Trostyanskaya, E. B., Venkova, E. S., and Mikhailin, Yu. A., Russ, Pat. 272,551 (1970); *Chem. Abstr.* **73**, 110503f (1970).
545. Ulmer, W. W., U.S. Patent 3,485,595 (Anaconda Wire and Cable Co.) (1969).

546. Ulmer, W. W., U.S. Patent 3,544,504 (Anaconda Wire and Cable Co.) (1970); *Chem. Abstr.* **74**, 43206s (1971).
547. Usai, A., Jap. Pat. 69/20,633 (Japan Bureau of Industrial Technology) (1969); *Chem. Abstr.* **72**, 3942f (1970).
548. Varlas, M., *SAMPE Quart.* **1**, 52 (1970); *Chem. Abstr.* **74**, 43180d (1971).
549. Vasil'eva, I. V., Teleshov, E. N., and Pravednikov, A. N., Russ. Pat. 275,387 (1970); *Chem. Abstr.* **74**, 4206x (1971).
550. Vaughan, G. B., Rose, J. C., and Brown, G. P., *Polym. Prepr., Amer. Chem. Soc., Div. Polym. Chem.*, **11** (1), 339 (1970).
551. Vaughan, M. F., S. Afr. Pat. 67/06,311 (National Res. Development Corp.) (1969); *Chem. Abstr.* **71**, 102451f (1969).
552. Vaughan, M. F., and Jones, J. I., Fr. Pat. 1,550,077 (National Res. Development Corp.) (1968); *Chem. Abstr.* **71**, 39767w (1969).
553. Vaughan, R. W., Jones, J. F., Lubowitz, H. R., and Burns, E. A., *Nat. SAMPE (Soc. Aerosp. Mater. Process Eng.) Symp. Exhib., 15th, 1969* pp. 59–65 (1969); *Chem. Abstr.* **72**, 122291c (1970).
554. Vayson de Pradenne, H. P., U.S. Pat. 3,532,661 (Société Generale de Constructions Electriques et Mécaniques, Alsthom) (1970).
555. Victorius, C., U.S. Pat. 3,441,532 (E. I. duPont de Nemours & Co.) (1969); *Chem. Abstr.* **71**, 13841r (1969).
556. Vinogradova, S. V., Slonimskii, G. L., Vygodskii, Ya. S., Askadskii, A. A., Mzhel'skii, A. I., Churochkina, N. A., and Korshak, V. V., *Vysokomol. Soedin., Ser. A* **11**, 2725 (1969); *Chem. Abstr.* **72**, 55999z (1970).
557. Vinogradova, S. V., Vygodskii, Ya. S., and Korshak, V. V., *Vysokomol. Soedin., Ser. A* **12**, 1987 (1970); *Chem. Abstr.* **74**, 13478q (1971).
558. Vlasova, K. N., Chernova, A. G., Dobrokhotova, M. L., Tanunina, P. M., Gershkokhen, S. L., Pilyaeva, V. F., Emel'yanova, L. N., Bublik, L. S., Shcherba, N. S., and Chechik, A. I. *Sov. Plast.* No. 5, p. 11 (1968).
559. Vlasova, K. N., Tanunina, P. M., and Palladina, T. I., *Plast. Massy* No. 2, p. 25 (1969); *Chem. Abstr.* **70**, 115621r (1969).
560. Wall, L. A., and Straus, S., *Polym. Prepr., Amer. Chem. Soc., Div. Polym. Chem.* **12** (1), 781 (1971).
561. Wallach, M. L., *J. Polym. Sci., Part A-2* **7**, 1435 (1969); *Chem. Abstr.* **71**, 113361c (1969).
562. Wallach, M. L., *J. Polym. Sci., Part A-2* **7**, 1995 (1969); *Chem. Abstr.* **72**, 32362x (1970).
563. Wilson, G. R., Ger. Pat. 1,808,126 (Monsanto Co.) (1970); *Chem. Abstr.* **73**, 56759d (1970).
564. Wilson, G. R., U.S. Pat. 3,520,837 (Monsanto Res. Corp.) (1970); *Chem. Abstr.* **73**, 78116u (1970).
565. Wrasidlo, W., *Polym. Prepr., Amer. Chem. Soc., Div. Polym. Chem.* **11** (2), 1159 (1970).
566. Wrasidlo, W., *J. Polym. Sci., Part A-1* **8**, 1107 (1970); *Chem. Abstr.* **73**, 4330n (1970).
567. Wrasidlo, W. J., *Polym. Prepr., Amer. Chem. Soc., Div. Polym. Chem.* **12** (1), 755 (1971).
568. Wrasidlo, W., and Augl, J. M., *J. Polym. Sci., Part A-1* **7**, 321 (1969); *Chem. Abstr.* **71**, 3697j (1969).
569. Wrasidlo, W., and Augl, J. M., *J. Polym. Sci., Part A-1* **7**, 1589 (1969); *Chem. Abstr.* **71**, 71037u (1969).
570. Wrasidlo, W., and Augl, J. M., *Macromolecules* **3**, 544 (1970); *Chem. Abstr.* **73**, 131366d (1970).
571. Yoda, N., Jap. Pat. 68/13,076 (Toyo Rayon Co., Ltd.) (1968); *Chem. Abstr.* **70**, 29602d (1969).

572. Yoda, N., Jap. Pat. 69/07,954 (Toyo Rayon Co., Ltd.) (1969); *Chem. Abstr.* **71**, 125222w (1969).
573. Yoda, N., *Encycl. Polym. Sci. Technol.* **10**, 682–690 (1969); *Chem. Abstr.* **72**, 44146t (1970).
574. Yoda, N., Baba, Y., and Ikeda, K., Jap. Pat. 70/08,993 (Toyo Rayon Co.) (1970); *Chem. Abstr.* **73**, 46060e (1970).
575. Yoda, N., Hagihara, Y., Ideda, K., and Kurihara, M., Jap. Pat. 70/01,631 (Toyo Rayon Co., Ltd.) (1970); *Chem. Abstr.* **72**, 133651w (1970).
576. Yoda, N., Ikeda, K., Kurihara, M., and Hagiwara, Y., Jap. Pat. 69/32,437 (Toyo Rayon Co., Ltd.) (1969); *Chem. Abstr.* **72**, 91029s (1970).
577. Yoda, N., Kubota, T., Kurihara, M., Dogoshi, N., and Yoshii, T., Ger. Pat. 2,009,626 (Toray Industries, Inc.) (1970); *Chem. Abstr.* **73**, 131562q (1970).
578. Yoda, N., Kurihara, M., Dogoshi, N., Ikeda, K., Toyama, S., and Nakanishi, Y., Jap. Pat. 69/19,874 (Toyo Rayon Co., Ltd.) (1969); *Chem. Abstr.* **71**, 113443f (1969).
579. Yoda, N., Kurihara, M., Dogoshi, N., Nakanishi, Y., Ikeda, K., and Toyama, S., Jap. Pat. 69/19,875 (Toyo Rayon Co., Ltd.) (1969); *Chem. Abstr.* **71**, 113444g (1969).
580. Yoda, N., Kurihara, M., Dogoshi, N., Toyama, S., and Ikdea, K., Jap. Pat. 69/19,877 (Toyo Rayon Co., Ltd.) (1969); *Chem. Abstr.* **71**, 125221v (1969).
581. Yoda, N., Kurihara, M., Toyama, S., Dogoshi, N., Hagiwara, Y., Itoga, M., Fujita, S. and Yamoto, H., Ger. Pat. 1,811,588 (Toyo Rayon Co., Ltd.) (1970); *Chem. Abstr.* **73**, 77843s (1970).
582. Yoda, N., Kurihara, M., Toyama, S., Dogoshi, N., Hagiwara, Y., Itoga, M., Fujita, S., and Yamoto, H., Jap. Pat. 70/19,074 (Toray Industries, Inc.) (1970); *Chem. Abstr.* **73**, 77867c (1970).
583. Yoda, N., Kurihara, M., Toyama, S., Dogoshi, N., Ikeda, K., and Nakanishi, R., Jap. Pat. 70/13,350 (Toray Industries, Inc.) (1970); *Chem. Abstr.* **73**, 46065k (1970).
584. Yoda, N., Kurihara, M., Toyama, S., Dogoshi, N., Ikeda, K., and Nakanishi, R., Jap. Pat. 70/13,351 (Toray Industries, Inc.) (1970); *Chem. Abstr.* **73**, 67098p (1970).
585. Yoda, N., Kurihara, M., Toyama, S., Dogoshi, N., Ikeda, K., and Nakanishi, R., Jap. Pat. 70/13,352 (Toray Industries, Inc.) (1970); *Chem. Abstr.* **73**, 46053e (1970).
586. Yoda, N., Kurihara, M., Toyama, S., Dogoshi, N., Ikeda, K., and Nakata, Y., Jap. Pat. 70/20,115 (Toray Industries, Inc.) (1970); *Chem. Abstr.* **73**, 99604s (1970).
587. Yoda, N., Nakanishi, R., Kubota, T., Kurihara, M., and Ikeda, K., U.S. Pat. 3,468,851 (Toyo Rayon Co., Ltd.) (1969).
588. Yoda, N., Nakanishi, R., Kurihara, M., Dogoshi, N., Toyama, S., and Ikeda, K., Jap. Pat. 70/22,954 (Toray Industries, Inc.) (1970); *Chem. Abstr.* **74**, 23248d (1971).
589. Yoda, N., Toyama, S., Dogoshi, N., Kurihara, M., Ikeda, K., and Nakanishi, R., Jap. Pat. 70/13,116 (Toray Industries, Inc.) (1970); *Chem. Abstr.* **73**, 56727s (1970).
590. Yoda, N., Toyama, S., Kurihara, M., Ikeda, K., and Nakanishi, R., Jap. Pat. 69/27,677 (Toyo Rayon Co., Ltd.) (1969); *Chem. Abstr.* **72**, 56109w (1970).
591. Yoda, N., Yoshii, T., Fujita, S., Mochizuki, H., Yumoto, H., Kurihara, M., Dogoshi, N., and Tanaka, C., Ger. Pat. 2,019,602 (Toray Industries, Inc.) (1970); *Chem. Abstr.* **74**, 23356n (1971).
592. Zakoschikov, S. A., Ignat'eva, I. N., Nikolaeva, N. V., and Pomerantseva, K. P., *Vysokomol. Soedin., Ser. A* **11**, 2487 (1969); *Chem. Abstr.* **72**, 67438d (1970).
593. Zakoshchikov, S. A., Ignat'eva, I. N., and Tanunina, P. M., *Vysokomol. Soedin., Ser. B* **11**, 106 (1969); *Chem. Abstr.* **70**, 106943v (1969).
594. Zakoshchikov, S. A., Pomerantseva, K. P., and Nikolaeva, N. V., *Vysokomol. Soedin., Ser. B* **11**, 483 (1969); *Chem. Abstr.* **71**, 102260t (1969).
595. Zecher, W., and Merten, R., U.S. Pat. 3,560,446 (Farbenfabriken Bayer, A. G.) (1971).

596. Zilinskaite, L., Maciulis, A., Misevicius, P., Shamraev, G. M., Liogon'kii, B. I., and Berlin, A. A., *Vysokomol. Soedin.*, *Ser. A*-**11**, 1214 (1969); *Chem. Abstr.* **71**, 81969p (1969).
597. Zurakowska-Orszagh, J., and Kobiela, S., *Polimery* **15**, 344 (1970); *Chem. Abstr.* **74**, 32124n (1971).

## Volume B, Part 2 Supplementary Reference List

## Chapter II

*General References and Reviews* (40, 42).

SECTION A (p. 291).

Polymerization of Dialdehydes (5–9, 37, 50, 52, 53, 59, 72).

SECTION C (p. 299).

Polymerization of Divinyl Acetals and Ketals (1, 3, 4, 17, 18, 20, 30, 34, 43–45, 68, 69).

SECTION D (p. 306).

Polymerization of Unsaturated Esters (14, 45, 46, 57, 58, 63, 64, 70, 71).

SECTION E (p. 317).

Polymerization of Diunsaturated Anhydrides (10, 25, 26, 48).

SECTION G (p. 322).

Polymerization of Diunsaturated Ammonium Salts, Amine Oxides and Amines (11, 12, 15, 27, 49, 60, 75).

SECTION H (p. 328).

Polymerization of Diunsaturated Amides (13, 28, 29, 38, 47, 50, 51, 61, 62, 65–67, 71).

SECTION I (p. 332).

Polymerization of Diisocyanates (21, 22, 35, 36).

SECTION J (p. 335).

Polymerization of Dinitriles (55, 56, 73).

SECTION K (p. 337).

Polymerization of Phosphorus-Containing Compounds (16, 27, 31, 32).

SECTION L (p. 339).

Polymerization of Sulfur-Containing Compounds (39, 41, 74).

SECTION M (p. 346).

Silicon Derivatives (2, 16, 19, 23, 24, 33).

SECTION N (p. 350).

Polymerization of Miscellaneous Compounds (54).

## REFERENCES

1. Allen, V. R., and Turner, S. R., *J. Macromol. Sci., Chem.* **5**, 2079 (1971); *Chem. Abstr.* **73**, 121002t (1970).
2. Andreev, D. N., Sokolova, N. P., Alekseeva, D. N., and Pavlovskaya, V. E., *Vysokomol. Soedin., Ser. A* **11**, 492 (1969); *Chem. Abstr.* **70**, 115595k (1969).
3. Ardis, A. E., Dietrich, H. J., Raymond, M. A., and Urs, V. S., U.S. Pat. 3,514,435 (Olin Mathieson Chem. Corp.) (1970); *Chem. Abstr.* **73**, 15718w (1970).
4. Aso, C., Kunitake, T., and Ando, S., *J. Macromol. Sci., Chem.* **5**, 2019 (1971); *Chem. Abstr.* **73**, 120931h (1970).
5. Aso, C., Kunitake, T., Miura, M., and Koyama, K., *Makromol. Chem.* **117**, 153 (1968); *Chem. Abstr.* **70**, 456, 4673q (1969).
6. Aso, C., Kunitake, T., Sasaki, M., and Koyama, K., *Kobunshi Kagaku* **27**, 260 (1970); *Chem. Abstr.* **73**, 99266h (1970).
7. Aso, C., and Tagami, S., *Macromolecules* **2**, 414 (1969); *Chem. Abstr.* **71**, 71019q (1969).
8. Aso, C., Tagami, S., and Kunitake, T., *J. Polym. Sci., Part A-1* **7**, 497 (1969); *Chem. Abstr.* **71**, 13439r (1969).
9. Aso, C., Tagami, S., and Kunitake, T., *Polym. J.* **1**, 395 (1970); *Chem. Abstr.* **73**, 110212k (1970).
10. Baines, F. C., and Bevington, J. C., *Polymer* **11**, 647 (1970); *Chem. Abstr.* **74**, 31983e (1971).
11. Boothe, J. E., Flock, H. G., and Hoover, M. F., *J. Macromol. Sci., Chem.* **4**, 1419 (1970); *Chem. Abstr.* **73**, 45870g (1970).
12. Boothe, J. E., Hoover, M. F., and Flock, H. G., *Polym. Prepr., Amer. Chem. Soc., Div. Polym. Chem.* **10**(2), 922 (1969).
13. Boyarchuk, Yu. M., *Vysokomol. Soedin., Ser. A* **11**, 2161 (1969); *Chem. Abstr.* **72**, 32400h (1970).
14. British Patent 1,129,230 (Air Reduction Co., Ltd.) (1968); *Chem. Abstr.* **69**, 10066, 107270q (1968).
15. British Patent 1,178,371 (Allied Chemical Corp.) (1970); *Chem. Abstr.* **72**, 67495v (1970).
16. Butler, G. B., U.S. Pat. Reissue 26,407 (Peninsular Chem. Research, Inc.) (1968).
17. Butler, G. B., *J. Macromol. Sci., Chem.* **5**, 219 (1971); *Chem. Abstr.* **73**, 120933k (1970).
18. Butler, G. B., and Campus, A. F., *Polym. Prepr., Amer. Chem. Soc., Div. Polym. Chem.* **9**(2), 1266 (1968).
19. Butler, G. B., and Campus, A. F., *J. Polym. Sci., Part A-1* **8**, 523 (1970); *Chem. Abstr.* **72**, 111892y (1970).

20. Butler, G. B., and Campus, A. F., *J. Polym. Sci., Part A-1* **8**, 545 (1970); *Chem. Abstr.* **72**, 111891x (1970).
21. Butler, G. B., and Corfield, G. C., U.S. Clearinghouse Fed. Sci. Tech. Inform., AD 1969, AD-694123 (1969); *Chem. Abstr.* **72**, 79712b (1970).
22. Butler, G. B., and Corfield, G. C., *J. Macromol. Sci., Chem.* **5**, 1889 (1971); *Chem. Abstr.* **73**, 131385j (1970).
23. Butler, G. B., and Iachia, B., *J. Macromol. Sci., Chem.* **3**, 803 (1969); *Chem. Abstr.* **71**, 30793u (1969).
24. Butler, G. B., and Iachia, B., *J. Macromol. Sci., Chem.* **3**, 1485 (1969); *Chem. Abstr.* **71**, 124979m (1969).
25. Butler, G. B., and Kimura, S., *J. Macromol. Sci., Chem.* **5**, 181 (1971); *Chem. Abstr.* **73**, 120932j (1970).
26. Butler, G. B., Kimura, S., and Baucom, K. B., *Polym. Prepr., Amer. Chem. Soc., Div. Polym. Chem.* **11**(1), 48 (1970).
27. Butler, G. B., and Miller, W. L., *J. Macromol. Sci., Chem.* **3**, 1493 (1969); *Chem. Abstr.* **71**, 124983h (1969).
28. Butler, G. B., and Myers, G. R., *J. Macromol. Sci., Chem.* **5**, 1957 (1971); *Chem. Abstr.* **73**, 120935n (1970).
29. Butler, G. B., and Myers, G. R., *J. Macromol. Sci., Chem.* **5**, 1987 (1971); *Chem. Abstr.* **73**, 120936p (1970).
30. Butler, G. B., and Sharpe, A. J., Jr., *Polym. Prepr., Amer. Chem. Soc. Div. Polym. Chem.* **11**(1), 42 (1970).
31. Butler, G. B., Skinner, D. L. Bond, W. C., and Rogers, C. L., *Polym. Prepr., Amer. Chem. Soc., Div. Polym. Chem.*, **10**(2), 923 (1969).
32. Butler, G. B., Skinner, D. L., Bond, W. C., Jr., and Rogers, C. L., *J. Macromol. Sci., Chem.* **4**, 1437 (1970); *Chem. Abstr.* **73**, 45900s (1970).
33. Butler, G. B., and Stackman, R. W., *J. Macromol. Sci., Chem.* **3**, 821 (1969); *Chem. Abstr.* **71**, 50559a (1969).
34. Butler, G. B., and Zeegers, B., *Polym. Prepr., Amer. Chem. Soc., Div. Polym. Chem.* **12**(1), 420 (1971).
35. Corfield, G. C., and Crawshaw, A., *J. Polym. Sci., Part A-1* **7**, 1179 (1969).
36. Corfield, G. C., and Crawshaw, A., *J. Macromol. Sci., Chem.* **5**, 1855 (1971); *Chem. Abstr.* **73**, 131386k (1970).
37. Corfield, G. C., and Crawshaw, A., *J. Macromol. Sci., Chem.* **5**, 1873 (1971); *Chem. Abstr.* **73**, 131387m (1970).
38. Crawshaw, A., and Jones, A. G., *J. Macromol. Sci., Chem.* **5**, 1903 (1971); *Chem. Abstr.* **73**, 131624m (1970).
39. DeWitte, E., and Goethals, E. J., *Makromol. Chem.* **115**, 234 (1968); *Chem. Abstr.* **69**, 3422, 36494c (1968).
40. Gibbs, W. E., and Barton, J. M., *Vinyl Polym.* **1**, 59–138 (1967); *Chem. Abstr.* **70**, 88271t (1969).
41. Goethals, E. J., and DeWitte, E., *J. Macromol. Sci., Chem.* **5**, 1915 (1971); *Chem. Abstr.* **73**, 131382f (1970).
42. Guaita, M., Camino, G., and Trosarelli, L., *Makromol. Chem.* **130**, 243 (1969); *Chem. Abstr.* **72**, 44182b (1970).
43. Guaita, M., Camino, G., and Trosarelli, L., *Makromol. Chem.* **130**, 252 (1969); *Chem. Abstr.* **72**, 44181a (1970).
44. Guaita, M., Camino, G., and Trosarelli, L., *Makromol. Chem.* **131**, 237 (1970); *Chem. Abstr.* **72**, 79535w (1970).

45. Guaita, M., Camino, G., and Trosarelli, L., *J. Macromol. Sci., Chem.* **5**, 89 (1971); *Chem. Abstr.* **73**, 131377h (1970).
46. Higgins, J. P. J., and Weale, K. E., *J. Polym. Sci., Part A-1* **6**, 3007 (1968).
47. Higgins, J. P. J., and Weale, K. E., *J. Polym. Sci., Part B* **7**, 153 (1969); *Chem. Abstr.* **70**, 115596m (1969).
48. Higgins, J. P. J., and Weale, K. E., *J. Polym. Sci., Part A-1* **8**, 1708 (1970); *Chem. Abstr.* **73**, 35798a (1970).
49. Hoover, M. F., Ger. Pat. 1,814,597 (Calgon Corp.) (1969); *Chem. Abstr.* **71**, 82113s (1969).
50. Ishida, S., Doctoral Dissertation, Brooklyn Polytechnical Institute (University Microfilms 67–13,397, Ann Arbor, Michigan) (1967).
51. Kaye, H., *Polym. Prepr., Amer. Chem. Soc., Div. Polym. Chem.* **11** (2), 1027 (1970).
52. Kitahama, Y., *J. Polym. Sci., Part A-1* **6**, 2309 (1968); *Chem. Abstr.* **69**, 6342, 67795f (1968).
53. Kobayashi, H., Ohama, H., and Kitahama, R., Jap. Pat. 70/25,313 (Asahi Chemical Industry Co.) (1970); *Chem. Abstr.* **74**, 4010d (1971).
54. Kunitake, T., Nakashima, T., and Aso, C., *J. Polym. Sci., Part A-1* **8**, 2853 (1970); *Chem. Abstr.* **73**, 131365c (1970).
55. Liepins, R., *Makromol. Chem.* **118**, 36 (1968); *Chem. Abstr.* **70**, 29406t (1969).
56. Liepins, R., Campbell, D., and Walker, C., *J. Polym. Sci., Part A-1* **6**, 3059 (1968); *Chem. Abstr.* **69**, 10050, 107108t (1968).
57. Matsumoto, A., and Oiwa, M., *Kogyo Kagaku Zasshi* **72**, 2127 (1969); *Chem. Abstr.* **72**, 21960q (1970).
58. Matsumoto, A., Takashima, K., and Oiwa, M., *Bull. Chem. Soc. Jap.* **42**, 1959 (1969); *Chem. Abstr.* **71**, 70965h (1969).
59. Moyer, W. W., Jr., U.S. Pat. 3,395,125 (Borg-Warner Corp.) (1968); *Chem. Abstr.* **69**, 5602, 59801q (1968).
60. Negi, Y., Harada, T., and Ishizuka, O., Jap. Pat. 70/01,457 (Nitto Boseki Co., Ltd.) (1970); *Chem. Abstr.* **72**, 91003d (1970).
61. Overberger, C. G., Montaudo, G., and Ishida, S., *J. Polym. Sci., Part A-1* **7**, 35 (1969).
62. Pyriadi, T. M., and Harwood, H. J., *Polym. Prepr., Amer. Chem. Soc., Div. Polym. Chem.* **11** (1), 60 (1970).
63. Rhum, D., and Moore, G. L., Brit. Pat. 1,129,229 (Air Reduction Co., Inc.) (1968); *Chem. Abstr.* **69**, 9907, 105926j (1968).
64. Rhum, D., and Weintraub, L., Brit. Pat. 1,129,228 (Air Reduction Co., Inc.) (1968); *Chem. Abstr.* **69**, 10065, 107255p (1968).
65. Sayadyan, A. G., and Simonyan, D. A., *Arm. Khim. Zh.* **21**, 1041 (1968); *Chem. Abstr.* **71**, 30720t (1969).
66. Sayadyan, A. G., and Simonyan, D. A., *Arm. Khim. Zh.* **22**, 528 (1969); *Chem. Abstr.* **71**, 102273z (1969).
67. Sokolova, T. A., and Osipova, I. N., *Vysokomol. Soedin., Ser. B* **10**, 384 (1968); *Chem. Abstr.* **69**, 2616, 27924r (1968).
68. Stackman, R. W., *J. Macromol. Sci., Chem.* **5**, 251 (1971); *Chem. Abstr.* **73**, 120934m (1970).
69. Sultanov, K., Ilkhamov, M. Kh., Askarov, M. A., and Tadzhibaev, A., Russ Pat. 203,901 (Scientific-Research Institute of the Chemistry and Technology of Cotton Cellulose) (1967); *Chem. Abstr.* **69**, 1877, 19801h (1968).
70. Trosarelli, L., and Guaita, M., *Polymer.* **9**, 233 (1968); *Chem. Abstr.* **69**, 3421, 36480v (1968).

71. Trosarelli, L., Guaita, M., and Priola, A., *J. Polym. Sci., Part C*, No. 16, Part 8, 4713 (1969); *Chem. Abstr.* **71**, 22388a (1969).
72. Vogl, O., *High Polym.* **23**, 419 (1968); *Chem. Abstr.* **69**, 10083, 107447c (1968).
73. Woehrle, D., and Manecke, G., *Makromol. Chem.* **138**, 283 (1970); *Chem. Abstr.* **74**, 3907q (1971).
74. Yamaguchi, T., and Ono, T., *Chem.* Ind. (*London*) p. 769 (1968); *Chem. Abstr.* **69**, 2611, 27875a (1968).
75. Zeh, H. J., Jr., Ger. Pat. 1,930,647 (Calgon Corp.) (1970); *Chem. Abstr.* **72**, 79682s (1970).

## Volume B, Part 2 Supplementary Reference List

### Chapter III (p. 441).

Polymerization of Acrolein (1, 5, 8, 11, 13–17, 19–23, 25).
Reactions of Polyacrolein (12, 15, 24).
Acrolein Copolymers (6, 9, 10, 18).
Poly ($\alpha$-alkyl Acroleins) (2–4, 8, 13).
Poly ($\alpha$-haloacroleins) (7).

## REFERENCES

1. Acosta, F., and Mateo, J. L., *Rev. Plast. Mod.* **19**, 693 (1968); *Chem. Abstr.* **70**, 88299h (1969).
2. Andreeva, I. V., Koton, M. M., Getmanchuk, Yu. P., Madorskaya, L. Ya., Pokrovskii, E. I., and Kol'tsov, A. I., *Khim. Atsetilena* p. 386 (1968); *Chem. Abstr.* **70**, 115658h (1969).
3. Andreeva, I. V., Koton, M. M., and Madorskaya, L. Ya., *Polym. Sci. USSR* **9**, 2825 (1967).
4. Andreeva, I. V., Koton, M. M., and Madorskaya, L. Ya., *Zh. Prikl. Khim.* (*Leningrad*) **41**, 2269 (1968); *Chem. Abstr.* **70**, 29375g (1969).
5. Andreeva, I. V., Koton, M. M., and Medvedev, Yu. V., *Vysokomol. Soedin., Ser. A* **11**, 1269 (1969); *Chem. Abstr.* **71**, 70970f (1969).
6. Annenkova, V. Z., Annenkova, V. M., and Gaitseva, E. A., *Khim. Atsetilena* p. 395 (1968); *Chem. Abstr.* **70**, 115612p (1969).
7. Annenkova, V. Z., Ivanova, L. T., and Ugryumova, G. S., *Khim. Atsetilena* p. 390 (1968); *Chem. Abstr.* **71**, 22381t (1969).
8. Bergman, E., Can. Pat. 764,558 (Shell Oil Company) (1967).
9. Buning, R., and Bier, G., U.S. Pat. 3,401,149 (Dynamit Nobel, A.-G.) (1968).
10. German Patent 1,445,331 (Dunlop, A.-G.) (1968).
11. German Patent 1,445,355 (Deutsche Gold-und Silber-Scheideanstalt) (1969).
12. Gole, J., and Calvayrac, H., *J. Polym. Sci., Part C*, No. 16, Part 7, 3765 (1968); *Chem. Abstr.* **70**, 29511y (1969).
13. Ishida, S., and Kitahama, R., Jap. Pat. 69/03,833 (Asahi Chemical Ind. Co., Ltd.) (1969); *Chem. Abstr.* **70**, 115702t (1969).
14. Japanese Patent 67/951 (Kokoku Rayon and Pulp Co., Ltd.) (1967).
15. Japanese Patent 67/4968 (Kureha Chem. Ind. Co., Ltd.) (1967).
16. Kekish, G. T., U.S. Pat. 3,438,941 (Nalco Chemical Co.) (1969); *Chem. Abstr.* **70**, 115729g (1969).

17. Kekish, G. T., U.S. Pat. 3,457,230 (Nalco Chemical Co.) (1969).
18. Kern, W., and Schulz, R., U.S. Pat. 3,404,133 (Deutsche Gold-und-Silber-Scheidean-stalt Vormals Roessler) (1968).
19. Kitahama, Y., and Ishida, S., *Makromol. Chem.* **119**, 64 (1968); *Chem. Abstr.* **70**, 29374f (1969).
20. Matsubara, Y., Sumitomo, H., and Maeshima, T., *Kogyo Kagaku Zasshi* **71**, 1726 (1968).
21. Schulz, R. C., *Vinyl Polym.* **1**, 403 (1967); *Chem. Abstr.* **70**, 88272u (1969).
22. Schulz, R. C., *Pure Appl. Chem.* **16**, 433 (1968).
23. Schulz, R. C., and Strobel, W., *Monatsh. Chem.* **99**, 1742 (1968); *Chem. Abstr.* **70**, 4752q (1969).
24. Sumitomo, Y., Hayakawa, A., and Tsuboshima, K., Jap. Pat. 69/30,988 (Kojin, Ltd.) (1969); *Chem. Abstr.* **72**, 101238a (1970).
25. Yamashita, N., Sumitomo, H., and Maeshima, T., *Kogyo Kagaku Zasshi* **71**, 1723 (1968).

## Supplementary Reference List For Part A

The following references supplement the reference lists at the end of the chapters in Part A. They are arranged by chapter and section. (See also the earlier supplementary reference list at the end of Part A, pp. 345–350).

### REVIEWS

1. Hojo, N., *Kobunshi Kako* **18**, 16 (1969); *Chem. Abstr.* **71**, 30847q (1969).
2. Jones, R. D. G., and Power, L. P., *Proc. Roy. Aust. Chem. Inst.* **35**, 338 (1968); *Chem. Abstr.* **70**, 78400v (1969).
3. Kiriyama, S., *Shikizai Kyokaishi* **42**, 221 (1969); *Chem. Abstr.* **71**, 91883r (1969).
4. Rode, V, V., *Progr. Polim. Khim.* pp. 347–374 (1969); (edited by V. V. Korshak); *Chem. Abstr.* **72**, 13081p (1970).

### CHAPTER 1 (p. 1).

1. Bach, H. C., U.S. Pat. 3,414,545 (Monsanto Co.) (1968).
2. Belgian Patent 725,122 (Glanzstoff, A.-G.) (1969).
3. Berlin, A. A., Liogon'kii, B. I., and Zelenetskii, A. N., *Vysokomol. Soedin.*, *Ser. A* **10**, 2076 (1968); *Polym. Sci.*, *USSR* **10**, 2415 (1968).
4. Berlin, A. A., Liogon'kii, B. I., and Zelenetskii, A. N., *Vysokomol. Soedin.*, *Ser. A* **10**, 2089 (1968). *Polym. Sci. USSR* **10**, 2431 (1968); *Chem. Abstr.* **70**, 12046c (1969).
5. Berlin, A. A., and Zelenetskii, A. N., *Bull Acad. Sci. USSR*, *Div. Chem. Sci.* p. 1734 (1968).
6. British Patent 1,092,824 (W. R. Grace & Co.) (1967).
7. British Patent 1,128,896 (Argus Chem. Corp.) (1968).
8. British Patent 1,130,265 (Dow Chem. Co.) (1968).
9. Chigir, A. N., Novikova, V. F., Kalikhman, I. D., Chumakov, Yu. I., Cherkashin, M. I., and Berlin, A. A., *Vysokomol. Soedin.*, *Ser. A* **11**, 1805 (1969); *Chem. Abstr.* **71**, 125042n (1969).
10. DeKoninck, L., and Smets, G., *J. Polym. Sci.*, *Part A-1* **7**, 3313 (1969); *Chem. Abstr.* **72**, 44223r (1970).
11. French Patent Addn. 94,186 (Rhone-Poulenc) (1969); *Chem. Abstr.*, **72**, 56108v (1970).
12. French Patent 2,000,055 (Glanzstoff, A.-G.) (1969); *Chem. Abstr.* **72**, 56093m (1970).
13. Gurney, J. A., and Hall, L. A. R., U.S. Pat. 3,577,476 (Geigy Chemical Co.) (1971).

14. Hermans, J. C., and Smets, G. J., Belg. Pat. 697,729 (Gevaert-Agfa, N. V.) (1967); *Chem. Abstr.* **69**, 10067, 107277x (1968).
15. Hirohashi, R., and Hishiki, Y., *Kogyo Kagaku Zasshi* **71**, 1744 (1968).
16. Hirohashi, R., Hishiki, Y., and Dohi, M., *Kogyo Kagaku Zasshi* **73**, 1450 (1970); *Chem. Abstr.* **73**, 121027e (1970).
17. Hirohashi, R., Hishiki, Y., and Ishii, S., *Kogyo Kagaku Zasshi* **72**, 1394 (1969); *Chem. Abstr.* **71**, 92020y (1969).
18. Hirohashi, R., Hishiki, Y., and Ishii, S., *Nippon Shashin Gakkai Kaishi* **33**, 95 (1970); *Chem. Abstr.* **74**, 42797y (1971).
19. Hoerhold, H. H., Graef, D., and Opfermann, J., *Plaste Kaut.* **17**, 84 (1970); *Chem. Abstr.* **72**, 101141p (1970).
20. Hoerhold, H. H., and Opfermann, J., *Makromol. Chem.* **131**, 105 (1970); *Chem. Abstr.* **72**, 90939b (1970).
21. Hoess, E., U.S. Pat. 3,431,221 (United States of America) (1969).
22. Japanese Patent 71/02,032 (Adeka Agasu Chem. Co. Ltd.) (1971).
23. Kabanov, V. A., Kargin, V. A., and Zubov, V. P., Russ. Pat. 134,862 (1970).
24. Kanbe, M., and Okawara, M., *Kogyo Kagaku Zasshi* **71**, 1276 (1968); *Chem. Abstr.* **70**, 457, 4682s (1969).
25. Kersten, H., Siggel, E., and Meyer, G., U.S. Pat. 3,577,387 (Glanzstoff A. G.) (1971).
26. Kossmehl, G., Haertel, M., and Manecke, G., *Makromol. Chem.* **131**, 15 (1970); *Chem. Abstr.* **72**, 101162w (1970).
27. Kossmehl, G., Haertel, M., and Manecke, G., *Makromol. Chem.* **131**, 37 (1970); *Chem. Abstr.* **72**, 101163x (1970).
28. Lapitskii, G. A., Makin, S. M., Lyapina, E. K., and Chebotarev, A. S., *Vysokomol. Soedin., Ser. B* **11**, 266 (1969); *Chem. Abstr.* **71**, 30745e (1969).
29. Lebsadze, T. N., Chkhaidze, L. T., Tabidze, B. A., Pavlenishvili, I. Ya., and Eligulashvili, I. A., *Soobshch. Akad. Nauk Gruz. SSR* **52**, 663 (1968); *Chem. Abstr.* **70**, 97274y (1969).
30. McGrath, J. J., and Spilners, I. J., U.S. Pat. 3,448,082 (Gulf R & D Co.) (1969).
31. Meyer, H. R., U.S. Pat. 3,427,261 (W. R. Grace and Co.) (1969); *Chem. Abstr.* **70**, 68889x (1969).
32. Netherlands Patent Application 67/00,441 (Stamicarbon, N. V.) (1968); *Chem. Abstr.* **69**, 10065, 107253m (1968).
33. Netherlands Patent Application 67/08,919 (Dow Chem. Co.) (1968).
34. Neuse, E. W., U.S. Pat. 3,437,634 (McDonnell Douglas Corp.) (1969); *Chem. Abstr.* **70**, 115727e (1969).
35. Oleinek, H., and Zugravescu, I., *Makromol. Chem.* **131**, 265 (1970); *Chem. Abstr.* **72**, 101166a (1970).
36. Quentin, J.-P., and Ruaud, M., U.S. Pat. 3,553,169 (Rhone-Poulenc) (1971).
37. Rembaum, A., and Singer, A., U.S. Pat. 3,538,053 (California Institute of Technology) (1970); *Chem. Abstr.* **74**, 32178h (1971).
38. Russian Patent 252,608 (1970).
39. Selezneva, E. N., Polak, L. S., and Shishkina, M. V., *Izv. Akad. Nauk SSSR, Ser. Khim.* p. 1086 (1969); *Chem. Abstr.* **71**, 50781s (1969).
40. Shostakovskii, M. F., Kryazhev, Yu. G., and Bestsenaya, N. V., Russ. Pat. 246,848 (1969).
41. Thomas, C. L., U.S. Pat. 3,505,353 (Sun Oil Co.) (1970); *Chem. Abstr.* **72**, 122181s (1970).
42. Thomson, D. W., U.S. Pat. 3,422,071 (U.S. Dept. of the Air Force) (1969); *Chem. Abstr.* **70**, 58414z (1969).
43. Wessling, R. A., and Zimmerman, R. G., U.S. Pat. 3,532,643 (Dow Chemical Co.) (1970); *Chem. Abstr.* **74**, 3994r (1971).

44. Wildi, B. S., U.S. Pat. 3,449,329 (Monsanto Co.) (1969); *Chem. Abstr.* **71**, 61892e (1969).
45. Yuminov, V. S., Bezborodova, S. P., Biragova, K. G., Bekichev, V. I., and Buder, S. A., *Vysokomol. Soedin., Ser. A* **11**, 640 (1969); *Chem. Abstr.* **71**, 13396z (1969).

CHAPTER II (p. 31).

1. Aso, C., and Kunitake, T., *Mem. Fac. Eng., Kyushu Univ.* **29**, 31 (1969); *Chem. Abstr.* **72**, 79533u (1970).
2. Aso, C., Kunitake, T., Khattak, R. K., and Sugi, N., *Macromol. Chem.* **134**, 147 (1970); *Chem. Abstr.* **73**, 25952g (1970).
3. Aso, C., Kunitake, T., Matsuguma, Y., and Imiazumi, Y., *J. Polym. Sci., Part A-1* **6**, 3049 (1968); *Chem. Abstr.* **69**, 10057, 107169p (1968).
4. Brodskaya, E. I., Kryazhev, Yu. G., Frulov, Yu. L., Kalikhman, I. D., and Yushmanova, T. I., *Vysokomol. Soedin., Ser. A* **11**, 655 (1969); *Chem. Abstr.* **70**, 115656f (1969).
5. Brodskaya, E. I., Kryazahev, Yu. G., Shergina, N. I., and Okladnikova, Z. A., *Vysokomol. Soedin., Ser. B* **10**, 895 (1968); *Chem. Abstr.* **70**, 38246t (1969).
6. Butler, G. B., and Jachia, B., *J. Macromol. Sci., Chem.* **3**, 803 (1969); *Chem. Abstr.* **71**, 30793u (1969).
7. Butler, G. B., Kimura, S., and Baucom, K. B., *Polym. Prepr., Amer. Chem. Soc., Div. Polym. Chem.* **11** (1), 48 (1970).
8. Butler, G. B., and Sharpe, A. J., Jr., *Polym. Prepr., Amer. Chem. Soc., Div. Polym. Chem.* **11** (1), 42 (1970).
9. Gibbs, W. E., and Barton, J. M., *Vinyl Polym.* **1**, 59–138 (1967); *Chem. Abstr.* **70**, 88271t (1969).
10. Guaita, M., Camino, G., and Trosarelli, L., *Makromol. Chem.* **130**, 243 (1969); *Chem. Abstr.* **72**, 44182b (1970).
11. Imanishi, Y., Matsuzaki, K., Kohjiya, S., and Okamura, S., *J. Macromol. Sci., Chem.* **3**, 237 (1969).
12. Kalikhman, I. D., Kryazhev, Yu. G., and Rzhepka, A. V., *Vysokomol. Soedin., Ser. B* **11**, 234 (1969); *Chem. Abstr.* **70**, 115664g (1969).
13. Kawai, W., and Katsuta, S., *J. Polym. Sci., A-1* **8**, 2421 (1970); *Chem. Abstr.* **73**, 99262d (1970).
14. Kryazhev, Yu. G., Cherkazhin, M. I., Yushmanova, T. I., Kalikhman, I. D., Baibovo-dina, E. N., and Shostakovskii, M. F., *Vysokomol. Soedin., Ser. A* **11**, 700 (1969); *Chem. Abstr.* **71**, 22379y (1969).
15. Kryazhev, Yu. G., Okladnikova, Z. A., Rzhepka, A. V., Brodskaya, E. I., and Shosta-kovskii, M. F., *Vysokomol. Soedin., Ser. A* **10**, 2366 (1968); *Chem. Abstr.* **70**, 29396q (1969).
16. LaFont, P., and Vivant, G., U.S. Pat. 3,528,952 (Rhone-Poulenc) (1970).
17. Matsoyan, S. G., Saakyan, A. A., and Pogosyan, G. M., Russ. Pat. 238,165 (Institute of Organic Chemistry, Academy of Sciences, Armenian SSR) (1969); *Chem. Abstr.* **71**, 50811b (1969).
18. Olson, S. G., U.S. Pat. 3,435,020 (Hercules, Inc.) (1969); *Chem. Abstr.* **70**, 97398s (1969).
19. Shostakovskii, M. F., Kryazhev, Yu. G., Okladnikova, Z. A., and Rzhepka, A. V., Russ. Pat. 240,236 (Irkutsk Institute of Organic Chemistry) (1969); *Chem. Abstr.* **71**, 50812c (1969).
20. Shostakovskii, M. F., Kryazhev, Yu. G., Okladnikova, Z. A., Rzhepka, A. V., and Komarov, N. V., *Vysokomol. Soedin., Ser. B* **11**, 174 (1969); *Chem. Abstr.* **71**, 3773f (1969).

21. Shostakovskii, M. F., Kryazhev, Yu. G., Rzhepka, A. V., and Okladnikova, Z. A., *Izv. Sib. Otd. Akad. Nauk SSSR, Ser. Khim. Nauk* p. 134 (1969); *Chem. Abstr.* **72**, 90907q (1970).
22. Takahashi, T., *J. Polym. Sci., Part A-1* **6**, 3327 (1968).
23. Trossarelli, I., Guaita, M., and Priola, A., *J. Polym. Sci., Part C*, No. 16, 4713 (1969); *Chem. Abstr.* **71**, 22388a (1969).
24. Yokota, K., and Takada, Y., *Kobunshi Kagaku* **26**, 317 (1969); *Chem. Abstr.* **71**, 22369v (1969).

CHAPTER III (p. 99).

1. Craven, J. M., U.S. Pat. 3,435,003 (E. I. duPont de Nemours & Co.) (1969).
2. Harris, F. W., and Stille, J. K., *Macromolecules* **1**, 463 (1968); *Chem. Abstr.* **70**, 457, 4690t (1969).
3. Otsu, T., Aoki, S., and Shimizu, A., *Kagaku* (Kyoto), **21**, 464 (1966); *Chem. Abstr.* **69**, 9123, 97178z (1968).
4. Rakutis, R. O., *Diss. Abstr. B* **29**, 2014 (1968); *Chem. Abstr.* **70**, 68792k (1969).
5. Schilling, C. L., Jr., Reed, J. A., and Stille, J. K., *Macromolecules* **2**, 85 (1969); *Chem. Abstr.* **71**, 3757d (1969).
6. Stevens, M. P., and Musa, Y., *Polym. Prepr., Amer. Chem. Soc., Div. Polym. Chem.* **12** (1), 615 (1971).
7. Stille, J. K., *J. Macromol. Sci., Chem.* **3**, 1043 (1969); *Chem. Abstr.* **71**, 91905d (1969).
8. Stille, J. K., and Noven, G. K., *J. Polym. Sci., Part B* **7**, 525 (1969); *Chem. Abstr.* **71**, 81787c (1969).
9. Stille, J. K., Noven, G. K., and Green, L. L., *J. Polym. Sci., Part A-1* **8**, 2245 (1970); *Chem. Abstr.* **73**, 77732e (1970).
10. Stille, J. K., Rakutis, R. O., Mukamal, H., and Harris, F. W., *Macromolecules* **1**, 431 (1968); *Chem. Abstr.* **69**, 10056, 107167m (1968).

CHAPTER IV (p. 121).

1. Berlin, A. A., Cherkashin, M. I., Chauser, M. G., and Shifrina, R. R., *Polym. Sci. USSR* **9**, 2510 (1967).
2. Corfield, G. C., and Crawshaw, A., *J. Macromol. Sci., Chem.* **5**, 1873 (1971); *Chem. Abstr.* **73**, 131387m (1970).
3. De Shryver, F. C., Geast, W. J., and Smets, G., *J. Polym. Sci., Part A-1* **8**, 1939 (1970); *Chem. Abstr.* **73**, 77805f (1970).
4. Gilliams, Y., and Smets, G., *Makromol. Chem.* **117**, 1 (1968); *Chem. Abstr.* **69**, 10055, 107160d (1968).
5. Guzzi, A., and Bosio, R., U.S. Pat. 3,418,282 (Ferrania S.p.A.) (1968).
6. Hasegawa, M., *Kobunshi Kagaku* **27**, 337 (1970); *Chem. Abstr.* **73**, 131391h (1970).
7. Kennedy, J. P., and Isaacson, R. B., U.S. Pat. 3,418,259 (Esso Research and Engineering Co.) (1969).
8. Neville, R. G., AD 677,111 (Boeing Scientific Research Labs.) (1968).
9. Neville, R. G., and Rosser, R. W., *Makromol. Chem.* **123**, 19 (1969).
10. Rinehard, R. E., Ger. Pat. 1,943,624 (Uniroyal, Inc.) (1970); *Chem. Abstr.* **72**, 101242x (1970).
11. Sosin, S. L., Korshak, V. V., and Frunze, T. M., *Dokl. Chem.* (*English transl.*) **179**, 345 (1968).

CHAPTER V (p. 139).

1. Marapov, R., Usmanov, A., Makhkamov, K., Kalontarov, I. Ya., and Nikitin, V. I., *Dokl. Akad. Nauk SSSR* **187**, 100 (1969); *Chem. Abstr.* **71**, 102255v (1969).

CHAPTER VI

SECTION A (p. 145).

1. Liepins, R., Verma, G. S. P., and Walker, C., *Macromolecules* **2**, 419 (1969).
1a. Poddar, S. N., and Podder, N. G., *J. Ind. Chem. Soc.* **45**, 562 (1968).
2. Trofimenko, S., U.S. Pat. 3,418,260 (E. I. duPont de Nemours & Co.) (1968).

SECTION C (p. 150).

3. Japanese Patent 69/2091 (Nippon Telegraph & Telephone Public Corp.) (1969).
4. Manecke, G., and Woehrle, D., *Makromol. Chem.* **120**, 192 (1968).
5. Rougee, M., and Andre, C., *J. Polym. Sci., Part C*, No. 16, Part 6, 3167 (1968).
6. Storbeck, I., *J. Polym. Sci., Part C*, No. 16, Part 8, 4361 (1969); *Chem. Abstr.* **71**, 22500f (1969).

SECTION D (p. 154).

7. Berezin, B. D., and Shormanova, L. P., *Vysokomol. Soedin., Ser. B* **10**, 784 (1968); *Chem. Abstr.* **70**, 20405j (1969).
8. Berezin, B. D., and Shormanova, L. P., *Vysokomol. Soedin., Ser. A.* **11**, 1033 (1969); *Chem. Abstr.* **71**, 39556b (1969).
9. Berlin, A. A., and Sherle, A. I., *Inorg. Macromol. Rev.* **1**, 235 (1971).
10. British Patent 1,185,975 (Nippon Oil Co., Ltd.) (1970).
11. Hamann, C., and Schmidt, H., *Plaste Kaut.* **16**, 85 (1969).
12. Manecke, G., and Woehrle, D., *Makromol. Chem.* **120**, 176 (1968).
13. Shormanova, L. P., and Berezin, B. D., *Polym. Sci. USSR* **12**, 782 (1970).
14. Shormanova, L. P., and Berezin, B. D., *Vysokomol. Soedin.* **12**, 692 (1970).

CHAPTER VII

SECTION A (p. 173).

1. D'Alelio, G. F., Hofman, E. T., and Zeman, J. R., *J. Macromol. Sci., Chem.* **3**, 959 (1969).
2. Gabe, I., and Leonte, D., *Rev. Roum. Chim.* **14**, 1535 (1969); *Chem. Abstr.* **73**, 4325q (1970).
3. Hartwell, I. O., *Diss. Abstr. B* **30**, 3079 (1970).
4. Oh, J. S., and Cho, K. O., *Daehan Hwahak Hwoejee* **13**, 309 (1969); *Chem. Abstr.* **73**, 99308y (1970).

SECTION B (p. 175).

5. Brown, J. E., Tryon, M., and Horowitz, E., *J. Appl. Polym. Sci.* **13**, 1937 (1969); *Chem. Abstr.* **71**, 113367j (1969).
6. Horowitz, E., *Advan. Chem. Ser.* No. 85, 82 (1968); *Chem. Abstr.* **70**, 38325t (1969).

SECTION C (p. 180).

7. Lebsadze, T. N., Chkhartishvili, K. A., Pavlenishvili, I. Ya., and Gugava, M. T., *Vysokomol. Soedin., Ser. B* **10**, 609 (1968); *Chem. Abstr.* **69**, 107154e (1968).
8. Marcu, M., and Dima, M., *Rev. Roum. Chim.* **15**, 1365 (1970); *Chem. Abstr.* **74**, 32024e (1971).

9. Poddar, S. N., and Saha, N., *J. Ind. Chem. Soc.* **47**, 255 (1970).
10. Zelentsov, V. V., and Suvorova, K. M., *J. Gen. Chem. USSR* **38**, 474 (1968).

SECTION E (p. 184).

11. Hojo, N., Shirai, H., and Suzuki, A., *Kogyo Kagaku Zasshi* **72**, 2040 (1969); *Chem. Abstr.* **72**, 32285z (1970).
12. Hojo, N., Shirai, H., and Suzuki, A., *Kogyo Kagaku Zasshi* **73**, 1438 (1970); *Chem. Abstr.* **73**, 131428a (1970).
13. Hojo, N., Shirai, H., and Suzuki, A., *J. Chem. Soc. Jap., Ind. Chem. Sect.* **73**, 2708 (1970).
14. Lidorenko, N. S., Ravich, I. Ya., Gindin, L. G., Kagan, A. S., Koval'skii, A. E., Toroptseva, T. N., and Zhigareva, N. K., *Dokl. Akad. Nauk SSSR* **182**, 1087 (1968); *Chem. Abstr.* **70**, 29637u (1969).

CHAPTER VIII

SECTION B (p. 212).

1. British Patent 1,132,573 (Imperial Smelting Corp.) (1968).
2. Hartman, R. D., Kanda, S., and Pohl, H. A., *Proc. Okla. Akad. Sci.* **47**, 246 (1968); *Chem. Abstr.* **70**, 97503x (1969).
3. Shetty, P. S., and Fernando, Q., *Anal. Chem.* **41**, 685 (1969); *Chem. Abstr.* **70**, 107036p (1969).

SECTION C (p. 214).

4. Marcu, M., and Dima, M., *Rev. Roum. Chim.* **13**, 359 (1968); *Chem. Abstr.* **70**, 29426z (1969).

CHAPTER IX

SECTION A (p. 227).

1. De Charentenay, F., and Teyssie, P., *C. R. Aacd. Sci., Ser. C* **269**, 814 (1969); *Chem. Abstr.* **72**, 3892q (1970).

SECTION B (p. 237).

2. Srivastava, K. C., and Banerji, S. K., *J. für Prakt. Chem.* [4] **38**, 327 (1968).
3. Tedesco, P. H., and Walton, H. F., *Inorg. Chem.* **8**, 932 (1969).

SECTION C (p. 242).

4. Marcu, M., and Dima, M., *Rev. Roum. Chim.* **15**, 1365 (1970); *Chem. Abstr.* **74**, 32024e (1971).

SECTION D (p. 249).

5. Block, B. P., U.S. Clearinghouse, Fed. Sci. Tech. Inform., AD 1968, AD-678324 (1968); *Chem. Abstr.* **70**, 78439q (1969).
6. Block, B. P., *Coord. Chem. Proc., John C. Bailar, Jr. Symp., 1969* p. 241, *Chem. Abstr.* **72**, 32298f (1970).
7. Block, B. P., *Inorg. Macromol. Rev.* **1**, 115 (1970); *Chem. Abstr.* **72**, 90895j (1970).
8. Block, B. P., and Dahl, G. H., U.S. Pat. 3,403,176 (Pennsalt Chemicals Corp.) (1968).
9. Block, B. P., and Dahl, G. H., U.S. Pat. 3,415,762 (Pennsalt Chemicals Corp.) (1968).
10. Block, B. P., and Dahl, G. H., U.S. Pat. 3,415,781 (Pennsalt Chemicals Corp.) (1968); *Chem. Abstr.* **70**, 38323r (1969).

11. Block, B. P., and Dahl, G. H., U.S. Pat. 3,457,195 (Pennsalt Chemicals Corp.) (1969); *Chem. Abstr.* **71**, 71227f (1969).
12. Block, B. P., Ocone, L. R., and Simkin, J., Can. Pat. 763,492 (1967).
13. Brückner, S., Calliganis, M., Nardin, G., Randaccio, L., and Ripamonti, A., *Chem. Commun.* p. 474 (1969).
14. Dahl, G. H., AD 671,882 (Pennsalt Chemicals Corp.) (1968).
15. Drew, M. G. B., Lewis, D. F., and Walton, R. A., *Chem. Commun.* p. 326 (1969).
16. Fetter, N. R., and Grieve, C. M., *J. Polym. Sci., Part A-1* **8**, 1337 (1970); *Chem. Abstr.* **73**, 15388p (1970).
17. Giancotti, V., Giordano, F., and Ripamonti, A., *Makromol. Chem.* **120**, 96 (1969).
18. Giancotti, V., and Ripamonti, A., *J. Chem. Soc., A* p. 706 (1969); *Chem. Abstr.* **70**, 78470t (1969).
19. King, J. P., Saraceno, A. J., and Block, B. P., U.S. Pat. 3,483,143 (Pennsalt Chemicals Corp.) (1969).
20. Korshak, V. V., Danilov, V. G., Vinogradova, O. V., and Suchkova, M. D., *Izv. Akad. Nauk SSSR, Ser. Khim.* p. 2586 (1969); *Chem. Abstr.* **72**, 67601b (1970).
21. Korshak, V. V., Krukovskii, S. P., Knyazeva, E. K., and Danilov, V. G., *Vysokomol. Soedin., Ser. A* **11** 3 (1969); *Chem. Abstr.* **70**, 68805s (1969).
22. Kuchen, W., and Hertel, H., *Angew, Chem., Int. Ed. Engl.* **8**, 89 (1969).
23. Ladwig, G., and Grunze, H., *Monatsber. Deut. Akad. Wiss. Berlin* **10**, 120 (1968); *Chem. Abstr.* **70**, 106944w (1969).
24. Maguire, K. D., U.S. Pat. 3,444,103 (Pennsalt Chemicals Corp.) (1965); *Chem. Abstr.* **71**, 39631x (1969).
25. Monroe, R. F., and Schmidt, D. L., U.S. Pat. 3,497,464 (Dow Chemical Co.) (1970); *Chem. Abstr.* **72**, 122177v (1970).
26. Owens, C., Pytlewskii, L. L., Karayannis, N. M., Wysoganski, J., and Labes, M. M., *J. Polym. Sci., Part B* **8**, 81 (1970); *Chem. Abstr.* **73**, 4243m (1970).
27. Rose, S. H., U.S. Pat. 3,440,186 (Pennsalt Chemicals Corp.) (1969); *Chem. Abstr.* **71**, 3904z (1969).
28. Saraceno, A. J., U.S. Pat. 3,483,142 (Pennsalt Chemicals Corp.) (1969); *Chem. Abstr.* **72**, 44588p (1970).
29. Schmidt, D. L., and Flagg, E. E., *J. Polym. Sci., Part A-1*, **7**, 865 (1969).
30. Schmidt, D. L., and Flagg, E. E., U.S. Pat. 3,538,136 (Dow Chemical Co.) (1970); *Chem. Abstr.* **74**, 23234w (1971).
31. Slota, P. J., Grieve, C. M., Fetter, N. R., and Bilbo, A. J., *J. Polym. Sci., Part A-1* **7**, 2051 (1969); *Chem. Abstr.* **71**, 102326m (1969).
32. Zhinkina, L. N., Polyakova, L. L., and Tarasov, E. V., *Plast. Massy* p. 28 (1969); *Chem. Abstr.* **71**, 13641a (1969).

CHAPTER X

SECTION A (p. 301).

1. Belgian Patent 725,552 (Badische Anilin) (1969).
2. Heller, K-H., Nentwig, J., and Schnell, H., U.S. Pat. 3,390,160 (Farbenfabriken Bayer) (1968).
3. Wöhrle, D., and Manecke, G., *Makromol. Chem.* **140**, 137 (1970).

SECTION B (p. 305).

4. Gagliardini, E., U.S. Pat. 3,412,117 (SIPCAM) (1968).
5. Windel, H., and Pommer, E-H., U.S. Pat. 3,441,581 (BASF) (1969).
6. Yokoi, H., and Isobe, T., *Bull. Chem. Soc. Jap.* **41**, 1489 (1968).

## CHAPTER XI

SECTION A (p. 317).

1. Andrianov, K. A., *Dokl. Akad. Nauk SSSR* **191**, 347 (1970); *Chem. Abstr.* **73**, 15352x (1970).
2. Andrianov, K. A., and Makarova, N. N., Russ. Pat. 231,813 (Institute of Heteroorganic Compounds, Academy of Sciences, USSR) (1968); *Chem. Abstr.* **70**, 58506f (1969).
3. Andrianov, K. A., and Makarova, N. N., *Izv. Akad. Nauk SSSR, Ser. Khim.* p. 625 (1969); *Chem. Abstr.* **71**, 4041c (1969).
4. Andrianov, K. A., and Makarova, N. N., *Vysokomol. Soedin., Ser. A* **12**, 663 (1970); *Chem. Abstr.* **73**, 4466m (1970).
5. Andrianov, K. A., Slonimskii, G. L., Genin, Ya. V., Gerasimov, V. I., Levin, V. Yu., Makarova, V. N., and Tsvankin, D. Ya., *Dokl. Akad. Nauk SSSR* **187**, 1285 (1969); *Chem. Abstr.* **72**, 22174y (1970).
6. Andrianov, K. A., Tikhonov, V. S., and Astapov, B. A., *Vysokomol. Soedin., Ser. B* **12**, 577 (1970); *Chem. Abstr.* **73**, 110197j (1970).
7. Avilova, T. P., Bykov, V. T., Zolotar, G. Ya., and Lesheheva, G. V., *Vysokomol. Soedin., Ser. B* **10**, 425 (1968); *Chem. Abstr.* **69**, 4918, 52536y (1968).
8. Brown, J. F., Jr., U.S. Pat. 3,390,163 (General Electric Co.) (1968).
9. Guyot, A., Cuidard, R., and Bartholin, M., *J. Polym. Sci., Part C* **22**, Part 2, 785 (1967); *Chem. Abstr.* **71**, 102406v (1969).
10. Krueger, G., Buettner, M., Oliew, G., and Thilo, E., *Z. Anorg. Allg. Chem.* **360**, 70 (1968); *Chem. Abstr.* **69**, 6343, 67810g (1968).
11. Petrashko, A. I., *Dokl. Chem. (English transl.)* **180**, 542 (1968).
12. Speier, J. L., U.S. Pat. 3,445,425 (Dow Corning Corp.) (1969); *Chem. Abstr.* **72**, 3940d (1970).
13. Tsvetkov, V. N., Andrianov, K. A., Okhrimenko, G. I., Shtennikova, I. N., Fomin, G. A., Vitovskaya, M. G., Pakhomov, V. I., Yarosh, A. A., and Andreev, D. N., *Vysokomol. Soedin., Ser. A* **12**, 1892 (1970); *Chem. Abstr.* **73**, 110262b (1970).
14. Tsvetkov, V. N., Andrianov, K. A., Shtennikova, I. N., Okhrimenko, G. I., Andreyeva, L. N., Fomin, G. A., and Pakhomov, V. I., *Polym. Sci. USSR* **10**, 636 (1968).
15. Voronkov, M. G., Romadane, I., Petsunovich, V. A., and Mazeika, I., *Khim. Geterotsikl. Soedin.* p. 972 (1968); *Chem. Abstr.* **70**, 68820t (1969).
16. Wu, T. C., U.S. Pat. 3,400,145 (General Electric Co.) (1968); *Chem. Abstr.* **69**, 9140, 97360c (1968).

SECTION B (p. 322).

17. Andrianov, K. A., and Kononov, A. M., *Kremniiorg. Soedin., Tr. Soveshch.* p. 35 (1967); *Chem. Abstr.* **69**, 5586, 59613e (1968).
18. Aylett, B. J., Brit. Pat, 1,123,252 (Minister of Technology, London) (1968); *Chem. Abstr.* **69**, 6352, 67901n (1968).
19. British Patent 1,156,881 (Monsanto Co.) (1969).
20. Canadian Patent 831,093 (Monsanto Co.) (1969).
21. Fink, W., U.S. Pat. 3,422,060 (Monsanto Co.) (1969).
22. Fink, W., U.S. Pat. 3,431,222 (Monsanto Co.) (1969).
23. Fink, W., U.S. Pat. 3,578,693 (Monsanto Co.) (1971).
24. German Patent 1,570,912 (Monsanto Co.) (1969).

Section C (p. 325).

25. Andrianov, K. A., Slonimskii, G. L., Zhdanov, A. A., and Levin, V. Yu., *Vysokomol. Soedin., Ser. A* **10**, 2102 (1968).
26. Andrianov, K. A., Slonimskii, G. L., Zhdanov, A. A., and Levin, V. Yu., *J. Macromol. Sci., Chem.* **2**, 959 (1968); *Chem. Abstr.* **69**, 9142, 97376n (1968).
27. Avilova, T. P., Bykov, V. T., Zolotar, G. Ya., Marinin, N. P., and Shapkin, N. P., *Izv. Akad. Nauk SSSR, Neorg. Mater.* **4**, 2188 (1968); *Chem. Abstr.* **70**, 47955a (1969).

# Author Index

Numbers in parentheses are reference numbers and indicate that an author's work is referred to, although his name is not cited in the text. Numbers in italics show the page on which the complete reference is listed.

## A

Abbott, L. S., 96(1), 97(1), 98(1), 254(1), 255(1), 256(1), 258(1), *262*

Abe, A., 299(1), 301(1), 302(1), 306(1), 363(1), *432*

Abkin, A. D., 443(36), 446(36), 448(36), 454(36), *469*

Abramo, J. G., 22(2), *262*

Abramo, S. V., 29(622), 30(622), 33(622), 34(622), 41(621, 622), 124(621), 126(621), 128(621), 130(621), 133(621), 135(621), *282*

Adachi, M., 81(335), 230(335), *272*

Adelson, D. E., 443(1), *468*

Adrova, N. A., 2(4), 7(4), 8(224, 405), 9(112), 30(609), 33(377, 609), 34(377), 39(377), 41(377, 565), 109(224), 110(224), 133(405), 136(5), 137(5, 378, 379), 145(224), 146(224), 147(6, 224), 148(6, 224), 158(7, 224), 159(3), 163(224), 164(224), 165(5, 380, 609), 166(8), 196(4), 197(4), *262, 265, 269, 274, 275, 280, 282*

Agolini, F., *262*

Aito, Y., 292(18), 293(16, 17, 18), 294(16, 24), 295(16, 20, 21), 296(19, 23, 25), 297(19, 25), 298(19, 25), 299(20), 352(2, 22, 24), 353(16, 17, 18, 20, 21), 354(20, 21), 355(19, 25), 356(19, 25), *432*

Akopyan, L. M., 300(175), 302(170, 171), 360(175), 361(175), 362(175), 363(170, 171, 172), 364(170, 171), 365(171), 366(172), 367(170, 172), *437*

Albrecht, W. M., 170(540), *279*

Alder, K., 451(2), 460(2), *468*

Allen, S. J., 85(10), 239(10), 240(10), *262*

Al-Sufi, H. H. M. A., 4(361), 9(361), 159(361), 160(361), *273*

Amborski, L. E., 10(14), 22(13, 14), 28(11, 12), 29(11, 12, 643), 39(11, 12), 133(13, 14), *262, 283*

Amerik, V. V., 459(3), *468*

Anand, C., 336(3), *432*

Andreeva, I. V., 442(5), 444(5, 71), 446(5), 454(5), 456(4, 72, 73), 457(5), 460(72), 463(5, 73), 464(70, 73), 466(5, 70, 73), 467(70), *468, 470*

Angelo, R. J., 7(15), 10(15, 27), 12(18, 20), 13(28), 14(29), 20(26, 27, 29), 22(15, 19, 26, 29), 23(25), 32(26), 40(27), 67(15, 19, 25, 29), 68(29), 75(16, 17), 76(22), 77(16, 21), 93(24), 95(23, 24), 128(15, 18, 20), 129(26), 133(28), 134(27), 162(27), 200(19), 201(19, 29), 213(16, 17), 214(16, 17, 21, 22), 215(21, 22), 216(16, 17, 22), 217(16, 17), 218(16, 17, 21, 22), 219(16, 17), 221(16, 17, 21, 22), 249(23, 24), 250(24), 252(24), *262, 263,* 287(72), 322(72, 73), 323(72), 382(73), 383(73), 384(72), 386(72), 392(73), *434*

Annenkova, V. M., 455(6), *468*

Annekova, V. Z., 455(6), 458(18, 19, 20), 459(7, 18, 20), *468,* 480(31), 481(31), *494*

Anonymous, 29(33, 34), 39(30), 41(30, 31), 42(35), 51(35), 84(33), 89(32), *263*

Arai, K., 322(125, 126), 324(125, 126), 395(126), 396(126), 397(125, 126), 398(126), *435*

Arbuzova, I. A., 299(7, 11, 12), 300(11, 12), 301(267), 307(4, 5, 240), 308(6, 9, 10), 309(6, 10), 315(8, 194), 358(11, 12), 359(11, 12), 360(12), 362(11), 363(7, 11), 368(4, 5, 240), 369(11), 370(194), 371(6, 10, 11), 372(10), 376(6, 9, 194), 377(11), 378(8), *432, 438, 439, 440*

Ardashnikov, A. Ya., 2(36), *263*

Arnold, F. E., 82(626), 83(37), 84(626, 637), 88(38), 233(637), 234(637), 236(637), 237(626), 238(626), 243(38), *263, 282, 283*

Asahara, T., 14(40), 28(39), 102(40), 103(40), 104(39, 40), 105(39), 106(39), *263*

Asano, N., 92(595), 193(341), 245(595), 246(595), 248(595), *272, 281*

Aso, C., 288(29), 290(15), 291(28, 29), 292 (18, 29), 293(16, 17, 18), 294(24, 28, 29), 295(16, 20, 21, 29), 296(19, 23, 25, 26, 27), 297(19, 25), 298(19, 25), 299 (20), 305(30, 32), 308(13), 352(2, 22, 24, 26, 27), 353(16, 17, 18, 20, 21), 354 (20, 21, 28, 29), 355(19, 25), 356(19, 25), 357(30, 31), 380(14), *432, 433*

Atlas, S. M., 456(74), *470*

Augl, J. M., 82(687, 688, 689), 84(687, 688, 689), 232(687, 689), 233(687, 688, 689), 234(687, 689), 236(687, 688, 689), 237(687), *285*

Avchen, B., 289(229), 290(229), 341(229), 342(229), 345(229, 269), 423(229), 424 (229), *439*

Avetyan, M. G., 299(177), 300(174, 175, 177), 303(177), 306(177), 358(174, 177), 359(176, 177, 178, 179), 360(173, 175, 176, 177, 178), 361(175, 176, 179), 362 (175), *437*

Azarov, V. I., 6(371), 127(371, 372), 129 (371, 372), *273*

**B**

Baba, Y., 72(693), 209(693), *285*, 492(1), *493*

Babbitt, G. E., 26(601), 108(601), *282*

Bäder, E., 443(8), 448(96), *468, 471*

Baginski, J., 492(7), *493*

Bailey, W. J., 96(43, 672), 97(43), 98(672), 256(43, 672), *263, 284*

Balaev, G. A., 298(216), 355(216), *438*

Banihashemi, A., 83(44), 230(44), 233(44), 235(44), 237(44), *263*

Barinova, A. N., 308(239), 369(239), *439*

Barito, R. W., 5(45), *263*

Barnett, M. D., 289(32, 33, 93), 308(32), 310(32, 33, 93), 371(32), 373(32, 33, 93), 374(32, 33), 375(33), 376(33), *433, 434*

Barsamyan, S. T., 300(35, 157), 358(34, 35, 157), 359(34, 35, 157), 360(34, 35, 157), *433, 436*

Barton, J. M., 290(36), 304(37), 357(37), 358(37), *433*

Baruba, Y., 93(713), 95(713), 250(713), 251(713), *286*

Batalin, O. E., 107(395), 111(395), *274*

Bauer, R. S., 298(38), 299(38), 355(38), 356(38), *433*

Bekasova, N. I., 474(2, 21), 475(2, 21), *493*

Bell, E. R., 433(17), *468*

Bell, V. L., 11(97), 62(98), 133(97), 204(98), *264*

Belohlav, L. R., 91(99), 248(99), *264*

Belova, G. V., 63(100), 100(101), 204(100), *264*

Belyaev, V. I., 443(141), 444(142, 143), 447 (142), 449(141), 454(142), 458(18, 19, 20), 459(18, 20), *468, 472*

Benina, I. N., 2(4), 7(4), 196(4), 197(4), *262*

Bergman, E., 443(17), 447(21), *468*

Berlin, A. A., 63(100), 69(104, 105), 70 (104, 105, 106), 71(176, 177), 99(102, 302), 100(101, 417), 117(103), 118 (103), 204(100), 205(104, 105, 106, 176, 177), 260(101, 102, 302, 417), *264, 265, 267, 271, 275*

Berlin, A. M., 481(22), *493*

Berlin, K. D., 337(44, 45), 338(44, 45), 417 (44, 45), 418(44, 45), 419(44), *433*

Bernier, G. A., 29(107), 37(107), 134(107), *265*

Berr, C. E., 29(622), 30(622), 33(622), 34 (290, 309, 622), 37(309), 39(309), 41 (621, 622), 124(621), 126(621), 128 (621), 130(621), 133(621), 135(621), *271, 282*

Bessonov, M. I., 8(405), 9(112), 28(567), 30(609), 33(377, 609), 34(377), 39 (377), 41(377, 565), 105(381), 106(381, 382), 107(382), 133(384, 405, 566), 134 (602), 137(378, 379), 159(3), 165(380, 609), *262, 265, 274, 275, 280, 282*

Beynon, K. I., 325(48), 326(48), 337(47), 402(48), 417(47), 419(46), 420(46), *433*

Binsack, R., 75(108), 218(108), 219(108), *265*

Blacet, F. E., 443(22), *468*

Black, W. B., 41(544), 138(541, 542, 543, 545), 139(541, 542, 543), 140(541, 542, 543), 141(546), 142(541, 542, 543), 143 (546), 144(546), 149(545), 152(546), *279, 280*, 300(205), 332(49, 203), 411 (49, 50, 203), *433, 438*

Blomstrom, D. C., 11(109), 154(109), *265*

Blout, E. R., 308(51), 311(51, 91), 312(51, 91), *433, 434*

Bock, W., 455(157), *472*
Bogert, M. T., 11, *265*
Bogomol'nyi, V. Ya., 349(52), 429(52), *433*
Bogomolova, T. B., 210(392), *274*
Boiko, L. V., 6(374), 30(374), 131(374), 132(374, 667), *274, 284*
Boldebuck, E. M., 13(111), *265*
Boldyrev, A. G., 9(112), *265*
Bolton, B. A., 42(113), 45(114), 185(114), *265*
Bombeke, J., 341(118), 422(118), *435*
Bond, W. C., Jr., 337(53), 338(53), 417(53), 420(53), *433*
Borodin, L. I., 455(6), *468*
Bottenbruch, L., 75(108), 76(557), 77(115), 216(115), 218(108, 115), 219(108, 115, 557), 220(115), 221(115), *265, 280*
Bottex, P., 80(116, 117), 227(116, 117), 228(116, 117), *265*
Bottger, H., 55(118), *265*
Bovey, F. A., 317(270), 320(270), *440*
Bower, G. M., 3(121, 122), 8(121, 122), 14(275), 29(166), 37(234, 275), 38(234), 40(166, 235, 275), 42(121, 122, 274), 43(121, 122, 275), 47(121, 122), 49(234), 50(122, 234), 51(119, 120, 122), 53(121, 122), 64(275), 65(275), 124(121, 122), 126(121, 122), 128(121, 122, 273), 130(121, 122, 273), 133(121, 122, 234), 134(121, 122, 273), 135(121, 122, 273), 138(275), 143(275), 146(166, 234), 147(275), 148(234), 149(275), 150(275), 151(166, 275), 152(275), 174(121, 122, 273), 175(121, 122, 234, 273, 274), 176(121, 122, 273, 275), 177(121, 122, 273, 274), 178(121, 122, 273, 274), 179(274), 180(274), 196(121, 122, 273), *265, 267, 269, 270*
Bradshaw, J. S., 117(123), *265*
Brady, W. T., 296(54, 55), 352(54, 55), *433*
Braeutigam, J., 489(28), *494*
Brake, W., 486(3), *493*
Braun, D., 321(56), *433*
Brebner, D. L., 11(124), 121(124), *265*
Breed, L. W., 98(125), *265*
Bresler, S. E., 317(57), 319(57), 320(57), 380(57), *433*
Brey, W. S., Jr., 320(204), *438*
Bristow, G. M., 308(58), *433*
Brooks, T. W., 288(74), 319(74), 322(75), 383(75), *434*

Bruck, S. D., 34(157, 158, 159, 160, 161, 162, 163, 164), 35(157, 158, 159, 160, 161, 162, 163, 164), 36(160, 163), 37(157, 158), 133(157, 158, 159, 161, 162), 134(160, 163, 164), *266, 267*
Brunovskaya, L. A., 325(223), 402(223), 403(223), 404(223), 405(223), *439*
Bublik, L. S., 29(669), *284*
Buckley, D. H., 41(165), *267*
Burgman, H. A., 14(275), 29(166), 37(275), 40(166, 235, 275), 43(275), 51(120), 64(275), 65(275), 138(275), 143(275), 146(166), 147(275), 149(275), 150(275), 151(166, 275), 152(275), 176(275), *265, 267, 269, 270*
Burkett, J. L., 490(25), 491(25), *493*
Burks, R. E., Jr., 98(167), *267*
Butler, G. B., 287(63, 72, 76, 77), 288(64, 70, 74, 82), 289(32, 33, 93), 290(64, 67), 304(37, 71, 85), 308(32), 310(32, 33, 93), 317(94, 95), 318(94, 95), 319(74), 320(66, 90, 204), 322(69, 72, 73, 75, 76, 77, 78, 79, 80), 323(69, 72), 337(44, 45, 65), 338(44, 45), 342(64, 68), 345(64, 71, 81), 346(65, 71, 84), 347(64, 71), 349(84), 350(68), 351(65), 357(37, 61, 62, 64, 71, 85), 358(37), 371(32), 373(32, 33, 93), 374(32, 33), 375(33), 376(33), 379(94, 95), 382(69, 73), 383(69, 73, 75, 76, 77), 384(69, 72, 76, 77), 385(69, 76, 77), 386(69, 72), 387(69), 388(69), 389(69), 390(69), 391(69), 392(69, 73), 393(69), 406(96), 417(44, 45, 65), 418(44, 45), 419(44), 420(83), 424(64, 68), 425(68), 426(62, 64, 71, 81), 427(62, 64, 71), 428(65, 84), 429(83, 84), 431(63, 65, 68), *433, 434, 438*

## C

Caldwell, D. S., 123(580), *281*
Calvayroc, H., 450(46), 454(46), *469*
Campanile, V. A., 443(18), 455(28), *468*
Campbell, D., 335(163), 336(163), 415(163), 416(163), *436*
Capps, D. B., 97(171), 254(171), 255(171), *267*
Carleton, P. S., 24(223), 25(223), 152(223), *269*
Carpenter, W. R., 475(4), 476(4), *493*

Chalmers, J. R., 3(172), 7(173, 174), 41 (175), 148(175), *267*
Chang, E., 288(86), 325(86), 326, 327(86), 328(86), 400(86), 401(86), *434*
Chang, H.-C., 308(88), 319(102), 346(87), 428(87), *434, 435*
Chapin, E. C., 304(37), 357(37), *433*
Chechak, J. J., 317(206), *438*
Chechik, A. I., 29(669), *284*
Chekulaeva, I. A., 315(250), *439*
Cherdron, H., 443(29, 30, 113, 134), 455 (60, 114, 115), 463(134), *468, 471, 472*
Cherkashina, L. G., 71(176, 177), 205(176, 177), *267*
Chernova, A. G., 29(669), 41(668), *284*
Chernysheva, T. I., 346(123, 124), 347(124), 428(123), 429(123), 430(124), *435*
Chiang, T.-C., 371(89), 372(89), 374(89), *434*
Chikenina, L. V., 43(393), 178(393), *274*
Cholakyan, A. A., 330(181), 339(181), 358 (181, 191), 359(191, 192), 360(191, 192), 361(191, 192), 362(191, 192), 369(181), 407(181), 421(181), *437*
Christie, P. A., 476(5, 11), *493*
Chu, C.-C., 81(691), 230(691), *285*
Cïhodaru, S., 473(35), 474(35), 476(6), *493, 494*
Clark, H. G., 320(90), *434*
Cohen, C., 90(297), 246(297), 248(297), *271*
Cohen, S. G., 311(91), 312(91), *434*
Cohen, S. M., 96(179, 180), 97(179), 98 (178, 179), 254(178, 179, 180), 255 (179, 180), 257(178, 179, 180), *267*
Collins, W. E., 133(181), *267*
Cooper, S. L., 37(182), 38(182), 49(182), 50(182), 133(182), 146(182), 148(182), 175(182), *267*
Corfield, G. C., 332(92), 335, 414(92), *434*
Costanza, J. R., 91(99), 248(99), *264*
Courtright, J. R., 29(183), *267*
Covington, E. R., 98(167), *267*
Cowan, J. C., 97(449), 258(449), *276*
Crawshaw, A., 287(76, 77), 289(33, 93), 310(33, 93), 317(94, 95), 318(94, 95), 322(73, 76, 77), 325(96), 332(92), 335, 373(33, 93), 374(33), 375(33), 376 (33), 379(94, 95), 382(73), 383(73, 76, 77), 384(76, 77), 385(76, 77), 392(73), 403(96), 404(96), 414(92), *433, 434*

Culbertson, B. M., 47(586), 181(586), 182 (586), 187(586), *281*, 297(263), 355 (263), *440*
Cummings, W., 125(184), *267*

**D**

Dabrowski, R., 492(7), *493*
D'Alelio, G. F., 61(185), 62(185), 204(185), *267*, 455(31), *469*, 476(8), 477(8), *493*
Daly, W. H., 341(225), 423(225), *439*
Dannenberg, H., 443(1), *468*
Darms, R., 479(9, 10), *493*
Davis, A. C., 18(186), 122(186), *267*
Davydova, S. L., 290(158), 321(159), 349 (159), 429(159), *436*
Dawans, F., 87(187), 243(187), *267*
De Brunner, R. E., 12(190), 29(188, 189), 41(190), *267*
de Gaudemaris, G., 9(551), 22(551), 32 (551), 73(576, 577), 74(576, 577), 80 (116, 117), 82(204), 84(607), 90(604, 605), 92(552), 95(549, 550, 551), 96 (604, 607), 127(551), 138(573), 173 (574, 575), 174(575), 179(575), 210 (576, 577), 227(116, 117), 228(116, 117), 235(204), 244(552, 604), 245(552, 604), 246(552, 604, 605, 606, 608), 247 (552, 604, 605, 606, 608), 248(604, 605, 606, 608), 249(549, 550, 551), 251(549, 550, 551), 252(549, 550, 551), *265, 268, 280, 281, 282*, 483(32, 33), *494*
De Gaudemaris, G., 84(191), 90(193), 233 (191, 192), 234(191, 192), 236(192), 245 (193), 246(193), 247(193), *268*
Delman, A. D., 41(194), *268*
Delmonte, D. W., 316(97), *434*
Derevyanchenko, V. P., 116(553), *280*
De Schryver, F., 78(442), 79(442), 83(195), 225(442), 226(442), 227(442), 228 (442), 229(442), 231(195), *268, 276*
Deshpande, A. B., 336(3), *432*
De Shryver, F. 83(343), 233(343), 234(343), *272*
Deval, N., 317(253, 254), 319(253), 320 (253), 379(253), 380(253), 381(253), *439*
DeWinter, W., 83(343), 233(343), 234(343), *272*, 290(98), *434*, 442(32), *468*
De Winter, W., 41(196), 84(196), 96(196), 141(546), 143(546), 144(546), 152 (546), *268, 280*

De Witte, E., 341(118), 422(118), *435*
Dieterle, R. J., 42(113), *265*
DiLeone, R. R., 29(197), *268*
Dine-Hart, R. A., 3(198, 200), 4(198, 200), 5(200), 10(198, 200), 15(199), 17(199), 40(201), 53(198, 200), 143(199), 171(199), *268*
Dobrokhotova, M. L., 29(669), *284*
Dogoshi, N., 95(711), 250(711), *285*
Donati, M., 289(99), *434*
Donleavy, J. J., 444(44), 451, *469*
Dreiman, N. A., 298(216), 354(216), *438*
Drewitt, J. G. N., 85(10), 239(10), 240(10), *262*
Dubnova, A. M., 136(5), 137(5, 378, 379), 159(3), 165(5), *262*, *274*
Dubnova, H. M., 165(380), *274*
Duenwald, W., 55(453), 189(453), *276*
Dufraisse, C., 444(76, 77, 78), *470*
Dugar'yan, S. G., 429(211, 271), *438*, *440*
Dunphy, J. F., 41(202, 203), *268*
Durif-Varambon, B., 82(204), 235(204), *268*
Dvorko, G. F., 320(100, 284), 321(100, 284), 379(100, 284), *434*, *440*
Dyankov, S. S., 429(271), *440*
Dyer, E., 96(205), *268*, 476(11), *493*
Dzhagalyan, A. O., 325(182), 403(182), 404(182), 405(182), *437*

**E**

Edwards, W. M., 2(208), 3(207, 208, 213), 5(206, 207, 208, 213), 7(208), 11(124, 212, 213, 214, 216), 14(208), 20(208, 209, 210), 22(210), 217(212, 213, 215), 29(622), 30(622), 33(622), 34(622), 41(209, 621, 622), 121(124, 212, 213, 214, 216), 122(214, 216), 123(212, 213, 215), 124(621), 126(209, 621), 128(209, 621), 130(209, 621), 133(209, 211, 621), 135(209, 621), 162(209), 164(209), *265*, *268*, *282*
Efremova, V. N., 307(4, 5), 368(4, 5), *432*
Eifert, R. L., 444(33), 462(34, 35), 463(35), 466(35), *468*
Eliazyan, M. A., 299(180), 300(175), 303(180, 183), 323(183), 337(183), 339(183), 360(175), 361(175, 180), 362(175, 180), 363(180), 367(183), 393(183), 402(183), 418(183), *437*
Eliseeva, A. G., 307(5), 368(5), *432*

Elzer, P., 449(116), *471*
Emel'yanova, L. N., 29(669), 135(716), *284*, *286*
Endrey, A. L., 2(218), 6(218), 8(388), 10(217), 13(221), 20(218, 222), 22(219), 29(622), 30(622), 33(622), 34(622), 39(220), 41(621, 622), 120(218), 121(218), 124(218, 220, 221, 621), 126(218, 621), 128(217, 218, 221, 621), 130(218, 221, 621), 133(211, 218, 221, 621), 135(218, 621), 148(221), *268*, *269*, *274*, *282*
Erickson, J. G., 307(101), *435*
Ermolaeva, T. I., 321(159), 349(159), 429(159), *436*

**F**

Fabbro, D., 83(44), 230(44), 233(44), 235(44), 237(44), *263*
Farino, M., 289(99), *434*
Farrissey, W. J., 24(223), 25(223), 152(223), *269*
Faulkner, D., 96(1), 97(1), 98(1), 254(1), 255(1), 256(1), 258(1), *262*
Fauth, H., 449(117, 118), *471*
Fedorova, E. F., 8(224, 225), 28(567), 106(225), 107(225), 109(224), 110(224), 145(224), 146(224), 147(224), 148(224), 158(224), 163(224), 164(224), *269*, *280*, 308(6), 309(6), 315(8), 370(194), 371(6), 376(6, 194), 378(8), *432*, *438*
Fedorova, I. P., 328(249), 330(249), 407(249), *439*
Fedorova, L. S., 15(366, 368), 16(366, 368), 44(367), 120(368), 124(366), 127(366), 173(367), 174(367), *273*
Fedotova, O. Ya., 2(360), 4(361), 9(361), 126(364), 128(364), 133(364), 159(361), 160(361), 163(363), 164(363), 186(362), *273*, 488(18, 19, 20), 489(20, 26), *493*
Feit, B., 444(158), *472*
Feng, H.-P., 346(87), 428(87), *434*
Feng, H.-T., 308(88), 319(102), 328(103), 346(87), 371(89), 372(89), 374(89), 407(103), 428(87), *434*, *435*
Fick, H. J., 40(226), *269*
Fielding, G. H., 443(22), *468*
Filatova, V. A., 456(73), 463(73), 464(73), 466(73), *470*

Fincke, J. K., 12(190), 40(227), 41(190), *267, 269*

Finkelshtein, E. I., 443(36), 446(36), 448 (36), 454(36), *469*

Fischer, R. F., 447(21, 37), 448(37), *468, 469*

Fisher, J. W., 85(228), 239(228), 240(228), *269*

Fisher, N. G., 98(229), 259(229), *269*

Flavell, W., 41(230), 84(230), *269*

Fleming, W. A., 288(139), 317(139), 319 (139), 379(139), 380(139), *436*

Fleshler, E., 308(116), *435*

Florinskii, F. S., 8(405), 9(112), 30(609), 33(377, 609), 34(377), 39(377), 41 (377), 106(382), 107(382), 133(384, 405, 566), 165(380, 609), *265, 274, 275, 280, 282*

Fomin, G. V., 70(106), 205(106), *265*

Forbes, J. W., 452(52), *469*

Fox, T. G., 320(235), *439*

Frankel, M., 444(158), *472*

Frazer, A. H., 342(104, 106), 343(105, 106), 424(104, 105, 106), *435*

Freeburger, M. E., 79(628, 632), 80(628), 83(627, 632), 84(627), 225(628, 632), 230(231, 627, 632), 231(632), 232(632), 238(627), *269, 282, 283*

Freeman, J. H., 14(275), 29(166), 37(234, 275), 38(234), 40(166, 232, 235, 275), 42(233), 43(275), 49(234), 50(234), 51 (120, 233), 52(233), 64(275), 65(275), 133(234), 138(275), 143(275), 146(166, 234), 147(275), 148(234), 149(275), 150 (275), 151(166, 275), 152(275), 175 (234), 176(275), *265, 267, 269, 270*

Freeman, R. R., 100(236), *269*

French, J. C., 29(197), *268*

Frenkel, S. Ya., 41(383), 106(382), 107 (382), 133(383), 137(379, 383), 158 (383), 159(383), *274*

Fritz, C. G., 10(269, 270), 133(269, 270), 146(269, 270), 148(269, 270), 153(269, 270), *270*

Frosch, C. J., 181(271), *270*

Frost, L. W., 3(121, 122), 4(276, 277), 6 (276, 277), 8(121, 122), 14(275), 29 (166), 37(234, 275), 38(234), 40(166, 235, 275), 42(121, 122, 274), 43(121, 122, 275), 47(121, 122), 49(234), 50 (122, 234), 51(120, 122), 53(121, 122),

64(275), 65(275), 124(121, 122), 126 (121, 122), 128(121, 122, 273), 130 (121, 122, 273), 133(121, 122, 234), 134 (121, 122, 273), 135(121, 122, 273), 138 (275), 143(275), 146(166, 234), 147 (275), 148(234), 149(272, 275), 150 (275), 151(166, 275), 152(275), 174 (121, 122, 273), 175(121, 122, 234, 273, 274), 176(121, 122, 273, 275), 177(121, 122, 273, 274), 178(121, 122, 273, 274), 179(274), 180(274), 196(121, 122, 273), *265, 267, 269, 270*

Fuji, S., 308(210), 309(210), 369(210), *438*

Fujita, S., 173(278), *270*

Fujiwara, T., 97(279), 254(279), *270*

Fukami, A., 34(513, 514, 515, 516, 517), 42(518), 120(513, 514), 178(518), *278, 279*

Fukuda, A., 41(280), *270*

Fukui, K., 444(41), *469*

Fukui, M., 14(40), 28(39), 102(40), 103 (40), 104(39, 40), 105(39), 106(39), 142 (340), *263, 272*

Fukui, S., 81(335), 280(335), *272*

Furue, M., 346(109, 110), 348(109, 110), 349(109), 427(109), *435*

Furukawa, I., 308(131), 418(132), 419(131), *435*

Furukawa, J., 292(221), *438*

Furuta, A., 294(24), 352(22, 24), *432*

**G**

Gaertner, R. F., 5(282), 128(282), 146(281, 282), 147(281), 148(282), 188(282), 189(282), *270*

Gaitseva, E. A., 455(6), *468*

Gall, W. G., 10(283, 284, 286, 287), 124 (286), 128(283, 284, 287), 130(283, 284), 133(283, 284, 287), 134(285), 145(286), 146(284, 286, 287), 147(287), 148(283, 284, 287), 149(283, 284), 157(283, 284), 158(283, 284), 160(283, 284, 286), 161 (286), 162(283, 284), 163(286), 165 (283, 284, 286), 166(283, 284, 286), 169 (283, 284), *270, 271*

Gallard, J., 40(288), 124(288), 126(288), *271*

Gallard-Nechtschein, J., 40(289), *271*

Gardner, J. M., 41(610), *282*

Garrison, W. E., 331(166), *431*

Gay, F. P., 8(388), 34(290), *262, 271, 274*

Gebelein, G. G., 339(111), *435*
George, N. J., 55(291), *271*
Gerow, C. W., 10(294, 295), 29(643), *271*, *283*
Gershkokhen, S. L., 29(669), *284*
Getmanchuk, Yu. P., 444(71), 456(4, 72, 73), 462(72), 463(73), 464(70, 73), 466 (70, 73), 467(70), *468*, *470*
Gevorkyan, E. Z., 299(180), 303(180), 359 (192), 360(192), 361(180, 192), 361 (180, 192), 362(180), 363(180), *437*
Gibbs, W. E., 15(296), 21(296), 159(296), *271*, 317(112, 113), 319(112, 113), 320 (112, 113), 325(114, 115), 328(114, 115), 380(112, 113), 405(114, 115), *435*
Gilbert, E. E., 444(44), 451, *469*
Gindin, L., 308(116), *435*
Giuliani, P., 90(297), 246(297), 248(297), *271*
Goeltner, W., 455(45, 103), *469*, *471*
Göltner, W., 455(47), *469*
Goethals, E. J., 289(117), 290(117), 341 (117, 118), 422(117, 118), *435*
Goette, R. L., 322(78, 79), *434*
Götzen, F., 328(122), 329(122), 331(122), 407(122), 408(122), *435*
Goins, O. K., 15(298), 21(298), 159(298), *271*
Gole, J., 450(46), 454(46), *469*
Goltsova, R. G., 98(539), *279*
Golubeva, I. A., 84(537), 230(537), *279*
Golubeva, N. A., 132(667), *284*
Goodman, M., 299(1), 301(1), 302(1), 306 (1), 363(1), *432*
Gordon, M., 308(119, 120, 121), *435*
Gorokhovskaya, A. S., 443(48), *469*
Goto, K., 307(224), *439*
Greber, G., 491(13), *493*
Green, D. E., 6(300), 41(299), 123(300), 158(300), 162(300), 167(300), 169 (300), 171(300), *271*
Gresham, W. F., 161(301), 163(301), *271*
Grev, D. A., 294(209), 353(209), *438*
Guaita, M., 290(273, 274, 275, 277, 278, 279, 280), 311(276), 312(275, 277, 278), 330(273, 274, 275, 278, 279, 280), 369 (276), 370(275, 277, 278), 401(273, 274, 275, 277, 278, 279, 280), *440*
Gurov, A. A., 99(102, 302), 100(101, 417), 260(101, 102, 302, 417), *264*, *271*, *275*

Gusel'nikova, L. E., 346(123, 124), 347 (124), 428(123), 429(123), 430(124), *435*

**H**

Hachihama, Y., 47(596), 72(596), *281*
Haggerty, W. J., 98(125), *265*
Hagiwara, Y., 74(303), 211(303), *271*, 477 (14), 478(14), *493*
Hait, P. W., 41(304), *271*
Hall, R. W., 126(305), 128(305), 133(305), 135(305), *271*
Hallensleben, M. L., 491(13), *493*
Haller, J. R., 10(306), *271*
Hank, R., 442(49), 443(49), 447(49), *469*
Hansch, F., 55(583), *281*
Harada, S., 322(125, 126, 127, 212), 323 (212), 324(125, 126, 127, 128), 325 (150), 383(212), 384(212), 385(212), 386(212), 388(212), 389(212), 395 (126, 127), 396(126), 397(125, 126), 398(126), 402(150), *435*, *436*, *438*
Harries, C., 292(129, 130), *435*
Harris, F. W., 79(632), 83(632), 225(632), 230(632), 231(632), 232(632), *283*
Hart, R., 313(230), 314(230), 377(230), *439*
Hartmann, H., 311(247), 328(247), 330 (247), 369(247), 409(247), *439*
Hartzler, J., 96(205), *268*
Harvey, J., 98(125), *265*
Hasegawa, M., 44(656), 174(656), *283*
Hashimoto, S., 11(307), 155(307), 156(307), *271*, 308(131), 418(132), 419(131), *435*
Haward, R. N., 308(133), *436*
Hay, J. M., 292(134), 352(134), *436*
Hayano, F., 86(339), 240(339), 241(339), *272*
Hayashi, K., 41(308), *271*, 292(221), 317 (219, 220), *438*
Hays, J. T., 316(97), *434*
Hayward, E. J., 325(48), 326(48), 402(48), *433*
Heacock, J. F., 34(309), 37(309), 39(309), *271*
Heathcock, J., 29(643), *283*
Heine, E., 34(310), 37(310), *271*
Heisenberg, E., 444(56), *469*
Hendrix, W. R., 10(312, 314), 20(311, 313), 22(312, 314), 128(311, 313), 133(311, 312, 313, 314), *271*

Henglein, A., 443(50), 446(50), 448(50), 454(50), 456(50), *469*

Hergenrother, P. M., 8(690), 34(316), 40(321), 82(315, 318), 83(315, 317), 84(316, 317, 319, 320), 146(316), 233(315), 234(316, 317, 318, 320), 235(315), 236(316, 317, 318, 320), 237(317, 318, 320), *271, 272, 285*

Hermans, P. H., 9(322), 11(322), 27(322), 120(322), 121(322), *272*

Heslinga, H., 34(323), 37(323), *272*

Heuer, W., 288(260), *440*

Hickam, W. M., 29(582), 34(581, 582), 50(581, 582), 175(581, 582), 178(582), *281*

Hillman, J. J., 297(264), 298(264), *440*

Hirsch, S. S., 37(324), 157(324), *272*

Ho, T.-C., 81(691), 230(691), *285*

Hoegger, E. F., 20(325), 67(325), 133(325), 146(326), 147(326), 200(326), 202(326), *272*

Hofbauer, E. I., 2(360), 4(361), 9(361), 93(327), 126(364), 128(364), 133(364), 159(361), 160(361), 163(363), 164(363), 186(362), *272, 273*

Hoffmeir, H., 76(557), 219(557), *280*

Hoglen, J. J., 289(272), 303(272), 323(272), 324(272), 349(272), 367(272), 393(272), 394(272), 430(272), *440*

Holländer, R., 449(119), *471*

Hollis, C. E., 96(1), 97(1), 98(1), 254(1), 255(1), 256(1), 258(1), *262*

Holt, T., 308(135, 251, 252), *436, 439*

Holub, F. F., 5(282, 328), 18(330), 45(331), 47(329, 332), 128(282), 145(328, 330), 146(282, 328), 147(328), 148(282, 328), 185(331), 188(282, 329, 332), 189(282), *270, 272*

Horner, L., 444(51), *469*

Hornyak, J., 7(547), *280*

Hous, P., 317(253, 254), 319(253), 320(253), 379(253), 380(253), 381(253), *439*

Howard, E., Jr., 339(111), *435*

Howard, E. G., Jr., 87(333), 242(333), *272*

Hrivik, A., 346(201), 428(201), *438*

Huemmer, T. F., 455(31), *469*

Hughes, R. B., 53(426), 55(426), 191(426), *276*

Hung, H.-Y., 328(103), 407(103), *435*

Hunt, C. F., 41(407, 408, 409), 98(178), 133(407, 408, 409), 254(178), 257(178), *267, 275*

Hunter, L., 453(52, 53), *469*

Hunter, R. W., 473(36), *494*

Hwa, J. C. H., 288(139, 140), 308(136), 317(137, 138, 139, 140), 319(137, 139, 140), 320(137, 140), 379(138, 139), 380(137, 139, 140), 381(140), *436*

**I**

Ichikawa, K., 332(142, 144), 333(142), 334(142, 143), 355(142, 144), 411(142, 143, 144, 145), 412(142, 144, 145), 413(144, 145), *436*

Ida, N., 65(397), 92(397), 245(397), *275*

Idelson, A. L., 485(15, 16), 486(16), *493*

Ikeda, K., 73(704), 74(303, 694, 703, 706, 707), 75(694, 695, 705, 708, 709, 710), 93(713), 95(650, 711, 713, 714), 211(303, 695, 703, 704, 705, 706, 707, 709, 710, 712), 212(694, 695, 704, 705, 708, 710, 712), 213(704, 705), 250(650, 711, 713), 251(650, 713, 714), *271, 283, 285, 286*

Imai, Y., 44(334), 142(340), 173(334), 174(334), *272*

Imoto, E., 81(335), 83(336), 84(336), 230(335, 336), *272*

Imoto, M., 307(224), *439*

Ingley, F. L., 322(80), *434*

Inoue, H., 81(335), 83(336), 84(336), 230(335, 336), *272*

Irwin, R. S., 15(337), 34(338), 39(338), 133(337), 135(337), *272*

Ishida, S., 291(228), 293(226, 228), 294(226, 228), 328(141, 227), 329(141, 227), 353(228), 354(226), 407(141, 227), *436, 439*

Ishii, Y., 293(285, 286), *440*

Ishizuka, D., 322(212), 323(212), 383(212), 384(212), 385(212), 386(212), 388(212), 389(212), *438*

Ito, Y., 293(285), *440*

Itoga, M., 173(278), *270*

Ivanova, L. P., 147(6), 148(6), *262*

Ivanova, L. T., 443(141), 449(141), 458(18, 19, 20), 459(7, 18, 20), *468, 472*

Iwagaki, T., 320(243), *439*

Iwakura, Y., 44(334), 86(339), 142(340), 173(334), 174(334), 240(339), 241(339), *272*, 325(282), 326(282), 332(142, 144), 333, 334(142, 143), 335(144), 402(282),

411(142, 143, 144, 145), 412(142, 144, 145), 413(144, 145), *436*, *440*
Izawa, S., 86(339), 240(339), 241(339), *272*
Izumi, M., 92(595), 193(341), 245(595), 246(595), 248(595), *272*, *281*

**J**

Jackson, W. G., 83(342), 231(342), *272*
Jadamus, H., 83(343), 233(343), 234(343), *272*
Janssen, R., 313(230), 314(230), 377(230), *439*
Jeffreys, K. D., 126(305), 128(305), 133(305), 135(305), *271*
Joliot, F., 443(54), *469*
Jones, A. G., 325(96), 403(96), 404(96), 406(96), *434*
Jones, J. F., 317(148), 318(148), 379(147, 148), *436*
Jones, J. I., 3(351), 40(349), 41(350), 42(351), 77(350), 84(350), 96(350), 124(351), 133(351), *273*
Jones, J. W., 41(352), *273*
Joshi, R. M., 443(55), *469*
June, R. K., 443(99), 447(99), 448(99), 455(99), *471*
Jurgeleit, W., 444(51, 56), *469*

**K**

Kachi, H., 5(353), *273*
Kagiya, T., 444(41), *469*
Kaiser, E., 455(120), 466(120), *471*
Kallistova, E. V., 8(405), 133(405), *275*
Kamai, G., 337(149), 338, 419(149), *436*
Kamao, H., 296(25), 297(25), 298(25), 355(25), 356(25), *432*
Kamenetskaya, N. M., 308(162), *436*
Kapur, S. L., 336(3), *432*
Kardash, I. E., 2(36), 18(354), 126(354), *263*, *273*
Karle, D. W., 473(36), *494*
Kasat, R. B., 345(81), 426(81), *434*
Kass, R. E., 98(178), 133(410), 145(411), 146(411), 148(411), 254(178), 257(178), *267*, *275*
Katayama, M., 322(127), 324(127, 128), 325(150), 395(127), 402(150), *435*, *436*
Katon, J. E., 71(683), 205(683), *284*

Katsarova, R. D., 30(370), 127(370), 147(370), 163(370)
Kawai, W., 290(152), 308(151), 311(152), 312(152), 314(151), 317(153), 328(152), 330(152), 370(152), 381(153), 407(152), *436*
Kawazura, H., 292(208), *438*
Kekish, G. T., 447(57), *469*
Kerkmeyer, J. L., 34(316), 84(316), 146(316), 234(316), 236(316), *272*
Kern, W., 293(199), 294(199), 353(199), *438*, 442(58, 137), 443(30, 58, 113, 134), 444(58, 137, 138), 445(58), 446(137), 447(122), 448(137), 449(58, 59, 61, 62, 63, 117, 118, 119, 121, 123, 124, 126, 127, 128, 129, 130, 131, 132), 450(58), 452(58, 121), 453(137, 138), 455(60, 114, 115, 120), 456(138), 460(138), 462(58), 463(58, 134), 465(138), 466(120), *468*, *469*, *470*, *471*, *472*
Kerr, C. M. L., 292(134), 352(134), *436*
Kesse, I., 4(276, 277), *270*
Khomutov, A. M., 315(250), *439*
Khomutova, N. M., 315(250), *439*
Kieffer, H. E., 61(185), 62(185), 204(185), *267*
Kikukawa, K., 308(210), 309(210), 369(210), *438*
Kimura, S., 485(17), *493*
King, C., 332(154), 333(154), 335, 411(154), 412(154), 413(154), 414(154), *436*
Kinosaki, T., 141(355), 142(355), 150(355), 151(355), *273*
Kitahama, Y., 353(155), 354(156), *436*
Klebe, J. F., 13(111), *265*
Kline, D. E., 29(107), 37(107), 134(107), *265*
Klüpfel, K., 444(51), *469*
Kluiber, R. W., 56(356), 198(356), 199(356), *273*
Knyazeva, T. S., 133(384), *274*
Kobayashi, A., 92(399), *275*
Kobayashi, H., 240(357, 358), *273*, 353(155), 354(156), *436*
Kocharyan, N. M., 300(157), 358(157), 359(157), 360(157), *436*
Koehler, A. M., 29(359), *273*
Kolesnikov, G. S., 2(360), 4(361), 9(361), 126(364), 128(364), 133(364), 159(361), 160(361), 163(363), 164(363), 186(362), *273*, 290(158), 321(159), 349(159), 429

Kolesnikov, G. S. (cont.)
(159), *436*, 488(18, 19, 20), 489(20, 26), *493*
Kol'k, A. R., 465(64), *470*
Kol'stov, A. I., 456(73), 463(73), 464(73), 466(73), *470*
Koltsov, A. I., 456(4), *468*
Konishi, T., 12(590), 124(590), *281*
Konkin, A. A., 465(64), *470*
Konopczynski, A., 492(7), *493*
Konsaka, T., 182(365), 183(365), 184(365), *273*
Konya, S., 123(715), 137(715), 139(715), 140(715), *286*
Koral, J. N., 295(160, 161), 352(160), 353(160), 354(160, 161), *436*, 459(65, 66, 67, 69), 460(68), *470*
Kornienko, A. A., 325(223), 402(223), 403(223), 404(223), 405(223), *439*
Korshak, V. V., 6(371, 374, 375), 15(366), 16(366, 368), 30(370, 374), 34(369), 41(369), 42(369), 44(367), 84(369), 120(368), 125(366), 127(366, 370, 371, 372), 129(371, 372), 130(371, 372), 131(374, 375), 132(374, 375, 661, 662, 664, 665, 666, 667, 673), 147(370), 157(373), 163(370), 164(375), 165(375), 169(663), 170(663), 171(663), 173(367), 174(367), *273, 274, 284*
Korshak, V. V., 474(2, 21), 475(2, 21), 481(22), *493*
Korzhavin, L. N., 106(382), 107(382), 137(379), *274*
Kostikov, R. R., 299(7), 363(7), *432*
Koton, M. M., 2(4), 7(4), 8(224, 225), 9(112), 28(567), 30(609), 33(377, 609), 34(377), 39(377), 41(377, 383, 565), 98(376), 105(381), 106(225, 381, 382, 394), 107(225, 382, 394, 395), 109(224), 110(224), 111(395), 133(383, 384, 566), 134(602), 136(5), 137(5, 378, 379, 383), 145(224), 146(224), 147(6, 224), 148(6, 224), 158(7, 224, 383), 159(3, 383), 163(224), 164(224), 165(5, 380, 609), 166(8), 196(4), 197(4), *262, 265, 269, 274, 280, 282*, 308(162), 317(57), 319(57), 320(57), 329(241), 380(57), 407(241), 408(241), 409(241), *433, 436, 439*, 442(5), 444(5, 71), 446(5), 454(5), 456(4, 72, 73), 457(5), 462(72), 463(5, 73), 464(70, 73), 466(5, 70, 73), 467(70), *468*

Kotov, B. V., 2(36), *263*
Kovacs, J., 447(122), 449(123, 124), *471*
Kovaleva, K. A., 442(5), 444(5), 446(5), 454(5), 457(5), 463(5), 466(5), *468*
Kozyrev, V. G., 455(148, 149), *472*
Krasnova, N. M., 3(720), 4(720), *286*
Krentsel, B. A., 459(3), *468*
Kress, B. H., 96(385), 97(385), 98(385), 254(385), 255(385), 257(385), *274*
Kreuz, J. A., 8(388), 13(28), 20(386, 387), 23(387), 67(386, 387), 133(28), 200(386), 201(386), 202(386), 203(386, 387), *263, 274*
Krieger, R. B., Jr., 29(389), *274*
Kronganz, E. S., 481(22), *493*
Krongauz, E. S., 34(369), 41(369), 42(369), 84(369), *273*
Kropa, E. L., 98(390), 254(390), 257(390), *274*
Kubota, T., 211(712), 212(712), *286*
Kuckertz, H., 122(391), 137(391), *274*
Kudryavtsev, G. I., 43(393), 135(651), 178(393), 210(392), *274, 283*, 478(23), 479(23), *493*
Kudryavtsev, V. V., 8(224, 225), 105(381), 106(225, 381, 382, 394), 107(225, 382, 394, 395), 109(224), 110(224), 111(395), 145(224), 146(224), 147(224), 148(224), 158(224), 163(224), 164(224), *269, 274*
Kukhtin, V. A., 337(149), 338, 419(149), *436*
Kuleva, M. M., 133(384), *274*
Kurihara, M., 65(397), 72(697, 698), 73(396, 398, 400, 401, 699, 701, 702, 704), 74(303, 694, 703, 706, 707), 75(401, 694, 695, 705, 708, 709, 710), 91(700), 92(397, 399, 400, 403), 93(402, 458, 713), 95(402, 404, 458, 650, 696, 711, 713, 714), 209(697, 698), 210(396, 398, 400), 211(303, 396, 398, 400, 401, 695, 699, 701, 702, 703, 704, 705, 706, 707, 709, 710, 712), 212(396, 398, 400, 401, 694, 695, 704, 705, 708, 710, 712), 213(704, 705), 245(397, 700), 249(458), 250(402, 458, 650, 696, 711, 713), 251(402, 458, 650, 696, 713, 714), 252(404, 458, 696), 253(458, 696), *271, 275, 277, 283, 285, 286*, 477(14), 478(14, 37), 492(24), *493, 494*
Kuryaev, B. S., 458(19), *468*

Kuvshinskii, E. V., 9(112), 33(603), 34(603), 38(603), 134(603), 164(603), *265*, *282*
Kuzina, V. V., 70(106), 205(106), *265*
Kuznetsova, T. S., 480(31), 481(31), *494*

## L

Laius, L. A., 8(405), 133(384, 405), *274, 275*
Lateltin, P., 90(406), 96(406), 248(406), *275*
Lavin, E., 41(407, 408, 409), 52(412), 96(179, 180), 97(179), 98(179), 133(407, 408, 409, 410), 145(411), 146(411), 147(591), 148(411), 185(412), 254(179, 180), 255(179, 180), 257(179, 180), *267*, *275, 281*
Lawrence, R. V., 47(586), 181(586), 182(586), 187(586), *281*
Layus, L. A., 33(377), 34(377), 39(377), 41(377), *274*
Lee, H., 29(414), 40(414), 41(414), 42(413), 50(413), 51(413), 52(413), *275*
Leibler, K., 492(7), *493*
Leutner, R., 291(281), *440*
Levine, H. H., 8(690), 34(316), 40(321), 82(318), 83(317), 84(316, 317, 319, 320), 146(316), 234(316, 317, 318, 320), 236(316, 317, 318, 320), 237(317, 318, 320), *272, 285*
Liepins, R., 335(163), 336(163), 415(163), 416(163), *436*
Lindsey, W. B., 41(415, 416), 133(415, 416), *275*
Liogon'kii, B. I., 63(100), 99(102, 302), 100(101, 417), 117(103), 118(103), 204(100), 260(101, 102, 302, 417), *264, 265, 271, 275*
Lis, A. L., 307(240), 368(240), *439*
Litt, M. H., 485(15, 16), 486(16), *493*
Löflund, I., 449(61, 125, 126, 127), *471*
Lohnizen, O., 257(585), 258(585), *281*
Lokshin, B. V., 132(665, 666), *284*
Loncrini, D. F., 47(419, 420), 53(419, 420, 422, 426), 55(426), 56(422), 58(422), 60(422), 96(418), 188(423), 189(423), 190(422, 423, 424), 191(421, 422, 423, 424, 426), 192(422, 423), 193(422, 423), 194(422, 423), 195(422, 423, 424, 425), *275, 276*
Lopez, A. H., 12(427), 124(427), *276*
Loughran, G. A., 490(25), 491(25), *493*

Lucas, H. R., 29(197), *268*, 295(160), 352(160), 353(160), 354(160), *436*
Luck, R. M., 51(652, 653), 177(652, 653), 178(652, 653), *283*
Luessi, H., 289(164), 310(164), *436*
Lunin, A. F., 84(537), 230(537), *279*
Lynch, E. R., 125(184), *267*
Lyubova, T. A., 135(651), *283*

## M

Mabuchi, K., 444(152), 453(152), *472*
Machus, F. F., 84(537), 230(537), *279*
McKeown, J. J., 10(428), 133(428), *276*
Madorskaya, L. Ya., 456(4, 72, 73), 460(72), 463(73), 464(73), 466(73), *468, 470*
Mainen, E., *282*
Mainen, E. L., 79(632), 83(429, 629, 630, 631, 632), 225(632), 230(632), 231(429, 629, 631, 632), 232(429, 630, 631, 632) 235(631), 238(631), *276, 282, 283*
Mair, A. D., 37(182, 430, 431), 38(182), 49(182), 50(182), 133(182), 146(182), 148(182), 175(182), *267, 276*
Mallet, P., 90(406), 96(406), 248(406), *275*
Mallock, R. S., 29(649), *283*
Malluck, R. S., 29(643), *283*
Manaka, K., 34(432), 35(432), 36(432), *276*
Manecke, G., 71(433, 434), 72(434), 205(433), 207(434), 208(434), 209(434), *276*
Marantz, L. B., 473(36), *494*
Maric, B., 478(34), *494*
Mark, H. F., 456(74), *470*
Markhart, A. H., 41(407, 408, 409), 52(412), 90(592), 98(178), 133(407, 408, 409, 410), 145(411), 146(411), 147(591), 148(411), 185(412), 244(592), 245(592), 254(178), 257(178), *267, 275, 281*
Markov, Yu. Ya., 84(537), 230(537), *279*
Marks, B. M., 444(33), 462(34, 35), 463(35), 466(35), *468*
Marschik, I., 29(582), 34(582), 50(582), 175(582), 178(582), *281*
Marvel, C. S., 41(436, 440, 441), 78(442, 521, 522, 685), 79(442, 521, 522, 523), 83(44, 195, 343), 84(436, 437, 438, 440, 441), 87(187), 222(685), 223(685), 224(685), 225(442, 521, 686), 226(442, 521, 686), 227(442, 523), 228(442, 521, 522), 229(442, 521, 522), 230(44),

Marvel, C. S. (cont).
231(195), 233(44, 343), 234(343), 235 (44), 237(44), 243(187, 435, 439), *263, 267, 268, 272, 276, 279, 284*
Marvel, C. S., 290(165), 331(166), 349(167), 429(167), *437*
Marx, M., 311(247), 328(247), 330(247), 369(247), 409(247), *439*
Mass, K. A., 133(181), *267*
Matray, A., 10(443), *276*
Matsoyan, S. G., 290(169), 299(168, 177, 180), 300(157, 168, 174, 175, 177, 185, 190), 301(185), 302(170, 171), 303(168, 177, 180, 183), 306(168, 177, 193), 322(183), 325(182), 330(181), 337(183), 339(181, 183), 340(188), 341(189), 358(157, 168, 174, 177, 181, 190, 191), 359(157, 168, 176, 177, 178, 179, 191, 192), 360(157, 168, 173, 175, 176, 177, 178, 187, 191, 192), 361(168, 175, 176, 179, 180, 191, 192), 362(175, 180, 187, 191, 192), 363(170, 171, 172, 180), 364(170, 171, 184, 185), 365(171, 185), 366(172, 185), 367(170, 172, 183), 369(181), 392(186), 393(183, 186), 402(183), 403(182), 404(182), 405(182, 186), 406(186), 407(181), 418(183), 421(181, 189), 422(188, 189), *436, 437*
Matsumoto, S., 97(279), 254(279), *270*
Matsumura, S., 92(595), 193(341), 245(595), 246(595), 248(595), *272, 281*
Matsuo, T., 352(2), *432*
Matveeva, N. G., 69(104, 105), 70(104, 105), 205(104, 105), *265*
Matvelashvili, G. S., 488(18, 19, 20), 489(20, 26), *493*
Maxim, I., 491(27), *494*
Maxwell, V. W., 109(444), 110(444), 179(444), *276*
Measday, D. F., 29(359), *273*
Medvedev, S. S., 308(116), *435*
Medvedeva, L. I., 315(194), 370(194), 376(194), *438*
Meiya, N. V., 298(216), 355(216), *438*
Mercier, J., 290(195), 317(196), 318(196, 197), *438*
Merten, R., 24(452), 25(452), 48(452), 55(453), 189(453), *276*
Meyer, J. F., 55(445, 446), *276*
Meyers, R. A., 24(447), 25(447), 128(447), *276*

Meyersen, K., 293(198, 199), 294(199), 343(200), 353(199), 425(200), *438*, 449 (128, 129, 130), *471*
Michael, K. W., 350(266), 429(266), *440*
Mielke, K. H., 55(453), 189(453), *276*
Mikulasova, D., 346(201), 428(201), *438*
Milek, J. T., 41(448), *276*
Miller, H. C., 443(75), 446(75), 455(75), *470*
Miller, L., 288(139, 140), 317(139, 140), 319(139, 140), 320(140), 379(139), 380(139, 140), 381(140), *436*
Miller, W. L., 287(76, 77), 300(205), 317(202), 320(90, 202, 204), 322(76, 77), 332(49, 203), 383(76, 77), 384(76, 77), 385(76, 77), 411(49, 50, 203), *433, 434, 438*
Miller, W. R., 97(449), 258(449), *276*
Minkova, R. M., 308(6), 309(6), 315(8), 371(6), 376(6), 378(8), *432*
Minsk, L., 317(206), *438*
Minoura, Y., 299(205), *438*
Mita, T., 41(450), *276*
Mitake, T., 299(207), 300(207), 306(207), *438*
Mitoh, M., 299(205), *438*
Miura, M., 296(26, 27), 352(26, 27), *432*
Miyamichi, K., 325(150), 402(150), *436*
Moldovskii, B. L., 107(395), 111(395), *274*
Moore, S. T., 322(244), 325(244), 340(244), 384(244), 385(244), 393(244), 394(244), 402(244), 403(244), 405(244), 421(244), *439*
Morgan, R. A., 17, 119(633), *283*
Mori, K., 292(221), *438*
Morimoto, G., 292(208), *438*
Morishima, A., 97(279), 254(279), *270*
Moriwaki, T., 12(519), 133(519), 153(519), *279*
Morlyan, N. M., 300(175), 360(175), 361(175), 362(175), *437*
Morrill, D. H., 29(359), *273*
Mortillaro, L., 96(451), *276*
Moskvina, E. M., 158(7), 159(3), *262*
Moureau, C., 444(76, 77, 78), *470*
Moyer, W. W., Jr., 294(209), 353(209), *438*
Mozgova, K. K., 15(366, 368), 16(366, 368), 44(367), 120(368), 124(366), 127(366), 173(367), 174(367), *273*
Müller, E., 449(131, 132), *471*
Mueller, G., 24(452), 25(452), 48(452), 55(453), 189(453), *276*

Mukamal, H., 89(529, 530, 531, 532), *279*
Mulyar, P. A., 308(162), *436*
Murahashi, S., 308(210), 309(210), 346(109, 110), 348(109, 110), 349(109), 350(268), 369(210), 427(109), *435, 438, 440*
Murray, J. T., 317(112, 113), 319(112, 113), 320(112, 113), 380(112, 113), *435*
Mushegyan, A. V., 325(182), 403(182), 404(182), 405(182), *437*


N

Naarman, H., 99(455), 100(454, 456, 457), 260(455, 456), 261(454, 455, 457), *276, 277*
Nagai, Y., 459(80), 463(79), *470*
Nagasuna, Y., 11(307), 155(307), 156(307), *271*
Nakajima, T., 459(80), 463(79), *470*
Nakanishi, R., 75(695, 710), 93(458, 713), 95(458), 211(695, 710, 712), 212(695, 710, 712), 249(458), 250(458, 713), 251(458, 713), 252(458), 253(458), *277, 285, 286*
Nametkin, N. S., 346(123, 124), 347(124), 428(123), 429(123, 211, 271), 430(124), *435, 438, 440*
Naylor, M. A., 161(301), 163(301), *271*
Nazimova, N., 478(23), *493*
Negi, Y., 322(212), 323(212), 383(212), 384(212), 385(212), 386(212), 388(212), 389(212), *438*
Neher, H. T., 455(81), *470*
Nesterova, E. I., 93(327), 186(362), *272, 273*
Neville, K., 29(414), 40(414), 41(414), 42(413), 50(413), 51(413), 52(413), *275*
Nifantev, E. E., 98(539), *279*
Nikolaev, A. F., 298(216), 355(216), *438*
Nishizaki, S., 12(519), 14(509, 510), 34(513, 514, 515, 516, 517), 42(518), 120(509, 513, 514), 124(509), 133(519), 134(512), 153(519), 168(511), 169(510), 170(510), 171(510), 178(518), *278, 279*
Nisova, S. A., 84(537), 230(537), *279*
Nixon, A. C., 443(151), 445(151), 453(53), *469, 472*
Noma, K., 308(217), *438*
Nozakura, S., 308(210), 309(210), 346(109, 110), 348(109, 110), 349(109), 350(268), 369(210), 427(109), *435, 438, 440*

Nukada, K., 73(398), 210(398), 211(398), 212(398), *275*
Nyrkov, E. S., 29(520), 89(520), *279*


O

Obora, M., 47(596), 72(596), *281*
Obreshkov, A., 299(231), 302(231, 232), 367(231), *439*
Ochynski, W. F., 3(351), 42(350), 124(351), 133(351), *273*
Oda, R., 108(641), 109(641), 115(641), 116(641), 182(365), 183(365), 184(365), *273, 283*, 332(222), 333(222), 340(269), 346(269), 413(222), 421(269), *438, 440*, 486(29), *494*
Odnoralova, V. N., 210(392), *274*, 478(23), 479(23), *493*
Ogata, Y., 308(218), *438*
Ohama, H., 353(155), 354(156), *436*
Oiwa, M., 308(218), *438*
Okada, M., 78(442, 521, 522), 79(442, 521, 522, 523), 255(442, 521, 686), 226(442, 521, 686), 227(442, 523), 228(442, 521, 522), 229(442, 521, 522), *276, 279, 284*, 317(219, 220), *438*
Okamura, S., 292(221), 317(219, 220), *438*,
Okano, M., 486(29), *494*
Okladnikova, Z. A., 444(142, 143), 447(142), 454(142), *472*
Oku, A., 332(222), 333(222), 413(222), *438*
Olivier, K. L., 6(524), 29(622), 30(622), 33(622), 34(622), 41(621, 622), 124(621), 126(621), 128(621), 130(621), 133(621), 135(621), 164(524), *279, 282*
Omarov, O. Yu., 84(537), 230(537), *279*
O'Neal, H. R., 296(54, 55), 352(54, 55), *433*
O'Neill, W. P., 342(106), 343(106), 424(106), 425(105), *435*
Oromi, J. C., 128(526, 527), 148(525), *279*
Orth, H., 98(528), *279*
Os'minskaya, A. T., 317(57), 319(57), 320(57), 380(57), *433*
Ostberg, B. E., 308(51), 311(51, 91), 312(51, 91), *433, 434*
Ostroverkhov, V. G., 325(223), 402(223), 403(223), 404(223), 405(223), *439*
Ota, T., 459(89), *470*
Otsu, T., 307(224), *439*
Overbeger, C. G., 289(229), 290(229, 236, 237), 291(228), 293(226, 228), 294(226,

Overbeger, C. G. (cont.)
228), 328(227), 329(227), 341(225, 229, 236, 237), 342(229), 345(229, 269), 353(228), 354(226), 407(227), 421(237), 423(225, 229, 236), 424(229), *439*, 442 (90), *470*
Overberger, C. G., 89(529, 530, 531, 532), *279*
Ozaki, S., 89(530, 531, 532), *279*

**P**

Packham, D. I., 71(533, 534, 535), 72(533, 535), 206(533, 534, 535), *279*
Panayotov, I. M., 299(231), 302(231, 232), 367(231), *439*
Pannell, C. E., 443(91), 447(91), 448(91), *471*
Pao, H.-L., 371(89), 372(89), 374(89), *434*
Paris, A., 41(536), *279*
Parish, D. J., 41(202, 203), *268*
Passmann, W., 442(133), 444(133), 446 (133), 450(133), 453(133), 454(133), 458(133), 463(133), 464(133), *472*
Patalakh, I. I., 84(537), 230(537), *279*
Paushkin, Ya. M., 84(537), 230(537), *279*
Pavlov, A. I., 6(371), 127(371, 372), 129 (371, 372), 130(371, 372), 157(373), *273*
Pavlova, S. A., 6(374), 30(374), 131(374), 132(374, 667), *274*, *284*
Pecher-Reboul, A., 40(289), *271*
Penczek, P., 116(538), *279*
Petrov, K. A., 98(539), *279*
Pezzaglia, P., 443(100), *471*
Pikalova, V. N., 300(35, 157), 358(35, 157), 359(35, 157), 360(35, 157), *433*, *436*
Pilyaeva, V. F., 29(669), *284*
Plonka, Z. J., 170(540), *279*
Plotkina, S. A., 308(6, 9, 10), 309(6, 10), 315(8), 371(6, 10), 372(10), 376(6, 9), 378(8), *432*
Pogorelova, T. G., 157(373), *273*
Pogosyan, G. M., 300(185), 301(185), 303 (183), 323(183), 325(182), 330(181), 337(183), 339(181, 183), 358(181), 364 (184, 185), 365(185), 366(185), 367 (183), 369(181), 392(186), 393(183, 186), 402(183), 403(182), 404(182), 405(182, 186), 406(186), 407(181), 418 (183), 421(181), *437*

Pokrovskii, E. I., 8(224, 225), 28(567), 106 (225), 107(225), 109(224), 110(224), 145(224), 146(224), 147(224), 148 (224), 158(224), 163(224), 164(224), *269*, *280*, 456(4, 72, 73), 462(72), 463 (73), 464 (73), 466(73), *468*, *470*
Polak, L. S., 84(537), 230(537), *279*, 346(123, 124), 347(124), 428(123), 429(123), 430 (124), *435*
Politi, R. E., 29(389), *274*
Pomerantseva, K. P., 135(716), *286*
Popescu, I., 473(35), 474(35), *494*
Popov, A. G., 317(57), 319(57), 320(47), 380(57), *433*
Popov, N., 82(598, 599), 235(598, 599), 237(598, 599), *281*
Pravednikov, A. N., 2(36), 18(354), 126 (354), *263*, *273*
Preston, J., 41(544), 138(541, 542, 543, 545), 139(541, 542, 543), 140(541, 542, 543), 141(546), 142(541, 542, 543), 143 (546), 144(546), 149(545), 152(546), *279*, *280*
Prévé, J., 90(193), 233(192), 234(192), 236 (192), 245(193), 246(193), 247(193), *268*
Price, C. C., 288(86), 325(86), 326, 327(86), 328(86), 400(86), 401(86), *434*
Price, J. A., 322(244), 325(233, 244), 340 (244), 384(244), 385(244), 393(244), 394(244), 402(244), 403(244), 405(244), 421(244), *439*
Prigozhina, M. P., 474(2, 21), 475(2, 21), *493*
Prince, M. I., 7(547), *280*
Priola, A., 290(273, 274, 275, 277, 278, 279, 280), 311(276), 312(275, 277, 278), 330 (273, 274, 275, 278, 279, 280), 369 (276), 370(275, 277, 278), 407(273, 274, 275, 278, 279, 280), *440*
Prokhorova, L. K., 166(8), *262*
Propp, L. N., 299(7), 363(7), *432*
Pruckmayr, G., 12(548), *280*
Pryde, E. H., 97(449), 258(449), *276*

**R**

Rabilloud, G., 9(551), 22(551), 32(551), 92(552), 95(549, 550, 551), 127(551), 244(552), 245(552), 246(552), 247(552), 249(549, 550, 551), 251(549, 550, 551), 252(549, 550, 551), *280*

Rackley, F. A., 3(351), 42(351), 71(533, 534, 535), 72(533, 535), 124(351), 133 (351), 206(533, 534, 535), *273*, *279*
Rafikov, S. R., 116(553), *280*
Ralea, R., 491(27), *495*
Ramadier, M. F., 304(85), 357(85), *437*
Ramp, F. L., 344(234), *439*
Ray, J. D., 41(619), *282*
Ray, T. W., 98(167), *267*
Raymond, M. A., 288(82), *434*
Razvadovskii, E. F., 99(102), 260(102), *264*
Read, J., 254(554), *280*
Redman, E. G., 184(555), *280*
Redtenbacher, J., 442, *471*
Reese, E., 76(557), 219(557), *280*
Reichel, B., 7(671), 9(558), 31(558, 671), 87(187), 114(671), 115(558,) 243(187), *267*, *280*, *284*
Reimschuessel, H. K., 26(559, 601), 28 (559), 108(559, 601), *280*, *282*
Reinhard, K. H., 455(93), *471*
Reinhardt, H. F., 97(560), 254(560), 258 (560), 259(560), *280*
Reinmöller, M., 320(235), *439*
Renshaw, R. R., 11, *265*
Reynolds, R. J. W., 13(561), 133(561), *280*
Ried, W., 489(28), *494*
Ringsdorf, H., 289(229), 290(229, 236, 237), 291(228), 293(228), 294(228), 341(229, 236, 237), 342(229), 345(229, 269), 353 (228), 421(237), 423(229, 236), 424 (229), *439*
Rink, K. H., 443(8), 444(94, 95, 97, 98), 448(96), *468*, *471*
Robinson, I. M., 3(213), 11(124, 212, 213, 214, 216), 27(215), 121(124, 212, 213, 214, 216), 122(214, 216), 123(212, 213, 215), *265*, *268*
Rodia, J. S., 2(562), 148(562), 151(563), 152(563), *280*
Roe, R.-J., 308(120, 121), *435*
Roehm, W., 76(557), 219(557), *280*
Rogers, F. E., 32(564), 38(564), 130(564), 131(564), 147(564), 160(564), 161(564), 162(564), 164(564), 166(564), *280*
Rogovin, Z. A., 465(64), *470*
Rombrecht, H.-M., 55(583), *281*
Roof, J. G., 443(22), *468*
Roovers, J., 314(238), 377(238), 378(238), *439*
Rose, J. S., 24(223), 25(223), 152(223), *269*

Rostovskii, E. N., 299(11), 300(11), 307 (240), 308(239), 358(11), 359(11), 362 (11), 363(11), 368(240), 369(11, 239), 371(11), 377(11), *432*, *439*
Rothrock, H. S., 443(75), 446(75), 455(75), *470*
Rozhkov, V. S., 4(717), *286*
Rubinshtein, E. I., 107(395), 111(395), *274*
Rudakov, A. P., 9(112), 28(567), 30(609), 33(377, 609), 34(377), 39(377), 41(377, 383, 565), 106(382, 394), 107(382, 394, 395), 111(395), 133(383, 384, 566), 134 (602), 137(378, 379, 383), 158(383), 159(3, 383), 165(380, 609), *262*, *265*, *274*, *280*, *282*
Rudakov, V. V., 105(381), 106(381), *274*
Rudkovskaya, G. D., 328(255, 256), 329 (241, 255, 256), 407(241, 256), 408 (241, 255, 256), 409(241, 256), *439*, *440*
Rüden, E., 451(2), 460(2), *468*
Ruffing, C. R., 14(275), 29(166), 37(275), 40(166, 235, 275), 43(275), 51(120), 64(275), 65(275), 138(275), 143(275), 146(166), 147(275), 149(275), 150(275), 151(166, 275), 152(275), 176(275), *265*, *267*, *269*, *270*
Rusanov, A. L., 30(370), 127(370), 146 (370), 163(370), *273*
Russo, M., 41(568, 571), 52(569), 83(570), 231(570), 233(570), 234(570), 236(570), 237(570), *280*
Ruzhentseva, G. A., 3(720), 4(717, 718, 720), *286*
Ryder, E. E., Jr., 443(99, 100), 447(99), 448(99), 455(99), *471*

## S

Saakyan, A. A., 300(185), 301(185), 306 (193), 340(188), 341(189), 360(187), 322(187), 364(184, 185), 365(185), 366 (185), 421(189), 422(188, 189), *437*
Sackis, J. J., 322(242), 325(242), 383(242), *439*
Saegusa, T., 292(221), *438*
Saga, M., 47(572, 596), 72(596), *280*, *281*
Saini, G., 290(280), 330(280), 407(280), *440*
Saito, H., 73(398), 210(398), 211(398), 212 (398), *275*
Sakaguchi, Y., 320(243), *439*
Sakurada, I., 308(217), 320(243), *438*, *439*

Salle, R., 40(288), 73(576, 577), 74(576, 577), 124(288), 126(288), 138(573), 173 (574, 575), 174(575), 179(575), 210 (576, 577), *271, 280, 281*

Sambeth, J., 48(578), 185(578), *281*

Samoilova, M. Ya., 455(144, 145), *472*

Santangelo, J. G., 84(579), 133(579), *281*

Sato, S., 340(269), 346(269), 421(269), *440*

Saunders, F. C., 123(580), *281*

Sauter, J. O., 52(412), 185(412), *275*

Savitskaya, M. N., 317(57), 319(57), 320 (57), 380(57), *433*

Scala, L. C., 29(582), 34(581, 582), 50(581, 582), 175(581, 582), 178(582), *281*

Scherlin, S. M., 443(101), 450, *471*

Schilling, H., 447(102), 448(102), *471*

Schlack, P., 455(45, 47, 103), *469, 471*

Schmidt, K., 55(583), *281*

Schmidt, P. G., 41(584), *281*

Schnabél, W., 443(50), 446(50), 448(50), 454(50), 456(50), *469*

Schnell, H., 75(108), 76(557), 77(115), 216 (115), 218(108, 115), 219(108, 115, 557), 220(115), 221(115), *265, 280*

Schopov, I., 299(231), 302(231, 232), 367 (231), *439*

Schors, A., 257(585), 258(585), *281*

Schröder, G., 328(122), 329(122), 407(122), 408(122), *435*

Schroeder, W., 83(342), 231(342), *272*

Schuller, W., 47(586), 181(586), 182(586), 187(586), *281*

Schuller, W. H., 322(244), 325(244), 340 (244), 384(244), 385(244), 393(244), 394(244), 402(244), 403(244), 405(244), 421(244), *439*

Schulz, H., 98(587), *281*, 444(135), 451(135), *472*

Schulz, R. C., 41(588), *281*, 290(246), 293 (199), 294(199), 311(247), 328(247, 248), 330(247, 248), 331(248), 353(199), 369 (247), 410(247, 248), 411(248), *438, 439*, 441 (109, 110, 112), 442(58, 104, 106, 107, 112, 133, 137), 443(30, 50, 58, 104, 106, 113, 134), 444(58, 104, 133, 137, 138), 445(58, 104, 107), 446(50, 133, 137), 447(107, 122), 448(50, 104, 106, 107, 137), 449(58, 59, 61, 62, 63, 105, 106, 107, 108, 111, 112, 116, 117, 118, 119, 121, 123, 124, 125, 126, 127, 128,

129, 130, 131, 132), 450(58, 104, 106, 107, 112, 133), 451(106), 452(58, 106, 107, 112, 121), 453(112, 133, 136, 137), 454 (50, 133), 455(60, 112, 114, 115, 120), 456(50, 112, 138), 458(133), 460 (138), 462(58), 463(58, 133, 134), 464 (133), 465(112, 138), 466(120), *468, 469, 470, 471, 472*

Schweitzer, F. E., 29(649), *283*

Schweitzer, O., 444(97, 98), 449(62, 63), *470, 471*

Seddon, J. D., 13(561), 133(561), *280*

Sekiguchi, H., 5(353), 12(590), 26(589), 124 (590), 128(589), *273, 281*

Seno, T., 325(150), 402(150), *436*

Serebrennikova, E. V., 444(142), 447(142), 454(142), *472*

Serlin, I., 90(592), 147(591), 244(592), 245 (592), *281*

Shablygin, M. V., 210(392), *274, 478(23), 479(23), 493*

Shamraev, G. M., 63(100), 204(100), *264*

Shatskaya, N. A., 16(368), 120(368), *273*

Shcherba, N. S., 29(669), *284*

Shcherbina, F. F., 328(249), 330(249), 407 (249), *439*

Sheffer, H. E., 55(445, 446, 593), *276, 281*

Shelgaeva, V. G., 126(364), 128(364), 133 (364), 163(363), 164(363), *273*

Shelton, C. F., Jr., 41(594), *281*

Shen, M. C., 37(430, 431), *276*

Sherle, A. I., 70(106), 205(106), *265*

Shilov, E. A., 320(284), 321(284), 379(284), *440*

Shimazaki, N., 444(152), 453(152), *472*

Shimizu, M., 444(152), 453(152), *472*

Shimizu, T., 444(41), *469*

Shimojo, S., 486(29), *494*

Shishkina, M. V., 459(3), *468*

Shiura, K., 47(572), 92(640), 245(640), 248 (640), *280, 281, 283*

Shokal, E. C., 444(139, 140), *472*

Shono, T., 47(572, 596), 72(596), 92(595, 640), 182(365), 183(365), 184(365), 245 (595, 640), 246(595), 248(595, 640), *273, 280, 281, 283*, 332(222), 333(222), 413(222), *438*

Shopov, I., 82(598, 599), 235(598, 599), 237 (598, 599), *281*, 483(30), *494*

Shostakovskii, M. F., 315(250), *439*, 443 (141), 444(142, 143), 447(142), 449(141),

454 (142), 455(144, 145, 146, 147, 148, 149), *472*, 480(31), 481(31), *494*

Shulman, G. P., 7(600), *281*

Shultz, A. R., 308(245), *439*

Shunzoku, T., 93(713), 95(713), 250(713), 251(713), *286*

Sibilia, J. P., 26(601), 108(601), *282*

Sidorovich, A. V., 33(603), 34(603), 38 (603), 134(602, 603), 164(603), *282*

Sillion, B., 9(551), 22(551), 32(551), 73(576, 577), 74(576, 577), 80(116, 117), 82 (204), 84(191, 607), 90(193, 604, 605), 92(552), 95(549, 550, 551), 96(604, 607), 127(551), 138(573), 173(574, 575), 174(575), 179(575), 210(576, 577), 227 (116, 117), 228(116, 117), 233(191, 192), 234(191, 192), 235(204), 236(192), 244(552, 604), 245(193, 552, 604), 246 (193, 552, 604, 605, 606, 608), 247(193, 552, 604, 605, 606, 608), 248(604, 605, 606, 608), 249(549, 550, 551), 251(549, 550, 551), 252(549, 550, 551), *265, 268, 280, 281, 282*, 483(32, 33), *494*

Simalty, M., 478(34), *494*

Simpson, W., 308(135), 308(251, 252), *436, 439*

Skinner, D. L., 420(83), 429(83), *434*

Skinner, J. S., 184(555), *280*

Skvortsova, G. G., 455(144, 145, 146, 147, 148, 149), *472*

Smets, G., 314(238), 317(196, 253, 254), 318(196, 197), 319(253), 320(253), 377 (238), 378(238), 379(253), 380(253), 381(253), *438, 439*

Smirnova, V. A., 157(373), *273*

Smirnova, V. E., 30(609), 33(609), 165(609), *282*

Smith, R. E., Jr., 41(610), *282*

Smolin, E. M., 295(161), 354(161), *436*

Smullen, R. S., 15(337), 133(337), 135(337), *272*

Sobolev, I., 442(150), 444(150), 463(150), *472*

Sogabe, M., 305(31), 357(31), *433*

Sokolinskaya, T. A., 84(537), 230(537), *279*

Sokolova, O. V., 308(10), 309(10), 371(10), 372(10), *432*

Sokolova, T. A., 328(255, 256), 329(241, 255, 256), 407(241, 256), 408(241, 255, 256), 409(241, 256), *439, 440*

Solintseva, L. M., 98(539), *279*

Sorenson, W. R., 12(611, 612), 124(611, 612), 130(611, 612), 133(611, 612), 135 (611, 612), *282*

Spain, R. G., 41(619), *282*

Sparrow, D. B., 311(91), 312(91), *434*

Speier, J. L., 350(266), 429(266), *440*

Spooncer, W. W., 337(258, 259), 338(258, 259), 339(258, 259), 417(259), 418 (259), 419(258, 259), 420(258, 259), *440*

Squire, E. N., 11(124, 216), 121(124, 216), 122(216), *265, 268*

Sroog, C. E., 3(620), 4(620), 8(388), 27 (620), 29(620, 622), 30(620, 622), 33 (620, 622), 34(620, 622), 41(620, 621, 622), 120(620), 121(620), 124(621), 126(621), 128(621), 130(621), 133 (621), 135(621), *274, 282*

Stackman, R. W., 346(84), 349(84), 420(83), 428(84), 429(83, 84), *434*

Standage, A. E., 41(623), 133(623), *282*

Starkweather, H. W., 11(124), 121(124), *265*

Staudinger, H., 288(260), *440*

Stenner, R., 328(248), 330(248), 331(248), 409(248), 410(248), *439*

Stephens, J. R., 45(114, 624), 185(114, 624), *265, 282*

Stevens, M. P., 117(123), *265*

Stewart, A. T., 443(151), 445(151), *472*

Stewart, A. T., Jr., 447(37), 448(37), *469*

Stille, J. K., 17, 79(628, 632), 80(628), 82 (626), 83(627, 629, 630, 631, 632), 84 (625, 626, 627, 637), 119(633), 225(628, 632), 230(627, 632), 231(629, 631, 632), 232(630, 631, 632), 233(634, 635, 636, 637), 234(637), 235(631), 236(637), 237 (626), 238(626, 627, 631), *282, 283*, 297(261, 262, 263, 264), 298(264), 342 (265), 355(263), 424(265), *440*

Stober, M. R., 350(266), 429(266), *440*

Stoffey, D., 29(414), 40(414), 41(414), 42 (413), 50(413), 51(413), 52(413), *275*

Strauss, E. L., 29(638), *283*

Streef, J. W., 9(322), 11(322), 27(322), 120 (322), 121(322), *272*

Strul, M., 87(639), 242(639), 243(639), *283*

Strzelecki, L., 478(34), *494*

Stychenko, V. A., 84(537), 230(537), *279*

Suga, M., 92(640), 245(640), 248(640), *283*

Sultanov, K., 299(12), 300(12), 301(267), 358(12), 359(12), 360(12), *432, 440*

Sumi, M., 350(268), *440*

Sumitomo, Y., 444(41), *469*

Sun, K.-H., 371(89), 372(89), 374(89), *434*

Suzuki, S., 443(134), 463(134), *472*

Suzuki, Y., 293(286), *440*

Sweeny, W., 34(338), 39(338), *272*

**T**

Tabushi, I., 108(641), 109(641), 115(641), 116(641), *283*, 340(269), 346(269), 421(269), *440*

Tagami, S., 288(29), 291(28, 29), 294(28, 29), 295(29), 354(28, 29), *432, 433*

Tanaka, S., 444(152), 453(152), *472*

Tanimot, S., 486(29), *494*

Tanimura, N., 108(641), 109(641), 115(641), 116(641), *283*

Tank, L., 292(130), *435*

Tanunina, P. M., 29(669), *284*

Tatum, W. E., 12(642), 14(29), 20(29), 29(643), 67(29), 68(29, 642, 644), 200(642), 201(29, 642), *263, 283*

Teyssie, P., 40(288), 124(288), 126(288), *271*

Thomas, W. M., 98(390), 254(390), 257(390), *274*, 322(244), 325(244), 340(244), 384(244), 385(244), 393(244), 394(244), 402(244), 403(244), 405(244), 421(244), *439*

Thomson, D. W., 342(265), 424(265), *440*

Thornton, R. L., 68(644), *283*

Tiers, G. V. D., 317(270), 320(270), *440*

Ting, M.-H., 41(645), 84(645), *283*

Tobolsky, A. V., 37(182, 430, 431), 38(182), 49(182), 50(182), 133(182), 146(182), 148(182), 175(182), *267, 276*

Tocker, S., 12(647), 32(647), 125(647), 148(647), 154(646), *283*

Todd, N. W., 29(648, 649), *283*

Tohyama, S., 75(695), 95(650), 211(695), 212(695), 250(650), 251(650), *283, 285*

Tokarev, A. V., 43(393), 135(651), 178(393), *274, 283*

Tolapchyan, L. S., 200(35, 157), 358(35, 157), 359(35, 157), 360(35, 157), *432, 436*

Tolchinskii, I. M., 429(211), *438*

Tolparova, G. A., 133(384), *274*

Topchiev, A. V., 429(211, 271), *438, 440*

Toyama, S., 75(710), 92(399), 95(711, 714), 211(710), 212(710), 250(711), 251(714), *275, 285, 286*

Trautwein, H., 443(8), *468*

Travnikova, A. P., 481(22), *493*

Traynard, P., 40(288, 289), 124(288), 126(288), *271*

Traynor, E. J., 14(275), 29(166), 37(234, 275), 38(234), 40(166, 235, 275), 43(275), 49(234), 50(234), 51(120), 64(275), 65(275), 133(234), 138(275), 143(275), 146(166, 234), 147(275), 148(234), 149(275), 150(275), 151(166, 275), 152(275), 175(234), 176(275), *265, 267, 269, 270*

Traynor, E. J., Jr., 51(652, 653), 177(652, 653), 178(652, 653), *283*

Trifan, D. S., 289(272), 303(272), 323(272), 324(272), 349(272), 367(272), 393(272), 394(272), 430(272), *440*

Trosarelli, L., 290(273, 274, 275, 277, 278, 279, 280), 311(276), 312(275, 277, 278), 330(273, 274, 275, 278, 279, 280), 369(276), 370(275, 277, 278), 407(273, 274, 275, 278, 279, 280), *440*

Ts'ao, W.-H., 308(88), 319(102), *434, 435*

Tsatsos, W. T., 447(21), 455(28), *468*

Tschamler, H., 291(281), *440*

Tseitlin, G. M., 6(371), 127(371, 372), 129(371, 372), 130(371, 372), 157(373), *273*

Tsujimura, T., 44(657), 173(657, 658), *283*

Tsuruoka, K., 325(282), 326(282), 402(282), *440*

Tuemmler, W. B., 70(654), 71(654), 205(654), *283*

Tugushi, D. S., 30(370), 127(370), 147(370), 163(370), *273*

Turner, W. N., 41(623), 133(623), *282*

Tusing, C. R., 7(600), *281*

Tyler, G. J., 18(186), 122(186), *267*

**U**

Ugryumova, G. S., 458(18, 19, 20), 459(18, 20), *468*

Ungurenasu, C., 473(35), 474(35), 476(6), 491(27), *493. 494*

Unishi, T., 44(655, 656, 657), 173(657, 658), 174(655, 656), *283*

Uno, K., 44(334), 142(340), 173(334), 174(334), *272*, 325(282), 326(282), 332(142, 144), 333(142), 334(142, 143), 335(142, 144), 402(282), 411(142, 143, 144, 145),

412(142, 144, 145), 413(144, 145), *436*, *440*

Ushio, S., 305(30, 31), 357(30, 31), *433*

## V

Van Deusen, R. L., 15(298), 21(298), 88(38), 159(298), 243(38), *263*, *271*, 325(114, 115), 328(114, 115), 405(114, 115), *435*

Vanhaeren, G., 304(85), 357(85), *434*

van Paesschen, G., 313(230), 314(230), 377 (230), *439*

Van Strien, R. E., 45(624), 185(624), *282*

Vasil'ev, E. K., 459(7), *468*

Vasil'eva, L. V., 444(142, 143), 447(142), 454(142), *472*

Vayson de Pradenne, H., 19(659), 146(659), *283*

Veitch, J., 12(660), 111(660), 112(660), *284*

Victorius, C., 41(175), 148(175), *267*

Vinogradova, S. V., 6(374, 375), 30(374), 131(374, 375), 132(374, 375, 661, 662, 664, 665, 666, 667, 673), 164(375), 165 (375), 169(663), 170(663), 171(663), *274*, *284*

Vishnyakova, T. P., 84(537), 230(537), *279*

Vlasova, K. N., 3(720), 4(720), 29(669), 41(668), 166(719), *284*, *286*

Vollmert, B., 7(671), 31(671), 41(670), 114 (671), *284*

Volpe, A. A., 96(43, 672), 97(43), 98(672), 256(43, 672), *263*, *284*

Voskanyan, M. G., 299(177), 300(157, 175, 177, 190), 303(177), 306(177, 193), 358 (157, 177, 190, 191), 359(157, 177, 178, 179, 191, 192), 360(157, 175, 176, 177, 178, 191, 192), 361(175, 176, 179, 191, 192), 362(175, 191, 192), *436*, *437*

Vygodskii, Ya. S., 6(374, 375), 30(374), 131 (374, 375), 132(374, 375, 661, 662, 664, 665, 666, 667, 673), 164(375), 165(375), 169(663), 170(663), 171(663), *274*, *284*

## W

Wagner, H., 98(587), *281*, 444(135), 451 (135), *472*

Walker, C., 335(163), 336(163), 415(163), 416(163), *436*

Wallach, M. L., 7(674, 677), 21(675, 676, 678), 125(676), *284*

Wallenberger, F. T., 41(679), 84(679), *284*

Walling, C., 308(283), *440*

Walton, W. L., 53(426), 55(426), 191(426), *276*

Wang, J. Y. C., 343(200), 425(200), *438*

Wang, P.-J., 41(680), *284*

Washburn, R. M., 473(36), *494*

Wegner, G., 442(137), 444(137, 138), 446 (137), 448(137), 453(136, 137, 138), 456 (138), 460(138), 465(138), *472*

Weiler, W., 466(153), *472*

Weisenberger, W. P., 10(14), 22(13, 14), 133(13, 14), *262*

Welch, F. J., 443(154, 155), 448(156), *472*

Werntz, J. H., 13(681, 682), 153(681, 682), *284*

Wildi, B. S., 71(683), 205(683), *284*

Wiley, R. H., 98(229), 259(229), *269*

Williamson, J. R., 84(637), 233(634, 635, 636, 637, 684), 234(637), 236(637, 684), *283*, *284*

Witkiewicz, Z., 492(7), *493*

Wöhrle, D., 71(433, 434), 72(434), 205 (433), 207(434), 208(434), 209(434), *276*

Wolf, R., 78(685), 222(685), 223(685), 224 (685), 225(686), 226(686), *284*

Wolff, F. A., 29(649), *283*

Wolz, H., 455(157), *472*

Woodward, C. F., 455(81), *470*

Woolford, R. G., 349(167), 429(167), *437*

Wrasidlo, W. J., 8(690), 40(321), 82(687, 688, 689), 84(687, 688, 689), 232(687, 689), 233(687, 688, 689), 234(687, 689), 236(687, 688, 689), 237(687), *272*, *285*

Wright, C. D., 10(428), 133(428), *276*

Wright, W. W., 3(200), 4(200), 5(200), 10 (200), 40(201), 53(200), *268*

Wu, H.-C., 81(691), 230(691), *285*

## Y

Yakhimovich, R. I., 320(100, 284), 321 (100, 284), 379(100, 284), *434*, *440*

Yakovlev, B. I., 133(384), *274*

Yamaguchi, K., 240(357, 358), *273*

Yamashita, T., 240(357, 358), *273*

Yarsley, V. E., 41(230), 84(230), *269*

Yeomans, B., 12(660), 111(660), 112(660), *284*

Yoda, N., 41(692), 72(693, 697, 698), 73 (398, 400, 401, 699, 701, 702, 704), 74 (303, 694, 703, 706, 707), 75(401, 694, 695, 705, 708, 709, 710), 77(692), 84 (692), 91(700), 92(399, 400, 403), 93 (402, 458, 713), 95(402, 404, 458, 650, 696, 711, 713, 714), 96(692), 209(693, 697, 698), 210(398, 400), 211(303, 398, 400, 401, 695, 699, 701, 702, 703, 704, 705, 706, 707, 709, 710, 712), 212(398, 400, 401, 694, 695, 704, 705, 708, 710, 712), 213(704, 705), 245(700), 249 (458), 250(402, 458, 650, 696, 711, 713), 251(402, 458, 650, 696, 713, 714), 252 (404, 458, 696), 253(458, 696), *271, 275, 277, 283, 285, 286*, 477(14), 478(14), 486(37), 492(1, 24), *493, 494*

Yokota, K., 293(285, 286), *440*

Yokoyama, M., 123(715), 137(715), 139 (715), 140(715), *286*

Yoshie, Y., 292(208), *438*

Yosomiya, R., 308(217), *438*

Young, R. R., 141(355), 142(355), 150 (355), 151(355), *273*

Yudin, B. N., 6(375), 131(375), 132(375), 164(375), *274*

**Z**

Zakoshchikov, S. A., 3(720, 721), 4(717, 718, 720, 721), 135(716), 166(719), *286*

Zalewski, E. J., 55(445, 446), *276*

Zapadinskii, B. I., 117(103), 118(103), *265*

Zapunnaya, K. V., 455(144, 145, 146, 147, 148, 149), *472*

Zavalei, V. M., 444(143), *472*

Zetie, R. J., 308(252), *439*

Zhamkochyan, G. A., 392(186), 393(186), 405(186), 406(186), *437*

Zhubanov, B. A., 116(553), *280*

Zilkha, A., 444(158), *472*

Zinder, M. F., 307(5), 368(5), *432*

Zolotareva, G. M., 3(721), 4(721), 166 (719), *286*

Zubareva, G. M., 3(720, 721), 4(717, 720, 721), *286*

Zugravescu, I., 87(639), 242(639), 243 (639), *283*

# Subject Index

Page numbers in italics refer to table entries.

## A

Acetone, divinyl thioketal, polymers of, *422*
Acrolein
  copolymers with aldehydes, 456
  copolymers with 1,4-butynediol, 480
  copolymers with vinyl ethers, 455
  copolymers with vinyl monomers, 455
  grafting on poly(methyl methacrylate), 456
  polymerization of, initiators, *443–444*
  polymers from, 441
  polymers of, applications, 447
  polymers of, chemical reactivity, 449
  polymers of, crystallinity, 447
  polymers of, density, 447
  polymers of, infrared spectra, 448
  polymers of, mechanical properties, 447
  polymers of, mechanism of formation, 452
  polymers of, molecular weight, 442
  polymers of, softening points, 445
  polymers of, solubility, 445
  polymers of, structure, 450
  terpolymers with vinyl monomers, 455
Acrolein dimer, 451
  polymerization of, 296
  polymers of, *353*
Acrylic anhydride
  copolymers with unsaturated monomers, 317
  copolymers with vinyl monomers, *379*
  polymerization of, free radical, 317
  polymers of, *379*
Acrylic methacrylic anhydride, polymers of, *379*
Acrylonitrile, Ziegler polymerization of, 336
Adipinaldehyde
  copolymer with chloral, 295, *354*
  polymerization of, cationic, 295
  polymers of, *353*
N-Allyl acrylamide
  polymerization of, free radical, 330
  polymers of, *407*

Allyl acrylate
  polymerization of, free radical, 311
  polymers of, *369*
Allyl 3-butenoate, polymerization of, free radical, 310
Allyl citraconate, polymers of, *372*
Allyl crotonate
  polymerization of, free radical, 310
  polymers of, *371*
Allyl isopropenyl ether, copolymer with selenium dioxide, *431*
Allyl maleate, polymers of, *371*
N-Allyl methacrylamide
  polymerization of, free radical, 330
  polymers of, *407*
Allyl methacrylate
  polymerization of, free radical, 311
  polymers of, *370*
N-Allyl-N-methyl-2-vinylmorpholinium chloride, polymers of, *389*
N-Allyl-N-methyl-2-vinylpiperidinium chloride, polymers of, *389*
N-Allyl-N-methyl-2-vinylpyrrolidinium chloride, polymers of, *389*
Allyl phenyl allyl phosphonate
  copolymer with lauryl methacrylate, 339
  polymerization of, free radical, 337
Allyl vinyl acetate, polymer of, *371*
N-Allyl vinyl sulfonamide
  polymerization of, free radical, 341
  polymers of, *422*
Allyl vinyl sulfonate
  polymer of, *422*
  polymerization of, free radical, 341
Amide anhydrides, poly(amide-imides) from, *187–188*
Amides, diunsaturated
  polymerization of, 328
  polymerization of, effect of incipient ring size on, 331
Amines, diunsaturated
  polymerization of, 322

Amines, diunsaturated (cont.)
polymerization of, effect of electron-withdrawing, groups on, 325
polymers of, *403–405*
Amine oxides, diunsaturated
copolymers with vinyl compounds, 325
polymerization of, 322
Aminoanhydrides, polymerization to poly-amic acids and polyimides, 10
4-(*p*-Aminobenzoyl)phthalic anhydride, polyimide from, *153*
3-Aminophthalic anhydride, polyimide from, *153*
4-Aminophthalic anhydride, polyimide from, *153*
*N*-Aminophthalimide
condensation with benzoyl chloride, 54
thermal rearrangement of, 54
6-Aminosaccharin, condensation in poly-phosphoric acid, 476
Ammonia, polymerization with chloranil, 100
Ammonium salts, diunsaturated
copolymers with sulfur dioxide, 324, *395–399*
copolymers with vinyl compounds, 322, 324, 325
polymerization of, 322
polymers of, *382–389*
Anhydrides, diunsaturated, polymerization of, 317
Anilinephthalein, polypyromellitimide from, solubility of, 30
Anilinephthaleinimide, polypyromellitimide from, solubility of, 30
Azobenzene tetracarboxylic acid dianhy-dride, polyimides from, *167*
Azobenzene-3,3′,4,4′-tetracarboxylic acid dianhydride, polyimide with 1,3-di-aminobenzene, *168*
Azoxybenzene-3,3′,4,4′-tetracarboxylic acid dianhydride, polyimide with 1,3-diaminobenzene, *168*

**B**

Benzalazine, polymerization with bismale-imides, 17
Benzaldehyde, polymerization with *m*-phenylene diamine, 485

Benzaldehyde, divinyl acetal, polymers of, *363*
Benzaldehyde, substituted, divinyl acetals, polymers of, *364–366*
Benzene-1,2,3,4-tetracarboxylic acid di-anhydride, polyimide with 4,4′-di-aminodiphenyl ether, *153*
Benzophenone, divinyl ketal, polymers of, *366*
3,3′,4,4′-Benzophenone tetracarboxylic acid dianhydride, copolyimide with 3,4-dicarboxy-1,2,3,4-tetrahydro-1-naphth-alene succinic dianhydride and 4,4′-diamino diphenyl ether, solubility and melt-processibility of, *31*
poly(amide-imides) from, *179–180*
polyimides from, *145–152*
polyimide with 1,3-diaminobenzene, glass transition temperature of, *38*
polyimide with 4,4′-diaminodiphenyl ether, glass transition temperature of, *38*
polyiminolactones from, *202*
polymerization with polyaniline-polyiso-cyanate, papi, 25
*p*-Benzoquinone, polymerization with 3,3′-dicarboxybenzidine, 487
Benzoyl chloride, condensation with *N*-aminophthalimide, 54
Benzoyl hydrazine, condensation with phthalic anhydride, 54
Benzoyl pyromellitic dianhydride, poly-imides from, *154*
1,1′-Bicyclohex-2,2′-ene, copolymer with sulfur dioxide, *425*
Bicyclo[2,2,2]-oct-7-ene-2,3,5,6-tetracar-boxylic acid dianhydride
polyimides from, *114, 115*
polyimide with 4,4′-diamino-3,3′dicar-boxyphenylmethane, solubility and melt-processibility of, *31*
polyimide with 4,4′-diaminodiphenyl-methane, solubility and melt-processi-bility of, *31*
1,1′-Bicyclopent-2,2′-ene, copolymer with sulfur dioxide, *425*
2,2′,6,6′-Biphenyltetracarboxylic acid di-anhydride
polyimides from, *159*
polymerization with aromatic diamines, 21

2,3,5,6-Biphenyl tetracarboxylic acid dianhydride, polyimides from, *159–160*

3,3′,4,4′-Biphenyl tetracarboxylic acid dianhydride
  polyimides from, *158, 159*
  polyiminolactone with 4,4′-diaminodiphenyl ether, *202*

1,4-Bis(acetylamino)benzene, condensation with trimellitic anhydride, 47

4,4′-Bis(acetylamino)diphenylmethane, polymerization with 3,3′-dicarboxy-4,4′-diaminobiphenyl, 91

Bis(*o*-amino acids), aromatic, polymerization with bis(imino esters), 92

*N,N*′-Bis(3-aminobenzoyl)hydrazine, polymerization with pyromellitic anhydride, 64

1,4-Bis(aminomethyl)benzene, copolyimide with pyromellitic anhydride and 2,2-bis(3,4-dicarboxyphenyl)propane dianhydride, properties of, 27

Bis(4-aminophenyl ether) of hydroquinone, polyimide with bis(3,4-dicarboxyphenyl)ether dianhydride, crystalline behavior of, *33*

2,2-Bis(4-aminophenyl)hexafluoropropane, polyimide with 2,2-bis(3,4-dicarboxyphenyl)hexafluoropropane dianhydride, glass transition temperature of, *38*
  solubility and melt-processibility of, *32*

2,5-Bis(4-aminophenyl)-1,3,4-oxadiazole, polyimide with pyromellitic anhydride, 60

2,2-Bis(4-aminophenyl)propane
  polypyromellitimide from,
  crystalline behavior of, *33*
  solubility of, *29*
  thermal stability of, *35*

1,4-Bis(bromoacetyl)benzene, polyphenylenepyrazine with ammonia, 81

4,4′-Bis(bromoacetyl)diphenyl ether, polythiazine with 4,4′-diamino-3,3′-dimercaptobiphenyl, 80

3,3′-Bis(carbamido)-4,4′-diaminobiphenyl, polyimide with pyromellitic anhydride, rearrangement of, 22

1,3-Bis(chloroacetylamino)benzene, polymerization with phenyl glycidyl ether, 86

Bis(α-chloroallyl)dimethyl silane, polymer of, *428*

Bis(2-chloroallyl)methyl phosphine, polymer of, *417*

Bis(2-chloropropenyl)ether, polymer of, *358*

Bis(2-chloropropenyl)sulfide, polymer of, *421*

Bis(2-chloropropenyl)sulfone, polymer of, *421*

Bis(3,4-dicarboxyphenyl)ether dianhydride, polyimides from, *163–165*
  polyimide with bis(4-aminophenyl ether) of hydroquinone, crystalline behavior of, *33*
  polyimide with 4,4′-diaminodiphenyl ether,
  crystalline behavior of, *33*
  glass transition temperature of, *38*
  polyiminolactone with 4,4′-diaminodiphenyl sulfide, *203*

2,2-Bis(3,4-dicarboxyphenyl)hexafluoropropane dianhydride, polyimides from, *162*
  polyimide with 2,2-bis(4-aminophenyl)hexafluoropropane
  glass transition temperature of, *38*
  solubility and melt-processibility of, *32*

Bis(3,4-dicarboxyphenyl)phosphine oxide dianhydride, polyimide with bis(4-aminophenyl)ether of hydroquinone, *166*

2,2-Bis(3,4-dicarboxyphenyl)propane dianhydride, copolyimide with pyromellitic anhydride and 1,4-bis(aminomethyl)benzene, properties of, 27
  polyimides from, *161–162*

Bis(3,4-dicarboxyphenyl)sulfone dianhydride, polyimides from, *165–166*

Bis(dimethylamino)phenyl boron, polymerization with dihydrazides and dihydrazines, 474

1,1-Bis(dimethyl-1-butadienylsilyl)ferrocene, polymerization of, 491

*N,N*′-Bis(β-hydroxyethyl)-1,12-dodecamethylene diamine, polymerization of, 85

Bis(imino esters), polymerization with aromatic bis(*o*-amino acids), 92

Bis(isocyanatomethyl)ether, polymers of, *412*

Bis(β-keto esters), polymerization with diamidines, 87

Bis(maleimides), polymerization with benzalazine, 17
Bis(oxetane), polymerization with terephthalaldehyde, 98
Bis(trimellitic anhydride ester) of 1,4-bis-(hydroxymethyl)benzene, poly(ester-imide) with 4,4′-diaminodiphenyl ether, *193*
Bis(trimellitic anhydride ester) of bisphenol, poly(ester-imides) from, *194*
Bis(trimellitic anhydride ester) of 2,2′-dihydroxybiphenyl, poly(ester-imides) from, *193*
Bis(trimellitic anhydride ester) of 4,4′-dihydroxydiphenyl sulfone, poly(ester-imides) from, *195*
Bis(trimellitic anhydride ester) of ethylene glycol, poly(ester-imides) from, *189*
Bis(trimellitic anhydride ester) of hydroquinone, poly(ester-imides) from, *190–192*
Bis(trimellitic anhydride ester) of resorcinol, poly(ester-imide) with 1,4-diaminobenzene, *193*
1,2,3,4-Butanetetracarboxylic acid dianhydride, polyimides from, *104–106*
properties of, 28
α-Butylacrolein, polymer of, *467*
1,4-Butylene diisocyanate, polymer of, *412*
1,4-Butynediol, copolymerization with acrolein, 480

## C

3-Carboxybenzylsuccinic anhydride, poly-(amide-imides) from, *183–184*
4-Carboxybenzylsuccinic anhydride, poly-(amide-imides) from, *182–183*
β-(Carboxymethyl)-ε-caprolactam
polyimide from, *108*
polymerization of, 26
Chloral, copolymer with adipinaldehyde, 295
Chloranil
polymerization with ammonia, 100
polymerization with potassium sulfide, 100
polymerization with tetrahydroxy-*p*-benzoquinone, 99
α-Chloroacrolein
copolymers with vinyl ethers, 458

polymerization of, 458
*o*-Chlorobenzaldehyde, divinyl acetal, polymer of, *363*
*m*-Chlorobenzaldehyde, divinyl acetal, polymer of, *363*
*p*-Chlorobenzaldehyde, polymer of, *363*
2-Chlorobutane-1,4-diisocyanate, polymer of, *412*
Copolymers, heterocyclic, imide-containing, 60
Crotonaldehyde, polymerization of, 459
1,2-*trans*-3,4-Cyclobutanetetracarboxylic acid dianhydride, polyimides from, *108–109*
*trans*-Cyclohexane-1,2-dicarboxaldehyde
polymer of, *354*
polymerization of, cationic, 294
Cyclohexane-1,2-diisocyanate, polymer of, *413*
Cyclohexane-1,3-diisocyanate, polymer of, *414*
1,2,4,5-Cyclohexanone tetrapropionic acid dianhydride, polyimides from, *111–112*
1,3-Cyclohexylene diisocyanate, polymerization of, anionic, 334
*cis,cis*-1,5-Cyclooctadiene
copolymer with carbon monoxide, 343
copolymer with sulfur dioxide, 342, 343
copolymer with carbon monoxide and sulfur dioxide, *424*
1,2,3,4-Cyclopentanetetracarboxylic acid dianhydride, polyimides from, *109–110*
1,2,3,4-*cis,cis,cis,cis*-Cyclopentanetetracarboxylic acid dianhydride, polyimides from, *111*
1,2-*cis*-Cyclopropane diisocyanate
copolymers of, *413*
polymer of, *413*

## D

Deemulsifiers, 325
Dialdehydes, polymerization of, 291
Diallyl amine
copolymers with vinyl monomers, *402*
polymers of, *401–402*
Diallyl *N*-butyl phosphonate, copolymers of, *419*
*N,N*-Diallylcyanamide
polymer of, *402*
polymerization of, free radical, 325

Diallyl dialkyl silanes, polymers of, *428–429*

Diallyl diaryl silanes, polymers of, *428–429*

Diallyl diethyl ammonium bromide, polymerization of, free radical, 322

Diallyl diethyl germanium, Ziegler polymerization of, 321

Diallyl diethyl lead
polymer of, *431*
polymerization of, free radical, 351
Ziegler polymerization of, 351

Diallyl dimethyl tin
polymer of, *431*
polymerization of, free radical, 351

Diallyl lauryl phosphonate, copolymer with vinyl stearate, *420*

Diallyl maleate
polymer of, *376*
polymerization of, free radical, 308

*N,N*-Diallyl melamine
copolymers of, *405*
polymer of, *405*
polymerization of, free radical, 328

Diallyl methylamine, copolymers of, *403*

Diallyl methylamine *N*-oxide, copolymers of, *394*

Diallyl methyl sulfonium methosulfate copolymers of, 340

Diallyl muconate
polymer of, *378*
polymerization of, free radical, 315

Diallyl phenyl phosphate, copolymers of, *418*

Diallyl phenyl phosphine oxide
copolymers of, *417*
polymer of, *417*

Diallyl phenyl phosphonate
copolymers of, *419–420*
polymer of, *419*

Diallyl sulfide
copolymer with acrylonitrile, 340
polymer of, *421*
polymerization of, cationic, 346
polymerization of, free radical, 340
Ziegler polymerization of, 345

Diallyl tartrate, polymerization of, free radical, 314

*N,N*-Diallyl urea, copolymers with lauryl methacrylate, 326, *402*

Diamidines, polymerization with bis($\beta$-keto esters), 87

Diamines
poly(ester-imides) with trimellitic anhydride and diols, 56
polymerization with dianhydrides, 2
polymerization with tetraacids, 11
polypyrrolines with 1,2,4,5-tetrakis-(chloromethyl)benzene, 72

Diamines, aromatic
poly(imide-imidazopyrrolones) with aromatic tetraamines and dianhydrides, 62

Diaminoamides
poly(amide-imides) with pyromellitic anhydride, properties of, *49*

4,3'-Diaminobenzanilide
poly(amide-imide) with pyromellitic anhydride, mechanical properties of, *49*

1,3-Diaminobenzene
polyamic acid with pyromellitic anhydride, mechanism of cyclization of, 8
polyimide with 3,3',4,4'-benzophenone tetracarboxylic acid dianhydride, glass transition temperature of, *38*
polymerization with pyromellitonitrile, 71
polypyromellitimide from,
crystalline behavior of, *33*
hydrolytic stability of, 34
solubility of, *29*
thermal stability of, *35*

1,4-Diaminobenzene
polypyromellitimide from,
crystalline behavior of, *33*
solubility of, *29*
thermal stability of, *35*

3,3'-Diaminobenzidine, polyquinoxalines from, *233, 234, 235*

4,4'-Diaminobiphenyl
polypyromellitimide from,
crystalline behavior of, *33*
solubility of, *29*
thermal stability of, *35*

2,2'-Diamino-1,10-decanedioic acid, polymerization of, 85

2,7-Diamino-3,6-di(carbophenoxy)naphthalene, poly(1,3-oxazinone) with diphenyl terephthalate, 74

4,4'-Diamino-3,3'-dicarboxybiphenyl
polymerization with terephthaloyl chloride, 74
polypyrrolinone with *p*-xylylene dichloride, 72

4,4'-Diamino-3,3'-dicarboxydiphenyl methane
  polyimide with bicyclo[2,2,2]-oct-7-ene-2,3,5,6-tetracarboxylic acid dianhydride, solubility and melt-processibility of, *31*
  polymerization to poly(1,3-oxazinone), 73
1,3-Diamino-4,6-dihydroxybenzene, polyoxazine with 2,5-dihydroxybenzoquinone, 79
4,4'-Diamino-3,3'-dihydroxybiphenyl, polyimide with pyromellitic anhydride, rearrangement of, 18
4,4'-Diamino-3,3'-dimercaptobiphenyl, polythiazine with 4,4'-bis(bromoacetyl) diphenyl ether, 80
4,4'-Diaminodiphenyl ether
  copolyimide with 3,4-dicarboxy-1,2,3,4-tetrahydro-1-naphthalene succinic dianhydride and 3,3',4,4'-benzophenone tetracarboxylic acid dianhydride, solubility and melt-processibility of, *31*
  poly(amide-imide) with isophthaloyl chloride and dimethyl 2,4-di(chloroformyl)terephthalate, 43
  poly(amide-imide) with trimellitoyl chloride, 45
  polyimide with 3,3',4,4'-benzophenone tetracarboxylic acid dianhydride, glass transition temperature of, *38*
  polyimide with bis(3,4-dicarboxyphenyl)-ether dianhydride,
    crystalline behavior of, *33*
    glass transition temperature of, *38*
  polyimide with pyromellitic anhydride,
    crystalline behavior of, *33*
    hydrolytic stability of, 34
    properties of, 28
    solubility of, *30*
    thermal stability of, *35*
  poly(imide-1,2,4-oxadiazole) with trimellitoyl chloride and isophthalbis(amidoxime), 66
  polymerization with 3,3'-dicarboxy-4,4'-bis(benzoylamino)biphenyl, 91
  polymerization with 2,2'-dimethyl-[6,6'-bibenzoxazine]-4,4'-dione, 90
  polymerization with 3,3',4,4'-diphenyl ether tetracarboxylic acid dianhydride, 6
  polymerization with pyromellitic anhydride, 1

  polymerization with pyromellitonitrile, 70
4,4'-Diaminodiphenyl methane
  poly(amide-imides) with isomeric tricarboxy diphenyl ether derivatives, 46
  polyimide with bicyclo[2,2,2]-oct-7-ene-2,3,5,6-tetracarboxylic acid dianhydride, solubility and melt-processibility of, *31*
  polypyromellitimide from,
    crystalline behavior of, *33*
    solubility of, *29*
    thermal stability of, *35*
4,4'-Diaminodiphenyl sulfide
  polypyromellitimide from,
    crystalline behavior of, *33*
    hydrolytic stability of, 34
    solubility of, *30*
    thermal stability of, *35*
3,3'-Diaminodiphenyl sulfone, polypyromellitimide from, solubility of, *30*
4,6-Diamino-1,3-dithiophenol, polythiazines from, *229*
2,5-Diamino-1,3-dithiophenol, polythiazine from, *228*
1,6-Diaminohexane, poly(amide-imide) with maleopimaric acid monochloride, 46
4,6-Diaminoisophthalaldehyde, polymerization with diketones, 486
2,4-Diaminoisopropylbenzene, polymerization with pyromellitic anhydride, 21
1,9-Diaminononane, polypyromellitimide from, properties of, 27
N,N'-Diamino pyromellitimide, polymerization with pyromellitic anhydride, 17
α,α'-Diaminosebacic acid, polydiketopiperazine from, *239*
α,α'-Diaminosuberic acid, polydiketopiperazine from, *239*
2,4-Diaminotoluene, polyamide with isophthaloyl chloride, cyclization of, 478
Dianhydrides
  poly(amide-imides) from, *180*
  poly(ester-imides) with diisocyanates, 55
  polyimides from, *171–172*
  polymerization with diamines, 2
Dianhydrides, aromatic
  poly(imide-imidazopyrrolones) with aromatic tetraamines and diamines, 62
Dianhydride from furfural, diamine and maleic anhydride, polyimides from, *117–118*

Diazides, condensation with dinitriles, 475
3,4-Dibenzylglutaconaldehyde, polymers of, *352*
3,3'-Dicarbamyl-4,4'-diaminodiphenyl methane, polymerization with pyromellitic anhydride, 94
3,3'-Dicarbethoxy-4,4'-diaminobiphenyl, polypyromellitimide from, solubility and melt, processibility of, *32*
3,4-Dicarboxy-1,2,3,4-tetrahydro-1-naphthalene succinic dianhydride, copolyimide with 3,3',4,4'-benzophenone tetracarboxylic acid dianhydride and 4,4'-diaminodiphenyl ether, solubility and melt processibility of, *31*
3,3'-Dicarboxybenzidine
polymerization with *p*-benzoquinone, 487
polymerization with pyroboretetracetate, 476
3,3'-Dicarboxy-4,4'-bis(benzoylamino)-biphenyl, polymerization with 4,4'-diaminodiphenyl ether, 91
3,3'-Dicarboxy-4,4'-diaminobiphenyl
polymerization with 4,4'-bis(acetylamino)diphenylmethane, 91
polymerization with 1,4-diisocyanatobenzene, 93
3,4-Dicarboxy-1,2,3,4-tetrahydro-1-naphthalene succinic dianhydride, polyimides from, *112–113*
1,5-Dienes, copolymers with selenium dioxide, 350
Dienes, copolymers with sulfur dioxide, 342
Diepoxides, polymerization of, 296
*o*-Di(epoxyethyl)benzene
polymerization of, anionic, 298
polymerization of DL and meso forms, cationic, 297
polymerization of, free radical, 296
polymers of, *355, 356*
1,2,5,6-Diepoxyhexane
polymerization of, anionic, 298
polymerization of, cationic, 297
polymers of, *355*
1,2,4,5-Diepoxypentane, polymerization of, cationic, 297
Diethyl oxaldiimidate, polymerization with 2,2',6,6'-tetraamino-4,4'-bis(methoxycarbonyl)biphenyl, 490
Diethyl succinosuccinate, condensation with *m*-phenylene diamine, 484

*N,N*-Diglycidylaniline
polymerization of, anionic, 299
polymers of, *356*
2,5-Dihalo-3,6-dihydroxy-*p*-benzoquinone, polymerization of, 99
2,5-Dihydroxybenzoquinone, polyoxazine with 1,3-diamino-4,6-dihydroxybenzene, 79
4,4'-Dihydroxybiphenyl, polymerization with 2,5-dichloroterephthalic acid, 479
2,3-Dihydroxy-6,7-diamino-1,2-diazanaphthalene, polymerization of, 83
4,4'-Dihydroxy-3,3'-di(carbophenoxy)-biphenyl poly(1,3-oxazinedione) with 4,4'-diisocyanatodiphenyl ether, 76
Diisocyanates
poly(ester-imides) with dianhydrides, 55
polymerization of, 332
polymerization with trimellitic anhydride, 48
1,4-Diisocyanatobenzene, polymerization with 3,3'-dicarboxy-4,4'-diaminobiphenyl, 93
4,4'-Diisocyanatodiphenyl ether, poly(1,3-oxazinedione) with 4,4'-dihydroxy-3,3'-di(carbophenoxy)biphenyl, 76
4,4'-Diisocyanatodiphenyl methane
poly(amide-imide) with trimellitic anhydride, 48
polymerization with pyromellitic anhydride, 25
Diketones, polymerization with tetrakis-(mercaptomethyl)methane, 98
3,3'-Dimercaptobenzidine, polythiazines from, *227–229*
Dimethacrylamide
polymerization of, free radical, 329
polymers of, *407*
Dimethallyl amines, copolymers, *406*
Dimethallylethyl phosphine oxide, polymers of, *418*
Dimethallyllauryl phosphonate, polymers of, *420*
Dimethallylmethyl phosphine oxide, polymers of, *417*
Dimethallylphenyl phosphine oxide
copolymers, *418*
polymers of, *418*
2,2'-Dimethyl-[6,6'-bibenzoxazine]-4,4'-dione, polymerization with 4,4'-diaminodiphenyl ether, 90

Dimethyl bis(trimellitimidates), poly(ester-imides) from, *198–199*

4,4-Dimethyl-1,7-diaminoheptane, properties of polypyromellitimide from, 27

Dimethyl 2,4-di(chloroformyl)terephthalate, poly(amide-imide) with isophthaloyl chloride and 4,4′-diaminodiphenyl ether, 43

*N,N*-Dimethyl-2,5-divinylmorpholinium chloride, polymers of, *390*

*N,N*-Dimethyl-2,6-divinylpiperidinium chloride, polymers of, *391*

*N,N*-Dimethyl-2,5-divinylpyrrolidinium chloride, polymers of, *390*

3,7-Dimethyloctanal, divinyl acetal, free radical polymerization of, 301

Dinitriles
condensation with diazides, 475
polyaminopyrimidines from, 87
polymerization of, 335

Diols, poly(ester-imides) with trimellitic anhydride and diamines, 56

4,8-Diphenyl-1,5-diazabicyclo[3,3,0]octane-2,3,6,7-tetracarboxylic acid dianhydride
polyimides from, *119*
polymerization with aliphatic and aromatic diamines, 17

Diphenyl esters, polymerization with 1,4,5,8-tetraaminonapthalene, 87, 88

3,3′,4,4′-Diphenyl ether tetracarboxylic acid dianhydride
poly(ester-imides) from, *197*
polymerization with 4,4′-diaminodiphenyl ether, 6

3,4,4′-Diphenyl ether tricarboxylic acid, poly(amide-imides) from, *186*

3,3′,4,4′-Diphenylmethane tetracarboxylic acid dianhydride, polyimides from, *160*

2,2′,3,3′-Diphenylmethane tetracarboxylic acid dianhydride, polyimide with 4,4′-diaminodiphenyl sulfone, *160*

*N,N′*-Diphenyl-*p*-phenylene diamine, condensation with trimellitoyl chloride, 47

Diphenyl terephthalate, poly(1,3-oxazinone) with 2,7-diamino-3,6-di(carbophenoxy) napthalene, 74

Dipropargyl maletate
polymerization of, free radical, 315
polymers of, *376*

Di-n-propenylphosphine oxide, polymerization of, anionic, 339

Di-*iso*-propenylphosphine oxide, polymerization of, anionic, 339

Disacryle, 442

3,3′-Disulfonamidobenzidine, polymerization with pyromellitic dianhydride, 477

Divinyl acetals
copolymers with olefins, 303
polymerization of, 299
polymerization of, cationic, 306
polymers of, *361, 362*
polymers of, structure and mechanism, 301

Divinyl adipate, copolymerization with vinyl acetate, 308

*N,N*-Divinylaniline
copolymer with diethyl fumarate, 327
copolymers with vinyl compounds, 327, *400, 401*
polymerization of, free radical, 326
polymers of, *400*

Divinyl butanal
copolymers with vinyl monomers, *360–361*
polymerization of, 306
polymers of, *360*

Divinyl isobutanal
copolymers with vinyl acetate, *361*
polymers of, *361*

Divinyl carbonate
polymerization of, free radical, 309
polymers of, *369*

Divinyl cyclohexanal, polymers of, *362*

1,2-Divinylcyclohexane, copolymer with sulfur dioxide, *425*

Divinylcyclopentamethylene silane, copolymers, *427*

Divinyldimethyl silane, copolymers with vinyl compounds, 347, 348, *427*

Divinyldimethyl siloxane
polymerization of, cationic, 350
polymerization of, free radical, 348, 349

*S,S′*-Divinyl dithiolocarbonate, polymers of, *423*

Divinyl diureas, polymers of, *410*

Divinyl ethanal
copolymers with vinyl monomers, *359–360*
polymerization of, cationic, 306
polymers of, *359*

Divinyl ether
copolymer with acrylonitrile, 304

copolymer with fumaronitrile, 304
copolymer with maleic anhydride, 304
copolymer with N-phenylmaleimide, 304
copolymer with tetracyanoethylene, 304
copolymers with vinyl monomers, 357, *358*
copolymer with 4-vinylpyridine, 304
polymerization of, free radical, 305
polymers of, *357*
Divinyl formal
copolymers with vinyl monomers, *358–359*
polymerization of, cationic, 306
polymerization of, free radical, 300
polymers of, *358*
Ziegler polymerization of, 306
Divinyl heptanal, polymers of, *363*
Divinyl hexanal, polymers of, *362*
Divinyl ketals
copolymers with olefins, 303
polymerization of, 299
Divinyl ketal of acetone, polymers of, *360*
Divinyl ketal of methylethylketone, polymers of, *360*
Divinyl mercaptal, polymerization of, free radical, 340
Divinyloxydimethyl silane, polymers of, *427*
Divinyl pentanal, polymers of, *362*
Divinyl isopentanal
copolymers of, *362*
polymers of, *362*
Divinylphenylphosphine, copolymers, *417*
Divinyl propanal, polymers of, *360*
Divinyl sulfone, copolymers with vinyl compounds, 345, *426*
Divinyl thioacetals, polymerization of, free radical, 340
Divinyl thiobutanal, polymers of, *422*
Divinyl thioethanal, polymers of, *422*
Divinyl thioformal, polymers of, *421*
Divinyl thioketals, polymerization of, free radical, 340
N,N'-Divinylurea
polymerization of, free radical, 328
polymers of, *407*
Divinyl urethans, polymers of, *409, 410*

E

Electrically conductive polymers, 325
Ester-amide salts, polymerization to polyimides, 11

Ester dianhydrides
from hydroquinone and trimellitic anhydride, poly(ester-imide) with isophthalic dihydrazide, preparation of, 53, 54
Esters, diunsaturated, polymerization of, 307
Esters, unsaturated, polymerization of, 306
1,1,2,2-Ethanetetracarboxylic acid dianhydride, polyimides from, *102–104*
α-Ethylacrolein, polymer of, *466*
Ethyl 2-acryloxy acrylate, polymer of, *372*
N-Ethyldimethacrylamide, polymers of, *408*
3,6-*endo*-Ethylene-1,2,4,5-cyclohexanetetracaboxylic acid dianhydride, polyimides from, *114*
Ethylene diacrylate, copolymerization with methyl methacrylate, 308
1,2-Ethylene diisocyanate
polymerization of, anionic, 333, 334
polymerization of, free radical, 333
polymers of, *411*
Ethylene dimethacrylate, polymerization of, free radical, 308
Ethyl 2-methacryloyloxy acrylate, polymers of, *374*

F

Flocculating agents, 324
Fumaronitrile
polymerization of, free radical, 335
polymers of, *415, 416*
Furfural, divinyl acetal
polymerization of, free radical, 302
polymers of, *367*

G

Germanium derivatives, diunsaturated, polymerization of, 321
Glutaraldehyde
polyacetal from, *257*
polymerization of, cationic, 292, 294
polymerization of, free radical, 291
polymers of, *353*
Glyceryl triisocyanate, polymers of, *414*
Glycidyl acrylate, polymers of, *368*
Glycidyl crotonate, polymers of, *368*
Glycidyl methacrylate
polymerization of, anionic, 307
polymerization of, cationic, 306
polymers of, 368

Glyoxal
  polyacetal from, *257*
  polymerization of, anionic, 296
  polymerization of, free radical, 291
  polymers of, *352*


## H

1,5-Hexadiene, copolymer with sulfur dioxide, 342, *424*
Hexamethylene bis(iminoacetic acid)
  polydiketopiperazine from, *240*
  polymerization of, 86
Hydrazine, polymerization with pyromellitic anhydride, 15
Hydrazines, arylene, di-, polyimides from, 16
Hydroquinone diacetate, condensation with trimellitic anhydride, 53


## I

Intra-intermolecular polymerizations, *see also*,
  cyclopolymerizations, 287
  mechanism of, 289
Isophthalaldehyde, polymerization of, 288
Isophthalbis(amidoxime), poly(imide-1,2,4-oxadiazole) with trimellitoyl chloride and 4,4'-diaminodiphenyl ether, 66
Isophthalic dihydrazide
  poly(ester-imide) with the ester dianhydride from hydroquinone and trimellitic anhydride, 54
  polymerization with oxy-bis[*N*-(4'-phenylene)-4-(chlorocarbonyl)phthalimide], 65
Isophthaloyl chloride, poly(amide-imide) with dimethyl 2,4-di(chloroformyl)-terephthalate and 4,4'-diaminodiphenyl ether, 43
Isoprene oxide
  polymerization of, cationic, 298
  polymers of, *355*
4-Isopropyl-1,3-diaminobenzene
  polypyromellitimide from, solubility and melt-processibility of, *32*


## M

Malealdehyde, polymers of, *352*

Maleonitrile
  polymerization of, free radical, 335
  polymers of, *415*
Maleopimaric acid, poly(amide-imides) from, *181–182*
Maleopimaric acid monochloride, poly-(amide-imide) with 1,6-diaminohexane, 46
*meso*-1,2,3,4-Butanetetracarboxylic acid dianhydride, polyimides from, *106, 107*
Methacrolein
  copolymers with vinyl monomers, 456, *465–466*
  polymerization of, 456
  polymers of, *462–465*
Methacrolein dimer
  polymerization of, 296
  polymers of, *354*
Methacrylic anhydride
  copolymers with unsaturated monomers, 317, 320, *380, 381*
  polymerization of, 288
  polymerization of, free radical, 319
  polymers of, *380*
Methacrylic crotonic anhydride, polymer of, *381*
*N*-Methallyl-*N*-*p*-acetylphenyl-2-vinyl-morpholinium chloride, polymers of, *392*
Methallyl 3-butenoate, polymerization of, free radical, 310
Methallyl crotonate
  polymerization of, free radical, 310
  polymers of, *374*
Methallyl vinyl acetate, polymers of, *374*
Methyl 2-acryloyloxy acrylate, polymers of, *371*
Methyl allyl fumarate, polymers of, *373*
Methyl allyl maleate, polymers of, *373*
Methyl 2-butenyl fumarate, polymers of, *375*
Methyl 3-butenyl fumarate, polymers of, *376*
Methyl 2-butenyl maleate, polymers of, *374*
Methyl 3-butenyl maleate, polymers of, *375*
2-Methyl-3-chlorosuberaldehyde, polymers of, *354*
*N*-Methyldiacrylamide, polymers of, *407*
3-Methyl-1,7-diaminoheptane, polypyromellitimide from, properties, 27
*N*-Methyldimethacrylamide
  polymerization of, anionic, 331
  polymers of, *408*

N-Methyl-*sym*-divinyltetramethyldisilazane
  polymerization of, anionic, 350
  polymers of, *429*
Methylene diisocyanate, polymers of, *411*
3-Methylglutaraldehyde
  polyacetal from, *257*
  polymerization of, cationic, 292
  polymers of, *353*
2-Methylpropane-1,3-diisocyanate, polymers of, *413*
Monoallyl citraconate, polymerization of,
  free radical, 309
Monoallyl maleate, polymerization of, free
  radical, 309


**N**

α-Napthaldehyde, divinyl acetal, polymers
  of, *367*
1,4,5,8-Napthalene tetracarboxylic acid dianhydride
  poly(imide-imidazopyrrolone) with 1,2,4-
    triaminobenzene, 62
  polyimides from, *169–171*
2,3,6,7-Napthalene tetracarboxylic acid dianhydride, polyimides from, *169*


**O**

1,2,3,4,5,6,7,8-Octahydrophenanthrene-
  1,2,5,6-tetracarboxylic acid dianhydride, polyimides from, *113*
1,2,4-Oxadiazole-imide copolymers, preparation, 65
1,3,4-Oxadiazole-imide copolymers, preparation, 64
oxy-*bis*[N-(4′-phenylene)-4-(chlorocarbonyl)phthalimide]
  polymerization with isophthalic hydra-
    zide, 65


**P**

Paper strengthening agents, 324
Pentaerythritol
  polyacetals from, *254–257*
  polyketals from, *255–256*
  polymerization with 1,1,5,5-tetraethoxy-
    pentane, 96

Phenanthrene-10-carboxy-1,2,7,8-tetra-
  carboxylic acid, dianhydride, *171*
N-Phenyldimethacrylamide, polymers of,
  *409*
*p*-Phenylene bis(benzil), polyquinoxaline
  with 1,2,4,5-tetraaminobenzene, 82
*m*-Phenylene diamine, polymerization with
  benzaldehyde, 485
*p*-Phenylene diboronic acid, polymerization
  with hydrazine, 473
3-Phenylglutaraldehyde
  polymerization of, cationic, 292
  polymers of, *353*
Phenyl glycidyl ether, polymerization with
  1,3-bis(chloroacetylamino)benzene, 86
Phosphine oxides, diunsaturated, polymerization of, free radical, 337
Phosphines, diunsaturated, polymerization
  of, free radical, 337
Phosphonates, diunsaturated, polymerization of, free radical, 337
Phosphonium salts, diunsaturated, polymerization of, free radical, 337
  polymers of, *420*
Phosphonium salts, unsaturated, polymerization of, effect of incipient ring size on,
  338
Phosphorus compounds, diunsaturated
  copolymers with vinyl compounds, 337
  polymerization of, 337
*o*-Phthalaldehyde
  polymerization of, anionic, 295
    cationic, 294
    free radical, 291
  polymers of, 288, *354*
  Ziegler polymerization of, 292
Phthalic anhydride
  condensation with benzoylhydrazine, 54
  condensation with dihydroxyphenols,
    492
Piperidinium salt, polymers of, 287
Polyacetals, *254–259*
  by acid-catalyzed transacetalization, 97
  by metal oxide-catalyzed ester exchange
    reaction, 97
  preparation of, 96
  properties of, 98
Polyacridines, 485
Poly(acrylic acid), 317
Poly(acrylic anhydride), stereochemistry of,
  318

Poly(amic acids)
  chemical dehydration to polyimides, 20
  cyclization to polyiminolactones, 20
  preparation from aminoanhydrides, 10
  preparation in bulk, 7
  preparation of, 2
  preparation of, as solvent-free powders, 5
  preparation of organosols, 7
  thermal dehydration to polyimides, 7
  use in preparation of polyimide coatings, 5
  use with silicone polymers for solvent
    resistant polyimide coatings, 10
Polyamic acids, ammonium salts of, de-
    hydration to polyimides, 13
Poly(amide-imides), 44
  cyclization to polyimides, 14
  electrical applications of, 51
  methods for preparation of, 42
  preparation by chemical dehydration of
    polyamic acids, 48
  preparation by thermal dehydration of
    polyamic acids, 42
  properties and applications, 49
Poly(amide-indoles), 478
Poly(aminotetrazoles), 476
Polyaniline-polyisocyanate, papi, polymer-
    ization with 3,3′,4,4′-benzophenone-
    tetracarboxylic acid dianhydride, 25
Polyanthrazolines, 486
Polybenzboroxazinones, 476
Polycarbodiimides, addition of hydrazoic
    acid to, 476
Poly(diallylphenyl phosphine oxide), 339
Polydiazadiphosphetidines, 473
Poly(4,9-diazapyrenes), 488
Polydiazepines, 489
Polydiketopiperazines, *239–241*
Polydioxins, 78
  from tetraphenols and tetrachloroquinox-
    alines, 78
Polydisalicylides, 490
Poly(ester-amides)
  polymerization to polyimides, 11
  preparation of, 12
Poly(ester-imides), 52
  as electrical insulating materials, 55
  block polymers, preparation of, 19
  preparation by thermal dehydration of
    polyamic acids, 52
  properties of, 56–60
  tensile properties of films, 58

  thermal stability of, 60
Polyferrocenes, 491
Poly(hydroxyquinolines), 483
Polyimides
  applications of, 40
  aromatic, mechanical properties of, 39
  aromatic, properties and applications of,
    29
  aromatic, solubility and melt-processabi-
    lity of, 29
  crystallinity of, 30
  fibers of, 15
  filled compositions, 10
  foams, 10
  from aliphatic diamines, properties of, 26
  from aromatic diamines, properties of, 28
  glass transition temperatures of, 37
  hydrolytic stability of, 34
  preparation and properties of, 1
  preparation by chemical dehydration of
    polyamic acids, 20
  preparation by dehydration of ammon-
    ium salts of polyamic acids, 13
  preparation by isomerization of poly-
    iminolactones, 20
  preparation by thermal dehydration of
    polyamic acids, 7
  preparation from aminoanhydrides, 10
  preparation from diamines and tetra-
    nitriles, 23
  preparation from diisocyanates and di-
    anhydrides, 24
  preparation from ester-amide salts, 11
  preparation from poly(amide-amides), 14
  preparation from poly(ester-amides), 11
  preparation from poly(iminolactones), 22
  thermal stability of, 34
Polyimides, elastomeric, preparation of, 18
Polyimides, filled compositions, preparation
    of, 22
Polyimides, foams, preparation of, 22
Poly(imide-imidazopyrrolones), 60, *204*
  preparation by thermal dehydration of
    polyamic acids, 61
  preparation in polyphosphoric acid, 62
  properties of, 63
Poly(imide-iminolactone) copolymers, *203*
Polyiminolactones, 66
  isomerization to polyimides, 20, 22, 67
  preparation by chemical dehydration of
    polyamic acids, 66

preparation by cyclization of polyamic acids, 20

reactions with alcohols, ammonia, amines and mercaptans, 67, 68

reaction with hydrazoic acid, 68

Polyindoloquinoxaline, preparation from tetraoxobiindole and, 3,3',4,4'-tetraaminobiphenyl, 82

Poly(isoindolonebenzothiadiazine dioxides), 477

Poly(isoindoloquinazolinediones), from aromatic bis(o-amino amides) and dianhydrides, 94

Polyketals, 254–259

preparation of, 96

properties of, 98

Polymer H

glass transition temperature of, 37

mechanical and electrical properties of, 39

preparation of, 1

properties of, 28

thermal degradation of, 35

Polymers, pyrrole-containing, 69–73

Poly(methacrylic acid),

pyrolysis product from, 320

Poly(methacrylic anhydride), 288, 319

stereochemistry of, 320

structure of, 320

Poly(1,3-oxazinediones), 214–221

by cyclization of ester-substituted polyurethanes, 76

from carbonate-blocked bis(hydroxyacids) and diamines, 76, 77

from carbonate-substituted polyamides, 77

from diisocyanates and bis(hydroxyacids), 75

from hydroxy-substituted polyamides and phenyl chloroformate, 77

Polyoxazines, 78, 225–226

from diaminodiphenols and quinones, 79

from diaminodiphenols and tetrachloro compounds, 79

from diaminodiphenols and tetrahydroxy compounds, 79

Poly(1,3-oxazinones), 210–213

by oxidative cyclization of methylated polyamides, 75

preparation by cyclization of carboxypolyamides, 74

preparation from o-aminoaromatic carboxylic acids, 73

Polyphenanthridines, 488

Polyphenylenepyrazine, from 1,4-bis(bromoacetyl)benzene and ammonia, 81

Poly(4-phosphoniapyran salts), synthesis, 478

Polyphthalocyanines, 70

Polypiperazines, 239–241

from bis(β-hydroxyethyl)diamines, 85

Poly(propiolic acid), 321

Polypyrazine, from silver cyanide, 81

Polypyrimidines, 242–243

preparation of, 87

Polypyrimidones, preparation of, 89

Polypyromellitimides, 120–144

Polyquinacridones, 486

Polyquinazolinediones, 249–253

from diisocyanates and aromatic bis-(o-aminoacids), 93

properties of, 95, 96

Polyquinazolones, 244–248

preparation of, 89–92

properties of, 95, 96

Polyquinolines, synthesis, 481

Polyquinolones, synthesis from diaminodiesters and diketone bis(ketals), 483

Polyquinones, 260–261

preparation of, 98–101

Polyquinoxalines, 230–238

applications of, 84

from bis(o-diamines) and tetracarbonyl compounds, 81

by polycondensation in polyphosphoric acid, 83

properties of, 84

Polysulfimides, 476

Polytetrazadiborines, 473

Polytetrazoles, 475

Polytetraazapyrenes, 242–243

preparation of, 87

Polytetracyanoethylene, preparation and properties of, 69

Polythiazines, 78

from diaminodithiophenols and bis(α-haloalkyl aryl ketones), 80

from diaminodithiophenols and quinones, 79

from diaminodithiophenols and tetrachloro compounds, 78

Polythiazines (cont.)
from diaminodithiophenols and tetrahydroxy compounds, 78
Polythioketals, *259*
Polytitanocenes, 491
Poly(triazololactams)
preparation of, 63
properties of, 63
Polyxanthones, 479
Potassium sulfide, polymerization with chloranil, 100
Propane 1,2-diisocyanate, polymers of, *412*
1,2,3-Propanetricarboxylic acid, poly(amide-imide) with 1,10-diaminodecane, *181*
Propane 1,2,3-triisocyanate, polymerization of, anionic, 333
Propargyl aldehyde, polymerization of, anionic, 460
Propargyl crotonate
polymerization of, 315
polymers of, *370*
Propargylic anhydride
polymerization of, anionic, 320
polymers of, *379*
Propargyl methacrylate, polymerization of, 315
α-Propylacrolein, polymers of, *467*
N-n-Propyldimethacrylamide, polymers of, *408*
1,3-Propylene diisocyanate, polymers of, *411*
Pyrazine-containing polymers, 81
Pyrazinetetracarboxylic acid dianhydride, polyimides from, *157*
Pyridine-2,3,5,6-tetracarboxylic acid dianhydride, polyimides from, *155–156*
Pyroboretetraacetate
polymerization with 3,3'-dicarboxybenzidine, 476
polymerization with hydrazine, 474
Pyromellitic anhydride
copolyimide with 2,2-bis(3,4-dicarboxyphenyl)propane dianhydride and 1,4-bis(aminomethyl)benzene, properties of, 27
polyamic acid with 1,3-diaminobenzene, mechanism of cyclization, 8
poly(amide-imides) from, *173–178*
poly(amide-imides) with diaminoamides, properties of, *49*
poly(amide-imide) with 4,3'-diaminobenzanilide, mechanical properties of, *49*

poly(ester-imides) from, *196*
poly(imide-benzimidazole) with 1',3,5-triaminobenzanilide, 43
poly(imide-imidazopyrrolone) with 1,2,4-triaminobenzene, 61
polyimide with 2,5-bis(4-aminophenyl)-1,3,4-oxadiazole, 60
polyimide with 4,4'-diamino-3,3'-dihydroxybiphenyl, rearrangement of, 18
polyimide with 4,4'-diaminodiphenyl ether, properties, 28
polyiminolactones from, *200, 201*
polymerization with N,N'-bis(3-aminobenzoyl), hydrazine, 64
polymerization with 4,4'-diaminodiphenyl ether, 1
polymerization with 2,4-diaminoisopropylbenzene, 21
polymerization with N,N'-diaminopyromellitimide, 17
polymerization with 3,3'-dicarbamyl-4,4'-diaminodiphenyl methane, 94
polymerization with 3,3'-disulfonamidobenzidine, 477
polymerization with hydrazine, 15
Pyromellitonitrile
polymerization with 1,3-diaminobenzene, 71
polymerization with 4,4'-diaminodiphenyl ether, 70
polymerization to polyphthalocyanines, 70
Pyrrole-containing polymers, *205–209*

## S

Sebaconitrile, polypyrimidines from, *242*
Selenium dioxide, copolymers with 1,5-dienes, 350
Silicon compounds, diunsaturated, polymerization of, 346
Stilbene-3,3',4,4'-tetracarboxylic acid dianhydride, polyimide with α,ω-bis(4-aminophenyl)methylvinyltrisiloxane, *167*
Suberaldehyde
polymerization of, cationic, 295
polymers of, *354*
Suberonitrile, polypyrimidines from, *242*
Succinaldehyde
polymerization of, cationic, 292, 294
polymers of, *352*

Succinonitrile
  polymerization of, free radical, 335
  polymers of, *415*, *416*
Sulfobenzoic anhydride, condensation with
  dihydroxyphenols, 492
Sulfonamides, diunsaturated, polymerization of, free radical, 341
Sulfonates, diunsaturated, polymerization
  of, free radical, 341
Sulfur compounds, diunsaturated, polymerization of, 339
Sulfur-containing polyquinones, preparation of, 100, 101


T


Tetraacids, polymerization with diamines, 11
Tetraallylammonium bromide
  polymerization of, 289
  polymerization of, free radical, 324
  polymers of, *394*
Tetraamines, aromatic, poly(imide-imidazopyrrolones) with aromatic diamines
  and dianhydrides, 62
1,2,4,5-Tetraaminobenzene
  polymerization with 1,2,6,7-tetraketopy-
  rene, 83
  polyquinoxaline with *p*-phenylene bis-
  (benzil), 82
  polyquinoxalines from, *230, 231, 232*
3,3',4,4'-Tetraaminobiphenyl tetrahydro-
  chloride, polymerization of, 83
2,2',6,6'-Tetraamino-4,4'-bis(methoxycarbonyl)biphenyl, polymerization with
  diethyl oxaldiimidate, 490
2,3,6,7-Tetraaminodibenzo-*p*-dioxin, polyquinoxalines from, *238*
3,3',4,4'-Tetraaminodiphenyl ether, polyquinoxalines from, *235*
3,3',4,4'-Tetraaminodiphenyl sulfone, polyquinoxalines from, *237, 238*
1,4,5,8-Tetraaminonapthalene, polymerization with diphenyl esters, 87, 88
2,3,6,7-Tetrachloro-1,4,5,8-tetraazaanthra-
  cene, polydioxin with 3,3',4,4'-tetra-
  hydroxybiphenyl, 78
1,1,5,5-Tetraethoxypentane, polymerization
  with pentaerythritol, 96
1,2,5,6-Tetrahydroxyanthraquinone, polydioxins from, *223, 224*

1,2,4,5-Tetrahydroxybenzene, polydioxin
  with 2,3,7,8-tetrachloro-1,4,6,9-tetraaza
  anthracene, *222*
Tetrahydroxy-*p*-benzoquinone
  polydioxins from, *222*
  polymerization with chloranil, 99
3,3',4,4'-Tetrahydroxybiphenyl, polydioxins
  from, *223*
  polydioxin with 2,3,6,7-tetrachloro-
  1,4,5,8-tetraazaanthracene, 78
2,3,6,7-Tetrahydroxythianthrene, polydioxin with 2,2',3,3'-tetrachloro-6,6'-bis-
  (quinoxaline), *224*
1,2,6,7-Tetraketopyrene, polymerization
  with 1,2,4,5-tetraaminobenzene, 83
1,2,4,5-Tetrakis(chloromethyl)benzene,
  polypyrrolines with diamines, 72
Tetrakis(mercaptomethyl)methane, polymerization with diketones, 98
Tetramethylene diisocyanate, polymerization of, free radical, 332
Tetramethylene diisothiocyanate
  polymerization of, anionic, 335
  polymers of, *413*
Terephthalaldehyde
  polymerization of, 288
  polymerization with bis(oxetane), 98
Terephthaloyl chloride, polymerization with
  4,4'-diamino-3,3'-dicarboxybiphenyl, 74
Terephthaloyl diphthalic anhydride, polyimides from, *168*
Thiocarbamates, diunsaturated, polymerization of, free radical, 341
Thiocarbonates, diunsaturated, polymerization of, free radical, 341
Thiophene-2,3,4,5-tetracarboxylic acid dianhydride, polyimide with 2,11-di-
  aminododecane, *154*
*m*-Tolualdehyde, divinyl acetal, polymers
  of, *363*
*o*-Tolualdehyde, divinyl acetal, polymers of,
  *363*
*p*-Tolualdehyde, divinylacetal, polymers of,
  *364*
Triallylamine hydrobromide, polymers of,
  *393*
Triallylammonium hydrobromide, polymerization of, free radical, 323
Triallylethylammonium bromide
  polymerization of, free radical, 323
  polymers of, *393*

Triallylmethyl silane
  polymers of, *430*
  Ziegler polymerization of, 349
Triallyl orthoformate
  polymerization of, free radical, 303
  polymers of, *367*
Triallylphenyl silane, polymers of, *430*
1′,3,5-Triaminobenzanilide, poly(imide-benzimidazole) with pyromellitic anhydride, 43
1,2,4-Triaminobenzene
  poly(imide-imidazopyrrolone) with 1,4,5,8-napthalene tetracarboxylic acid dianhydride, 62
  poly(imide-imidazopyrrolone) with pyromellitic anhydride, 61
Tricyclo[4,2,2,0$^{2,5}$]dec-7-ene-7-alkyl-3,4,9,10-tetracarboxylic acid dianhydride, polyimides from, *117*
Tricyclo[4,2,2,0$^{2,5}$]dec-7-ene-3,4,9,10-tetracarboxylic acid dianhydride, polyimides from, *115–116*
Trimellitic anhydride
  condensation with 1,4-bis(acetylamino)-benzene, 47
  condensation with hydroquinone diacetate, 53
  poly(amide-imides) from, *184–185*
  poly(amide-imide) with 4,4′-diisocyanatodiphenylmethane, 48
  poly(ester-imides) with diamines and diols, 56
  polymerization with diisocyanates, 48
Trimellitoyl chloride
  condensation with N,N′-diphenyl-p-phenylene diamine, 47
  poly(amide-imide) with 4,4′-diaminodiphenyl ether, 45
  poly(imide-1,2,4-oxadiazole) with 4,4′-diaminodiphenyl ether and isophthalbis(amidoxime), 66
Trimethallyl amine hydrochloride, polymers of, *393*
Trimethylene diisocyanate, polymerization of, anionic, *333*
Triphenyl methane-bis(alkoxy)-diamines, polyimides from, *32*

Trivinyl orthoformate
  polymerization of, free radical, 303
  polymers of, *367*
Trivinyl phosphate, polymers of, *418*

V

Vinyl acrylate, chain transfer activity in olefin polymerization, 316
N-Vinyl allyl urethan
  polymerization of, free radical, 330
  polymers of, 409
Vinyl atropate, polymerization of, 310
S-Vinyl-N-n-butyl vinyl thiocarbamate, polymers of, *424*
trans-Vinyl cinnamate
  copolymer with vinyl acetate, 314
    methacrylonitrile, 314
    vinylpyrrolidone, 314
    styrene, 314
  copolymers with vinyl monomers, *377–378*
  polymerization of, free radical, 313
  polymers of, 377
Vinyl cinnamate, substituted, polymerization of, 314
Vinyl crotonate, polymers of, *369*
S-Vinyl-N-ethyl vinyl thiocarbamate, polymers of, *423*
Vinyl isocyanate, polymerization of, 89
Vinyl ketones, polymerization of, 460
Vinyl methacrylate
  polymerization of, free radical, 311, 312
  polymers of, *370*
S-Vinyl-N-methyl vinyl thiocarbamate, polymers of, *423*
S-Vinyl-N-vinylthiocarbamate, polymerization of, free radical, cationic, 289, 290
S-Vinyl-N-vinyl thiocarbamates
  polymerization of, cationic, 345, 346
  polymerization of, free radical, 341, 342

X

p-Xylylene dichloride, polypyrrolinone with 4,4′-diamino-3,3′-dicarboxybiphenyl, 72

# ORGANIC CHEMISTRY
## A SERIES OF MONOGRAPHS

### EDITORS

**ALFRED T. BLOMQUIST**
*Department of Chemistry*
*Cornell University*
*Ithaca, New York*

**HARRY WASSERMAN**
*Department of Chemistry*
*Yale University*
*New Haven, Connecticut*

1. Wolfgang Kirmse. CARBENE CHEMISTRY, 1964; 2nd Edition, 1971

2. Brandes H. Smith. BRIDGED AROMATIC COMPOUNDS, 1964

3. Michael Hanack. CONFORMATION THEORY, 1965

4. Donald J. Cram. FUNDAMENTALS OF CARBANION CHEMISTRY, 1965

5. Kenneth B. Wiberg (Editor). OXIDATION IN ORGANIC CHEMISTRY, PART A, 1965; PART B, *In preparation*

6. R. F. Hudson. STRUCTURE AND MECHANISM IN ORGANO-PHOSPHORUS CHEMISTRY, 1965

7. A. William Johnson. YLID CHEMISTRY, 1966

8. Jan Hamer (Editor). 1,4-CYCLOADDITION REACTIONS, 1967

9. Henri Ulrich. CYCLOADDITION REACTIONS OF HETEROCUMULENES, 1967

10. M. P. Cava and M. J. Mitchell. CYCLOBUTADIENE AND RELATED COMPOUNDS, 1967

11. Reinhard W. Hoffman. DEHYDROBENZENE AND CYCLOALKYNES, 1967

12. Stanley R. Sandler and Wolf Karo. ORGANIC FUNCTIONAL GROUP PREPARATIONS, VOLUME I, 1968; VOLUME II, 1971; VOLUME III, *In preparation*

13. Robert J. Cotter and Markus Matzner. RING-FORMING POLYMERIZATIONS, PART A, 1969; PART B, 1; B, 2, 1972

14. R. H. DeWolfe. CARBOXYLIC ORTHO ACID DERIVATIVES, 1970

15. R. Foster. ORGANIC CHARGE-TRANSFER COMPLEXES, 1969

16. James P. Snyder (Editor). NONBENZENOID AROMATICS, VOLUME I, 1969; VOLUME II, 1971

# ORGANIC CHEMISTRY
*A Series of Monographs*

**17.** C. H. Rochester. ACIDITY FUNCTIONS, 1970

**18.** Richard J. Sundberg. THE CHEMISTRY OF INDOLES, 1970

**19.** A. R. Katritzky and J. M. Lagowski. CHEMISTRY OF THE HETEROCYCLIC *N*-OXIDES, 1970

**20.** Ivar Ugi (Editor). ISONITRILE CHEMISTRY, 1971

**21.** G. Chiurdoglu (Editor). CONFORMATIONAL ANALYSIS, 1971

**22.** Gottfried Schill. CATENANES, ROTAXANES, AND KNOTS, 1971

**23.** M. Liler. REACTION MECHANISMS IN SULPHURIC ACID AND OTHER STRONG ACID SOLUTIONS, 1971

*In preparation*

J. B. Stothers. CARBON-13 NMR SPECTROSCOPY

Maurice Shamma. THE ISOQUINOLINE ALKALOIDS: CHEMISTRY AND PHARMACOLOGY

Walter S. Trahanovsky (Editor). OXIDATION IN ORGANIC CHEMISTRY, PART B

Samuel P. McManus (Editor). ORGANIC REACTIVE INTERMEDIATES